Integrated Solar Fuel Generators

Energy and Environment Series

Editor-in-chief:
Heinz Frei, *Lawrence Berkeley National Laboratory, USA*

Series editors:
Nigel Brandon OBE FREng, *Imperial College London, UK*
Roberto Rinaldi, *Imperial College London, UK*
Vivian Wing-Wah Yam, *University of Hong Kong, Hong Kong*

Titles in the series:
1: Thermochemical Conversion of Biomass to Liquid Fuels and Chemicals
2: Innovations in Fuel Cell Technologies
3: Energy Crops
4: Chemical and Biochemical Catalysis for Next Generation Biofuels
5: Molecular Solar Fuels
6: Catalysts for Alcohol-Fuelled Direct Oxidation Fuel Cells
7: Solid Oxide Fuel Cells: From Materials to System Modeling
8: Solar Energy Conversion: Dynamics of Interfacial Electron and Excitation Transfer
9: Photoelectrochemical Water Splitting: Materials, Processes and Architectures
10: Biological Conversion of Biomass for Fuels and Chemicals: Explorations from Natural Utilization Systems
11: Advanced Concepts in Photovoltaics
12: Materials Challenges: Inorganic Photovoltaic Solar Energy
13: Catalytic Hydrogenation for Biomass Valorization
14: Photocatalysis: Fundamentals and Perspectives
15: Photocatalysis: Applications
16: Unconventional Thin Film Photovoltaics
17: Thermoelectric Materials and Devices
18: X-Ray Free Electron Lasers: Applications in Materials, Chemistry and Biology
19: Lignin Valorization: Emerging Approaches
20: Advances in Photoelectrochemical Water Splitting: Theory, Experiment and Systems Analysis
21: Electrochemical Reduction of Carbon Dioxide: Overcoming the Limitations of Photosynthesis
22: Integrated Solar Fuel Generators

How to obtain future titles on publication:
A standing order plan is available for this series. A standing order will bring delivery of each new volume immediately on publication.

For further information please contact:
Book Sales Department, Royal Society of Chemistry, Thomas Graham House, Science Park, Milton Road, Cambridge, CB4 0WF, UK
Telephone: +44 (0)1223 420066, Fax: +44 (0)1223 420247
Email: booksales@rsc.org
Visit our website at www.rsc.org/books

Integrated Solar Fuel Generators

Edited by

Ian D. Sharp
Technical University of Munich, Germany
Email: sharp@wsi.tum.de

Harry A. Atwater
California Institute of Technology, CA, USA
Email: haa@caltech.edu

and

Hans-Joachim Lewerenz
California Institute of Technology, CA, USA
Email: lewerenz@caltech.edu

ROYAL SOCIETY OF **CHEMISTRY**

Energy and Environment Series No. 22

Print ISBN: 978-1-78262-555-1
PDF ISBN: 978-1-78801-031-3
EPUB ISBN: 978-1-78801-521-9
Print ISSN: 2044-0774
Electronic ISSN: 2044-0782

A catalogue record for this book is available from the British Library

The Royal Society of Chemistry is a charity, registered in England and Wales, Number 207890, and a company incorporated in England by Royal Charter (Registered No. RC000524), registered office: Burlington House, Piccadilly, London W1J 0BA, UK, Telephone: +44 (0) 207 4378 6556.

For further information see our web site at www.rsc.org

Printed in the United Kingdom by CPI Group (UK) Ltd, Croydon, CR0 4YY, UK

Preface

The prospect of sustainably generating fuels from sunlight *via* artificial photosynthesis has inspired scientists for decades. Over time, intensive research devoted to materials for catalysis and light harvesting provided a set of components individually optimized for specific steps of sunlight-to-fuel conversion, as well as a conceptual basis for understanding conversion mechanisms. These efforts also revealed critical gaps in properties of existing materials, in models of non-ideal function, and especially in our understanding of concerted processes occurring in integrated photosystems. Within this context, the Joint Center for Artificial Photosynthesis (JCAP) was established with the mission 'to demonstrate a scalable, manufacturable solar-fuels generator using Earth-abundant elements, that, with no wires, robustly produces fuel from the sun ten times more efficiently than (current) crops'. Intrinsic to this mission was the recognition that integration of components into functional systems introduces new scientific challenges that must be addressed to advance along the path to a technology. The drive towards integration forces the researcher to consider how components are physically, electronically, and chemically coupled over multiple length, time, and energy scales. Such coupling places dramatic restrictions on material synthesis processes, electrolyte environments, and system geometries. However, these considerations also provide a framework for defining research priorities and for conceiving practical device architectures.

This text describes recent developments in the field of solar water splitting that encompass major work at JCAP and its scientific environment. The Center was operated as an Energy Innovation Hub, a novel concept at that time, initiated by the former US Secretary of Energy Steven Chu. He modeled the hubs after Bell Laboratories, also known as 'The Idea

Energy and Environment Series No. 22
Integrated Solar Fuel Generators
Edited by Ian D. Sharp, Harry A. Atwater and Hans-Joachim Lewerenz

Published by the Royal Society of Chemistry, www.rsc.org

Factory,' where fundamental and applied science, extending to devices and systems, were integrated into a coherent research and development effort. This model led to the 'one-roof' concept of the Energy Innovation Hubs of the US Department of Energy. The hubs were considered smaller Bell Laboratories—affine units. The Hub concept entails the participation of researchers and engineers from various scientific disciplines in a proactive approach to conducting research and in managing activities towards applications. From the beginning, the focus was set on well-funded, accelerated development achieved by highly integrated research and development teams that could advance progress more rapidly than researchers working separately. In this way, the Energy Innovation Hubs were conceptualized and modeled after the way industrial laboratories operated during their most productive periods.

When the Hub concept was launched, the production of fuels from sunlight was considered as one of the grand challenges where interdisciplinary work would particularly benefit accelerated development. In 2010, the activity for artificial photosynthesis was started in a joint effort that involved the California Institute of Technology, the Lawrence Berkeley National Laboratory and the Stanford Linear Accelerator as major partners, including also the Universities of California at Irvine and San Diego under the leadership of Prof. Nathan Lewis, the Founding Director.

This book comprises the outcome of the first 5-year period of the Solar Fuels Hub and, also, puts the work into an international context by inclusion of chapters from renowned researchers in the field. In the spirt of the Hub model of scientific activity inspired by industrial-style research, we include and emphasize the prospective end-use application and its role in the development of integrated solar water splitting devices. Research successes are described alongside those that ultimately failed or were deemed impractical, especially at the device level. The aim is to provide a record that is useful to the practitioner by presenting not just major advancements, but also the path towards their realization.

This book is divided into five sections comprising a total of thirteen chapters.

The first section, *Introduction and System Considerations*, begins with a basic introduction by Lewerenz and Sharp to key concepts in photovoltaic, electrochemical, and photoelectrochemical energy conversion processes and materials. This first chapter provides a basis for understanding the more in-depth chapters within the book. The following chapter, by Greenblatt, describes a complete life cycle assessment of a prospective 1 GW solar hydrogen plant. The outcome of this assessment motivates research into robust and high efficiency solar water splitting devices and provides context to targeted research—from materials to devices—described in the remainder of the book.

The second section focuses on *Electrocatalysis* and spans from materials and mechanisms to advanced methods of characterization. The chapter by

Bell provides a detailed account of how the understanding of the function of water oxidation catalysts, particularly transition metal oxides and related compounds, has evolved and deepened in recent years. Much of this improved understanding has been enabled by advances in characterization and the subsequent chapter by Friebel describes state-of-the-art x-ray and electron spectroscopic methods that are used to probe catalysts, increasingly under realistic operating conditions. The section on electrocatalysis concludes with a chapter by McCrory *et al.*, which presents the importance of standardized benchmarking methods and protocols for assessing activities and stabilities of both oxygen and hydrogen evolution catalysts in reactive environments that are relevant to integrated solar water splitting devices.

The third section is dedicated to *Semiconductor Light Absorbers*, which are the engines of any solar fuel device. The chapter by Ager explores heterojunction concepts and configurations upon which integrated devices are based. A particular emphasis of this chapter is on the role solid–solid junctions in semiconductor stabilization, which represents one of the most significant breakthroughs in solar water splitting systems over the last decade and enables highly efficient and stable system constructions. In the next chapter, Osterloh provides a comprehensive overview of particle-based approaches to photochemical energy conversion. Such approaches offer significant opportunity for scalability compared to thin film architectures, but present another set of fascinating challenges in terms of understanding photochemical mechanisms, engineering materials, and integrating into systems. The section on semiconductors concludes with a chapter by Toma and co-workers that describes ubiquitous (photo)chemical instabilities of semiconductor light absorbers, how their mechanisms are determined, and approaches for developing robust photoelectrodes.

The fourth section, entitled *New Materials and Components*, provides two case examples of approaches to new materials discovery and component integration. In the chapter by Gregoire *et al.*, the modern approach to discovery of new materials *via* high throughput experimentation is presented. The complete pipeline, including material synthesis, screening, characterization, and data management is discussed and examples of how this pipeline has been implemented for discovery of electrocatalysts and photoelectrodes is provided. The next chapter, by Miller and Houle, turns towards the challenge of integrating membranes, which are necessary for creating operationally safe devices by separating gaseous hydrogen and oxygen products while simultaneously allowing ionic transport between the anolyte and catholyte. The membrane properties specific to integrated solar fuel generators are discussed and opportunities for engineering new materials for this application are presented.

The fifth and final section is devoted to *Devices and Modelling* and represents the heart of this book. The chapter by Xiang and colleagues describes the scientific and engineering efforts at JCAP devoted to prototype development and integrated device realization, with a true emphasis on the process. Descriptions of the challenges of moving from isolated components to integrated devices provide important insights that can aid those seeking to develop solar fuels generators. The next chapter, by May, Döscher, and Turner, describes the path to modern high efficiency integrated solar water splitting devices, as well as important considerations for accurate assessment of key performance indicators. Finally, the chapter by Singh, Haussener, and Weber, develops a continuum scale model of the function of integrated solar water splitting devices. As integrated systems are experimentally realized, optimization of device design, specification of material and component properties, and exploration of novel architectures are all greatly advanced by computational models.

Ian Sharp, Joachim Lewerenz, and Harry Atwater
Munich and Pasadena

Contents

Energy and Environment Series No. 22
Integrated Solar Fuel Generators
Edited by Ian D. Sharp, Harry A. Atwater and Hans-Joachim Lewerenz

Published by the Royal Society of Chemistry, www.rsc.org

Semiconductor Light Absorbers

New Materials and Components

Devices and Modelling

Introduction and System Considerations

CHAPTER 1

Concepts of Photoelectrochemical Energy Conversion and Fuel Generation

HANS JOACHIM LEWERENZ*[a,b] AND IAN D. SHARP[c,d]

[a] Division of Engineering and Applied Science, California Institute of Technology, Pasadena, CA 91125, USA; [b] Joint Center for Artificial Photosynthesis, California Institute of Technology, Pasadena, CA 91125, USA; [c] Walter Schottky Institut and Physik Department, Technische Universität München, 85748 Garching, Germany; [d] Joint Center for Artificial Photosynthesis, Lawrence Berkeley National Laboratory, Berkeley, CA 94720, USA

*Email: lewerenz@caltech.edu

1.1 Introductory Remarks

Photoelectrochemical energy conversion dates back to the seminal work of Fujishima and Honda, who demonstrated light-induced water splitting[1] 183 years after van Trostwijk and Deiman first split water using electricity in 1789.[2] Although photoeffects at the electrolyte contact were observed as early as 1835 by Becquerel,[3] energy conversion at the interface with electrolytes was preceded by the development of solid state photovoltaic solar cells.[4,5] In 1975, Gerischer published the concept of formation of rectifying junctions between semiconductors and redox electrolytes,[6] thereby providing the

Energy and Environment Series No. 22
Integrated Solar Fuel Generators
Edited by Ian D. Sharp, Harry A. Atwater and Hans-Joachim Lewerenz

Published by the Royal Society of Chemistry, www.rsc.org

conceptual and scientific basis for already ongoing, as well as for subsequently pursued, work.[7] The first efficient and, in part, stable photoelectrochemical solar cells were developed by the Bell Laboratories group of Heller.[8,9] The initial realization of half cells focused on the photovoltaic mode of energy conversion, but was followed rather soon by the first efficient light-induced hydrogen evolution at an InP photocathode.[10]

As stability was recognized early as a serious issue for these systems, which operate at a reactive solution junction, a framework for assessing the underlying thermodynamic stability was introduced by Gerischer and Wrighton.[11,12] It was noted that semiconductors that exhibit both cathodic and anodic *thermodynamic* stability did not exist. Accordingly, the issue of materials development and optimization, as well as the search for corrosion protection schemes, dominated the field from early on. However, the issue of stability seriously compromised technological developments and the hitherto only example of a technically realized photoelectrochemical solar cell is the dye sensitized cell introduced by O'Reagan and Graetzel.[13] Despite its limitations and the considerable struggle for increased robustness while maintaining sufficiently high efficiency,[14] this success was instrumental for the next phase of development in the field: the attempt to arrive at an integrated water splitting system, sometimes referred to as the *artificial leaf*.[15] Worldwide activities in this direction began more than a decade ago and culminated, in part, in the foundation of the Joint Center for Artificial Photosynthesis (JCAP), a United States Department of Energy (US DOE) Energy Innovation Hub, located in California.[16]

The Energy Innovation Hubs (or, Hubs for short) were the brainchild of former DOE Secretary Steven Chu and were originally modelled after the Bell Laboratory R&D approach, where work on fundamental and use-inspired research, devices, and applications existed alongside efforts dedicated to prototype development with deployment to industry, integrated in single facilities.[17] Consequently, the Hubs have been conceptualized to enable developments that reach a similar breadth in pursued projects and, as such, are highly multidisciplinary in an unprecedented manner within the sphere of publically funded research. Figure 1.1 shows a comparison of the different Hubs (including the Joint Bioenergy Institute (JBEI)), their approximate technological readiness, and the respective funding programs. JCAP, in its first phase of funding, spanned the largest range of activities: from basic to applied research, from fundamental understanding and discovery to the realization and scale-up of functional water splitting prototypes. In contrast to many of the other Hubs, with topics closely linked to existing industries, at JCAP's inception, no trained work force or industry existed in the field of photoelectrochemical water splitting.

In the following sections, we give an overview and background on photoelectrochemistry and light-induced energy conversion at the electrolyte contact, with an emphasis on water photolysis. This will allow the readers to follow most of the more expert-level chapters in this book without having to resort extensively to overview articles and the seminal papers in the respective fields.

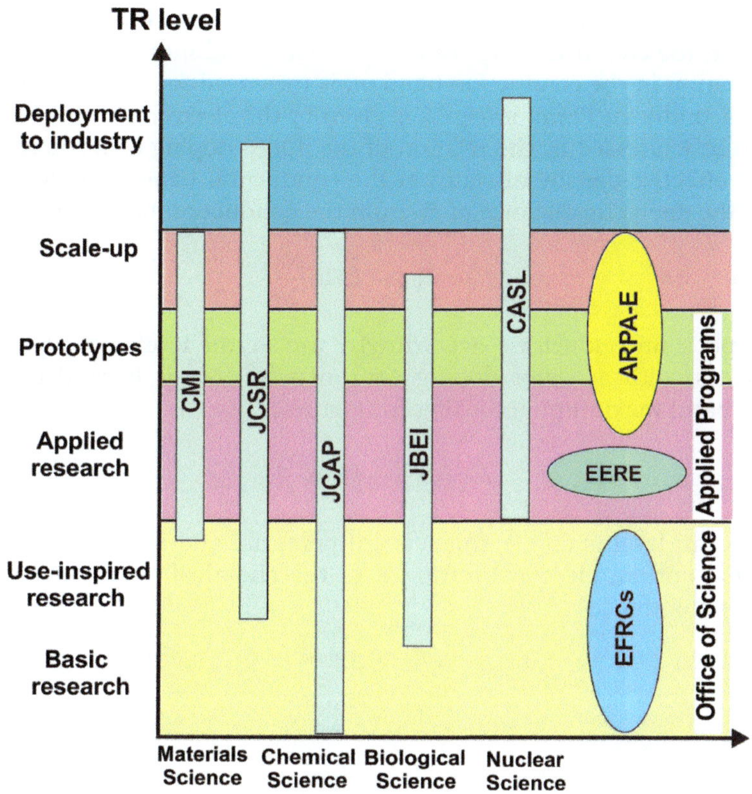

Figure 1.1 Technological readiness for energy innovation hubs versus R&D fields, including the Joint Bioenergy Institute (JBEI). Glossary—Centers: Critical Materials Initiative (CMI); Joint Center for Energy Storage Research (JCSR); Joint Center for Artificial Photosynthesis (JCAP); Consortium for Advancement of Light Water Reactors (CASL); Joint Bioenergy Institute (JBEI). Glossary—Agencies: Advanced Research Projects Agency-Energy (ARPA-E); Office of Energy Efficiency & Renewable Energy (EERE); Energy Frontier Research Center (EFRC).

1.2 Semiconductor Junctions and Dark Electrochemical Processes

1.2.1 Concept of the Classical Silicon Solar Cell

Semiconductors are characterized by an absolute energy gap, where thermal excitation of electrons from the valence band to the conduction band results in a conductivity that depends on temperature and the energetic width of the bandgap. Doping with foreign atoms allows the conductivity to be tuned over a potentially broad range if the energy levels of the dopants, $E_{D,A}$, are located close enough to the band edges. In this case, thermal excitations of electrons from E_D (donors) to the conduction band of an n-type semiconductor or from the valence band to E_A (acceptors) in a p-type semiconductor increases the majority carrier concentration and thus the conductivity. For an n-type semiconductor, the conductivity is given by $\sigma = e\, n\, \mu$, and increases as the carrier concentration in the conduction band, n, is increased for a given mobility, μ. The semiconductor Fermi level, E_F, is given by the electroneutrality condition and can be expressed by the relation of the donor doping concentration, N_D, and the effective density of states at the conduction band edge, N_{CB}, which defines the energetic distance of E_F from the conduction band edge, E_{CB}:

$$E_F = E_{CB} - kT\ln\frac{N_{CB}}{N_D} \tag{1.1}$$

for n-type semiconductors; accordingly, the Fermi level for p-type semiconductors, with acceptor doping concentration N_A, is located above the valence band maximum according to (Figure 1.2):

$$E_F = E_{CB} - kT\ln\frac{N_{VB}}{N_A} \tag{1.1a}$$

The Fermi level is an electrochemical potential and is the sum of a concentration and an electrical term, *i.e.* of the chemical potential μ and the Galvani potential φ:

$$\mu^* = \mu + e\varphi \tag{1.2a}$$

$$\mu = kT \ln c^0 \tag{1.2b}$$

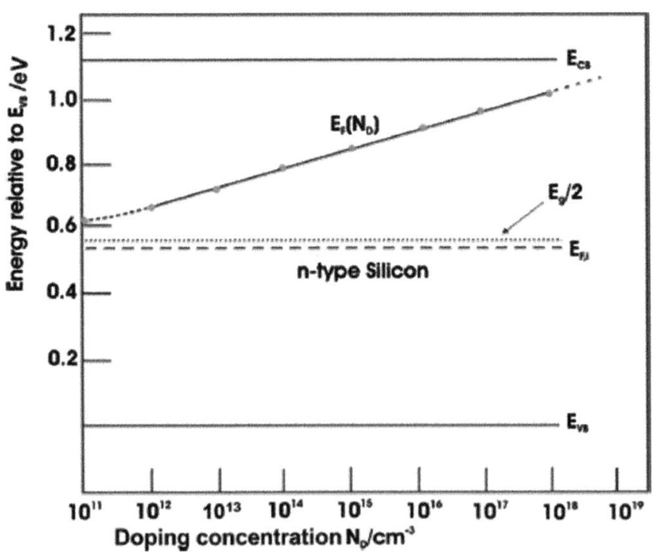

Figure 1.2 Shift of the energetic position of the Fermi level in Si for n-type doping with the donor doping concentration N_D. Also shown is the position of the intrinsic Fermi level (undoped Si), located slightly above the middle of the energy gap because the effective density of states at the conduction- and valence band edges differ ($N_{VB} = 1.83\times10^{19}\,\text{cm}^{-3}$, $N_{CB} = 3.2\times10^{19}\,\text{cm}^{-3}$).

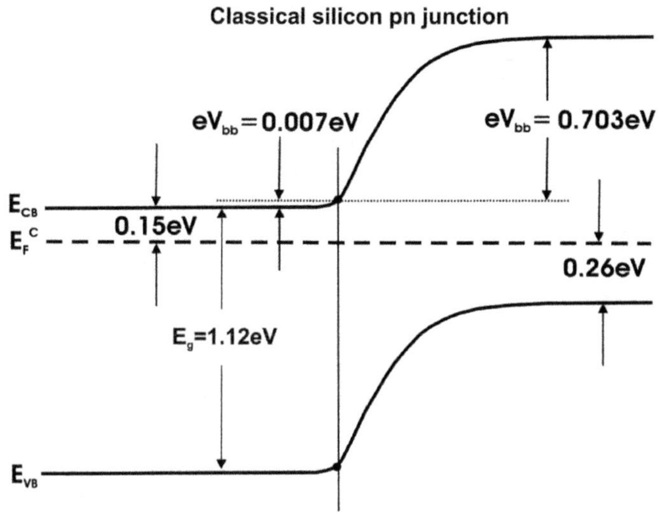

Figure 1.3 Si pn junction such as used in the classical asymmetrically doped crystalline Si solar cell; the doping concentrations here are $10^{15}\,cm^{-3}$ for the p-type side and $10^{17}\,cm^{-3}$ for the n-type side of the junction. From the doping concentrations, the positions of the Fermi levels away from the junction have been determined using eqn (1.1), (1.1a) to be 0.15 eV below the CB edge (n-type) and 0.26 eV above the VB edge (p-type). The total contact potential difference of 0.71 eV drops, however, almost exclusively in the lower doped p-type part, where a pronounced space charge layer is formed.

The Galvani term arises because equilibria between charged phases are considered and the depletion of, for example, electrons on one side of a contact results in its positive charging, whereas the electron-receiving phase will be negatively charged. The energetic position of the Fermi level (or electrochemical potential) in two phases that are contacted defines the flow of charges upon contact formation. The phase with larger electrochemical potential (${}^{1}E_{F} > {}^{2}E_{F}$ or ${}^{1}\mu^{*} > {}^{2}\mu^{*}$) will provide the electrons during contact formation. This charge exchange continues until the redistribution of charges is compensated by the built-up electrical field. Accordingly, both phases are charged. For an asymmetric semiconductor p–n junction in which the n-type doping concentration is higher than the p-type doping concentration, an energy schematic is shown in Figure 1.3.

1.2.2 The Semiconductor–redox Electrolyte Contact

Whereas for solid state devices, contact and equilibrium formation is rather straightforward and has a long history,[18,19] the establishment of equilibria at the semiconductor–electrolyte junction was less explored. In fact, the first observation of light-induced water splitting with TiO_2[1] can be viewed as an empirical result since a conceptual basis for operation had not yet been developed. The fundamental concept uses the definition of the solution redox potential according to the Marcus–Gerischer theory,[20–22] the consideration of the interface between a semiconductor and a redox solution, and the partition of the contact potential difference between electrolyte and semiconductor. The redox potential or energy has been derived using the picture of fluctuating energy levels for electrons on ions in (aqueous) solution. The influence of a polar solvent such as water on the electronic levels is basically twofold: first, the solvent becomes polarized due to the charged ions, resulting in an altered electronic energy on the ion upon formation of the solvation shell; second, the thermal fluctuation of the dipoles at a given temperature induces a type of 'breathing' of the solvation shell on the timescale of nuclear motion, *e.g.* picoseconds,[23] which results in a fluctuation of the electron energy around its most likely value. The derivation of the concept was classical in its early version[20–22] and advanced later[24,25] and the solvation shell can be viewed as a type of quantum confinement of the ion with periodic fluctuation of its size. Based on a harmonic oscillator approximation, Gerischer developed the occupation probability, $w^{G}(E)$, for oxidized and reduced ionic species in solution as

$$w^{G}_{R,O}(E) = \frac{1}{\sqrt{4\pi\lambda kT}} \cdot e^{-\frac{(E_{R,O}-E)^2}{4\lambda kT}} \tag{1.3}$$

Eqn (1.3) describes a Gaussian behavior of the energy levels on the ion. The thermal fluctuation shifts the electronic energy from its equilibrium values E_{ox} or E_{red} for the oxidized or reduced species, respectively, to a value E. The reorganization energy, λ, is the energy needed to change the nuclear configuration of the solvent from the donor to the acceptor state or *vice versa*. Another picture is to envisage an outer sphere charge transfer of an electron from the solid to the oxidized state of the redox couple, for example, Fe^{3+}_{aq}, which, as an ultrashort tunneling process, occurs without structural changes in solution (Franck Condon principle[26]). On mostly a ps timescale, the CT process is followed by relaxation of the nuclear coordinates from the acceptor state to the donor state in solution, and the corresponding free energy change $E_{ox} - E_{red}$ is 2λ.

For contact formation, the equilibrium potential of the redox couple in solution has to be considered; it is given by the condition $D_{Ox}(E) = D_{Red}(E)$, where D denotes the density of electronic states on the ions. The functional dependence of $D(E)$ is defined by $w^{G}(E)$ and the concentration of the respective species in solution: $D_{R,O} = c_{R,O}\, w^{G}_{R,O}(E)$. The situation is graphically depicted in Figure 1.4 where the electron energy relations also include the semiconductor at flatband condition, *i.e.* before contact formation and without surface states. The main parameter that defines the formation of the junction is the contact potential difference (CPD), *i.e.* the energetic difference between the electrochemical potential in solution and in the semiconductor, hence CPD $= E_F - E_{R,O}$. In the simplest case, one can approximate the potential distribution across the junction by considering the respective

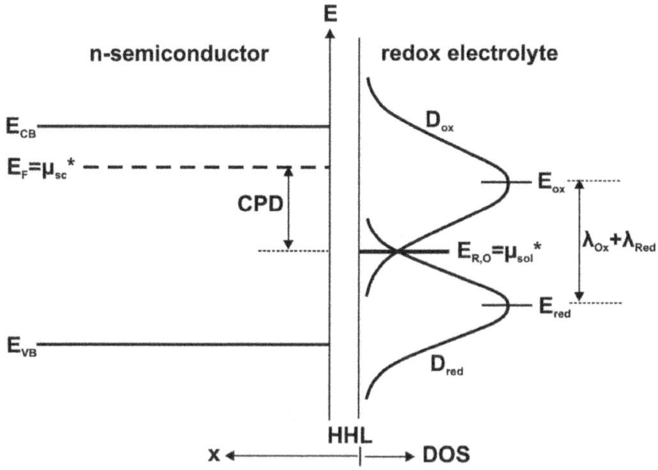

Figure 1.4 Energy scheme for the semiconductor–redox electrolyte junction before contact formation. Note the difference between the redox energy of the electrolyte and the Fermi level of the semiconductor, with these values corresponding to the electrochemical potentials μ^*_{sol} and μ^*_{sol} for the electrolyte and the semiconductor, respectively. For generality, the density of states in solution (DOS) for the oxidized and the reduced species have been drawn with different energetic width. Accordingly, the difference $E_{Ox} - E_{Red}$ is $\lambda_{Ox} + \lambda_{Red}$ instead of 2λ. The spatial location of the redox couple has been assumed to be at the outer part of the interfacial Helmholtz layer (HHL) since electron transfer is assumed as an outer sphere diabatic tunneling process (see below). The contact potential difference, CPD, determines the energetics of the semiconductor-electrolyte junction.

capacitance in each phase; in semiconductors, $C = \varepsilon\ \varepsilon_0/W$ where the space charge layer width is given by $W = (2\varepsilon\varepsilon_0 V_{bi}/qN)^{1/2}$ and depends on the dopant concentration (N), the built-in voltage, V_{bi}, and the static dielectric constant ε. For non-degenerately doped semiconductors, typical space charge layer widths are in the range between 100 nm and 1 μm. The corresponding space charge layer capacitance is of the order of 10 nF. The Helmholtz double layer capacitance for solutions with ionic strength around 0.1–1 M is about 1–3 μF,[27] *i.e.* larger by a factor of ~100. The relative voltage drop on either side of the junction is $V = Q\ (1/C_{HH} + 1/C_{sc})$ and the larger part of the CPD drops across the semiconductor surface, forming a space charge layer.

Figure 1.5 depicts the situation shown in Figure 1.4 after contact formation. Equilibrium is established by equalizing the electrochemical potentials on both sides of the junction, $E_F = E_{R,O}$. In this ideal case (absence of surface states, non-degenerate semiconductor, moderate built-in voltage), most of the CPD drops within the semiconductor, exhibiting a linear-parabolic behavior with distance, x, below the surface. The figure also shows a small potential drop in the HHL, which has been included for

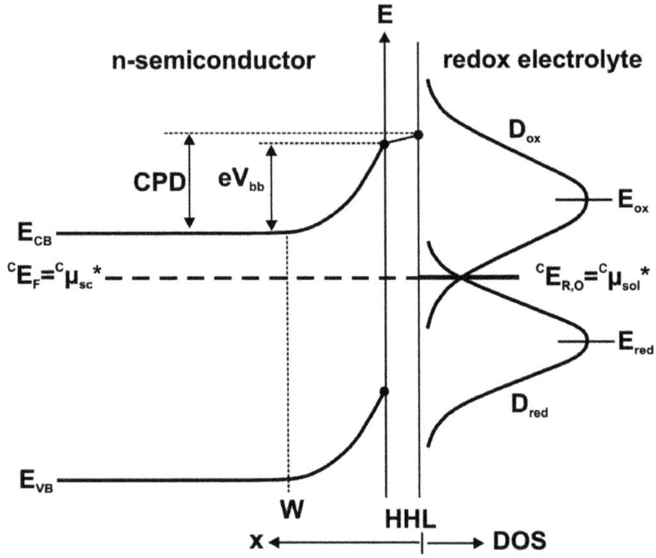

Figure 1.5 Semiconductor–redox electrolyte junction in contact. Note the equilibration of the electrochemical potentials on either side of the junction. The total CPD is divided into a smaller part that drops across the HHL (overemphasized in the figure) and across the semiconductor. The resulting bending of the energy bands, eV_{bb}, is indicated along with the width of the space charge layer w (see text).

clarity. Depending on the doping level of the semiconductor (which in turn changes the capacitance) the branching between semiconductor band bending eV_{bb} and HHL potential eV_{HH} changes. Increased doping reduces eV_{bb} relative to eV_{HH}, and *vice versa*. The energetic situation shown represents a rectifying contact.

1.2.3 Dark Currents at the Semiconductor–electrolyte Boundary

In its functional behavior, the dark current at the redox–electrolyte junction resembles that of a Schottky diode.[18] However, the factors/parameters that determine the current are considerably different from the thermionic emission model[28] of a metal–semiconductor junction. Although one sometimes encounters the expression 'Tafel equation for semiconductors' for the corresponding characteristic, this is seriously misleading because the Tafel equation is an approximation of the Butler–Volmer equation[29] for larger overpotentials. However, the Butler–Volmer equation has been derived for the metal–electrolyte case, where the total potential drops across the HHL, resulting in an electrochemical Stark effect where the potential alters the energetic position of an activated complex (transition state theory[30]). The charge transfer is adiabatic and the tunneling probability for outer sphere processes is 1 in the Butler–Volmer approach.

In the case of semiconductors, the distance dependence of the tunneling process cannot be neglected since the charge transfer reaction is not based on the potential-induced alteration of the energy of the transition state. The dark forward current for an n-type semiconductor is given by

$$j_d^C(V_a) = ek^C \cdot n_s(V_a) \cdot D_{Ox} = ek_0^C e^{-\frac{eV_{bb}}{kT}} \cdot n_0 e^{-\frac{eV_a}{kT}} \cdot D_{Ox} \tag{1.4}$$

which gives an exponential increase of the cathodic dark current $j_d^C(V_a)$ with applied voltage. This similarity in behavior has led to the confusion of the concepts of adiabatic and diabatic electron transfer at the electrode–electrolyte phase boundary.[31] The fundamental differences between the two approaches are contained in the rate constant for the cathodic (forward) reaction, k^C. Whereas it carries the potential dependence of the reaction in the activated complex theory framework that leads to the Butler–Volmer equation for metal–electrolyte systems, it is ideally potential-independent for the semiconductor case. The potential dependence is given by the surface concentration of electrons, n_s. k_0^C contains a term that includes the tunneling probability for electron transfer summed over all distances.[32] Since the potential drop in the HHL can moderately change with applied potential, the rate constant is also a function of V. In the (typical) case when surface states are present, the according capacitance depends on their density of states and, for a high density, a substantial part of the CPD drops across the HHL, resulting in a mixed situation where $n_s = n_s(V_a)$ and $k^C = k^C(V_a)$.

The reverse current from the electrolyte to the semiconductor (anodic component) is, in this ideal consideration, potential independent and has the same value as the forward current without applied voltage, j_0, although the transfer mechanism is different. Accordingly, the total current at the semiconductor–electrolyte junction follows the functional behavior of the well-known relation for rectifying junctions:[33]

$$j(V_a) = j_0 e^{-\frac{eV_a}{kT}} - j_0' \tag{1.5}$$

The parameters j_0 and j_0' in eqn (1.5) represent different charge transfer processes in the forward and reverse directions. In the reverse (anodic) process, the reservoir of the electrons on the solvated ions differs from the Boltzmann term for the surface electron concentration of the semiconductor and the current, j_0', is given by

$$J_0' = qk_0'^c \cdot N_{CB} \cdot D_{Red} \tag{1.6}$$

where N_{CB} denotes the effective density of states in the semiconductor conduction band.

1.2.4 The Role of Surface States at the Electrolyte Boundary

Surface states are ubiquitous at semiconductor interfaces. One distinguishes between intrinsic surface states that arise from the termination of the solid in one direction and extrinsic ones, which result from surface interactions, such as exposure to atmospheres, as well as contact to other solids and to liquids. The capacitance of surface states depends on their density and can substantially exceed that of the space charge layer. In that case, with a surface state DOS in the range above $10^{13}\,cm^{-2}\,eV^{-1}$, the semiconductor behaves in a hybrid manner: it shows partial or complete Fermi level pinning (FLP).[34] Partial FLP exemplifies the situation most clearly: the total CPD now drops across both the HHL and the semiconductor space charge region because the potential drops follow the capacitance relationship. Complete FLP is typically observed for a DOS $> 10^{14}\,cm^{-2}\,eV^{-1}$. Figure 1.6 depicts the situation for partial FLP where the CPD is split between a drop in the HHL and in the space charge region. Applying a positive potential will then increase both the HHL potential drop and that of the space charge. In that case, the band edges shift and the band bending increases but the applied potential only partially increases the semiconductor band bending.

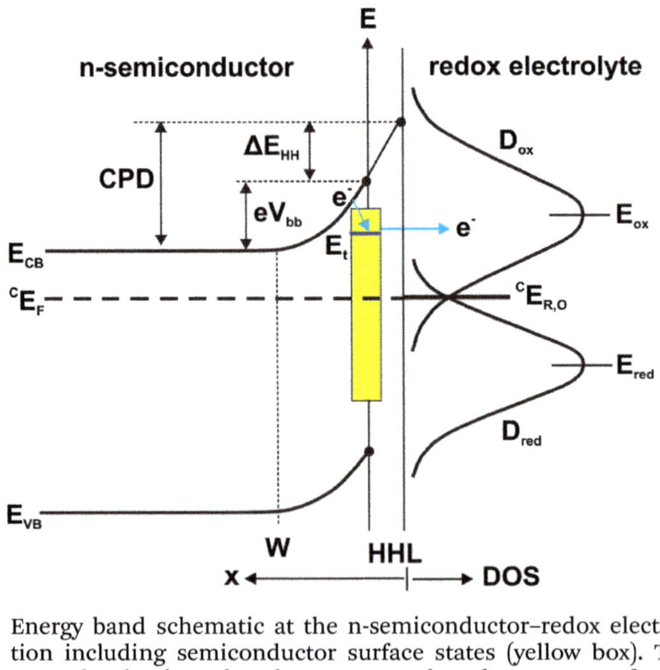

Figure 1.6 Energy band schematic at the n-semiconductor–redox electrolyte junction including semiconductor surface states (yellow box). The surface state density here has been assumed to be constant for simplicity, though it is generally a function of energy. ΔE_{HH} denotes the energy drop in the HHL, with $CPD = eV_{bb} + \Delta E_{HH}$. The blue arrows indicate first an inelastic and subsequently an isoenergetic charge transfer over a trap state with energy E_t (see text).

In cases where the energetic overlap between the semiconductor band edges and the DOS of the redox couple is small, the explanation of the experimental results has to invoke the participation of surface states in the charge transfer process.[35,36] The overall transfer process consists of (i) transport of carriers in the semiconductor bands to the surface, (ii) inelastic transfer of carriers from the bands to surface states and (iii) the charge transfer to the oxidized redox species (see blue arrows in Figure 1.6). It becomes obvious from the figure that the overlap with the DOS of the oxidized redox species is increased, resulting in higher currents. In this scenario, eqn (1.4) and (1.5) have to be modified in two ways: first, the electron flux to the surface is given by

$$f_e = n_s \cdot N_t^e \cdot \sigma_n \cdot \nu_{th}^{CB} \tag{1.7}$$

where the terms N_t^e, σ_n and ν_{th}^{CB} denote the trap density of empty surface states, the electron capture cross section (ranging from 10^{-12} cm^2 for Coulomb attraction to 10^{-18} cm^2 for repulsion), and the thermal velocity of electrons in the conduction band (typically 10^6–10^7 cm s^{-1}), respectively. Second, the charge transfer process to the redox species does not follow the ideal non-adiabatic behavior. Here, theoretical development is needed that connects adiabatic and diabatic electron transfer processes in relation to the surface state density.

Figure 1.7 shows the influence of an applied negative potential on the energetic relations at the semiconductor–electrolyte contact based on the

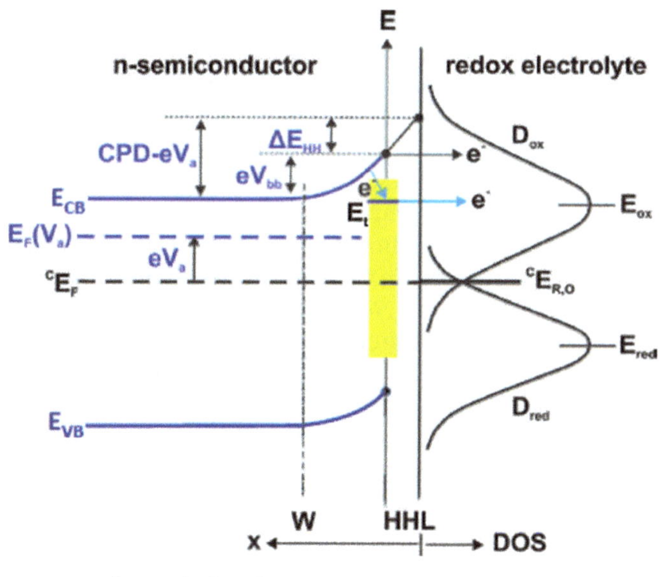

Figure 1.7 Energy schematic for charge transfer *via* conduction band (black arrow) and *via* surface states, E_t, (blue arrow) upon applying a negative potential V_a to the n-type semiconductor. Note that both the band bending, eV_{bb}, and the HHL drop, ΔE_{HH}, have decreased because the original CPD is smaller by eV_a. The relative potential drop between semiconductor and HHL has been assumed to be of similar size.

schematic presented in Figure 1.6. One observes that a forward (negative) voltage results in a reduction of the semiconductor band bending and, also in a reduced potential drop within the HHL. Accordingly, the band edges shift upward. In this situation, the overlap between the redox DOS and the conduction band minimum is reduced and processes *via* surface states gain more importance due to the Gaussian reduction of the DOS with a linear shift of the semiconductor energy levels.

Up to this point, the fundamental interfacial properties of semiconductors in contact with electrolyte, as well as corresponding dark charge transfer characteristics, have been presented. Before discussing redox reactions at illuminated semiconductor interfaces, we provide a general overview of the various types of charge-separating semiconductor junctions and how they are commonly assembled into integrated solar fuels systems.

1.3 Semiconductor Junctions for Solar Energy Conversion

1.3.1 Overview of Junction Types

Semiconductor light absorbers lie at the heart of integrated solar fuel generators, though a wide range of different internal configurations for generating photovoltage and photocurrent are used in practice. These include both solid–solid photovoltaic junctions and solid–liquid photoelectrochemical junctions. Figure 1.8 gives a general overview of semiconductor-based rectifying junctions. Further below, the scope will be extended to address the taxonomy of solar energy converting solid–liquid junctions which are represented by two examples in Figure 1.8. *Schottky solar cells* are limited in their efficiency due to light attenuation by the metal, where sheet resistance, metal thickness, and absorption length have to be balanced. Recently, however, ultra thin graphene sheets have been used as metallic front contact in an efficient Schottky cell configuration.[37] The typical example of a *pn homojunction cell* is the classical crystalline Si solar cell[38] that has been introduced in 1954.[5] *Heterojunction solar cells*, where the semiconductors are made from different materials, are typically found in thin film solar cells, such as the ternary chalcopyrite-based material class that includes the more recently developed kesterites.[39,40] The exploitation of the electron excess energy, which otherwise is lost by thermalization, is pursued by structures with dual or multiple absorbers where the respective energy gaps are chosen to optimize the efficiency. These *tandem cells* or *multi-junction cells* generate higher voltages and the maximum attainable photocurrent is given by a judicious balance between energy gap, absorption behavior and thickness of the subcells (absorbers) of the stack. Due to the increase in photovoltage, such structures have become increasingly interesting for application in light-induced water splitting.[41–44] Since multi-junction solar cells, if monolithically integrated, need tunnel junctions between their component cells, the preparation of devices with more than

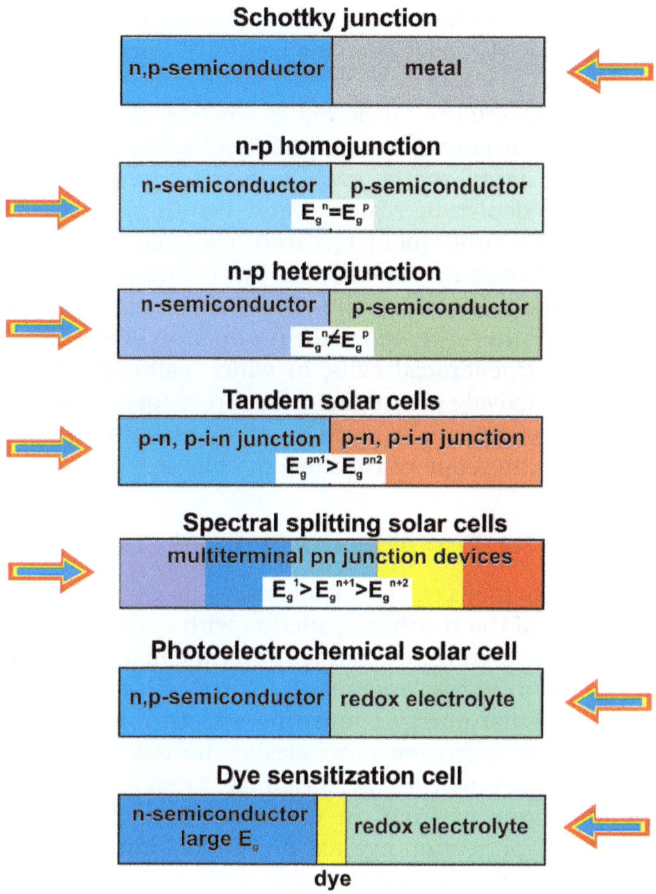

Figure 1.8 Overview of rectifying junctions for solar energy harvesting using semiconductor-based structures and systems (see text); E_g^n, E_g^p, $E_g^{pn1,2}$ denote the energy gaps of the n- and p-type semiconductors, respectively, and the last term stands for the energy gap of the pn homojunctions 1 and 2 of a tandem structure. Here, the difference in energy gaps shall indicate different materials. However, in III–V and II–VI semiconductor families, identical energy gaps with different materials can, in principle, be prepared. The arrows indicate the illumination direction.

three absorbers is challenging. In particular, abrupt junction doping profiles have to be established, which imposes stringent demands on materials control at highly doped interfacial regions. Another approach uses the *spectral splitting* method where the incoming solar light is split into a multitude of spectral beams that each illuminate a solar cell, optimized for the given spectral irradiation.[45] These multi-terminal devices operate as solar concentrators.[46]

Photoelectrochemical solar cells are, in principle, based on the formation of a rectifying junction between a semiconductor and a redox electrolyte. Both photovoltaic, *i.e.* current generating, and photoelectrosynthetic cells, which can generate fuels (*e.g.* H_2), have been realized[41–44,47] and their details will be discussed below.

The *dye sensitization cell* can be viewed as a hybrid as its function is based on the asymmetry of the chemical potential for charge carriers between a dye on the surface of a large energy gap semiconductor and the host semiconductor, where a depletion region (band bending) exists. The original work and most of the subsequent research has been done on Ru-dyes on TiO_2.[13,48] The redox reaction replenishes the oxidized dye, whose excited electron has been injected into the substrate. While dye sensitized cells have most commonly been investigated for the production of electrical power, dye sensitized photoelectrochemcial cells, in which catalysts are incorporated into the device, are actively investigated for direct fuels generation.[49] However, these intriguing systems, as well as many elaborate molecular architectures, lie outside the scope of the current volume.

1.3.2 Junctions for Photoelectrochemical Energy Conversion

A classification of photoelectrochemical solar cells can be developed by assessing the location of the rectifying junction with respect to the electrolyte. Figure 1.9 presents an overview of some hitherto considered systems. Since the rectifying junction is the 'engine' of a solar cell, its formation and its position within the solar energy converting structure is of interest, in particular because (photo)corrosion processes at the reactive electrolyte interface can seriously limit the device performance. Therefore, the displacement of the active junction away from the electrolyte boundary can protect this junction. Additional protection against decomposition can be achieved by interfacial layers and also by compact and surface covering electrocatalyst films if the catalysts are stable in the respective electrolyte. Naturally, these interfaces and interfacial films affect the interfacial energetics and, in particular, the energy band alignment.[50]

Figure 1.9(a) shows the formation of a rectifying junction between a redox solution and an n-type semiconductor. The half cell is protected by a passivation layer and for photoelectrocatalysis applications, electrocatalysts reside at the electrolyte boundary. Without catalysts, a classical photoelectrochemical junction would exist.[6] However, the presence of the metallic catalyst can alter the rectification properties. Rectification can depend on the size and thickness of the catalyst layer. The influence of smaller catalytic deposits (nano-particle layers) on the junction energetics is a topic of ongoing investigation. In Figure 1.9(b), the photovoltage is generated between an asymmetrically doped n^+- and a p-type semiconductor (see perpendicular dashed lines that indicate the space charge layer extension) in a photocathode heterojunction structure (compare also Figure 1.8). The contact of the catalyst/n^+–p structure to the electrolyte has been assumed to be ohmic.

Approaches that generate higher photovoltages are pursued because of the efficiency limitation for single junction photocatalysis cells: for unassisted

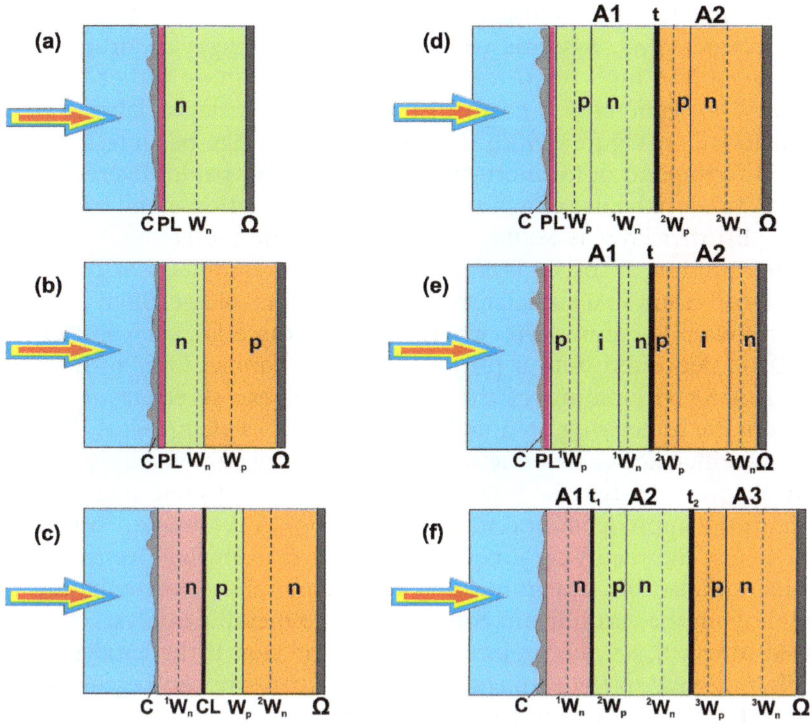

Figure 1.9 Schematic overview of rectifying junctions envisioned/used for light-induced water splitting. Thin full lines represent separation of differently doped regions, bold full lines indicate tunneling layers, and dashed lines show space charge regions. Abbreviations are as follows: C, catalyst; PL, passivating layer; Ω, ohmic (conductive) back contact; W, space charge layer width; subscripts n and p refer to n- and p-type semiconductors, respectively; CL, contact layer, for example FTO or ITO (fluorine doped tin oxide/indium tin oxide); A1–A3, absorber 1-absorber 3; t_i, tunnel junctions in the tandem cells consisting of highly (degenerately) doped p^+ and n^+ regions; i denotes intrinsic, *i.e.* a undoped part of a semiconductor in a tandem configuration (see text).

efficient water splitting with a photocurrent of >10 mA cm^{-2}, the thermodynamic value of 1.23 V has to be exceeded substantially because of reaction and working overpotentials. Considering radiative recombination in the Shockley–Queisser detailed balance limit,[51] the minimum photovoltage that has ideally to be generated is ∼1.6 V.[52] For the oxygen evolution reaction (OER), a 4-electron transfer process, an additional potential of 0.3–0.4 V is needed to sustain the current density. Including a series of additional loss processes, the photovoltage of an *efficient system* should be ∼2.2 eV, which places the maximum power point close to 2 V for a system with high fill factor. The resulting solar-to-hydrogen (STH) efficiency using a single semiconductor light absorber is then limited to a theoretical efficiency of 15.1% for 1 sun (AM 1.5G spectrum) illumination.[52] The theoretical approach includes reflection losses and absorption by the (metallic) highly active electrocatalyst. Systems with more than a single absorber are employed to achieve high photovoltages while maintaining rather large photocurrents. For generality, a corrosion protection interlayer (labeled PL) has been included in Figure 1.9(a), (b), (d) and (e). It has become a standard approach in the search for more robust systems. For oxide covered surfaces (Figure 1.9(c), (e)) the use of a larger energy gap transition metal oxide (TMO) as front absorber layer is assumed to provide some protection.

Adding another absorber layer on top of, for example, a Si or GaAs pn junction, can result in a sufficiently increased photovoltage. Such a structure has been drawn in Figure 1.9(c) where an n-type material, such as TiO_2, WO_3 or $BiVO_4$ is deposited onto a pn junction. The contact layer provides tunneling and recombination of the light-induced excess electrons with the holes from the underlying pn junction. These devices are presently limited to rather low efficiency values due to the large energy gap of the typically employed absorber materials.[53] The search for better suited materials uses combinatorial approaches and first advances have been made.[54] Whereas in Figure 1.9(b), the rectifying junction is removed from the electrolyte interface, the top absorber in Figure 1.9(c) can form a photoelectrochemical cell (depending on the equilibrium condition of the metallic catalyst) coupled to a buried junction, *i.e.* the pn-junction in its interior. If the catalyst is energetically 'coupled' to the semiconductor, the structure can be considered a hybrid PEC and pn junction cell.

The dual tandem structure (Figure 1.9(d)) gains its power from the combination of two buried pn-junction solar cells with four space charge regions. Only the larger extended space charge layers contribute notably to the total achievable photovoltage because the systems are asymmetrically doped, as also shown in Figure 1.3. The use of p–i–n (or n–i–p) semiconductor structures has become a standard whenever the product of excess carrier lifetime, τ, and mobility, μ is small.[55] Such a condition results in a minority carrier diffusion length of $L=(D\ \tau)^{1/2}$, where $D=(kT/q)\ \mu$, that is small compared to the photon absorption length of $x_\alpha=3\alpha^{-1}$ ($x_\alpha>L$). By creating systems in which the electric field extends throughout the intrinsic part of the semiconductor, transport of photogenerated carries occurs *via* drift and charge extraction competes favorably with bulk recombination processes. This concept has been originally employed to develop efficient solar cells with amorphous silicon.[56] The p–i–n design is shown in Figure 1.9(e). Such cell structures have been used to develop efficient water photolysis systems. The photoactive junctions are removed from the electrolyte contact and such devices represent buried junction photocatalysis cells. Figure 1.9(f) shows the equivalent structure of Figure 1.9(c) for a dual tandem junction. Here again, the system is of hybrid nature, consisting of a buried and a possible semiconductor–redox electrolyte junction.

A variety of architectures have been used to assemble semiconductor light harvesting elements into integrated solar fuels generators. A comprehensive treatment is given in the chapter by Xiang *et al.* Here, a major consideration

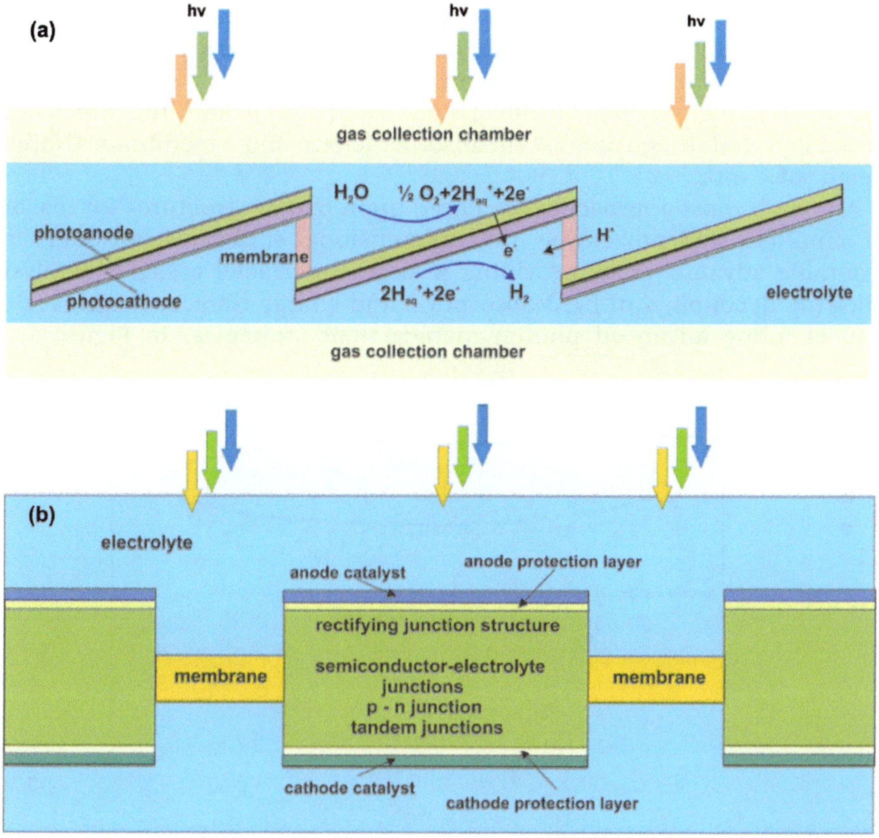

Figure 1.10 Illustrations of monolithically integrated artificial leaf designs. (a) Louvered-design, where the arrangement of the membrane is not obstructing the path of the incoming light. The photooxidation reaction generates electrons that migrate to the cathode through the solid state absorber structure, protons are transferred *via* the membrane, which also inhibits gas crossover. (b) Planar design in which the membrane area reduces the maximum attainable efficiency. The structure shows details of the absorber systems that comprise catalysts, protection layers against photocorrosion on the photo-anode and -cathode side. The approaches used to form rectifying junctions are also indicated.

is the geometric placement of not just the semiconductor but also the ion-conducting membrane. As described in the chapter by Miller and Houle, such membranes are essential elements that separate generated products, while allowing transport of ions between the catholyte and anolyte. Product separation ensures reduced back reaction rates (*i.e.* oxygen reduction and hydrogen oxidation), which are loss processes in solar water splitting devices, and ensures that pure product streams with no explosive mixtures are created. In Figure 1.10, two examples of geometries for monolithic integration into a solar fuels device (*i.e.*, an "artificial leaf") are depicted. The structures have been modeled to obtain the dimensions necessary to minimize losses due to ion transport, membrane conductivity, and motion of light-induced carriers.[57] For photocurrents of 10 mA cm^{-2}, corresponding to 12.3% STH efficiency (see Section 1.4 and eqn (1.16) below), the dimensions of the louvered design are: 1.5 cm absorber length and a membrane (Nafion) height of 4 mm.

Although most approaches are based upon planar structures for reasons of simplicity and processing, three-dimensional structures can offer considerable advantages by providing a larger interfacial area for catalysis, allowing decoupling of light absorption and charge transfer length scales, and enabling advanced photon management strategies. In Figure 1.11,

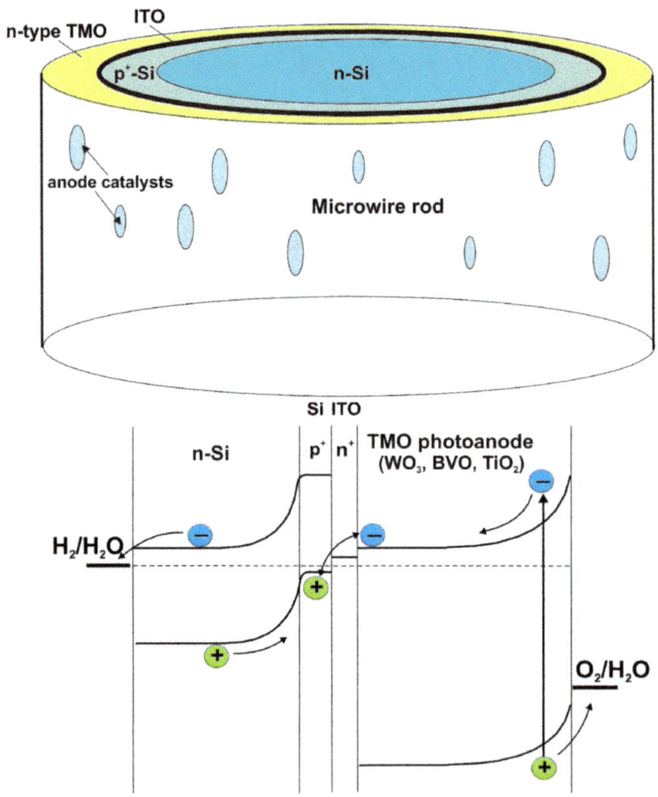

Figure 1.11 Top: schematic of a silicon microwire rod structure that includes the various semiconductor layers and the electrocatalysts at the photoanode; illumination is from above. Bottom: energy band schematic that shows the operation under illumination under short circuit conditions and that describes the role of each part of the structure for collection of carriers at the catalyst sites (due to multiple reflections depending on the lateral distance of the rods, light impinges between perpendicular and inclined directions). Note that excessive catalyst loading can result in reflection in case of metal catalysts (see text).

a microwire design is shown where radial carrier collection occurs. Light absorption, parallel to the rod axis, is orthogonal to carrier separation. An asymmetric pn junction of Si is used that generates ~0.6 V photovoltage. To overcome the limit of 1.23 V for water splitting, the TMO is deposited onto its surface and needs to supply a voltage of at least 1 V in the ideal case discussed above. For TiO_2, the observed photovoltage exceeds 1.6 V. The structure thus shows unassisted water splitting albeit with reduced efficiency.[58] It should be noted that this design uses majority carrier electrons from n-Si to reduce water and light-induced minority carriers from the TMO for oxidation. The indium tin oxide interlayer (ITO) functions as a tunnel junction where the majority electrons from the TMO recombine with the minority holes from n-Si. An energy band schematic is shown in Figure 1.11.

1.4 Photocurrent Generation at Illuminated Semiconductor Junctions

We now turn from device architectures to the internal processes that govern the function of light harvesting elements of integrated solar fuels generators, beginning with light absorption and extending to redox reactions at solid–liquid interfaces.

1.4.1 Photon Absorption

Semiconductors are generally classified as either indirect or direct according to the nature of optical transitions across their fundamental energy gaps. The terminology for direct and indirect transitions results from consideration of the alignment of the conduction band minimum and valence band maximum in reciprocal space, *i.e.* in electron **k**-space. *Direct transitions* are typically shown as vertical transitions in the reduced zone scheme and are associated with high oscillator strength, leading to large absorption coefficients. It can be shown that, for nearly free electrons, the absorption coefficient near the fundamental absorption edge is given by[59]

$$\alpha(\omega) = A\frac{(\eta\hbar - E_g)^{1/2}}{\hbar\omega} \tag{1.8}$$

Thus, a plot of $[\alpha(\omega)\hbar\omega]^2$ *versus* the photon excitation energy, $\hbar\omega$, gives a linear relationship and allows extrapolation of the energy gap of the respective semiconductor.

For *indirect transitions*, the conduction band minimum and valence band maximum lie at different positions in **k**-space. Therefore, energy and momentum conservation dictates that optical transitions across the fundamental energy gap must include additional momentum sources, *e.g.* phonons, that, depending on their wave vector, $\mathbf{k}_{ph}$, and their nature (optical, acoustic) vary in energy. Accordingly, both the phonon energy and momentum have to be considered in these transitions and one obtains

$$E_f - E_i = h\nu + E_{ph} \text{ (phonon absorption)}$$
$$E_f - E_i = h\nu - E_{ph} \text{ (phonon emission)} \quad (1.8a)$$

and

$$\mathbf{k_f} = \mathbf{k_i} + \mathbf{k_{ph}} \text{ (momentum conservation)}$$

The photon energy dependence of the absorption coefficient can be determined for free-electron like semiconductors in the effective mass approximation.[60] The energy bands are represented by parabolae, where the crystal potential enters *via* the effective mass. The curvature of the parabola is given by the effective mass, leading to expressions such as heavy and light holes (in GaAs, for example). The energies in **k**-space assume a spheroidal shape since $E = \hbar^2 k^2/2m^*$ (m^* effective mass). Integration over initial and final states and including the electron-phonon interaction yields, for *indirect transitions*:[59]

$$\alpha(h\nu) = B^*(h\nu - E_g)^2 \quad (1.9)$$

Note, that the pre-factor A in eqn (1.8), which is on the order of 10^4, is approximately 5×10^2 times larger than B^* (for example, when comparing the absorption coefficients of the indirect semiconductor Ge with that of the direct band gap material GaAs).

In some materials, optical transitions are forbidden for $\mathbf{k} = 0$ due to selection rules,[61] such as for $s \rightarrow d$ ($\Delta l = 2$) excitations, which, however, become allowed for $\mathbf{k} \neq 0$ for hybridized bands that can be described by s, p and/or d band contributions. Such transitions are termed *forbidden direct* and the photon energy dependence of the absorption coefficient in such cases is given by

$$\alpha(h\nu) = A^*(h\nu - E_g)^{3/2} \quad (1.10)$$

More generally, the wavelength or photon energy dependence of the absorption coefficient is evaluated using a Tauc plot[62] according to the following equation:

$$\alpha(h\nu) \cdot h\nu = C^* \cdot (h\nu - E_g^{pot})^m \quad (1.11)$$

where $m = \frac{1}{2}$, 2, 3/2 and 3, depending on whether the excitation leads to a *direct*, *indirect*, *forbidden direct*, or *forbidden indirect* transition. The constant C^* contains the refractive index, the dc conductivity extrapolated to $T = 0$ K, the vacuum velocity of light, and the extent of the band tailing, a deviation from the idealized picture of a perfect semiconductor crystal, which plays a prominent role in disordered materials such as amorphous silicon.

With the advent of amorphous semiconductors for solar applications, the concepts developed for crystalline solids with 3D periodic atomic arrangement do not always apply and a modified formalism has been developed to describe absorption behavior.[63] It accounts for low energy transitions that involve defect states, given by an exponential behavior with photon energy that extends over several orders of magnitude of the absorption coefficient, along with a higher energy component with $\alpha \sim (h\nu - E_g)^2$. The determined optical gap and the mobility edge, where carrier transport is notable, differ in amorphous semiconductors. One observes that the absorption coefficient of amorphous hydrogenated silicon (a-Si : H) is significantly higher than the crystalline (c-Si) form, allowing the fabrication of thin film and tandem solar cells.[64]

1.4.2 Illuminated Rectifying Junctions

We consider, for example, Figure 1.5 under illumination. The energetic situation is depicted in Figure 1.12 where the redox energy is indicated without the DOS of the redox couple constituents. It is assumed that the CPD drops exclusively across the semiconductor space charge layer. Hence, the band bending eV_{bb} equals the CPD (Figure 1.5). Following photonic excitation (process 1), the light-induced excess carriers either rapidly thermalize (process 2) or, if generated very close to the surface, are transferred as hot carriers. The electrons (process 3) and holes (process 6) *drift* in the electric field of the space charge region. Outside *W*, the electrons and

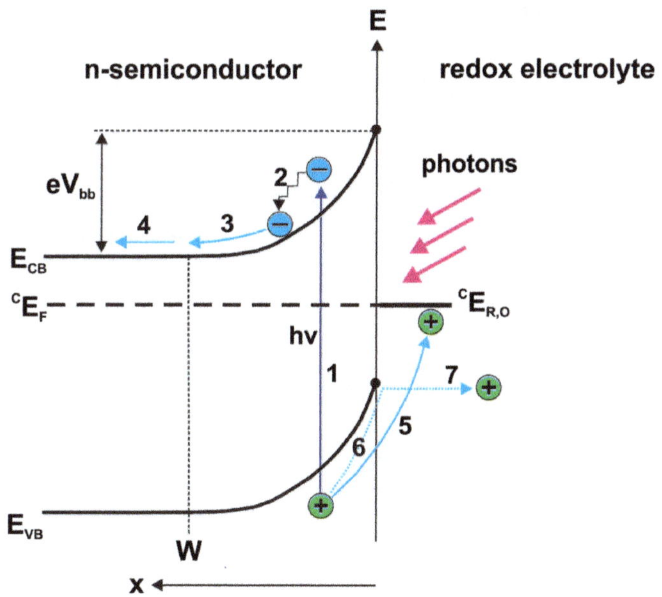

Figure 1.12 Photoexcitation processes for an n-type semiconductor–electrolyte junction; 1: optical electron transition from the valence to the conduction band; 2: thermalization to the band edge; 3: excess carrier drift; 4: excess carrier diffusion; 5: simplified scheme for hole transfer at the electrolyte interface; 6: drift of excess holes to the absorber surface; 7: outer sphere charge transfer from the valence band edge to the reduced species (DOS not shown in the figure) of the redox couple (electron transfer from the redox couple to the valence band).

holes *diffuse* in the neutral regions of the absorber (process 4). The excess holes, which are minority carriers for an n-type semiconductor, drift to the surface (process 6) where they are transferred to the reduced species of the redox couple in an outer sphere tunneling charge transfer (process 7). The arrow indicating the hole transfer to $E_{R,O}$ (labeled 5) is an often-found but somewhat misleading graphic representation of the light-induced excess carrier hole transfer. The electron transfer occurs from the reduced part of the redox couple to the surface. Here, we depict a situation where holes at the valence band edge become neutralized by electron transfer (iso-energetic tunneling from E_R to the semiconductor). In the simplest approximation, the light-induced current density can be written as the product of the number of absorbed photons ($h\nu > E_g$) with $A = 1 - R$ (reflectivity):

$$j_L(h\nu) = qn_{Ph}(h\nu \geq E_g) \cdot (1 - R) \tag{1.12}$$

In this simple approximation, the photocurrent–voltage characteristic assumes the shape of the dark j–V curve.

The photocurrent–voltage curve shown in Figure 1.13 follows from the simple superposition principle where the dark current density and the light-induced component have opposite signs $j_{Ph}(V) = j_D(V) - j_L$ $j_{ph}(V) = j_D(V) - j_L$ using eqn (1.4) and (1.5):

$$j_{Ph}(V) = j_D(V) - j_L = \left(j_0 e^{-\frac{eV}{kT}} - j_0^{'}\right) - j_L \tag{1.13}$$

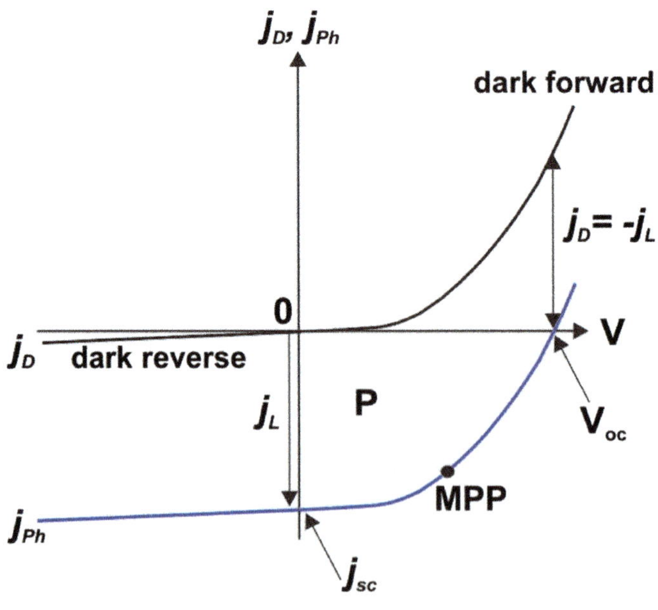

Figure 1.13 Dark and photocurrent characteristics for an illuminated rectifying junction (see text for parameters).

In this simplest picture of the voltage dependence of the photocurrent for a rectifying junction, one subtracts the light-induced current j_L (eqn (1.12)) from the dark current–voltage curve $j_D(V)$ as shown in Figure 1.13. The main solar cell parameters are included in the figure; the output power P is defined by the largest rectangle in the fourth quadrant of the j–V diagram. It is given by the current and voltage at the maximum power point, which is defined by $d(j_{Ph}V)/dV = dP/dV = 0$. The open circuit voltage, V_{oc}, is obtained for $j_{Ph} = 0$ or $j_D = -j_L$. The latter defines the short circuit current at $V = 0$. The solar-to-electrical conversion efficiency is given by

$$\eta = \frac{j_{MP} \cdot V_{MP}}{I} = ff \cdot \frac{j_{SC} \cdot V_{OC}}{I} \tag{1.14}$$

where the fill factor ff defines the rectangularity of the photocurrent–voltage characteristic: $ff = (j_{MP} \cdot V_{MP})/(j_{SC} \cdot V_{SC})$. The open circuit voltage is obtained from the condition $j_{Ph} = 0$ in eqn (1.13):

$$V_{oc} = \frac{kT}{q} \ln\left(\frac{j_L}{j_0} + j_0'\right) \tag{1.15}$$

Hence, in this simple approximation, the photovoltage varies logarithmically with the light intensity $I(x)$, where x denotes the direction perpendicular to the surface as shown in Figure 1.12, because j_L is proportional to $I(x)$.

1.5 Photoelectrochemical Water Splitting

In Section 1.2, we discussed junctions for photoelectrochemical energy conversion (Figure 1.9) and related devices (Figures 1.10 and 1.11). Here, we present the governing equations for combined photovoltaic–electrocatalyst systems and derive their basic behaviour. Water dissociation is characterized by a thermodynamic energy threshold as mentioned above in Section 1.2.2. The according voltage is given by $V_0 = -\Delta G^0/nF$ with $\Delta G^0 = \Delta H^0 - T\Delta S^0$ which yields for the water splitting reaction $H_2O \rightarrow H_2 + \frac{1}{2}O_2$ a voltage of $V_0 = -1.23$ V at 25 °C for pH 0. Hence, in contrast to solid state photovoltaics, a minimum voltage is necessary to achieve water dissociation into hydrogen and oxygen gas. This cut-off affects the achievable solar-to-hydrogen conversion efficiency and, besides stability issues, puts an additional constraint on the materials selection. Materials such as hematite are characterized by a comparably low theoretical water photolysis efficiency.

The low overpotentials selected in Figure 1.14 are given by the activation overpotential and the reaction overpotential which is defined by the current passed at the respective heterogeneous catalysts. It has been established in benchmarking studies to use a current density of 10 mA cm^{-2} since a photocurrent of 10 mA cm^{-2} has been considered a norm for efficient PEC water splitting. Overvoltages of 0.4 V as shown in Figure 1.14 result from highly

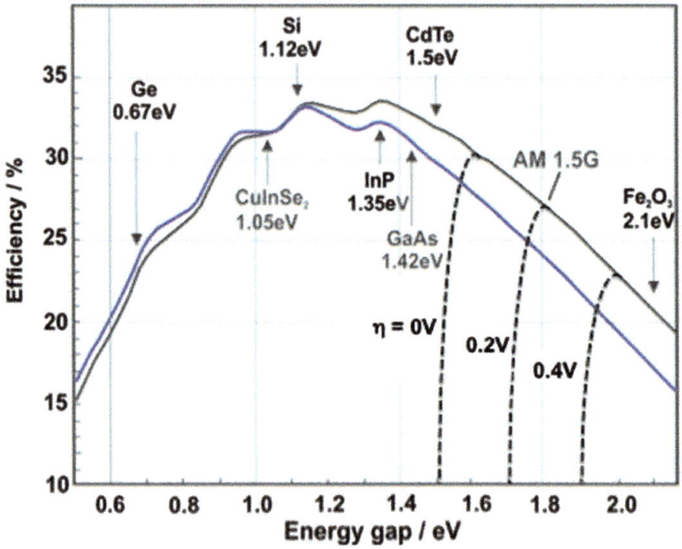

Figure 1.14 Dependence of the theoretical efficiency of photovoltaic energy conversion for AM 1.5G irradiation (black curve) and for AM 1.5d (blue curve) including solar water splitting for the ideal case (thermodynamic limit, $\eta=0$) and for two overpotentials η as indicated; energy gaps for various semiconductors are given.
Reproduced from ref. 65 with permission from the Royal Society of Chemistry.

active electrocatalysts, such as Pt for the hydrogen evolution reaction (HER) and RuO_2, IrO_2, or $NiFeO_x$ for the oxygen evolution reaction (OER). Catalyst activities are typically described in the form of so-called Volcano plots, where either the exchange current density (see below) or the reaction overpotential is plotted *versus* the metal_hydrogen bond energy (HER) or *versus* the standard enthalpy change from lower to higher oxide $MO_x \rightarrow MO_{x+1}$, as introduced by Trasatti.[66]

Currents at metallic electrocatalysts have been described by the Butler–Volmer equation.[29,67] This equation is based on the transition state theory,[30] where the activated complex between the initial and the final state of an electrochemical reaction becomes voltage dependent due to the high electric field in the compact Helmholtz layer at the metal/electrolyte interface. The current–voltage behaviour is characterized by the anodic and cathodic reaction branches. At low overpotentials, the Butler–Volmer behaviour is linear with potential. For increased potential, one branch dominates, described by the Tafel approximation that shows an exponential increase of the current.[68] This increase is defined by the exchange current density,[69] a measure of the kinetic activity of a material in a given electrolyte.

Solar-to-fuel efficiency can be defined for the two basic situations, *i.e.* when solar testing is performed with half-cells under applied bias or for unassisted water splitting. For integrated solar hydrogen generators, in which spontaneous overall water splitting occurs without external electrical power applied, the solar-to-hydrogen (STH) efficiency is calculated by the limiting (thermodynamic) voltage for water splitting $V_{H_2} = 1.23$ V:

$$\eta_{STH} = \frac{j_{op} \cdot V_{H_2} \cdot f_{FE}}{I} \tag{1.16}$$

where j_{op} denotes the operating current density of the device, V_{H_2} is the thermodynamically defined voltage of 1.23 V, I is the incident light power density, and f_{FE} the faradaic efficiency. If, for instance, side reactions, related to (photo)corrosion are present, the faradaic efficiency drops below 1.

For half cells, operating with applied potential, the achieved conversion efficiency can be described by the energy saved due to illumination, compared to the unassisted case[70]

$$\eta_{STH}^{HC} = \frac{j_{MP} \cdot (1.23V - V_a) \cdot f_{FE}}{I} \tag{1.17}$$

V_a is the voltage applied between the hydrogen evolving (photo)cathode and the oxygen evolving (photo)anode and j_{MP} denotes the photocurrent measured at the maximum power point.

The photocatalytic current j_{op}, j_{MP} in eqn (1.16) and (1.17) can be calculated for both a pn junction with deposited electrocatalysts[71,72] and tandem junctions.[52] The according photocurrent–voltage characteristics are written in inverted form with $V(j)$ to allow analytic representation. The expression for the total voltage of the system $V(j)$ (eqn (1.18)) combines the catalytic dark current, described by the Tafel approximation of the Butler–Volmer equation for cathodic currents with the light-induced part of the photocurrent, j_L, of a diode

$$V(j) = \pm\frac{n_d kT}{q}\ln\left(\pm\frac{j_L - j}{j_{0,PV}}\right) - \frac{kT}{q\alpha_C n_e}\ln\left(-\frac{j}{j_{0,cat}}\right) + jR_s \tag{1.18}$$

Here, the dark current $j_{0,PV}$ corresponds to the reverse saturation current of a Schottky diode or a pn junction, n_d is the junction ideality factor that extends the description of the photocurrent to behavior beyond purely thermionic emission; α_C, n_e and $j_{0,cat}$ denote the cathodic charge transfer coefficient, the number of transferred electrons and the exchange current density of the metal–electrolyte junction. R_s is the total series resistance that is used as a floating parameter to simulate experimental systems.

An energy band schematic of a combined PV-electrocatalyst structure is shown in Figure 1.15. Here, quasi-Fermi levels (QFL) are used to show the operation condition at the maximum power point where, according to Figure 1.13, the voltage V_{MP} can be large and close to V_{oc} for systems with large fill factors. QFLs describe the stationary energy content of the system due to the light-induced excess carrier concentration with $n^*(x,h\nu) = n_0 + \Delta n(x,h\nu)$. The latter term denotes the excess carrier concentration that depends on a

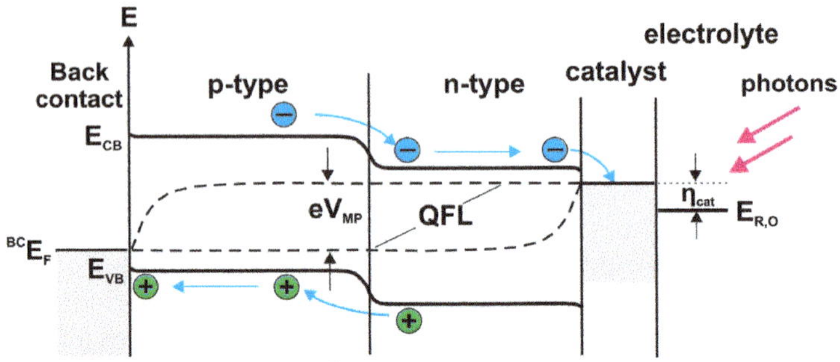

Figure 1.15 Energy schematic for a PV pn junction in contact with an electrocatalyst and a redox electrolyte operating at the maximum power point; note that despite the large split in the QFLs, almost the maximum photocurrent is passing through the device (see text).

spatial coordinate because of the absorption profile and the subsequent carrier diffusion. QFLs $({}_{n,p}E_F^*(x))$ can be schematically derived from the exponential Boltzmann distribution of the carrier concentration (eqn (1.1) and (1.1a)) by replacing N_D and N_A by $n^*(x,h\nu)$ and $p^*(x,h\nu)$, respectively:

$$n^*(x) = N_{CB}\exp\left[-\frac{E_{CB} - {}_nE_F^*(x)}{kT}\right]; \quad p^*(x) = N_{VB}\exp\left[-\frac{{}_pE_F^*(x) - E_{VB}}{kT}\right] \quad (1.19)$$

Eqn (1.19) yields

$${}_{n,p}E_F^*(x) = E_F \pm kT\ln\frac{n(p)^*(x)}{n(p)} = E_F \pm kT\ln\left(1 + \frac{\Delta n(p)(x)}{n(p)}\right) \quad (1.20)$$

The difference between QFLs for electrons and holes at the surface is the maximum attainable photovoltage, hence

$$eV_{ph}^{max} = |{}_nE_F^*(x=0) - {}_pE_F^*(x=0)| \quad (1.21)$$

Figure 1.16 shows a half cell where the photocurrent–voltage characteristics of a photovoltaic pn junction and that of the pn junction modified with an electrocatalyst are compared. The exponential onset of the photocurrent near V_{oc} (blue curve) is due to the (exponential) Tafel behavior of the electrocatalyst and indicates an electrocatalytically limited system. This limitation depends on the interplay of electrocatalytic exchange current density and Tafel slope, as well as on the photodiode fill factor and the total device current. One notes a shift of the MPP to lower potentials and slightly lower currents for the catalyst loaded device. This shift is caused by the subtraction of the catalytic overvoltage from the quasi-Fermi Level splitting in the PV diode. Hence, the operating voltage is reduced and, for systems with low fill factor, the corresponding photocurrent is also lowered. In the example presented here, the fill factor is rather large, which results in only a small lowering of j_{ph}. Upon inclusion of a series resistance, an efficient p–InP half

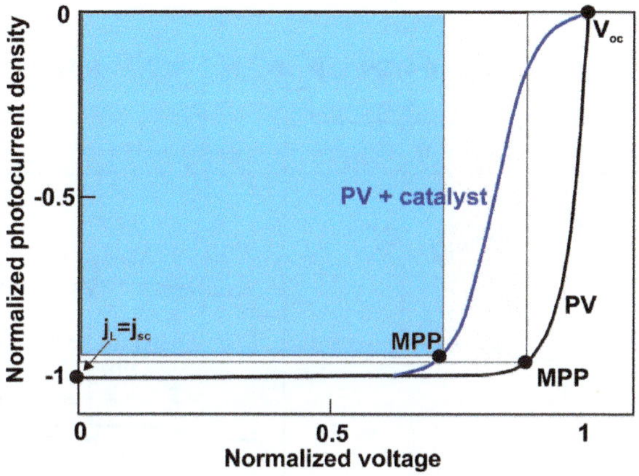

Figure 1.16 Modelled photocurrents for a PV (solid state) photodiode and a coupled PV–electrocatalyst system;[71] normalized currents (j/j_L) and voltages (V/V_{oc}) are displayed; parameters were $j_L = 35$ mA cm^{-2}, PV dark saturation current 5×10^{-11} mA cm^{-2}, exchange current density 0.25 mA cm^{-2}, $n_e = 2$, $n_d = 1$, $\alpha = 0.5$, $R_s = 0$.

cell with a thin film epitaxial (111) : (2×4)-oriented absorber could be modelled.[71]

1.6 Tandem Junction Water Splitting Cells

High efficiency unassisted water splitting has been reported for several device structures. Photoanode[73] as well as photocathode[42–44] designs have been realized. The photoactive core of these systems is a photovoltaic 'buried' tandem junction. Here, we provide a brief introduction to one such system, as well as a description of its functional characteristics and limiting efficiencies. Details and device related considerations can be found in the chapters by May *et al.* and by Xiang *et al.*

We consider tandem structures as shown in Figure 1.9(d), (e) and visualized in Figure 1.10. Figure 1.17 shows a design where the dual tandem consists of np junctions, formed between the higher doped n^+ regions and the p-type absorber. Here, a fully immersed artificial leaf is displayed, although without passivation layers and membrane. The 'engine' of the cell is the PV tandem, where carrier separation occurs. Energetic alignment with the electrocatalysts depends on their size, thickness, and distribution. For small deposit amounts (ultrathin layers or island formation), the energetics are dominated by the semiconductor.[74]

The maximum theoretical efficiencies achievable at the electrolyte contact with such systems has been calculated recently.[52,75] In Figure 1.18 theoretical efficiencies under realistic conditions with respect to the system components are displayed for a dual PV tandem structure coupled to electrocatalysts. Efficiencies have been calculated using the STH relationship of eqn (1.16).

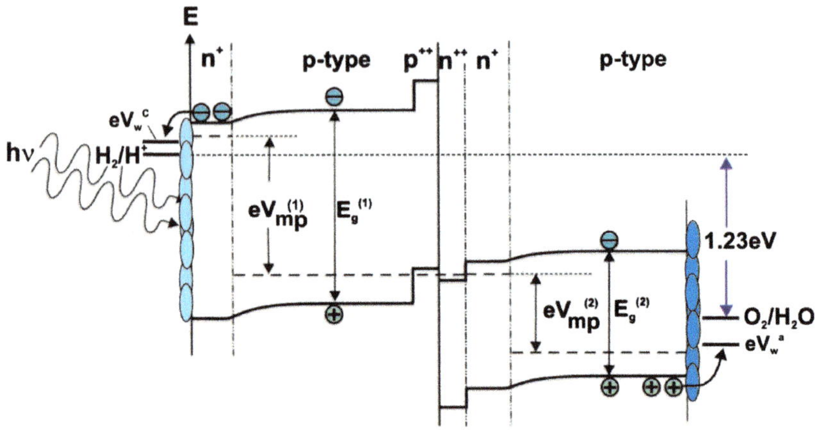

Figure 1.17 Energy band schematic of a dual junction tandem cell including tunnel junctions (doping p^{++} and n^{++}); $V_{mp}(1,2)$ denote the photovoltage at the maximum power point of the respective absorbers, *i.e.* the QFL splitting in that component, V_w^c, V_w^a are the cathodic and anodic working potentials at the given photocurrent (see text).

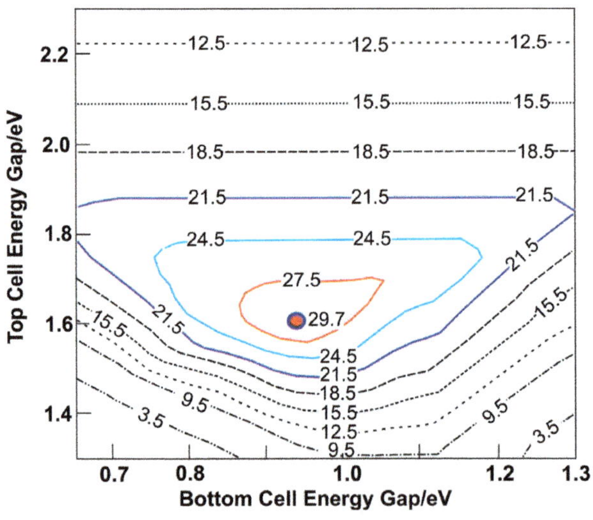

Figure 1.18 Iso-STH efficiency plots for coupled PV tandem-electrocatalyst systems. As catalysts, Pt (HER) and RuO_2 (OER) were chosen. The fill factor of the PV part of the device is $ff = 0.85$ and the solution resistance is assumed to be 5 $\Omega\,cm^{-2}$.
Reproduced from ref. 75 with permission from the Royal Society of Chemistry.

The maximum STH efficiency is reached for a tandem structure with $E_g^{(1)}$ (bottom subcell) = 0.95 eV and $E_g^{(2)}$ (top cell) = 1.60 eV.

Analytically, the current–voltage characteristic of integrated catalyst–tandem structures can be developed based on eqn (1.18) by expanding to a variable number of photodiodes and including two catalysts, for the anodic and cathodic reactions. The photoelectrochemical device output voltage, V_{PEC}, can then be written as

$$V_{\text{PEC}} = \sum_{i}^{n} V_{\text{PV},j}(j) + \eta_{\text{cat}} + \eta_{\text{cat}} + V_{\text{s}} \tag{1.22}$$

with the requirement that $V_{\text{PEC}} > V_0^{\text{rxn}}$ in order for the reaction to proceed; V_0^{rxn} denotes the thermodynamic value of the reaction and V_{S} the IR drop in the solution. Please note that for these ideal considerations, recombination terms have not been included. The individual PV tandem component voltage includes a factor that describes the optical transmission of the electrocatalyst because, in the integrated design, one catalyst layer will be on the illuminated side. The factor describes the light attenuation due to (often metallic) absorption by the catalyst. The first term of eqn (1.18) becomes

$$V_{\text{PV},i}(j) = \frac{n_{\text{d},i}kT}{q} \ln\left(\frac{j_{\text{L},i} \cdot f_{\text{T},i} - j}{j_{0,\text{PV},i}} + 1\right) \tag{1.23}$$

For the overvoltages, we use the expression from the Butler–Volmer equation instead of the Tafel approximation and include a surface area factor, f_{sa}, that modifies the catalytic exchange current density (c, a denote cathodic and anodic branches)

$$\eta_{\text{cat}}^{\text{a/c}}(j) = -\frac{2kT}{q \cdot n_{\text{e}}} \sinh^{-1}\left(\frac{j}{2j_{0,\text{cat}}^{\text{a/c}} \cdot f_{\text{sa}}^{\text{a/c}}}\right) \tag{1.24}$$

With eqn (1.22) one obtains the analytic expression for the output voltage of the coupled tandem-electrocatalyst device using $\alpha = 0.5$

$$V_{\text{PEC}}(j) = \sum_{i} \frac{n_{\text{d},i}kT}{q} \ln\left(\frac{j_{\text{L},i} \cdot f_{\text{T},i} - j}{j_{0,\text{PV},i}} + 1\right) - \sum_{a,c} \frac{2kT}{q \cdot n_{\text{e}}} \sinh^{-1}\left(\frac{j}{2j_{0,\text{cat}}^{\text{a/c}} \cdot f_{\text{sa}}^{\text{a/c}}}\right) + V_{\text{S}}(j) \tag{1.25}$$

Figure 1.19 shows the calculated behavior of a dual tandem ($i = 2$) for the ideal case ($f_{\text{T}} = 1, f_{\text{sa}} = 1$ and external radiative efficiency (ERE) 1 (see ref. 52)) and in the more realisitc scenario $f_{\text{T}} = 0.9$ and ERE $= 10^{-3}$ have been assumed. Further parameters regarding energy gaps and exchange current densities of the catalysts are given in the figure caption.

Reducing the transmission by 10% lowers the short circuit current by this amount and, accordingly a logarithmic reduction of the photovoltage in addition to the influence of the external radiative efficiency is noted. For complete transmission and optimized surface area, the tandem should reach its maximum performance. In real systems, however, the parameters are interlinked: for 100% transmission, the catalyst coverage must be minimized, which results in a low value for f_{sa} (eqn (1.24)). For increased f_{sa}, the transmission typically decreases. Judicious choice of catalyst nano- and

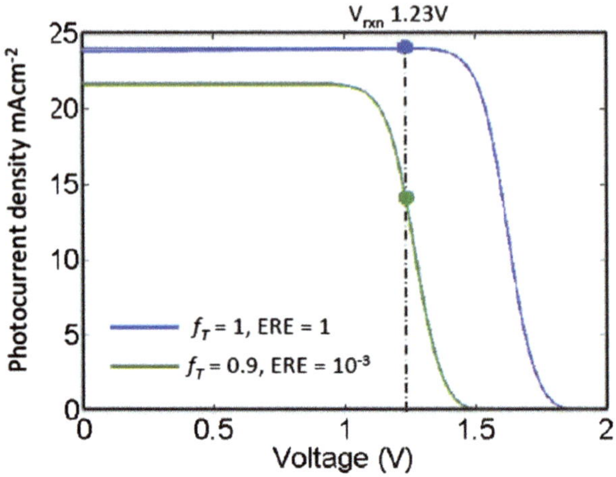

Figure 1.19 Modelled photocurrent–voltage characteristics for a dual junction photoelectrochemical device ($E_g^{(1)} = 1.5$ eV, $E_g^{(2)} = 1.0$ eV; j_{0c}(HER) = 1 mA cm^2, j_{0a}(OER) = 10^{-3} mA cm^2) overlaid with the reaction potential ($V_{rxn} = 1.23$ V) and a dashed–dotted line marking the operating point for ideal photodiodes (blue curve) and for non-ideal photodiodes (green curve).

microtopography on the surfaces, however, can result in higher transmission than in planar structures.[72]

The present costs of the tandem cores demand alternatives. For application of III–V semiconductors, epitaxial lift-off and spalling techniques can alleviate the situation.[76,77] The also costly material use of scarce noble metal catalysts can be reduced by advancing the design of earth abundant core–noble metal shell strutures, which has already found considerable attention in fuel cell research and development.[78]

1.7 New and Emerging Materials for Photoelectrochemical Energy Conversion

To have potential for making an impact on renewable energy conversion technologies, integrated solar water splitting devices must be simultaneously efficient, stable, and capable of being produced at scale. While impressive advances have been made with tandem photovolatic elements directly coupled to catalysts (see above), the III–V semiconductor material stacks used for these demonstrations remain expensive. Furthermore, corrosion protection strategies based on conformal coating of chemically sensitive semiconductors have allowed remarkable progress in recent years, but such interfaces are delicate and, for photoanode operations, have been limited to very small electrodes because of an increased probability for pinhole formation. For these reasons, the most robust protected photoelectrodes are limited to those, like silicon, that self-passivate.[79] Considerable current research is focused on reducing the cost of III–V tandems, as well as on improving the reliabiltiy of corrosion protection schemes. However, the discovery and development of thin film photoelectrodes that can be inexpensively produced and are intrinsically stable under reaction conditions remains a grand challenge in the field of artificial photosynthesis.

Historically, transition metal oxides have been considered as a promising class of materials for application as photoanodes. Indeed, the seminal experiment by Fujishima and Honda[1] was based on one such material, n-type TiO_2. While a variety of wide bandgap semicondctors, like TiO_2, exhibit the requisite stability and even desirable charge transport properties, their efficiencies are severly limited due to their transparency to photons that make up the majority of the solar spectrum (see Figure 1.14). On the other hand, numerous narrower bandgap transition metal oxides have been identified but these materials tend to suffer from optical absorption lengths that are significantly longer than their charge extraction lengths; the result is significant photocarrier recombination and poor overall efficiency. Nevertheless, considerable progress is being made on improving function of established materials, identifying new ones, and integrating such components into overall water splitting devices.

Among transition metal oxides, two materials—hematite (Fe_2O_3) and bismuth vanadate ($BiVO_4$)—have attracted considerable interest over the years. For the case of Fe_2O_3, the combination of a well-suited energy gap of 2.1 eV (Figure 1.14), high chemical stability, and very low prospective production costs are particularly alluring.[80] While it has long been known that the material possesses extremely short minority carrier diffusion lengths, on the order of 10 nm, significant progress has been made on creating nano- and meso-structured films with characteristic physical lengths that are well matched to charge transport lengths.[81] The result has been a pronounced improvement in the attainable photocurrent.[80] The intrinsic stability of hematite, as well as methods for its controlled synthesis, has enabled mechanistic *operando* studies that reveal how competitions between photocarrier recombination and chemical reaction define efficiency, the chemical nature of accumulated holes on its surface affect water oxidation, how its electronic strucuture impacts the generation of mobile charge carriers at different photoexcitation energies, and integration of catalyst on the surface serves to passivate surface sites or improve catalysis.[82–86] These mechanistic insights provide a stronger framework for understanding function and limtiations of new materials, an endeavor that will be further aided by the unprecedented availability of such *operando* characterization methods.

In many respects, $BiVO_4$ is complimentary to Fe_2O_3 as a material for studying fundamental processes in transition metal oxide photoanodes and for developing generalizable strategies for improving function.[87] With an indirect bandgap of 2.5 eV,[88] it is naturally incompatible with high efficiency integrated solar fuels generators. However, it presently stands as the best performing transition metal oxide photoanode with achieved water splitting efficiencies near 5%, as shown in Figure 1.20.[89,90] Therefore, it deserves special attention for mechanistic understanding to guide discovery and aid

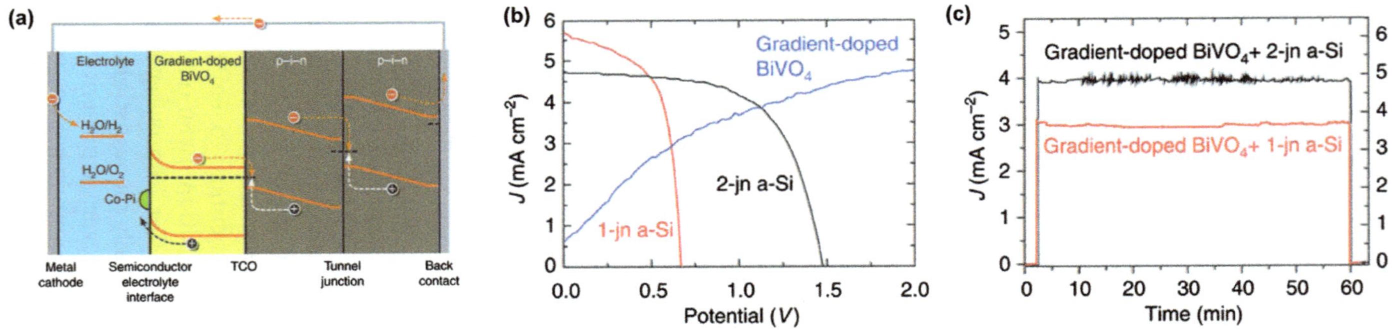

Figure 1.20 Integration of gradient-doped $BiVO_4$ with single and double junction amorphous silicon photovoltaic elements provides sufficient voltage for driving overall water splitting. (a) Schematic illustration of the device structure created by van de Krol and co-workers, (b) current density as a function of applied potential showing the operational point defined by the current matching conditon for the single and double junction device, (c) chronoamperometric characteristation of the devices shown in (b), with stable solar-to-hydrogen efficiencies approaching 5% achieved for $BiVO_4$ integrated with double junction amorphous silicon photovoltaic elements.
Reprinted by permission from Springer Nature: Nature Communications.[89] Copyright 2013 Nature Publishing Group.

in the evaluation of new semiconductors with smaller bandgaps and, thus, prospects for higher efficiency. In terms of electronic structure, $BiVO_4$ does not contain partially occupied d-orbitals, meaning that its bandgap does not arise specifically from electron correlation.[91] This feature may prove important for understanding the relatively rapid progress towards high quantum efficiencies compared to Mott- and charge-transfer insulators[92] that describe a substantial fraction of recently identified, yet poorly performing, oxide semiconductors. On the other hand, it has been determined that both electrons and holes in the material self-trap to form small polarons that transport *via* thermally activated hopping with low mobilties.[93,94] The fact that charges can still be efficiently extracted points to the need for improved characterization of barriers to recombination associated with local structural distortions, as well as for understanding of interactions between polarons and both native and impurity defect states.[87] This example highlights that although polaronic conduction is not a desirable feature, it may not intrinsically exclude materials from use as high performance photoelectrodes.

As described in the Chapter by Eichhorn *et al.*, stabiltiy of metal oxide semiconductors under reaction conditions cannot be assumed.[95] In addition to the Pourbaix stability, which describes the thermodynamiclly stable phases and ions as a function of electrochemical potential and pH, photocorrosion and kinetic processes can have a dramatic impact on the robust operation of light absorbers. For the case of $BiVO_4$, it has been found that in addition to vanadium leaching from the solid into solution, trapping of both electrons and holes at the surface can destabilize the lattice and promote photocorrosion.[95] While consideration of the Pourbaix diagram would suggest self-passivation by electochemically stable bismuth oxides, (photo)corrosion of $BiVO_4$ may leave a highly disorderd and atomically sparse layer that is kinetically hindered from transforming to a passivating stable surface layer.[95] Such an observation explains discrepencies in the literature regarding stability of $BiVO_4$, since different synthetic approaches may leave different surface compositions, including those with pre-formed stable passivation layers, which either favor or disfavor stabilization of the natively unstable $BiVO_4$.

Despite significant progress on improving photocurrent in visible light absorbing transition metal oxide semconductors, no demonstrations of integrated overall water splitting systems incorporating these materials with efficiencies exceeding 10% have been reported. In large part, this is because progress towards improved photovoltage in both Fe_2O_3 and $BiVO_4$ photoanodes has been slow. For the case of Fe_2O_3, the band edge energetics are not favorable for achieving low onset potentials for the oxygen evolution reaction. For the case of $BiVO_4$, future work should address the relative impacts of Fermi level pinning at surface defect states and energetic relaxation associated with formation of both electron and hole polarons, which would serve to reduce the effective bandgap of the material from the perspective of photovoltage.

The search for new materials that can replace Fe_2O_3 and $BiVO_4$ as workhorse photoanodes continues in parallel with developments in III–V based high efficiency systems. On the discovery front, theory has begun to make a significant impact in directing experiment, particularly with respect to high throughput surveys.[54,96] However, the discovery challenge remains great due to the strict combination of material properties that are simultansously desired. Identification of just one new material that possesses a bandgap in the range of 1.7–2.0 eV, favorable band edge positions for driving the desired reactions while generating maximum photovoltages, intrinsic (photo)-chemical stability, defect tolerance, and long photocarrier diffusion lengths would represent a dramatic andvance on the path towards practical integrated solar fuels generators.

1.8 Concluding Remarks

Decades of research into electrochemical, photovoltaic, and photoelectrochemical energy conversion provide the basis for modern sprints towards integrated, robust, and high efficiency solar fuels devices. This chapter presented a brief survey of some of the central concepts that underlie the function of these systems and their constituent components. In the remainder of this book, modern advances in the international solar fuels community, with an emphasis on how the Joint Center for Artificial Photosynthesis contributed to progress, are described.

References

1. A. Fujishima and K. Honda, Electrochemical Photolysis of Water at a Semiconductor Electrode, *Nature*, 1972, **238**, 37.
2. A. P. van Troostwijk and J. R. Deiman, Sur une maniere de decomposer l'Eau en Air in ammable et en Air vital, *Obs. Phys.*, 1789, **35**, 369.
3. E. Becquerel, Mémoire sur les effets électriques produits sous l'influence des rayons solaires, *C. R.*, 1839, **9**, 561.
4. C. E. Fritts, On a new form of selenium cell, and some electrical discoveries made by its use, *Am. J. Sci.*, 1883, **26**, 465.
5. D. M. Chapin, C. S. Fuller and P. L. Pearson, A New Silicon p-n Junction Photocell for Converting Solar Radiation into Electric Power, *J. Appl. Phys.*, 1954, **25**, 676.
6. H. Gerischer, Electrochemical photo and solar cell principles and some experiments, *J. Electroanal. Chem. Interfacial Electrochem.*, 1975, **58**, 263.
7. A. J. Nozik, Photoelectrochemistry: Applications to Solar Energy Conversion, *Annu. Rev. Phys. Chem.*, 1978, **29**, 189.
8. B. A. Parkinson, A. Heller and B. Miller, Enhanced photoelectrochemical solar-energy conversion by gallium arsenide surface modification, *Appl. Phys. Lett.*, 1978, **33**, 521.
9. A. Heller, B. Miller, H. J. Lewerenz and K. J. Bachmann, An efficient photocathode for semiconductor liquid junction cells: 9.4% solar conversion efficiency with p-InP/VCl_3-VCl_2-HCl/C, *J. Am. Chem. Soc.*, 1980, **102**(21), 6555.

10. A. Heller and R. G. Vadimsky, Efficient Solar to Chemical Conversion: 12% Efficient Photoassisted Electrolysis in the [p-type InP(Ru)]/HCl-KCl/Pt(Rh) Cell, *Phys. Rev. Lett.*, 1981, **46**, 1153.
11. H. Gerischer, Electrolytic decomposition and photodecomposition of compound semiconductors in contact with electrolytes, *J. Vac. Sci. Technol.*, 1978, **15**, 1422.
12. A. J. Bard and M. S. Wrighton, Thermodynamic Potential for the Anodic Dissolution of n-Type Semiconductors, *J. Electrochem. Soc.*, 1977, **124**(11), 1706.
13. B. O'Reagan and M. Graetzel, A low-cost, high-efficiency solar cell based on dye- sensitized colloidal TiO_2films, *Nature*, 1991, **353**, 737.
14. S. Y. Chae, S. J. Park, O.-S. Joo, Y. Jun, B. K. Min and Y. J. Hwang, Highly stable tandem solar cell monolithically integrating dye-sensitized and CIGS solar cells, *Sci. Rep.*, 2016, **6**, 30868.
15. D. G. Nocera, The artificial leaf, *Acc. Chem. Res.*, 2012, **45**(5), 767.
16. A. Eisenberg, The Answer Is (Artificially) Blowing in the Wind, *The New York* Times, 2011, May 22, p. BU3.
17. J. Gertner, *The Idea Factory: Bell Labs and the Great Age of American Innovation*, Penguin Books, New York, 2012.
18. W. Schottky, Zur Halbleitertheorie der Sperrschicht- und Spitzengleichrichter, *Z. Phys.A*, 1939, **113**, 813.
19. W. Shockley, *Electrons and Holes in Semiconductors*, Van Nostrand Company Inc., 1950.
20. H. Gerischer, Über den Ablauf von Redoxreaktionen an Metallen und an Halbleitern. I. Allgemeines zum Elektronenübergang zwischen einem Festkörper und einem Redoxelektrolyten, *Z. Phys. Chem. N. F.*, 1960, **26**, 223.
21. H. Gerischer, Über den Ablauf von Redoxreaktionen an Metallen und an Halbleitern. II. Metall-Elektroden, *Z. Phys. Chem. N. F.*, 1960, **26**, 325.
22. H. Gerischer, Über den Ablauf von Redoxreaktionen an Metallen und an Halbleitern. III. Halbleiterelektroden, *Z. Phys. Chem. N. F.*, 1961, **27**, 48.
23. J. M. Wiesenfeld and E. P. Ippen, Dynamics of electron solvation in liquid water, *Chem. Phys. Lett.*, 1980, **73**(11), 47.
24. R. Memming, *Semiconductor Electrochemistry*, Wiley-VCH, 2nd edn, 2015, ch. 6.3.1.
25. S. Tanaka and C.-P. Hsu, Theory of nonadiabatic electron transfer at electrode/liquid interfaces: Role of quantum effects,*J. Chem. Phys.*, 1999, **111**, 11117.
26. J. Franck, Elementary processes of photochemical reactions, *Trans. Faraday Soc.*, 1926, **21**, 536; E. U. Condon, A Theory of Intensity Distribution in Band Systems, *Phys. Rev.*, 1926, **28**, 1182.
27. H. Jungblut, S. L. Schweizer, J. Jakubowicz and H. J. Lewerenz, Mechanism of initial structure formation on highly doped Si(111), *J. Electroanal. Chem.*, 2002, **527**, 41.
28. O. W. Richardson, *Thermionic Emission from Hot Bodies*, Wexford College Press, 2003, p. 196.

29. T. Erdey-Grúz and M. Volmer, Zur Theorie der Wasserstoffüberspannung, *Z. Phys. Chem.*, 1930, **150A**, 203.
30. H. Eyring, The Activated Complex in Chemical Reactions, *J. Chem. Phys.*, 1935, **3**, 107.
31. W. Schmickler and E. Santos, *Interfacial Electrochemistry*, Springer, 2nd edn, 2010, ch. 10.
32. Y. Q. Gao, Y. Georgievskii and R. A. Marcus, The theory of electron transfer reactions at semiconductor electrode/liquid interfaces, *J. Chem. Phys.*, 2000, **112**(7), 3358.
33. P. Kaushal, S. Chand and J. Osvald, Current–voltage characteristics of Schottky diode simulated using semiconductor device equations, *Int. J. Electron.*, 2013, **100**(5), 686.
34. H. J. Lewerenz, Surface states and Fermi level pinning at semiconductor/electrolyte junctions, *J. Appl. Phys.*, 1993, **356**(1-2), 121.
35. L. Bertoluzzi, P. Lopez-Varo, J. A. J. Tejada and J. Bisquert, Charge transfer processes at the semiconductor/electrolyte interface for solar fuel production: insight from impedance spectroscopy, *J. Mater. Chem. A*, 2016, **4**, 2873.
36. M. Nishida, A theoretical treatment of charge transfer via surface states at a semiconductor-electrolyte interface: Analysis of the water photoelectrolysis process, *J. Appl. Phys.*, 1980, **51**, 1669.
37. X. Liu, X. W. Zhang, J. H. Meng, Z. G. Yin, L. Q. Zhang, H. L. Wang and J. L. Wu, High efficiency Schottky junction solar cells by co-doping of graphene with gold nanoparticles and nitric acid, *Appl. Phys. Lett.*, 2015, **106**, 233901.
38. C. Battaglia, A. Cuevas and S. De Wolf, High-efficiency crystalline silicon solar cells: status and perspectives, *Energy Environ. Sci.*, 2016, **9**, 1552.
39. P. D. Antunez, D. M. Bishop, Y. Luo and R. Haight, Efficient kesterite solar cells with high open-circuit voltage for applications in powering distributed devices, *Nat. Energy*, 2017, **2**, 884.
40. S. K. Wallace, D. B. Mitzi and A. Walsh, The Steady Rise of Kesterite Solar Cells, *ACS Energy Lett.*, 2017, **2**(4), 776.
41. O. Khaselev and J. A. Turner, A Monolithic Photovoltaic-Photoelectrochemical Device for Hydrogen Production via Water Splitting, *Science*, 1998, **280**(11), 425.
42. M. M. May, H. J. Lewerenz, F. Dimroth, D. Lackner and T. Hannappel, Efficient direct solar-to-hydrogen conversion by *in situ* interface transformation of a tandem structure, *Nat. Commun.*, 2015, **6**, 8286.
43. J. L. Young, M. A. Steiner, H. Doescher, R. M. France, J. A. Turner and T. G. Deutsch, Direct solar-to-hydrogen conversion via inverted metamorphic multi-junction semiconductor architectures, *Nat. Energy*, 2017, **2**, 17028.
44. W.-H. Cheng, M. H. Richter, M. M. May, J. Ohlmann, D. Lackner, F. Dimroth, T. Hannappel, H. A. Atwater and H. J. Lewerenz, *ACS Energy Lett.*, 2018, **3**, 1795.

45. A. Polman and H. A. Atwater, Photonic design principles for ultrahigh-efficiency photovoltaics, *Nat. Mater.*, 2012, **11**, 174.
46. C. Maragliano, M. Chiesa and M. Stefancich, Point-focus spectral splitting solar concentrator for multiple cells concentrating photovoltaic system, *J. Opt.*, 2015, **17**(10), 105901.
47. M. G. Walter, E. L. Warren, J. R. McKone, S. W. Boettcher, Q. Mi, E. A. Santori and N. S. Lewis, Solar Water Splitting Cells, *Chem. Rev.*, 2010, **110**(11), 6446.
48. M. Law, L. E. Greene, J. C. Johnson, R. Saykally and P. Yang, Nanowire dye-sensitized solar cells, *Nat. Mater.*, 2005, **4**, 455.
49. J. Brillet, J.-H. Yum, M. Cornuz, T. Hisatomi, R. Solarska, J. Augustynski, M. Graetzel and K. Sivula, Highly efficient water splitting by a dual-absorber tandem cell, *Nat. Photonics*, 2012, **6**, 824.
50. D. O. Scanlon, C. W. Dunnill, J. Buckeridge, S. A. Shevlin, A. J. Logsdail, S. M. Woodley, C. R. A. Catlow, M. J. Powell, R. G. Palgrave, I. P. Parkin, G. W. Watson, T. W. Keal, P. Sherwood, A. Walsh and A. A. Sokol, Band alignment of rutile and anatase TiO_2, *Nat. Mater.*, 2013, **12**, 798.
51. W. Shockley and H. J. Queisser, Detailed Balance Limit of Efficiency of *p-n* Junction Solar Cells, *J. Appl. Phys.*, 1961, **32**, 510.
52. K. T. Fountaine, H. J. Lewerenz and H. A. Atwater, Efficiency limits for photoelectrochemical water splitting, *Nat. Commun.*, 2016, **7**, 13706.
53. J. Su, L. Guo, N. Bao and C. A. Grimes, Nanostructured $WO_3/BiVO_4$ Heterojunction Films for Efficient Photoelectrochemical Water Splitting, *Nano Lett.*, 2011, **11**(5), 1928.
54. Q. Yan, J. Yu, S. K. Suram, L. Zhou, A. Shinde, P. F. Newhouse, W. Chen, G. Li, K. A. Persson, J. M. Gregoire and J. B. Neaton, Solar fuels photoanode materials discovery by integrating high-throughput theory and experiment, *Proc. Natl. Acad. Sci.*, 2017, **114**(12), 3040.
55. J. M. Asensi, J. Merten, C. Voz and J. Andreu, Analysis of the role of mobility-lifetime products in the performance of amorphous silicon p-i-n solar cells, *J. Appl. Phys.*, 1999, **85**, 2939.
56. B. Rech and H. Wagner, Potential of amorphous silicon for solar cells, *Appl. Phys. A*, 1999, **69**(2), 155.
57. K. Walczak, Y. Chen, C. Karp, J. W. Beeman, M. R. Shaner, J. Spurgeon, I. D. Sharp, X. Amashukeli, W. West, J. Jin, N. S. Lewis and C. Xiang, Modeling, Simulation, and Fabrication of a Fully Integrated, Acid-stable, Scalable Solar-Driven Water-Splitting System, *ChemSusChem*, 2015, **8**(3), 544.
58. M. R. Shaner, M. T. Dowell, A. Pien, H. A. Atwater and N. S. Lewis, Si/TiO_2 Tandem-Junction Microwire Arrays for Unassisted Solar-Driven Water Splitting, *J. Electrochem. Soc.*, 2016, **163**(5), H 261.
59. J. I. Pankove, *Optical Processes in Semiconductors*, Dover, New York, 1971.
60. S. M. Sze and K. M. Kwok, *Physics of Semiconductor Devices*, Wiley Interscience, 3rd edn, 2006.
61. W. Eberhardt and F. J. Himpsel, Dipole selection rules for optical transitions in the fcc and bcc lattices, *Phys. Rev. B*, 1980, **21**, 5572.

62. J. Tauc, Optical properties and electronic structure of amorphous Ge and Si, *Mater. Res. Bull.*, 1968, **3**, 37.
63. S. K. O'Leary, S. Zukotynski and J. M. Perz, Optical absorption in amorphous semiconductors, *Phys. Rev. B*, 1995, **52**, 7795.
64. D. E. Carlson and C. R. Wronski, Amorphous silicon solar cell, *Appl. Phys. Lett.*, 1976, **28**, 671.
65. A. J. Nozik, Novel Approaches to Water Splitting by Solar Photons, in *Photoelectrochemical Water Splitting*, ed. H. J. Lewerenz and L. M. Peter, RSC Publishing, 2013, ch. 13.
66. S. Trasatti, Work function, electronegativity, and electrochemical behaviour of metals: III. Electrolytic hydrogen evolution in acid solutions, *J. Electroanal. Chem.*, 1972, **39**, 163.
67. J. A. V. Butler, Studies in heterogeneous equilibria. Part II. The kinetic interpretation of the Nernst theory of electromotive force, *Trans. Faraday Soc.*, 1924, **19**, 729.
68. J. Tafel, Über die Polarisation bei kathodischer Wasserstoffentwicklung, *Z. Phys. Chem.*, 1905, **50A**, 641.
69. T. Shinagawa, A. T. Garcia-Esparza and K. Takanabe, Insight on Tafel slopes from a microkinetic analysis of aqueous electrocatalysis for energy conversion, *Nat. Commun. Sci. Rep.*, 2015, **5**, 13801.
70. M. G. Walter, E. L. Warren, J. R. McKone, S. W. Boettcher, Q. Mi, E. A. Santori and N. S. Lewis, Solar Water Splitting Cells, *Chem. Rev.*, 2010, **110**(11), 6446.
71. M. R. Shaner, K. T. Fountaine and H. J. Lewerenz, Current-voltage characteristics of coupled photodiode-electrocatalyst devices, *Appl. Phys. Lett.*, 2013, **103**, 143905.
72. K. T. Fountaine, H. J. Lewerenz and H. A. Atwater, Interplay of light transmission and catalytic exchange current in photoelectrochemical systems, *Appl. Phys. Lett.*, 2014, **105**, 173901.
73. E. Verlage, S. Hu, R. Liu, R. J. R. Jones, K. Sun, C. Xiang, N. S. Lewis and H. A. Atwater, A monolithically integrated, intrinsically safe, 10% efficient, solar-driven water-splitting system based on active, stable earth-abundant electrocatalysts in conjunction with tandem III–V light absorbers protected by amorphous TiO_2 films., *Energy Environ. Sci.*, 2015, **8**, 3166.
74. V. Subramanian, E. E. Wolf and P. V. Kamat, Green Emission to Probe Photoinduced Charging Events in ZnO-Au Nanoparticles. Charge Distribution and Fermi-Level Equilibration, *J. Phys. Chem. B*, 2003, **107**, 7479.
75. S. Hu, C. Xiang, S. Haussener, A. D. Berger and N. S. Lewis, An analysis of the optimal band gaps of light absorbers in integrated tandem photoelectrochemical water-splitting systems, *Energy Environ. Sci.*, 2013, **6**, 2984.
76. C.-W. Cheng, K.-T. Shiu, N. Li, S.-J. Han, L. Shi and D. K. Sadana, Epitaxial lift-off process for gallium arsenide substrate reuse and flexible electronics, *Nat. Commun.*, 2013, **4**, 1577.

77. D. Shahrjerdi, S. W. Bedell, C. Ebert, C. Bayram, B. Hekmatshoar, K. Fogel, P. Lauro, M. Gaynes, T. Gokmen, J. A. Ott and D. K. Sadana, High-efficiency thin-film InGaP/InGaAs/Ge tandem solar cells enabled by controlled spalling technology, *Appl. Phys. Lett.*, 2012, **100**, 053901.
78. S. T. Hunt and Y. Roman-Leshkov, Principles and Methods for the Rational Design of Core–Shell Nanoparticle Catalysts with Ultralow Noble Metal Loadings, *Acc. Chem Res.*, 2018, **51**, 1054.
79. K. Sun, F. H. Saadi, M. F. Lichterman, W. G. Hale, H.-P. Wang, X. Zhou, N. T. Plymale, S. T. Omelchenko, J.-H. He, K. M. Papadantonakis, B. S. Brunschwig and N. S. Lewis, Stable solar-driven oxidation of water by semiconducting photoanodes protected by transparent catalytic nickel oxide films, *Proc. Natl. Acad. Sci.*, 2015, **112**, 3612.
80. K. Sivula, F. Le Formal and M. Grätzel, Solar Water Splitting: Progress Using Hematite (α-Fe_2O_3) Photoelectrodes, *ChemSusChem*, 2011, **4**, 432.
81. S. C. Warren, K. Voïtchovsky, H. Dotan, C. M. Leroy, M. Cornuz, F. Stellacci, C. Hébert, A. Rothschild and M. Grätzel, Identifying champion nanostructures for solar water-splitting, *Nat. Mater.*, 2013, **12**, 842.
82. F. Le Formal, E. Pastor, S. D. Tilley, C. A. Mesa, S. R. Pendlebury, M. Grätzel and J. R. Durrant, Rate Law Analysis of Water Oxidation on a Hematite Surface, *J. Am. Chem. Soc.*, 2015, **137**, 6629.
83. O. Zandi and T. W. Hamann, Determination of photoelectrochemical water oxidation intermediates on haematite electrode surfaces using operando infrared spectroscopy, *Nat. Chem.*, 2016, **8**, 778.
84. G. Segev, H. Dotan, D. S. Ellis, Y. Piekner, D. Klotz, J. W. Beeman, J. K. Cooper, D. A. Grave, I. D. Sharp and A. Rothschild, The Spatial Collection Efficiency of Charge Carriers in Photovoltaic and Photoelectrochemical Cell, *Joule*, 2018, **2**, 210.
85. G. Segev, C.-M. Jiang, J. K. Cooper, J. Eichhorn, F. M. Toma and I. D. Sharp, Quantification of the loss mechanisms in emerging water splitting photoanodes through empirical extraction of the spatial charge collection efficiency, *Energy Environ. Sci.*, 2018, **11**, 904.
86. A. Shavorskiy, X. Ye, O. Karslıoğlu, A. D. Poletayev, M. Hartl, I. Zegkinoglou, L. Trotochaud, S. Nemšák, C. M. Schneider, E. J. Crumlin, S. Axnanda, Z. Liu, P. N. Ross, W. Chueh and H. Bluhm, Direct Mapping of Band Positions in Doped and Undoped Hematite during Photoelectrochemical Water Splitting, *J. Phys. Chem. Lett.*, 2017, **8**, 5579.
87. I. D. Sharp, J. K. Cooper, F. M. Toma and R. Buonsanti, Bismuth Vanadate as a Platform for Accelerating Discovery and Development of Complex Transition-Metal Oxide Photoanodes, *ACS Energy Lett.*, 2017, **2**, 139.
88. J. K. Cooper, S. Gul, F. M. Toma, L. Chen, Y.-S. Liu, J. Guo, J. W. Ager, J. Yano and I. D. Sharp, Indirect Bandgap and Optical Properties of Monoclinic Bismuth Vanadate, *J. Phys. Chem. C*, 2015, **119**, 2969.
89. F. F. Abdi, L. Han, A. H. M. Smets, M. Zeman, B. Dam and R. van de Krol, Efficient solar water splitting by enhanced charge separation in a bismuth vanadate-silicon tandem photoelectrode, *Nat. Commun.*, 2013, **4**, 2195.

90. L. Han, F. F. Abdi, R. van de Krol, R. Liu, Z. Huang, H.-J. Lewerenz, B. Dam, M. Zeman and A. H. M. Smets, Efficient Water-Splitting Device Based on a Bismuth Vanadate Photoanode and Thin-Film Silicon Solar Cells, *ChemSusChem*, 2014, **7**, 2832.
91. J. K. Cooper, S. Gul, F. M. Toma, L. Chen, P.-A. Glans, J. Guo, J. W. Ager, J. Yano and I. D. Sharp, Electronic Structure of Monoclinic $BiVO_4$, *Chem. Mater.*, 2014, **26**, 5365.
92. M. N. Huda, M. M. Al-Jassim and J. A. Turner, Mott insulators: An early selection criterion for materials for photoelectrochemical H_2 production, *J. Renewable Sustainable Energy*, 2011, **3**, 053101.
93. A. J. E. Rettie, W. D. Chemelewski, D. Emin and C. B. Mullins, Unravelling Small-Polaron Transport in Metal Oxide Photoelectrodes, *J. Phys. Chem. Lett.*, 2016, **7**, 471.
94. L. Zhang, X. Ye, M. Boloor, A. Poletayev, N. A. Melosh and W. C. Chueh, Significantly enhanced photocurrent for water oxidation in monolithic $Mo:BiVO_4/SnO_2/Si$ by thermally increasing the minority carrier diffusion length, *Energy Environ. Sci.*, 2016, **9**, 2044.
95. F. M. Toma, J. K. Cooper, V. Kunzelmann, M. T. McDowell, J. Yu, D. M. Larson, N. J. Borys, C. Abelyan, J. W. Beeman, K. M. Yu, J. Yang, L. Chen, M. R. Shaner, J. Spurgeon, F. A. Houle, K. A. Persson and I. D. Sharp, Mechanistic insights into chemical and photochemical transformations of bismuth vanadate photoanodes, *Nat. Commun.*, 2016, **7**, 12012.
96. A. Shinde, S. K. Suram, Q. Yan, L. Zhou, A. K. Singh, J. Yu, K. A. Persson, J. B. Neaton and J. M. Gregoire, Discovery of Manganese-Based Solar Fuel Photoanodes via Integration of Electronic Structure Calculations, Pourbaix Stability Modeling, and High-Throughput Experiments, *ACS Energy Lett.*, 2017, **2**, 2307.

CHAPTER 2

Photo-electrochemical Hydrogen Plants at Scale: A Life-cycle Net Energy Assessment

JEFFERY B. GREENBLATT†

Joint Center for Artificial Photosynthesis, Lawrence Berkeley National Laboratory, Berkeley, CA 94720, USA
Email: jbgreenblatt@lbl.gov

2.1 Introduction

The primary motivation for developing photo-electrochemical (PEC) hydrogen (H_2) systems is to reduce global greenhouse gas (GHG) emissions, as required by the 2015 Paris Agreement[1] and recently ratified by the threshold number of countries required for entry into force.[2] While numerous approaches exist that can result in large GHG reductions,[3] H_2 represents an energy carrier with many attractive features,[4] including feedstocks for its production, relative ease of storage and transport, high efficiency of conversion into electricity, and clean combustion without any particulates or CO_2 emissions. Moreover, it is a useful intermediate chemical for many industrial purposes, including fuel upgrading or Fischer–Tropsch hydrocarbon synthesis,[5] ammonia production,[6] hydrochloric acid and methanol production,[7] silicon manufacturing,[8] and as a reducing agent for iron oxide.[9] It can also be used in atomic hydrogen welding, as a rotor coolant in electric

†Present address: Emerging Futures, LLC, Berkeley, CA 94710, USA. Email: jeff@emerging-futures.com

Energy and Environment Series No. 22
Integrated Solar Fuel Generators
Edited by Ian D. Sharp, Harry A. Atwater and Hans-Joachim Lewerenz

Published by the Royal Society of Chemistry, www.rsc.org

generators, and as a leak detector.[7] Global H_2 production was 48 million metric tons (Mt) in 2006 and >50 Mt in 2014, with 40% produced in the United States.[10,11] Of the total US hydrogen production, petroleum refining consumed 64%, and chemical manufacturing (mainly ammonia production) consumed 33%. Global estimates indicate a potential demand of 300 Mt H_2 per year in 2050,[12] with an energy equivalent of ~300 billion gallons of gasoline, or ~20% of global liquid fuel demand in 2013.[13]

Currently, the lowest-cost methods of producing hydrogen depend on fossil fuels such as natural gas,[14] and therefore emit large quantities of GHGs at scale. The main alternatives for producing low net GHG hydrogen are coal gasification with CO_2 capture and sequestration,[15] biomass,[16] and water electrolysis.[17] The first approach adds considerable complexity and cost to hydrogen production (as well as still emitting ~15% of the CO_2 in coal),[15] while biomass has its own set of complexity and cost challenges, in addition to being an inherently limited resource that must compete with demands for hydrocarbon fuel production, direct combustion for electricity, and non-energy uses (*e.g.*, agriculture or ecosystem preservation). Electrolysis, while in principle possible to be supplied by zero net GHG electricity, does not offer attractive economics because of the relatively high cost of electricity.

For all these reasons, PEC conversion *via* water splitting is a promising alternative because it has the potential to directly convert solar radiation into an energy-rich, storable chemical (hydrogen) at efficiencies many times greater than that of natural photosynthesis.[18] While many PEC designs have been explored,[19,20] nearly all approaches involve the use of semiconductor absorbers in combination with catalysts to generate H_2 and O_2 products, using only sunlight and water as inputs.[21] The approach taken at the Joint Center for Artificial Photosynthesis (JCAP), as well as others in the field, has been to develop systems that operate spontaneously, without the need for external electrical bias, and to combine the photoabsorption and water-splitting functions together in intimate proximity in the same cell, without the need for 'wires' (transporting electrical charge over macroscopic distances), as described in other chapters of this book. Laboratory demonstrations have achieved solar-to-hydrogen (STH) efficiencies in the 5–18% range,[22–28] and ongoing research aims to improve performance, increase lifetime and reduce cost to allow eventual large-scale implementation.[29–33]

Solar-to-hydrogen technologies are far less developed than other solar-to-fuel routes, such as biofuels. As a result, prospective modeling of such systems at an early stage in the research and development cycle can help identify key challenges and opportunities for improvement, and provide insight into their potential competitiveness. Pioneering studies by James *et al.*[19] and Pinaud *et al.*[20] explored four different configurations of PEC devices: two variations of PEC nanoparticle colloidal suspensions, a flat plate collector, and a concentrating collector. These studies developed economic metrics to compare the technologies at scale. A small number of commercial firms (*e.g.*, Sun Catalytix[34]) have also investigated the economics of these systems, but most have concluded thus far that they are not economical. However, the technology remains at an early stage and meaningful economic analysis may need to await improved designs, enhanced performance and novel manufacturing approaches.

Our foray into life-cycle assessment was driven by limited information on the macroscale energetic flows associated with PEC H_2 systems in the literature. Rather than developing a comprehensive life-cycle analysis that took into account all possible materials and impacts, our study was limited to a net energy analysis, which provides a fundamental basis for comparing energy systems, particularly renewable technologies and their contributions to sustainability. (We did include cursory evaluations of the impacts of a large-scale system on land use, local water supplies, and possible spill hazards, but it was far from comprehensive.) The advantage of a net energy focus is that it identifies, separate from cost or detailed performance concerns, those materials or processes that require the most energy on a life-cycle basis, which often points to important areas of potential improvement, and thus can guide future research and stimulate innovation in critical areas. Energy consumption is also often a proxy for cost, and thus reducing energy use usually translates into reductions in cost, so a low net energy system has a greater opportunity to become cost-competitive.

The net energy balance of a system is the energy available for 'societal use,' after subtracting the energy inputs required to produce and operate the technology, and ultimately dispose of it at end of life (EOL). It differs from solar-to-hydrogen (STH) efficiency, which only looks at the conversion energy ratios; see Figure 2.1 for a visual illustration. If energy inputs to the system approach or exceed the resulting energy harvested by it, its contribution to societal wellbeing becomes questionable, and in the long term, unsustainable. Thus, net energy life-cycle analysis concerns itself with accounting for all energy inputs and outputs for a technology with well-defined system

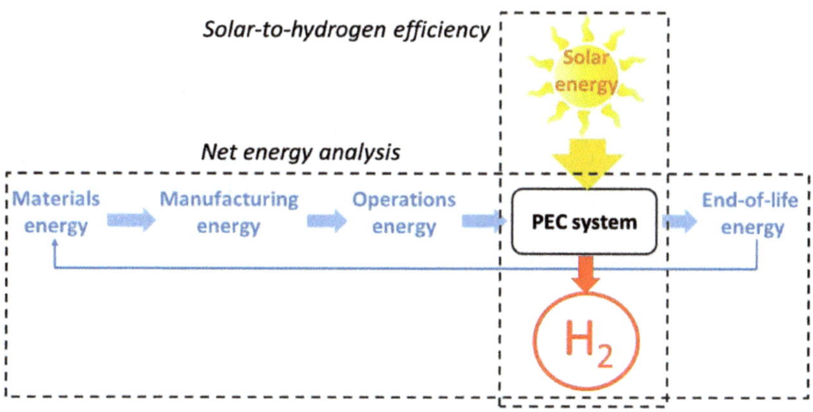

Figure 2.1 STH efficiency compares the solar energy incident on a PEC cell to the chemical energy of the produced hydrogen. Net energy analysis compares the total energy inputs to the system, including materials, manufacturing, operations, and end-of-life management, to the energy of produced H_2. Reproduced from ref. 37 with permission from The Royal Society of Chemistry.

boundaries and plant lifetime and can be used to determine a baseline for viability or identify specific component or system-level energy costs that may constrain future implementation.

Our prior work has analyzed the net energy balance of an individual PEC cell (including photoelectrodes, catalysts, and encapsulation),[35] as well as balance of system (BOS) requirements of a large-scale PEC H_2 facility, including structural supports, pipes, pumps, compressors, storage tanks, roads and monitoring systems.[36] Our latest analysis re-examined the PEC cell, identifying energy-saving design as well as manufacturing changes that could dramatically improve the net energy balance.[37] Current work being prepared for publication has identified even greater energy savings through the utilization of an unexpected resource: waste heat.

The output from an analysis such as the one presented here is valuable, because it identifies parameters and/or design elements that most strongly affect the net energy balance of the total system over its entire life cycle. This information can be used to guide future research needs and impacts all areas of work described throughout this book, ranging from studies of basic phenomena and discovery of catalysts and light absorbers, to development and assessment of prototypes.

2.2 Methods

2.2.1 Modeling Approach

The approach we have taken is generally described as prospective life-cycle assessment (LCA), meaning an LCA that examines the potential impacts of technologies and infrastructure systems that do not yet exist at commercial scale (or at all). It often utilizes scenario analysis, which is an evaluation of potential future pathways that differ in important respects such as demand for different types of energy and products; costs of energy, materials and technologies; and existence or absence of key technologies and policies.[38]

Our hypothetical hydrogen plant is conceived of at production-level scale, comparable to a large industrial facility or power plant, in order to capture any economies of scale inherent in such a system. (We reason that if such a plant is not sufficiently energy- and cost-efficient at this scale, it will not be commercially competitive at smaller scales, though there may be niche applications, *e.g.*, in remote areas or under special conditions, where it could still offer advantages.) We assume H_2 production equivalent to 1 GW on an annual average basis: using the higher heating value (HHV) of H_2, 1 GW equals 610 000 kg day^{-1} or 6.8 million Nm^3 day^{-1} of H_2).[‡] However, the scale we have chosen may be excessively large to capture requisite economies of scale, because the plant is almost fully modular at a scale above

[‡]The annual H_2 production of the modeled facility corresponds to 1 GW continuous power output, averaged across diurnal and seasonal cycles. 1 GW continuous power for one year is equivalent to 31.5 PJ of energy. Peak H_2 production rate in full sunlight is 3.6 times the average output rate. Nm^3 is normal cubic meters, at 20 °C temperature and 1 atm pressure.

~380 kg day^{-1} H_2 (~630 kW)—the size of a single 'field' (see below). The exceptions to this modularity are the operations building (housing personnel, repair equipment, and spare parts), and connecting infrastructure between the plant and municipal water supplies, grid electricity, and hydrogen pipelines (which consist of large single infrastructure elements requiring far less material—and hence embodied energy—per unit of H_2 flow capacity than at smaller scales). An interesting topic for future research is the suitability of smaller modular facilities for deployment in ways that are similar to how solar PV is installed at military, commercial, and residential scales.[39]

The PEC plant is configured on five hierarchical levels: cells, modules, panels, fields, and facility; each is described in detail in subsequent sections. In particular, we investigate the significance of six key energy characteristics of PEC systems:

1. Embodied energy of active cell materials (semiconductors, catalysts, transparent conductive oxides (TCO), conductive substrate, protective layers, electrolyte, and membrane materials)
2. Embodied energy of inactive module materials (window, back cover, frame, and ribs)
3. Energy intensity of active cell fabrication (thin-film deposition processes and membrane manufacture)
4. Energy intensity of inactive module fabrication (cell and module assembly)
5. Initial energy use for production of BOS components
6. Ongoing energy use for operation and end-of-life of BOS components

The first four characteristics pertain to the PEC cells and modules, with the first two describing the embodied energy of material supplies, accounting for energy used during extraction, processing, and transportation to the PEC factory. The next two characteristics describe the energy intensity of the fabrication processes within the factory to produce finished cells and modules. The final two characteristics pertain to the energy use of the BOS during production, operation and end-of-life, and concern the panel, field, and facility levels of the analysis.

To model the facility, we apply fundamental principles such as mass and energy balance to determine essential flows. We also conduct a preliminary engineering design of the facility, based on heuristic methods, to establish configurations, equipment characteristics, materials, and masses of materials sufficient to provide reliable performance. We then use a parametric model describing system-wide energy flows to determine the following three net energy metrics:

1. Life-cycle *primary energy balance:* Describes the usable energy the facility provides to society during its lifespan. It is calculated as the total energy output minus the total energy input:

$$\text{Energy Balance} = \{T \times E_\text{H}\} - \{E_\text{P} + (T \times E_\text{O}) + E_\text{D}\} \tag{2.1}$$

where

T = Service life of the facility (years)
E_H = Energy (HHV) in hydrogen produced in 1 year (PJ per year)
E_P = Energy used to produce the facility (PJ)
E_O = Energy used to operate the facility for 1 year (PJ per year)
E_D = Energy used to decommission the facility (PJ)

2. *Energy return on energy invested* (EROEI, or sometimes EROI): Describes the usable energy delivered relative to energy inputs. It is a dimensionless metric, calculated as the total energy output divided by the total energy input:

$$\text{EROEI} = \frac{T \times E_H}{E_P + (T \times E_O) + E_D} \tag{2.2}$$

3. *Energy payback time*: Describes how long the facility must operate in order to deliver sufficient H_2 to offset the energy consumed during manufacturing, construction, operations, and decommissioning. Note that energy payback time is not a true life-cycle metric, as it does not consider energy delivered after the payback time is reached, nor does it include the facility service life. It is calculated as the fixed energy inputs divided by the annual net energy output under full-scale, steady state operation:

$$\text{Energy Payback Time} = \frac{E_P + E_D}{E_H - E_O} \tag{2.3}$$

2.2.2 Uncertainty

Our system model includes base-case values for each of the six characteristics, as well as low and high energy input values indicative of current uncertainty regarding the actual value of the parameters in a future large-scale system. Parameter valuation is based on literature review, proxy data on analogous processes, and informed estimates. The range of cell characteristics are described in Table 2.1 for the most current published plant design, using a planar cell configuration with thin-film coatings based on a GaAs or amorphous (a-) Si photoanode and InGaP, $BiVO_4$, or Fe_2O_3 photocathode.[37] (In Sathre *et al.*[36] we assumed a Si nanowire configuration based on our initial PEC cell analysis.[35])

In addition to these parameters, we vary two key performance parameters: the lifetime of PEC cells (as opposed to the facility lifetime, which is much longer), and the STH conversion efficiency of the cells. We assume base-case values of 10 years and 10% efficiency, respectively, with the latter parameter suggested by Chen *et al.*,[40] based on the Gibbs free energy[§] of the H_2

[§]STH efficiency is defined in terms of Gibbs free energy of H_2, per the convention in the PEC community.[33] We define our 1 GW facility based on the higher heating value (HHV) of H_2, because HHV is an energy measurement convention commonly used in US energy system analysis. Our analysis applies appropriate conversion factors. One mole of H_2 has Gibbs free energy of 237.1 kJ, lower heating value (LHV) of 242.3 kJ, and HHV of 286.6 kJ.

Table 2.1 Summary description of high energy input, base-case, and low energy input characteristics of active cell, inactive module, and balance-of-system components.[a]

	High energy input	Base-case	Low energy input
1. Active cell materials	• 250 nm GaAs photoanode • 250 nm InGaP photocathode • 120 nm TiO_2 protective layer • 1 nm Pt HER catalyst • 1 nm IrO_x OER catalyst • 100 nm ITO TCO layer • 20% material utilization • 2:1 precursor energy • 0.2 mm stainless steel substrate • 70 μm membrane • electrolyte	• 250 nm a-Si photoanode • 250 nm $BiVO_4$ photocathode • 60 nm TiO_2 protective layer • 1 nm Pt HER catalyst • 1 nm IrO_x OER catalyst • 50 nm FTO TCO layer • 50% material utilization • 1:1 precursor energy • 0.1 mm stainless steel substrate • 50 μm membrane • electrolyte	• 250 nm a-Si photoanode • 250 nm Fe_2O_3 photocathode • 30 nm TiO_2 protective layer • 1 nm Fe_3P HER catalyst • 1 nm NiFeOx OER catalyst • 10 nm FTO TCO layer • 80% material utilization • 0.5:1 precursor energy • 0.05 mm stainless steel substrate • 30 μm membrane • electrolyte
2. Inactive module materials	• 2 × 3 mm glass • large PVC frame • PVC support ribs	• 2 × 2 mm glass • medium PVC frame • PVC support ribs	• 2 × 1 mm glass • small PVC frame • PVC support ribs
3. Active cell fabrication	Current typical thin-film coating	Current state-of-the-art inline thin-film coating	Solution coating at atmospheric pressure and room temperature
4. Inactive module fabrication	Base-case × 200%	From ref. 35	Base-case×50%
5. Initial BOS production	Upper 90% confidence interval from Monte Carlo modeling	Adapted from ref. 36, see text	Lower 90% confidence interval from Monte Carlo modeling
6. BOS operation and EOL	Upper 90% confidence interval from Monte Carlo modeling	Adapted from ref. 36, see text	Lower 90% confidence interval from Monte Carlo modeling
Key performance metrics	• 5-year PEC cell lifetime • 5% STH efficiency	• 10-year PEC cell lifetime • 10% STH efficiency	• 20-year PEC cell lifetime • 20% STH efficiency

[a]Abbreviations: amorphous silicon (a-Si), fluorine-doped tin oxide (FTO), H_2 evolution reaction (HER), indium tin oxide (ITO), O_2 evolution reaction (OER), photo-electrochemical (PEC), polyvinyl chloride (PVC), solar-to-H_2 (STH), transparent conductive oxide (TCO), balance of system (BOS), end of life (EOL). (Reproduced from ref. 37 with permission from The Royal Society of Chemistry.).

produced under AM1.5G solar radiation, and a number of PEC devices demonstrated to operate at or above this value. Due to a paucity of robust data, we are unable to quantitatively link the variations in cell production energy and cell performance, but such linkages could be introduced as research advances to provide convergence on realistic data.

We use a simplified approach to account for the shorter service life of PEC cells, by assuming steady-state conditions in which a fixed percentage of panels are replaced each year. For instance, we assume that for cells with a 10-year life span, 10% of panels are replaced annually. We do not model a facility ramp-up period, but instead base our calculations on eventual steady-state conditions. Furthermore, we assume cell performance degrades linearly over time, with panel replacement occurring when STH efficiency falls to 80% of its original value. This is simulated through a simple 'degradation factor' so that the average H_2 output over the lifetime of the panel corresponds to 90% of the nominal STH efficiency. We also assume a further 10% solar transmittance loss due to the combined effects of absorption and reflection by surface dust, encapsulation material, and electrolyte.

We conduct a sensitivity analysis by varying individual parameters from their base-case values to their low input and high input values. Within parameter categories, we also conduct Monte Carlo simulations to estimate uncertainty introduced by interactions between individual parameters, using triangular probability distributions based on low, base-case and high values. Based on the outcome distribution of 10 000 simulations, we used 90% confidence intervals to estimate uncertainty ranges of results.

We consider two forms of energy used to construct and operate the facility: fossil fuels and electricity. Comprehensive energy accounting must be done using comparable units, thus different energy carriers and end-uses are generally considered in terms of their 'primary energy' use. Primary energy is in a form found in nature, prior to any conversion or transformation process. For raw fossil fuels, their primary energy is defined by their heat content. For other derived energy carriers, such as electricity and refined fuels, primary energy includes all energy used upstream for *e.g.* extraction, transportation, processing, conversion and distribution of the energy carriers. For diesel fuel used in the facility, we estimate its primary energy as the higher heating value (HHV) of the fuel plus 13% fuel cycle inputs.[41] The primary energy of electricity depends on how it is generated; for example, electricity from a coal-fired power plant typically has a primary energy conversion efficiency of about 33%, meaning that about 3 J of primary energy are needed for each J of electricity produced. Estimating the primary energy of non-thermal electricity sources, such as hydroelectric and PV, is less straightforward.[42] Furthermore, a particular challenge in performing prospective energy analysis is to estimate primary energy use associated with future electricity generation, for example the electricity used to operate the H_2 production facility modeled here.

Grid electricity, which comes from a range of fossil and renewable sources, is likely to change markedly in composition over the next several decades.

We consider the hydrogen gas produced by the facility as primary energy, equivalent to its higher heating value. We convert end-use electricity to primary energy, assuming a conversion efficiency of 50% based on current H_2 fuel cell performance.[43] In other words, we estimate the primary energy of electricity inputs to the facility, as if that electricity were produced from the hydrogen made by the facility. We use this energy accounting technique merely to derive comparable primary energy values for different energy carriers; all the H_2 produced is actually delivered for external use, and electricity is a separate input to the facility. Despite the modest electrical requirements of the system, the impact on total net primary energy is sensitive to this choice, which we explore in a sensitivity analysis.

2.2.3 Externally-supplied *versus* On-site Electricity

The electricity required for on-site operations, such as gas compression and electronic monitoring, is assumed to be generated externally and provided as an input to the facility from the electricity grid. On the other hand, one could consider a case where additional solar PV cells are included in the facility design to directly supply electricity to meet these operational demands.

In fact, our most recent work suggests that such PV cells could be integrated into the existing PEC cell design with very little additional manufacturing energy, providing nearly twice as much electrical energy (on a primary energy-equivalent basis) as hydrogen for the base and low case scenarios, with negligible performance loss in the PEC cells. Only about 12% of this electrical energy would be needed to run the plant equipment; the rest (about 820 MW of electricity) could be exported along with the hydrogen. Moreover, because the electricity is produced on-site rather than being imported, it is subtracted from the net electricity output as well as from the operational energy, which significantly reduces the denominator in eqn (2.2). As a result, the EROEI would increase from 2.34 to 11.66, a nearly fivefold increase (Segev *et al.*, submitted), with dramatic improvements in other net energy metrics as well. While this performance has yet to be demonstrated in the laboratory, it would have significant implications for environmental sustainability as well as economic viability.

2.2.4 PEC Cell and Module Design

We have chosen to model the flat panel format described in James *et al.*[19] and Pinaud *et al.*,[20] because it is at a more mature developmental stage than the alternatives. It consists of a planar thin-film device in an aqueous electrolyte, with the photoanode (O_2-producing reaction) facing the sun, and the photocathode (H_2-producing reaction) on the other side of the electrolyte membrane. See Figure 2.2, which shows cross-sections of the PEC cell and module assembly at different scales. The overall module is similar in appearance to a dual pane window, with dimensions of 1 m$\times$1.2 m$\times$11 mm. A ribbed PVC frame is laminated between the glass panes, supporting the

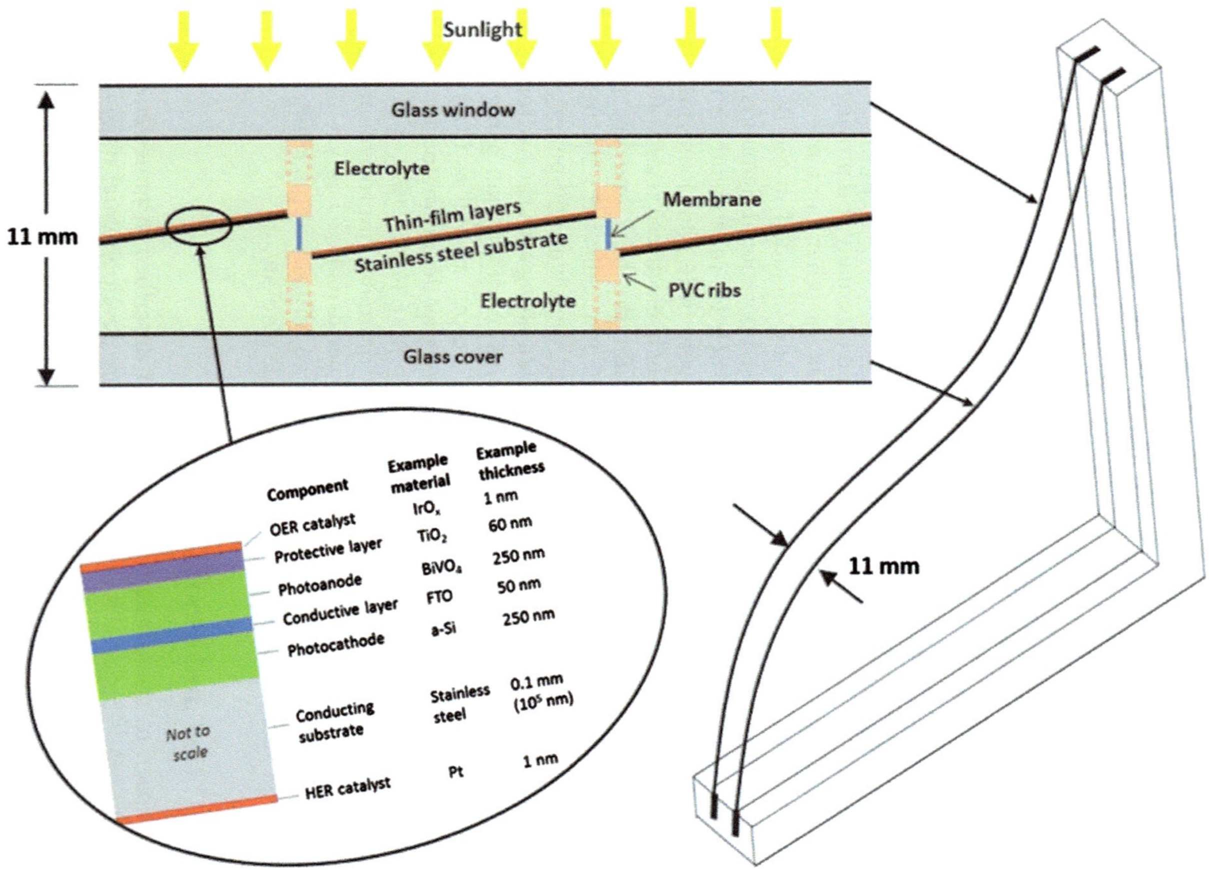

Figure 2.2 Cross-sections of PEC cell and module, including dual pane glass module encapsulation (right), louvered active cell configuration (top left), and thin-film layers of active cell materials (bottom left).
Reproduced from ref. 37 with permission from The Royal Society of Chemistry.

louvered active cell assembly. Active thin-film layers are angled within the louvered cell geometry, with an ion-exchange membrane positioned vertically to minimize transport distances and solar blockage.[44] The thin-film device layers are also shown (not to scale), and include a tandem photo-absorber, transparent conductive oxide (TCO) layer, protective anti-corrosion layer, as well as electrocatalyst layers facilitating the redox reactions.

2.2.4.1 *Active Cell Materials Energy*

We estimated the energy used for extraction, primary processing and transportation to the cell factory gate for the active device materials or their precursors. (Fabrication energy was a separate step detailed below in Section 2.4.2) Materials include semiconductors, catalysts, transparent conductive oxide (TCO), conductive substrate, electrolyte, and membranes. For the louvered geometry, we adopt a generic electrochemical design but consider a range of potential active materials that vary in embodied energy: the primary energy used for extraction, processing and transportation of materials. Embodied energy does not include the energy used for thin film deposition, which is considered part of fabrication energy. We also do not consider correlations between materials and deposition energy intensity, product yield, STH efficiency, or cell life span, as information was mostly lacking; this potentially important issue should be pursued in the future as more information becomes available.

Quantifying the embodied energy of chemical compounds accurately remains a challenge throughout the LCA community. Standard life-cycle inventory (LCI) databases contain such estimates for common materials, but specialty chemicals produced and used in smaller amounts remain out of reach to both LCI databases and related literature. Unfortunately, many of the precursor and/or final materials used in the thin films modeled here fall into this latter category. To overcome this challenge, we have developed a simplified approach to estimating the embodied energy of precursor materials for thin film deposition:

$$\text{Thin Film Material Energy} = \sum_{\text{layers}}^{i} \sum_{\text{elements}}^{j} \frac{M_\mathrm{s} \times E_\mathrm{s} \times R_\mathrm{p}}{F_\mathrm{m}} \tag{2.4}$$

This approach sums over i layers of thin film, each containing up to j elements, where the embodied energy of each element is quantified using four factors: M_s is the mass per thin film area (kg m^{-2}), or alternatively, the material density of the thin film material (kg m^{-3}) times the film thickness (m) times the mass fraction of element j in the thin film material. E_s is the embodied primary energy (that is, extraction, processing, transportation and refinement energy) of the refined element per unit mass (MJ kg^{-1}). R_p is the dimensionless ratio of embodied energy of the refined element to that of the precursor material used in the deposition process. F_m is the dimensionless material utilization efficiency, or the mass in the final thin film per mass in the precursor material. The overall embodied energy therefore has units of MJ of primary energy per m^2 of active cell area.

The embodied energy E_s of most refined elements is available from standard LCI databases. We base our calculations largely on embodied energy data from the Ecoinvent database,[45] supplemented with data from Nuss and Eckelman.[46] The embodied energy of an element depends on its purity, as processing energy increases significantly with increasing purity.[47] In this analysis, we assume the default purity levels provided by the Ecoinvent database, which typically represent the purest form that is widely used in representative industries.[48]

In some thin film deposition processes, the final deposited material is the same as the process raw material. For example, magnetron sputtering often uses targets made of the same metal as the resulting thin film. However, in many cases, the material composition of the thin film differs from that of the precursor materials. For example, when making a TiO_2 layer, precursors may consist of titanium isopropoxide or tetrakis(dimethylamido)titanium, laid down using atomic layer deposition. Another example is InP, which is deposited by metal–organic chemical vapor deposition from precursors such as trimethylindium and phosphine. Atmospheric pressure depositions also frequently use non-elemental or oxide precursors; *e.g.* $BiVO_4$ is deposited from $Bi(NO_3)_3$ and vanadium acetylacetonate. Even amorphous silicon is typically deposited from H_2 and SiH_4 using plasma-enhanced chemical vapor deposition. We account for this ambiguity using R_p to express the ratio of embodied energy of the refined element to that of the precursors, both measured per unit mass of the element. Our base-case assumes a value of 1.0, and we consider a range from 2.0 to 0.5.

The material utilization efficiency F_m is determined from the material application yield, defined as the fraction of raw material that is contained in the finished thin film (assuming the remainder is not recovered; however, see below). This factor includes the factory product yield (fraction of thin film area present in finished PEC modules; the remainder represents defective products, assumed scrapped) and the material recycling rate (the fraction of wasted material that is recovered and reused). In practice, F_m can vary widely among deposition processes and materials; in general, gaseous precursor chemical reaction processes are the least material efficient (as low as 10%), while electrodeposition processes are the most efficient (up to 95%).[49–52] We assume a base-case value of 50%, with low and high energy inputs cases of 80% and 20% respectively. (We assume the factory product yield is ~100% for all cases since business viability requires that defective products be minimal.)

For the base-case, we assume amorphous silicon (a-Si) for the photocathode material, though other potential materials include InP and Cu_2O. The base-case photoanode material is assumed to be $BiVO_4$, a moderate-bandgap material, though other potential materials include a-Si, WO_3, hematite (Fe_2O_3), and GaAs. (Amorphous silicon is a buried junction device, meaning in principle it can be incorporated as a photoanode or photocathode.) Our high energy input case considers GaAs and InGaP photoelectrodes, and our low energy input case uses a-Si and Fe_2O_3 photoelectrodes. While not all combinations of materials considered have the required properties (*e.g.*, sufficient bandgap) for effective PEC cells, or adequate stability against corrosion, we use them as representative materials and fabrication processes that span a reasonable range of embodied energies.

Potential H_2 evolution reaction (HER) catalysts for acidic or basic electrolyte solutions include Pt, NiMo, Co_2P, and Fe_3P, while potential O_2 evolution reaction (OER) catalysts for basic solutions include NiO_x, CoO_x, and $NiFeO_x$,[53] while only IrO_x is currently available under acidic conditions, as described in the chapter by McCrory within this book. Our base-case analysis assumes a Pt HER catalyst and IrO_x OER catalyst. (In our earlier work, we assumed a Co or Pt HER catalyst, and no catalyst or a Pt catalyst for the OER,[35,36] which was unrealistic.) The same materials are used in the high energy input case, while the low energy input case considers a Fe_3P HER catalyst and $NiFeO_x$ OER catalyst.

The cell stack also requires protective layers to prevent photoelectrode corrosion; we assume TiO_2.[54–56] (In our earlier work,[35,36] before corrosion mitigation strategies were well-developed, we assumed no such layers.) In our base-case we use 60 nm of TiO_2 to protect the photocathode and vary the thickness between 30 and 120 nm for the low and high input cases, respectively. For the photoanode, we assume a 50 nm thick transparent conductive oxide (TCO) layer made of fluorine-doped tin oxide (FTO) in our base-case. For the low and high energy input cases, we examine a 10 nm thick FTO layer and a 100 nm thick indium tin oxide (ITO) layer, respectively.

We include the thin film substrate, upon which thin film layers are deposited, as an active material because electrical charge must be transported through it. We select stainless steel because of its electrical conductivity, flexibility for use in roll-to-roll processing, and chemical resistance in both acidic and basic environments. However, different types of stainless steel are more resistant under acidic or basic conditions, so an appropriate type must be selected depending on the electrolyte solution. As information is lacking on the requisite thickness to use, we choose 0.1 mm for our base-case and consider 0.2 and 0.05 mm thicknesses in our high and low energy input cases, respectively. (For the Si nanowire design,[35,36] no substrate was utilized because the membrane acts as support.)

Membranes for PEC devices must be nearly impermeable to H_2 and O_2, yet conductive to cations (*e.g.*, H^+) or anions (*e.g.*, OH^-), as described in the chapter by Miller and Houle within this book. The most commonly used cation membrane material is perfluorosulfonic acid (PFSA), available commercially as Nafion®.[57] The required PFSA thickness depends on device design and is assumed to be 50 μm in our base-case. For our low and high energy input cases, we vary this thickness between 30 and 70 μm respectively. Based on Zhai *et al.*,[35] the primary energy for PFSA raw materials is

140 MJ m^{-2} of membrane. Due to the geometry of the louvered device (Figure 2.2), which is optimized for ionic transport,[44] the membrane area is only 8.8% of the cell area, resulting in a much lower embodied energy for membrane raw materials of 12 MJ m^{-2} of cell area.

We assume a 1 M H_2SO_4 electrolyte in our modeling, though we note that some electrode and catalyst materials are not stable in such an environment. Alternatively, a 1 M KOH electrolyte could be used to create a basic environment. However, we learned that production of KOH requires over 10 times the energy as production of H_2SO_4 per unit mass. Nonetheless, this would have little impact on overall net energy, because 1 M H_2SO_4 is responsible for only 0.07% of total energy in the initial construction of the facility; increasing this to 0.7% would have little consequence on the overall energy balance.

2.2.4.2 Active Cell Fabrication Energy

Fabrication energy for thin-film deposition as well as membrane fabrication is included in this category. Data for thin-film PV manufacturing provide the most relevant information, since the same processes are expected to be applicable to solar fuels technologies, though materials used would be somewhat different. A variety of thin film deposition techniques are available, including physical vapor deposition methods[58] such as evaporation, sputtering, laser ablation, and arc-based emission; chemical vapor deposition methods[59] such as thermal, plasma enhanced, and combustion; and solution-based methods[60,61] including spin-coating, inkjet printing, spray pyrolysis, and gravure printing. These techniques consume energy for various purposes, such as vacuum pumping, ion generation, heating, and cooling, resulting in a wide range of process energy intensities.[62] Process energy use may vary depending on required pressure differential, pumping speed, temperature, film thickness (directly correlated with processing time), precursor or source material phase, *etc.* In larger facilities, economies of scale can be realized, requiring much less energy per unit deposited area,[63] placing constraints on how useful energy intensity data can be obtained for certain processes. For example, theoretical values of vacuum pumping energy use can be calculated from first principles, but at industrial scale, load locks and transfer chambers are used in place of repeated venting-evacuation cycles, and efficiencies are gained through differential vacuum pumping of small volumes over short durations.[64] Moreover, the climate at a particular geographic location of a manufacturing plant may also affect energy intensities.[65]

In consultation with a vapor deposition expert (A. Anders, pers. commun., 2014–2015), we settled on a state-of-the-art integrated inline coater for our base-case, where multiple layers are deposited sequentially in the same vacuum chamber, minimizing total vacuum pumping energy use. Reviews of thin-film PV technology LCA studies[51,66] show that direct process deposition energy ranges from 129 to 1150 MJ m^{-2}, with most examples falling between

300 and 500 MJ m^{-2}. Although using PV manufacturing as a proxy for PEC device fabrication does not fully capture likely processes, it provides an initial process energy estimate; we select 400 MJ m^{-2} as our base-case value, representing current best practice in an integrated inline deposition process. For our high energy input case, we select 800 MJ m^{-2} representing current average practice (*e.g.*, involving non-integrated processes with separate sequential deposition lines). For our low energy input case, we select 60 MJ m^{-2}, representing a room temperature, atmospheric pressure, solution-based deposition method based on a recent LCA study for perovskite solar modules.[67] This low energy input case also considers the use of a-Si electrodes made with a solution-phase process, which is currently at an early development stage.[68] As currently practiced, it requires strictly water and air-free chemical handling and process conditions, increasing the complexity and energy intensity of the method. Nevertheless, we adopt this option to bound the energy consumption of potential future system improvements.

Our membrane fabrication energy estimate was retained from our earliest work,[35] 166 MJ m^{-2} of membrane. Due to the geometry of the louvered device, the membrane area is reduced by 91%, resulting in fabrication energy of 15 MJ m^{-2} of cell area.

2.2.4.3 *Inactive Component Materials Energy*

Module materials include window glass, back cover, perimeter frame, and cell support ribs. A module is assumed to have dimensions of 1 m×1.2 m, encapsulated on the front and back by 2 mm thick panes of float glass. The thickness was chosen after an investigation into the strength of glass, whose elastic modulus was found to be much more strongly affected by surface defects than overall thickness.[69] We found that the minimum recommended thickness for architectural glass was 2 mm when spanning an open area of ~1.4 m^2,[70,71] while also resisting both the typical three-second maximum gust wind speeds (<90 mph) found in the desert US southwest, away from coastal hurricane regions,[72,73] and a maximum hail stone damage of ~19 J impact energy (<1.75 inch diameter at terminal velocity), found outside the Great Plains region.[74,75]

We select glass as the encapsulation material due to its rigidity, transparency (a requirement only for the side facing the sun), resistance to most chemicals, and low embodied energy compared to metals or plastics. Properties of various potential encapsulation materials are shown in Table 2.2. Thicknesses of 1 and 3 mm are explored in the low and high energy input cases, respectively: 1 mm glass may be possible, because the glass is not freestanding but is adhered to periodic ribbed supports, reducing the maximum span and making the entire module a rigid structural element. 3 mm glass is considered more standard for architectural glass and in the PV industry, but we deem it overly conservative given the above considerations.

Table 2.2 Properties of six potential encapsulation materials for PEC modules.

Material	Density $g\,cm^{-3}$	Elastic modulus GPa	Specific embodied energy of 2 mm thick sheet $MJ\,m^{-2}$
Float glass	2.5	~70	65
PVC	1.4	~3	170
Polycarbonate	1.2	~2.2	260
Poly(methyl methacrylate)	1.2	~2.6	340
Aluminum	2.7	~70	1050
Stainless steel	7.9	~180	1230

For configurations requiring basic conditions, glass cannot come into direct contact with the electrolyte because the glass will etch over time. However, glass may be easily protected with a base-resistant thin coating, such as PVC, polypropylene, or ethylene-propylene diene monomer.[76] The embodied energy of a 0.1 mm thick polypropylene layer is about 7 $MJ\,m^{-2}$. For comparison, a 2 mm glass pane has an embodied energy of 64 $MJ\,m^{-2}$, so adding such a coating only modestly increases the total encapsulation energy. As an alternative, polycarbonate could be used instead of glass, but its embodied energy is significantly higher, ~260 $MJ\,m^{-2}$ for 2 mm thickness.

A molded PVC perimeter frame surrounds the module, securing the glass panes and acting as an impermeable barrier to the electrolyte. We consider three alternative designs for the perimeter frame corresponding to low energy, base-case, and high energy inputs. A molded PVC ribbed framework supports the membrane and the thin-film substrate and is adhered to the two glass panes (see Figure 2.2).

2.2.4.4 *Inactive Component Fabrication Energy*

This category includes encapsulation fabrication (*e.g.*, cutting and pressing), ancillary supplies (*e.g.*, miscellaneous process chemicals), and indirect industrial processes (*e.g.*, environmental controls of the manufacturing facilities including space heating, cooling, ventilation, and air filtration). Based on our earliest work,[35] we assume a base-case value for module fabrication of 241 $MJ\,m^{-2}$ of cell area. As literature on the energy use of large-scale PEC cell assembly is absent, this value has considerable uncertainty, and we therefore arbitrarily assume a broad range of values for our low and high energy input cases (50% and 200% of base-case values, respectively).

2.2.5 Balance of System (panel-, field- and facility-level) Design

The remaining three levels comprise the balance of system (Figure 2.3), consisting of balance-of-systems (BOS) components including structural

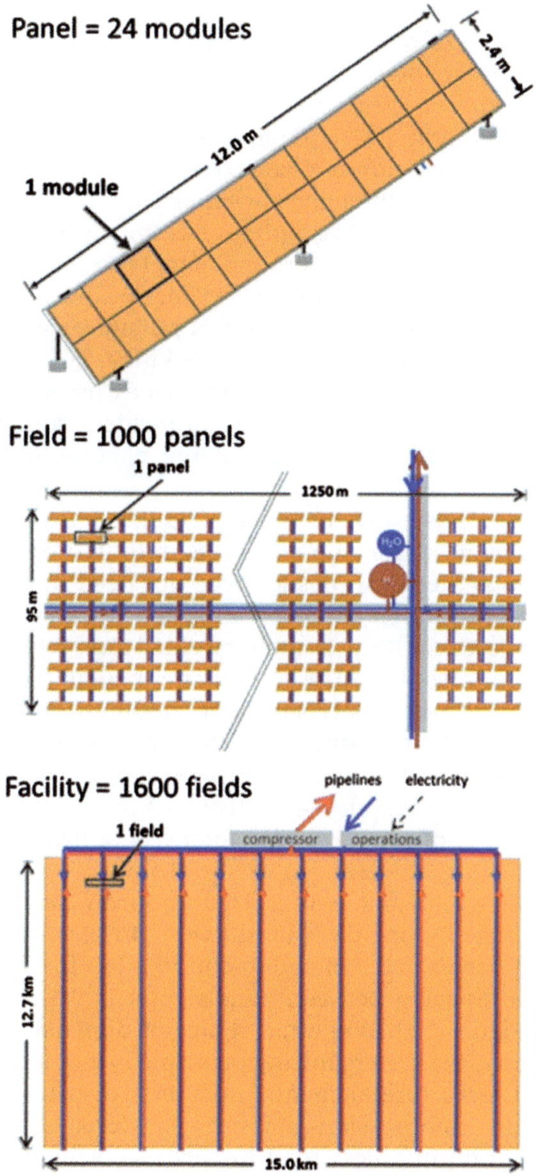

Figure 2.3 Three hierarchical levels (panel, field and facility) comprise the balance of system of the 1 GW hydrogen plant.
Reproduced from ref. 37 with permission from The Royal Society of Chemistry.

supports, manifolds and pipes, pumps, compressors, storage tanks, pipelines, roads and monitoring systems. Each level is described in detail below.

Twenty-four modules are mounted on a ~29 m^2 'panel' designed to fit in the interior of an intermodal shipping container (~12.0 m length by ~2.4 m height). With a container width of 2.4 m and panel thickness of ~20 cm including padding, ~12 panels would fit in each container. Panels are envisioned to be modular and self-contained, and include onboard monitoring and diagnostics sensors, electrical and data connections, and pipe manifolds connecting fluid supply (H_2O) and uptake (H_2, O_2) among modules. Panels and modules have standardized connections so they can be installed and removed easily for maintenance or replacement.

The total panel mass is ~510 kg without electrolyte and ~690 kg including electrolyte. When deployed on the ground, panels are tilted at an angle equal to the local latitude (assumed to be 34° for locations in the US desert southwest). While sun-tracking capability is possible, we have not assumed it here, as the additional solar energy captured must be balanced against the extra hardware, complexity and reliability issues associated with moving panels. Moreover, an appreciable fraction of incident sunlight is diffuse, so sun tracking does not improve the energy capture of this portion of sunlight.[77]

The panel design has evolved[37] since we published our first analysis in 2014.[36] First, we reduced the amount of steel required by redesigning panel structural elements: the newer design does not need to support as heavy a load, mainly due to thinner modules and less electrolyte, and replaces a single large steel section with two smaller welded steel sections, reducing the steel mass by 39%. Also, the inactive area, defined as the fraction of gross panel area that does not collect solar energy due to blockage by inactive materials, is doubled to 10% due to the presence of supporting ribs in the louvered PEC modules (these were not necessary in the Si microwire cell design[36]). The current design has an active area of 25.8 m^2 per panel, which increases the number of panels needed for 1 GW of hydrogen to 1.6 million (see below).

A 'field' contains 1000 panels in a fixed flat array format, plus gas compression and storage for one day's production of H_2, occupying 120 000 m^2. Spacing between panels is 0.5 m end-to-end to allow for access by cranes or other equipment. Spacing between panels rows is 9.5 m, based on a 10° shading angle (Figure 2.4). While the choice of dimensions for the field is somewhat arbitrary, the key organizing principle is that it includes a number of important infrastructure elements, including distributed water storage, and gas collection, compression and storage. Moreover, all panels are located within 50 m of an access road in order to facilitate replacement *via* mobile crane. In this way, the field represents the smallest fully self-contained unit in a scalable system, with the only higher-level elements being an operations building (with personnel, spare parts, common equipment, *etc.*), electrical grid tie-in, and connections to regional water, H_2 and O_2 pipelines (the latter being optional, depending on the assumed market for compressed O_2). It is also completely modular: while an overall facility site may have complex topography, preventing it from occupying a simple rectangular area, the available land could be covered with a series of fields that are completely regular and uniform, greatly simplifying design and

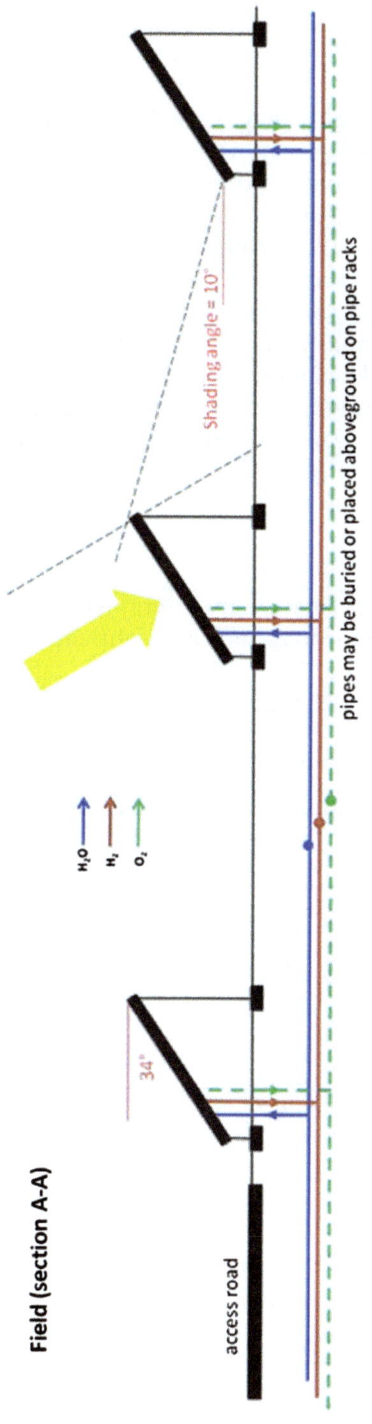

Figure 2.4 Cross sectional diagram of a PEC field, showing geometry of panel layout. Reproduced from ref. 36 with permission from The Royal Society of Chemistry.

construction.[78] In a final, optimized engineering design, there might be multiple levels of intermediate organization in a facility of this size, and the unit size could be considerably larger or smaller than the 1000-panel design presented here.

Finally, the 1 GW (annual average) 'facility' comprises 1600 fields occupying 190 km^2 in the base-case of 10% STH efficiency, with 90% capacity factor (*e.g.*, 90% of the panels are operating). The facility is assumed to be located in the southwest US in an area of high insolation. Monthly mean, flat-panel insolation data taken from NREL[79] are averaged over four sites: Phoenix, Daggett, Tucson, and Las Vegas, providing an annual average solar insolation of 276 W m^{-2}. A life span of 40 years is assumed for the facility, which is arbitrary but actually on the short side for large-scale energy infrastructure, which often lasts 60 years or longer. At EOL, the facility must be decommissioned (dismantled, parts recycled or disposed of, and land reconditioned in some way). Following the method used by NETL,[80] we assume that decommissioning requires 10% of initial BOS construction energy.

2.3 Results

We find a total of 981 MJ of primary energy per m^2 of active cell area is required in our base-case for materials and fabrication of cells and modules (*e.g.*, not including BOS), see Table 2.3. This value ranges from 1843 MJ m^{-2} in the high input case to 373 MJ m^{-2} in the low input case. Fabrication of

Table 2.3 Assumed energy inputs for PEC cell and module production. Reproduced from ref. 37 with permission from The Royal Society of Chemistry.

	High energy input	Base-case	Low energy input
Component	*MJ of primary energy per m^2 of active cell area*		
Active cell materials	194.3	93.3	40.6
Thin-film materials	*53.9*	*18.1*	*0.8*
Membrane materials	*15.6*	*12.2*	*7.6*
Conductive substrate	*123.4*	*61.7*	*30.9*
Electrolyte	*1.4*	*1.4*	*1.4*
Inactive module materials	348.5	231.8	143.0
Glass window[a]	*106.2*	*70.8*	*35.4*
Glass back cover[a]	*106.2*	*70.8*	*35.4*
Supporting ribs	*51.5*	*51.5*	*51.5*
Perimeter frame	*84.7*	*38.7*	*20.8*
Active cell fabrication	818.6	414.5	69.1
Thin-film deposition	*800.0*	*400.0*	*60.0*
Membrane fabrication	*18.6*	*14.5*	*9.1*
Inactive module fabrication	481.9	240.9	120.5
Total without BOS	1843	981	373

[a]The embodied energy values for glass shown here are greater than those discussed elsewhere, because these values are expressed per m^2 of active cell area, which is assumed to be 90% of total module area.

active cell components is the largest single energy contributor in the base and high energy input cases, mainly due to the thin film deposition process. However, for the low energy input case, our assumption of a room temperature and pressure solution-phase deposition process lowers this energy considerably, and the inactive module material becomes the dominant contributor to total energy. We highlight the thin film deposition processes as an important opportunity for lowering energy inputs and improving the net energy performance.

Several material inputs for cell or module production are finished products of earlier production processes (*e.g.* stainless steel, glass, PVC), each of which requires their own raw material inputs. Thus, the energy use breakdown begun in Table 2.3 could also include the embodied energy of these materials. For example, a typical soda-lime glass composition used for window glass contains ~60% silica derived from sand, ~20% sodium oxide derived from sodium carbonate (soda ash), and ~18% calcium oxide derived from calcium carbonate (limestone).[81] The sand, limestone and soda ash are combined and heated, typically using methane fuel, and the melted glass is poured on a float bed and cooled until solid, after which the glass sheets are cut to size, and finally annealed. Heating the raw materials to melting temperature is the most energy-intensive step in the process, contributing over half the total embodied energy of the finished product, while supply of raw materials, including extraction, primary processing and transport, contributes less than 15% (calculation from ref. 37, based on Worrell *et al.*[81] and Gaines and Mintz[82]).

A breakdown of the embodied energy of the precursor materials used for thin film layers is shown in Table 2.4. The variations among the cases are due to different material selections, layer thicknesses, and values for material utilization efficiency and the ratio of embodied energy of the refined element to that of the precursor materials. The totals in Table 2.4 are included in Table 2.3 under 'Thin film materials'.

Table 2.5 shows energy inputs for initial production, and ongoing operation and EOL, of the BOS under various combinations of cell life span, energy input case and STH efficiency. Note that these values are normalized by cell life span so they can be compared to the energy inputs shown in Table 2.3. In all cases, the ongoing operation and EOL energy inputs are found to be substantially greater than those for initial BOS production.

Figure 2.5 compares the production energy use of our current cell design—a louvered thin-film device[37]—to the silicon microwire cell design modeled in Zhai *et al.*[35] and used in Sathre *et al.*[36] The base-case value of the current design uses substantially less energy for production (981 MJ m^{-2}) than the silicon microwire design (2110 MJ m^{-2}). Major improvements are due to reduced energy for fabricating photoactive components, and reduced embodied energy of inactive materials, with significant consequences for overall net energy performance of the plant (see below). This demonstrates that innovative cell design and manufacture can improve net energy performance.

Table 2.4 Embodied energy of precursor materials to the thin film layers. Reproduced from ref. 37 with permission from The Royal Society of Chemistry.

	High energy input	Base-case	Low energy input
Thin film layer	*MJ primary energy per m² active cell area*		
OER catalyst	5.8	2.9	0.001
Protective layer	2.4	0.60	0.15
Photoanode	8.2	1.1	0.56
TCO	5.5	0.17	0.02
Photocathode	9.5	2.0	0.02
HER catalyst	22.6	11.3	<0.001
Total	53.9	18.1	0.75

Table 2.5 Energy inputs for initial and ongoing balance of system under different conditions of cell life span, energy input case, and STH efficiency. Reproduced from ref. 37 with permission from The Royal Society of Chemistry.

Cell life span:		5 year			10 year			20 year		
Energy input case:		High	Base	Low	High	Base	Low	High	Base	Low
STH efficiency	BOS stage	*MJ primary energy per m² active cell area per cell life span*								
5%	Initial	235	219	204	471	437	407	941	875	816
	Ongoing & EOL	735	602	474	1475	1204	948	2943	2410	1897
10%	Initial	259	238	220	517	476	439	1036	953	878
	Ongoing & EOL	1060	913	768	2126	1826	1536	4251	3649	3068
20%	Initial	309	276	248	618	553	496	1234	1104	989
	Ongoing & EOL	1732	1534	1339	3468	3068	2682	6907	6133	5348

Figure 2.6 shows our latest annual energy balance estimates for plant operations under base-case conditions. The produced H_2 has a primary energy content of 31.5 PJ per year, equivalent to 1 GW of continuous power. From this positive value we must subtract the primary energy needed to operate the system, including the energy to manufacture replacement parts. The largest single energy use is for gas compression (5.0 PJ per year), while cell replacement constitutes the second largest energy input (4.1 PJ per year assuming a 10-year cell life span). Note that cell replacement energy is less than half of what we estimated in our previous work (8.7 PJ per year),[36] due to the lower energy intensity of cell production (see Figure 2.5). Module heating to avoid electrolyte freezing on cold winter nights is the next largest energy drain (1.5 PJ per year) and is unchanged from what we estimated in Sathre *et al.*[36] Though the louvered design will have different internal convection coefficients, we assumed this would not change the heating requirement significantly. The need for module heating is discussed further in the Supplementary Information to Sathre *et al.*[37] Gas handling, water supply and

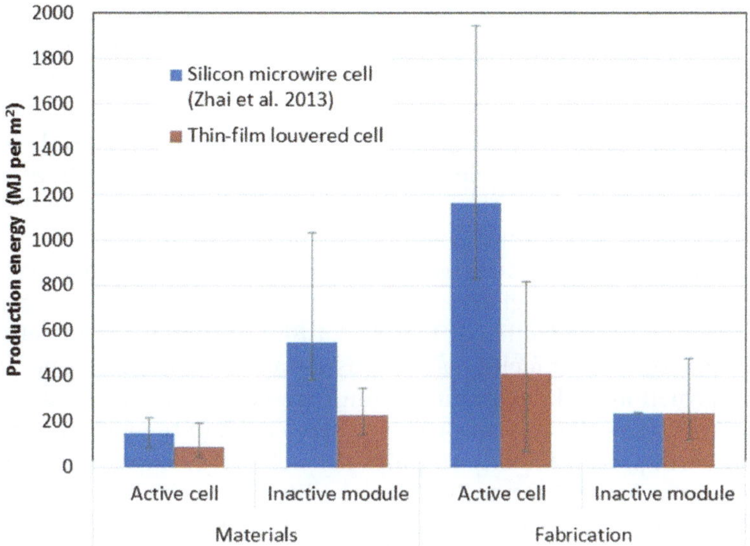

Figure 2.5 Production energy use for the louvered thin-film and silicon microwire cell designs.
Reproduced from ref. 37 with permission from The Royal Society of Chemistry.

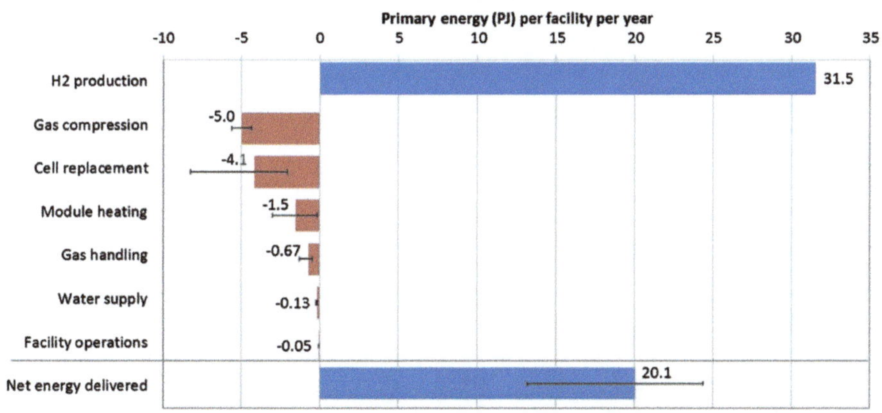

Figure 2.6 Annual energy balance (PJ per year) of a 1 GW annual average H_2 production facility under base-case conditions.
Reproduced from ref. 37 with permission from The Royal Society of Chemistry.

facility operations energy use is minor. In total, about 11.4 PJ per year are needed, resulting in a net primary energy delivery of 20.1 PJ per year under base-case conditions.

It may be possible to reduce the compression energy needed by designing cells to produce H_2 at higher pressure, using a higher overpotential; the thermodynamics of H_2 production requires an additional 0.06 V per atm.[83] While this is a modest requirement compared with the required water-splitting voltage (1.23 V), maintaining cells and panels above atmospheric pressure would entail trade-offs between the decreased energy requirements for gas compression and the higher performance needs of other components including module encapsulation and pipes. We have not explored these trade-offs to date.

Figure 2.7 shows one-time energy inputs required to construct and decommission the facility. Under base-case conditions, initial construction will need 72.9 PJ of primary energy, while the estimated energy input for decommissioning of the facility at EOL is 3.4 PJ. Compared to the net annual delivered energy of the facility, shown as a dashed line, these one-time inputs (construction and decommissioning) are equivalent to about 4.1 years of operations. Over the 40-year projected service life of the facility, therefore, the facility will produce nearly 10 times the net energy required to construct and decommission it. Put another way, the life-cycle primary energy balance is +720 PJ, almost ten times the initial construction energy. However, the energy return on energy investment (EROEI) is a lower ratio (2.34), because of the way it is calculated: the energy of the produced H_2 appears in the numerator, and all energy inputs, including operational inputs, appear in the denominator.

The net energy performance of the facility is substantially better than that we estimated in our previous silicon microwire-based design,[36] however,

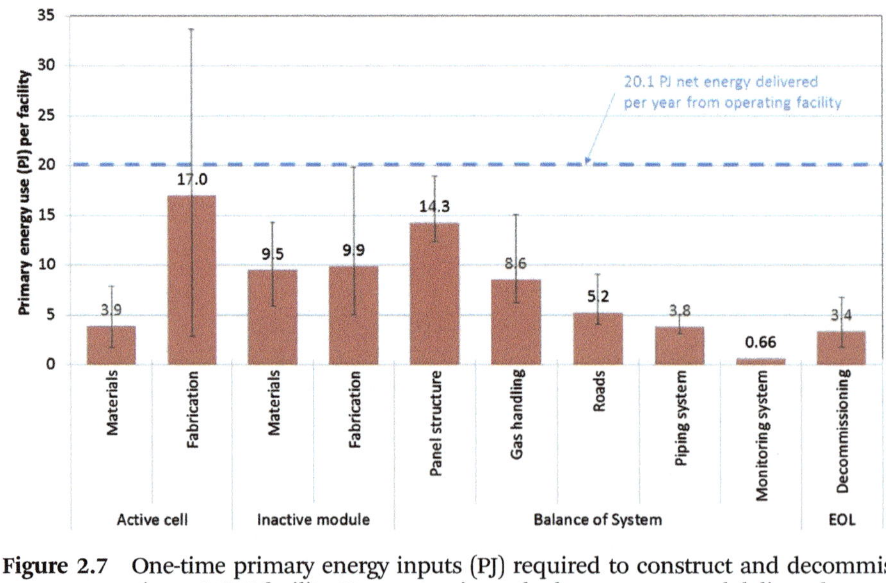

Figure 2.7 One-time primary energy inputs (PJ) required to construct and decommission a 1 GW facility. For comparison, the base-case annual delivered energy of the completed facility in operation is shown as a blue dashed line. Reproduced from ref. 37 with permission from The Royal Society of Chemistry.

which had an energy payback time of 8.1 years, an EROEI of 1.66, and an overall energy balance of +500 PJ. The improved net energy performance is largely due to the reduced energy inputs for cell and module production, which means lower initial energy use for facility construction (Figure 2.7), as well as lower ongoing energy use for cell replacement (Figure 2.6).

Thus far, our discussion of results has focused on the base-case. Figure 2.8 shows the change in life-cycle energy balance, EROEI, and energy payback time with variations between low and high values in eight individual parameters showing the most impact. Two describe cell performance (STH efficiency and cell life span), two describe the embodied energy of cell and module materials, two describe the energy intensity of cell and module fabrication processes, and two describe the initial and ongoing energy use of the BOS. Of these shown, STH efficiency is the parameter whose variation causes the most significant change in all three metrics. Variation in PEC cell life span is the second most significant parameter. The range of energy use for active cell fabrication, dominated by thin-film deposition energy use, is the third most significant parameter. Variation in ongoing energy use for BOS operation is the next most significant. The modeled range of variation is much less significant for other parameters, including inactive module materials and fabrication, active cell materials, initial BOS energy use, as well as additional parameters not shown in the figure.

As there appears to be a strong trade-off between cell efficiency and lifetime *versus* energy inputs for materials and fabrication, we compare three PEC systems, with all six characteristics (from Table 2.2) at the low input, base-case, and high input values, respectively. Furthermore, we assumed that the system with low energy inputs has lower performance (5% STH efficiency and five year life span), and the high input system has higher performance (20% STH efficiency and 20 year life span). The base-case system assumed unchanged performance values (10% STH efficiency and 10 year life span). Figure 2.9 shows that with these assumptions, the high input system with its higher performance displays much better overall energy metrics. The low input system, despite lower energy inputs than the other systems, delivers lower life-cycle net energy due to its lower efficiency and shorter life span. This suggests that high STH efficiency and long life spans are primary design requirements for PEC systems, even if such performance requires additional energy and material inputs for construction and operation.

2.3.1 Re-use of Materials

Our model already assumes re-use of panel support structure, manifold pipes, and sensors upon replacement of PEC modules. If inactive module materials (such as the glass window and back cover, and PVC support ribs and frame) are also re-used at the end of cell life span, there is an increase in system performance because less energy is required to produce replacement cell components. We find that the energy balance increases by 5% to +760 PJ, the EROEI increases 8% to 2.50, and the energy payback time decreases

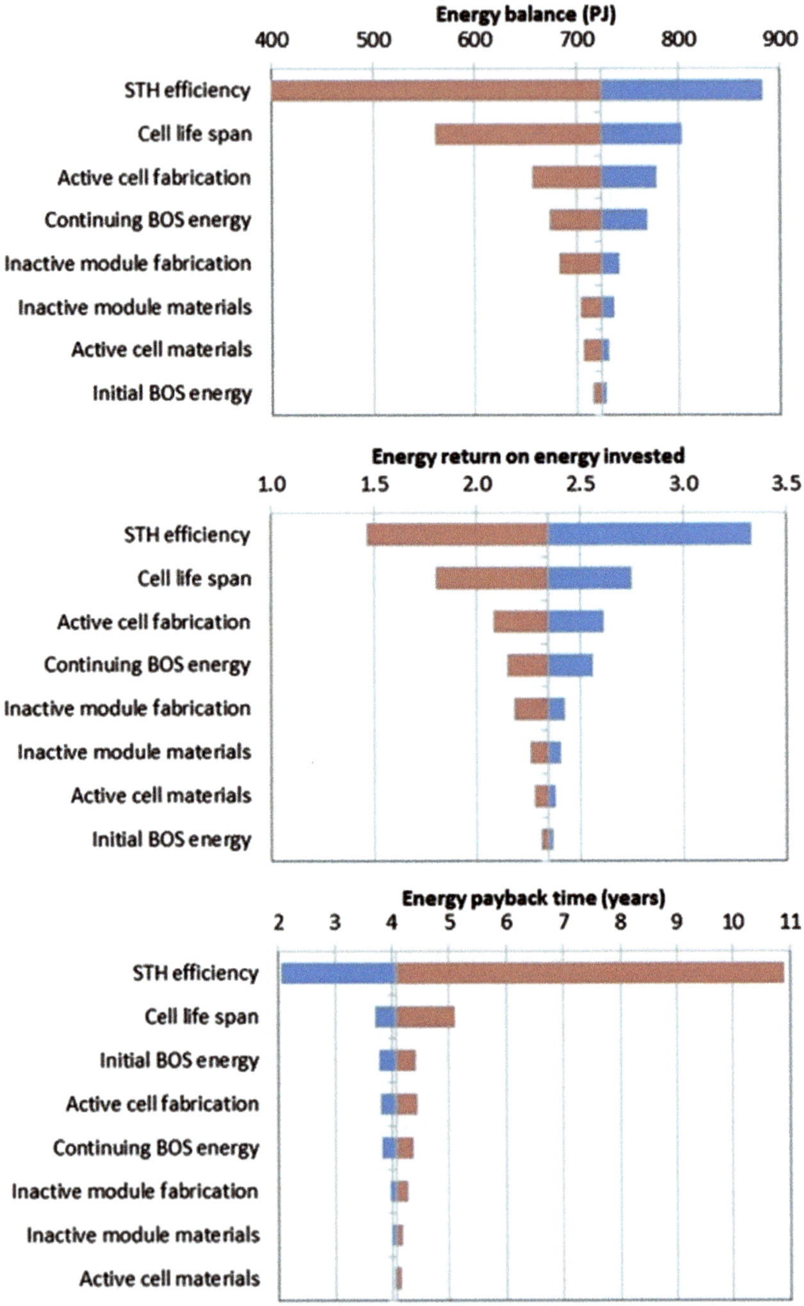

Figure 2.8 Effect of variation in key parameters on system performance metrics. Top image shows energy balance; middle image shows EROEI; bottom image shows energy payback time.
Reproduced from ref. 37 with permission from The Royal Society of Chemistry.

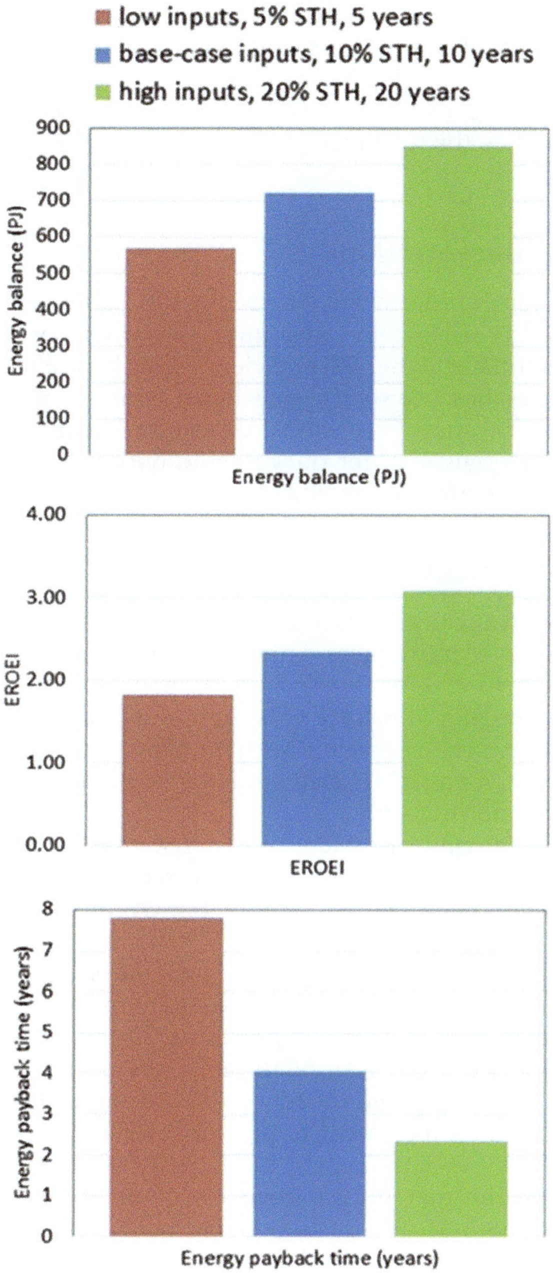

Figure 2.9 System performance metrics of three combinations of parameters: low input and low performance (left bars), base-case inputs and base-case performance (middle bars), and high input and high performance (right bars).
Reproduced from ref. 37 with permission from The Royal Society of Chemistry.

by 5% to 3.9 years. However, the suitability of re-using glass encapsulation materials may be limited due to possible degradation after years of exposure to the 'elements,' *e.g.*, etching by wind-blown sand. Unlike direct re-use of glass products (*e.g.*, glass containers), which has clear energy benefits, the energy benefits when the recovered glass must be melted and reconfigured is much lower.[82]

2.3.2 Solar Concentration

Further net energy improvements may be possible using concentrating PEC device designs.[84] Concentration may improve net energy performance because of the potentially higher STH efficiency and the use of a smaller active collection area, as has been suggested in an analysis of a PV-electrolyzer system for H_2 generation.[85] However, concentration mechanisms require more inactive materials, *e.g.*, for tracking and optics. A detailed net energy analysis should be performed to elucidate these trade-offs.

2.3.3 Scale-up Analysis

We also conducted a scale-up analysis to estimate required quantities of thin film materials and identify potential constraints in material availability if many 1 GW plants were to be constructed. We considered scale-up scenarios for H_2 demand, ranging from 10% of US light-duty vehicle demand to 100% of global light-duty vehicle demand in 2040 being satisfied by H_2 from our modeled facilities. We also included a middle scenario of 100% of US light-duty vehicle demand in 2040.

For the lowest demand case, we estimate that 14 GW (continuous annual average) would be required, *e.g.*, 14 facilities producing 1 GW annual average each. The middle scenario would require 142 GW of capacity, noting that some improvements in vehicle fuel cell efficiency would compensate for a larger overall travel demand. The highest demand scenario would require 850 GW of capacity. We then calculated the amount of thin film materials required annually for these hydrogen production facilities, assuming an STH efficiency of 10%, a cell life span of 10 years, and a material utilization efficiency of 50%, and compared the requirements to the 2014 global primary mine production of each element to provide a reference point for scale-up feasibility.

For many thin film materials, particularly abundant elements including Si, Fe, Ni, Cu, P, Sn, Ti and V, large-scale deployment of PEC hydrogen generation would require a very low proportion of current global mine production, suggesting that supply constraints are unlikely. At the lowest demand scale-up case, only Te (as part of CdTe photoelectrodes) and Ir (as an OER catalyst) showed likely supply constraints, each requiring over 30% of current production rates of the elements. For the middle demand case, Ga (as part of GaAs or InGaP photoelectrodes) and In (a component of InP and InGaP photoelectrodes as well as ITO TCO layers) begin to show likely supply constraints as measured by their increase relative to 2014 global production. For the highest demand case, Pt (as a HER catalyst) and Bi (as part of $BiVO_4$ photoelectrodes) supply would likely be constrained, and the quantities of Te, Ir, In and Ga required annually would exceed current primary production rates of these elements. More information is detailed in the Supplemental Information of Sathre *et al.*[37]

An important caveat when comparing against current mine production is that future production may differ substantially due to changes in economic demand or extraction and processing technology. Resource constraints for key materials may be encountered due to limited absolute quantities of materials, or to conflicts with other potential uses of the materials. Potential material constraints are yet to be fully incorporated into most technology assessments. The LCA framework traditionally includes a 'resource depeletion' impact category, though the longer time scale and non-linear response of that impact compared to some other categories makes objective comparison difficult. The emerging metric of 'material criticality' takes into account two aspects: the risks to reliable supply of a material, and vulnerability of a technological system to a potential supply disruption of the material,[86] and should eventually be applied to this question.

2.4 Conclusions

Prospective life-cycle assessment is an approach used to provide early insight into the eventual large-scale manufacturing, deployment, operation and decommissioning of emerging technologies. We developed and applied this approach to explore the net energy implications of a PEC H_2 production system at a facility scale of 1 GW annual average H_2 output (610 t H_2 day^{-1}). We focused on three indices of net energy performance: life-cycle primary energy balance, energy return on energy invested, and energy payback time.

Our analysis identified several priorities for continued PEC development. First, the STH conversion efficiency plays a strong role in determining the net energy viability of a PEC H_2 system, and it remains a primary system requirement. We find that STH efficiencies greater than 5% are necessary for net energy viability, and efficiencies of 10% can provide a substantial energy return on energy investment (*e.g.*, EROEI $\geq$ 2). Second, the cell life span also plays a strong role in net energy viability, and we find that cells with life spans of less than five years will be energetically challenged, and ~10 years are required for substantial positive net energy benefits. The third most important parameter is the energy intensity of cell fabrication, where thin-film deposition, with its much lower energy requirements, will improve performance. Improvements in STH efficiency or cell life span, even if they come at the cost of increases in cell production energy inputs, may be energetically favorable overall. We find that other system characteristics are less important, including inactive module fabrication, active cell materials, inactive module materials, and balance-of-system energy use.

Compared with our earlier work based on a silicon nanowire design,[35,36] our most recent work[37] has demonstrated that innovative design and manufacturing approaches can improve the net energy performance of PEC systems. Co-production of electricity in amounts exceeding H_2 production in primary energy terms could dramatically improve the EROEI and other system performance metrics, though this has yet to be demonstrated in the laboratory. Concentration may also improve the net energy performance because of the potentially higher STH efficiency and the use of a smaller active collection area, but these advantages must be balanced against the additional materials required to accomplish concentration.

Applying the same life-cycle approach to PEC plants producing hydrocarbon fuels from reduced CO_2 and H_2O is the focus of our current research, and we expect it to produce useful insights as well.

Acknowledgements

This material is based upon work performed by the Joint Center for Artificial Photosynthesis, a DOE Energy Innovation Hub, supported through the Office of Science of the US Department of Energy under Award Number DE-SC0004993. Portions of the text were adapted from Sathre *et al.* (2014)[36] and Sathre *et al.* (2016),[37] both published in *Energy and Environmental Science*.

References

1. Adoption of the Paris Agreement Conference of the Parties, Twenty-first session FCCC/CP/2015/L.9, United Nations Framework Convention on Climate Change, 2015. https://unfccc.int/resource/docs/2015/cop21/eng/l09.pdf.
2. Paris Agreement – Status of Ratification, United Nations Framework Convention on Climate Change, 2016. http://unfccc.int/paris_agreement/items/9444.php.
3. S. Pacala and R. Socolow, Stabilization Wedges: Solving the Climate Problem for the Next 50 Years with Current Technologies, *Science*, 2004, **305**, 968, DOI: 10.1126/science.1100103.
4. J. A. Turner, Sustainable hydrogen production, *Science*, 2004, **305**, 972.
5. C. Graves, S. D. Ebbesen, M. Mogensen and K. S. Lackner, Sustainable hydrocarbon fuels by recycling CO_2 and H_2O with renewable or nuclear energy, *Renewable Sustainable Energy Rev.*, 2011, **15**, 1.
6. P. Atkins, T. Overton, J. Rourke, M. Weller, F. Armstrong and M. Hagerman, Shriver & Atkins' Inorganic Chemistry, ISBN 978-1-42-921820-7, W. H. Freeman and Company, New York, 5th edn, 2010, ch. 15, p. 383. https://docs.google.com/file/d/0B4Ka5HSSrR_yeENFS0hvbjdUR00/view.
7. Uses of Hydrogen, 2016, http://www.usesof.net/uses-of-hydrogen.html.

8. Analyst silicon field trip, *Renewable Energy Corporation*, 28 March 2007. http://hugin.info/136555/R/1115224/203491.pdf.
9. D. Wagner, O. Devisme, F. Patisson, D. Ablitzer, A laboratory study of the reduction of iron oxides by hydrogen, *Proceedings of the 2006 Sohn International Symposium*, 27–31 August 2006, San Diego, *CA*, ed. F. Kongoli and R. G. Reddy, *The Minerals, Metals and Materials Society (TMS)*, vol. 2, pp. 111–120. https://arxiv.org/pdf/0803.2831.pdf.
10. National Hydrogen Association, *Hydrogen and Fuel Cells: The US Market Report*, 2010. Web-accessed at http://www.hydrogenassociation.org/.
11. E. Miller, Hydrogen Supply/Demand, *IEA North American Roadmap Workshop*, Fuel Cell Technologies Office, Office of Energy Efficiency and Renewable Energy, U.S. Department of Energy, 28 January 2014. https://www.iea.org/media/workshops/2014/hydrogenroadmap/7DOEEricMiller.pdf.
12. European Commission, World Energy Technology Outlook, 2006. Web-accessed at http://ec.europa.eu/research/energy/pdf/weto-h2_en.pdf.
13. Energy Information Administration, International Energy Outlook, 2013. Web-accessed at http://www.eia.gov/forecasts/ieo/.
14. Hydrogen production: Natural gas reforming, Office of Energy Efficiency and Renewable Energy, U.S. Department of Energy, no date. http://energy.gov/eere/fuelcells/hydrogen-production-natural-gas-reforming.
15. National Research Council; National Academy of Engineering; Division on Engineering and Physical Sciences; Board on Energy and Environmental Systems; Committee on Alternatives and Strategies for Future Hydrogen Production and Use, Chapter 7: Carbon Capture and Storage: The rationale of carbon capture and storage from hydrogen production, The Hydrogen Economy: Opportunities, Costs, Barriers, and R&D Needs, 2004, ISBN: 978-0-309-09163-3, DOI: 10.17226/10922. https://www.nap.edu/read/10922/chapter/9.
16. N. Parker, J. M. Ogden and Y. Fan, The role of biomass in California's hydrogen economy, *Energy Policy*, 2008, **36**, 3925–3939.
17. Hydrogen production: Electrolysis, Office of Energy Efficiency and Renewable Energy, U.S. Department of Energy, no date. http://energy.gov/eere/fuelcells/hydrogen-production-electrolysis.
18. N. S. Lewis and D. G. Nocera, Powering the planet: Chemical challenges in solar energy utilization, *Proc. Natl. Acad. Sci. U. S. A.*, 2006, **103**, 15729.
19. B. D. James, G. N. Baum, J. Perez and K. N. Baum, Technoeconomic Analysis of Photoelectrochemical (PEC) Hydrogen Production, Final Report, Direct Technologies, DOE Contract Number GS-10F-009J, 2009.
20. B. A. Pinaud, J. D. Benck, L. C. Seitz, A. J. Forman, Z. Chen, T. G. Deutsch, B. D. James, K. N. Baum, G. N. Baum, S. Ardo, H. Wang, E. Miller and T. F. Jaramillo, Technical and economic feasibility of centralized facilities for solar hydrogen production via photocatalysis and photoelectrochemistry, *Energy Environ. Sci.*, 2013, **6**, 1983.

21. M. G. Walter, E. L. Warren, J. R. McKone, S. W. Boettcher, Q. Mi, E. A. Santori and N. S. Lewis, Solar water splitting cells, *Chem. Rev.*, 2010, **110**, 6446.
22. R. E. Rocheleau, E. L. Miller and A. Misra, High-efficiency photoelectrochemical hydrogen production using multijunction amorphous silicon photoelectrodes, *Energy Fuels*, 1998, **12**, 3.
23. O. Khaselev, A. Bansal and J. Turner, High-efficiency integrated multijunction photovoltaic/electrolysis systems for hydrogen production, *Int. J. Hydrogen Energy*, 2001, **26**, 127.
24. N. A. Kelly and T. L. Gibson, Design and characterization of a robust photoelectrochemical device to generate hydrogen using solar water splitting, *Int. J. Hydrogen Energy*, 2006, **31**, 1658.
25. S. Licht, B. Wang, S. Mukerji, T. Soga, M. Umeno and H. Tributsch, Efficient solar water splitting, exemplified by RuO2-catalyzed AlGaAs/Si photoelectrolysis, *J. Phys. Chem. B*, 2000, **104**, 8920.
26. G. Peharz, F. Dimroth and U. Wittstadt, Solar hydrogen production by water splitting with a conversion efficiency of 18%, *Int. J. Hydrogen Energy*, 2007, **32**, 3248.
27. K. Fujii, S. Nakamura, M. Sugiyama, K. Watanabe, B. Bagheri and Y. Nakano, , Characteristics of hydrogen generation from water splitting by polymer electrolyte electrochemical cell directly connected with concentrated photovoltaic cell, *Int. J. Hydrogen Energy*, 2013, **38**(34), 14424.
28. T. J. Jacobsson, V. Fjallstrom, M. Sahlberg, M. Edoff and T. Edvinsson, A monolithic device for solar water splitting based on series interconnected thin film absorbers reaching over 10% solar-to-hydrogen efficiency, *Energy Environ. Sci.*, 2013, **6**, 3676.
29. S. Y. Reece, J. A. Hamel, K. Sung, T. D. Jarvi, A. J. Esswein, J. J. H. Pijpers and D. G. Nocera, Wireless solar water splitting using silicon-based semiconductors and earth-abundant catalysts, *Science*, 2011, **334**, 645.
30. J. Brillet, J.-H. Yum, M. Cornuz, T. Hisatomi, R. Solarska, J. Augustynski, M. Graetzel and K. Sivula, Highly efficient water splitting by a dual-absorber tandem cell, *Nat. Photonics*, 2012, **6**, 824.
31. F. F. Abdi, L. Han, A. H. M. Smets, M. Zeman, B. Dam and R. van de Krol, Efficient solar water splitting by enhanced charge separation in a bismuth vanadate-silicon tandem photoelectrode, *Nat. Commun.*, 2013, **4**, 2195.
32. M. R. Shaner, K. T. Fountaine, S. Ardo, R. H. Coridan, H. A. Atwater and N. S. Lewis, Photoelectrochemistry of core–shell tandem junction n–p + -Si/n-WO3microwire array photoelectrodes, *Energy Environ. Sci.*, 2014, **7**, 779.
33. C. Liu, J. Tang, H. M. Chen, B. Liu and P. Yang, A fully integrated nanosystem of semiconductor nanowires for direct solar water splitting, *Nano Lett.*, 2013, **13**, 298.
34. E. Wesoff, 2014. MIT Energy Storage Spinout Sun Catalytix's Assets Acquired by Lockheed Martin, Greentech Media, 26 August. http://www.greentechmedia.com/articles/read/MIT-Energy-Storage-Spinout-Sun-Catalytix-Assets-Acquired-by-Lockheed-Marti (accessed 12 September 2016).

35. P. Zhai, S. Haussener, J. Ager, R. Sathre, K. Walczak, J. Greenblatt and T. McKone, Net primary energy balance of a solar-driven photo-electrochemical water-splitting device, *Energy Environ. Sci.*, 2013, **6**, 2380.
36. R. Sathre, C. D. Scown, W. R. Morrow, J. C. Stevens, I. D. Sharp, J. W. Ager, K. Walczak, F. A. Houle and J. B. Greenblatt, Life-cycle net energy assessment of large-scale hydrogen production via photoelectrochemical water splitting, *Energy Environ. Sci.*, 2014, **7**, 3264–3278.
37. R. Sathre, J. B. Greenblatt, K. Walczak, I. D. Sharp, J. C. Stevens, J. W. Ager III and F. A. Houle, Opportunities to improve the net energy performance of photoelectrochemical water-splitting technology, *Energy Environ. Sci.*, 2016, **9**(3), 803–819.
38. M. Spielmann, R. W. Scholz, O. Tietje and P. de Haan, Scenario modelling in prospective LCA of transportation systems: Application of formative scenario analysis, *Int. J. LCA.*, 2005, **10**(5), 325.
39. E. Dahlgren, C. Gocmen, K. Lackner and G. Van Ryzin, Small modular infrastructure, *Eng. Econ.*, 2013, **58**, 231–264.
40. Z. Chen, T. F. Jaramillo, T. G. Deutsch, A. Kleiman-Shwarsctein, A. J. Forman, N. Gaillard, R. Garland, K. Takanabe, C. Heske, M. Sunkara, E. W. McFarland, K. Domen, E. L. Miller, J. A. Turner and H. N. Dinh, Accelerating materials development for photoelectrochemical (PEC) hydrogen production: Standards for methods, definitions, and reporting protocols, *J. Mater. Res.*, 2010, **25**, 3.
41. K. Coughlin, Projections of Full-fuel-cycle Energy and Emission Metrics, Report LBNL-6025E, Lawrence Berkeley National Laboratory, 2013.
42. M. Raugei, P. Fullana-i-Palmer and V. Fthenakis, The energy return on energy investment (EROI) of photovoltaics: Methodology and comparisons with fossil fuel lifecycles, *Energy Policy*, 2012, **45**, 576.
43. EERE (US DOE Energy Efficiency and Renewable Energy), Comparison of Fuel Cell Technologies, 2011, web accessed at http://www1.eere.energy.gov/hydrogenandfuelcells/fuelcells/pdfs/fc_comparison_chart.pdf.
44. K. Walczak, Y. Chen, C. Karp, J. W. Beeman, M. Shaner, J. Spurgeon, I. D. Sharp, X. Amashukeli, W. West, J. Jin, N. Lewis and C.. Xiang, Modeling, simulation, and fabrication of a fully integrated, acid-stable, scalable solar-driven water-splitting system, *ChemSusChem*, 2015, **8**, 544–551.
45. Ecoinvent, 2012, *Ecoinvent v2.2 Life Cycle Inventory (LCI) database*. Web-accessed at www.ecoinvent.org.
46. P. Nuss and M. J. Eckelman, Life cycle assessment of metals: A scientific synthesis, *PLOS One*, 2014, **9**(7), e101298.
47. E. Williams, N. Krishnan and S. Boyd, Ultrapurity and energy use: Case study of semiconductor manufacturing, Chapter 7, in *Thermodynamics and the Destruction of Resources*, ed. B. R. Bakshi, T. G. Gutowski and D. P. Sekulic, Cambridge University Press, 2014.
48. Ecoinvent, 2007, *Life Cycle Inventories of Chemicals*, Web-accessed at www.ecoinvent.org.

49. M. A. Kreiger, D. R. Shonnard and J. M. Pearce, Life cycle analysis of silane recycling in amorphous silicon-based solar photovoltaic manufacturing, *Resour., Conserv. Recycl.*, 2013, **70**, 44–49.
50. V. Fthenakis, Sustainability of photovoltaics: The case for thin-film solar cells, *Renewable Sustainable Energy Rev.*, 2009, **13**, 2746–2750.
51. E. Alsema, Energy requirements of thin-film solar cell modules: A review, *Renewable Sustainable Energy Rev.*, 1998, **2**, 387–415.
52. K. Zweibel, Thin film PV manufacturing: Materials costs and their optimization, *Sol. Energy Mater. Sol. Cells*, 2000, **63**, 375–386.
53. C. L. C. McCrory, S. Jung, J. C. Peters and T. F. Jaramillo, Benchmarking heterogeneous electrocatalysts for the oxygen evolution reaction, *J. Am. Chem. Soc.*, 2013, **135**(45), 16977–16987.
54. B. Seger, T. Pedersen, A. B. Laursen, P. C. K. Vesborg, O. Hansen and I. Chorkendorff, Using TiO_2 as a conductive protective layer for photocathodic H_2 evolution, *J. Am. Chem. Soc.*, 2013, **135**, 1057–1064.
55. Y. Lin, C. Battaglia, M. Boccard, M. Hettick, Z. Yu, C. Ballif, J. W. Ager and A. Javey, Amorphous Si thin film based photocathodes with high photovoltage for efficient hydrogen production, *Nano Lett.*, 2013, **13**(11), 5615–5618.
56. S. Hu, M. R. Shaner, J. A. Beardslee, M. Lichterman, B. S. Brunschwig and N. S. Lewis, Amorphous TiO_2 coatings stabilize Si, GaAs, and GaP photoanodes for efficient water oxidation, *Science*, 2014, **344**, 1005–1009.
57. *Fuel Cell Technology Handbook*, ed. G. Hoogers, CRC Press, Boca Raton, London, New York, Washington, DC, 2003.
58. S. M. Rossnagel, Thin film deposition with physical vapor deposition and related technologies, *J. Vac. Sci. Technol., A*, 2003, **21**(5), S74–S87.
59. M. J. Hampden-Smith and T. T. Kodas, Chemical vapor deposition of metals: Part 1: An overview of CVD processes, *Chem. Vap. Deposition*, 1995, **1**, 8–23.
60. R. M. Pasquarelli, D. S. Ginley and R. O'Hayre, Solution processing of transparent conductors: From flask to film, *Chem. Soc. Rev.*, 2011, **40**, 5406–5544.
61. S. J. Kim, S. Yoon and H. J. Kim, Review of solution-processed oxide thin-film transistors, *Jpn. J. Appl. Phys.*, 2014, **53**, 02BA02.
62. C. A. Wolden, J. Kurtin, J. B. Baxter, I. Repins, S. E. Shaheen, J. T. Torvik, A. A. Rockett, V. M. Fthenakis and E. S. Aydil, Photovoltaic manufacturing: Present status, future prospects, and research needs, *J. Vac. Sci. Technol., A*, 2011, **29**, 030801.
63. A. Shah, J. Meier, A. Buechel, U. Kroll, J. Steinhauser, F. Meillaud, H. Schade and D. Domine, Towards very low-cost mass production of thin-film silicon photovoltaic (PV) solar modules on glass, *Thin Solid Films*, 2006, **502**, 292–299.
64. R. J. Hill and S. J. Nadel, *Coated Glass Applications and Markets*, BOC Coating Technology, Fairfield, CA, 1999.
65. D. Yue, F. You and S. B. Darling, , Domestic and overseas manufacturing scenarios of silicon-based photovoltaics: Life cycle energy

and environmental comparative analysis, *Sol. Energy*, 2014, **105**, 669–678.

66. J. Peng, L. Lu and H. Yang, Review on life cycle assessment of energy payback and greenhouse gas emission of solar photovoltaic systems, *Renewable Sustainable Energy Rev.*, 2013, **19**, 255–274.
67. J. Gong, S. B. Darling and F. You, Perovskite photovoltaics: Life-cycle assessment of energy and environmental impacts, *Energy Environ. Sci.*, 2015, **8**, 1953–1968.
68. T. Bronger, P. H. Wöbkenberg, J. Wördenweber, S. Muthmann, U. W. Paetzold and V. Smirnov, et al., Solution-based Silicon in thin film solar cells, *Adv. Energy Mater.*, 2014, **4**, 1301871.
69. F. A. Veer, The strength of glass, a nontransparent value, *Heron J.*, 2007.
70. Cardinal IG Company, Glass Windload Tables, Technical service bulletin #IG03, 2008, http://www.cardinalcorp.com/wp-content/uploads/pdf/tsb/ig/IG03_05-08.pdf.
71. Cardinal IG Company, Technical Glass Guide, Version 2.0, 2014. http://www.cardinalcorp.com/wp-content/uploads/pdf/Technical_Glass_Guide_Web.pdf.
72. WindSpeedByZip, Wind Speed By Zip, 2015. http://www.windspeedbyzip.com/.
73. National Certified Testing Laboratories, Wind Velocity Chart, 2015. http://www.nctlinc.com/velocity-chart/.
74. SolarWorld Americas, The SolarWorld Standard, 2015. http://www.solarworld-usa.com/why-choose-solarworld/the-solarworld-standard.
75. S. A. Chagnon, D. Chagnon and S. D. Hilbert, Hailstorms across the Nation: An Atlas about Hail and Its Damages, Illinois State Water Survey, Contract Report 2009-12, 2009. http://www.isws.illinois.edu/pubdoc/CR/ISWSCR2009-12.pdf.
76. IPEX, Chemical Resistance Guide: Thermoplastic Piping Systems, 2001.
77. J. C. Stevens and A. Z. Weber, A Computational Study of Optically Concentrating, Solar-Fuels Generators from Annual Thermal- and Fuel-Production Efficiency Perspectives, *J. Electrochem. Soc.*, 2016, **163**(7), H475–H484.
78. S. Ong, C. Campbell, P. Denholm, R. Margolis and G. Heath, *Land-Use Requirements for Solar Power Plants in the United States*, National Renewable Energy Laboratory, Report NREL/TP-6A20-56290, 2013.
79. National Renewable Energy Laboratory (NREL), *PV-Watts calculator*, 2015. Web-accessed at http://pvwatts.nrel.gov/index.php.
80. National Energy Technology Laboratory (NETL), Life Cycle Analysis: Power Studies Compilation Report, National Energy Technology Laboratory Report DOE/NETL-2010/1419, 2010.
81. E. Worrell, C. Galitsky, E. Masanet and W. Graus, *Energy Efficiency Improvement and Cost Saving Opportunities for the Glass Industry*, Report LBNL-57335, Lawrence Berkeley National Laboratory, 2008.
82. L. L. Gaines and M. M. Mintz, *Energy Implications of Glass-Container Recycling*, Report ANL/ESD-18, Argonne National Laboratory, 1994.

83. J. S. Newman and K. E. Thomas-Alyea, *Electrochemical Systems*, Wiley, Hoboken, NJ, 2004.
84. Y. Chen, C. Xiang, S. Hu and N. S. Lewis, Modeling the performance of an integrated photoelectrolysis system with 10× solar concentrators, *J. Electrochem. Soc.*, 2014, **161**, F1101–F1110.
85. M. Dumortier and S. Haussener, Design guidelines for concentrated photo-electrochemical water splitting devices based on energy and greenhouse gas yield ratios, *Energy Environ. Sci.*, 2015, **8**, 3069–3082.
86. American Physical Society (APS) and Materials Research Society (MRS), *Energy Critical Elements: Securing Materials for Emerging Technologies*, 2011, http://www.aps.org/policy/reports/popa-reports/upload/elementsreport.pdf.

Electrocatalysis

CHAPTER 3

Understanding the Effects of Composition and Structure on the Oxygen Evolution Reaction (OER) Occurring on NiFeO$_x$ Catalysts

ALEXIS T. BELL[a,b]

[a] Joint Center for Artificial Photosynthesis, Lawrence Berkeley National Laboratory, Berkeley, CA 94720, USA; [b] Department of Chemical and Biomolecular Engineering, University of California, Berkeley, CA 94720-1462, USA
Email: alexbell@berkeley.edu

3.1 Introduction

The use of wind and solar energy to carry out the electrochemical splitting of water is a subject of considerable contemporary interest, since hydrogen produced by this means could be used to store renewable but intermittent electrical energy, to power fuel-cell vehicles, and to hydrodeoxygenate biomass or reduce CO_2 to carbon-containing fuels.[1–5] The challenge, therefore, is to develop energy-efficient systems utilizing renewable sources of electricity to achieve the electrochemical splitting of water. This chapter reviews the catalyst requirements for achieving this goal using earth-abundant elements. Particular attention is devoted to oxygen evolution, which occurs

Energy and Environment Series No. 22
Integrated Solar Fuel Generators
Edited by Ian D. Sharp, Harry A. Atwater and Hans-Joachim Lewerenz

Published by the Royal Society of Chemistry, www.rsc.org

at the electrolyzer anode, since this reaction requires a potential well above the minimum set by thermodynamics. We begin by summarizing the thermodynamics for the electrochemical splitting of water and then review what is known about catalysts based on earth-abundant elements that can be used to promote the oxygen evolution reaction. This overview will show that the most promising catalysts are based on oxides and oxyhydroxides containing Ni and Fe. We will next examine in some detail how Fe^{3+} cations incorporate into Ni oxyhydroxides (NiOOH) and what the role of these cations is in promoting the OER. The chapter will end with an examination of recent efforts to find additive elements other than Fe that could be used to enhance the activity of NiOOH and elements that might be added to $Ni_{1-x}Fe_xOOH$ to further enhance its activity.

3.2 Thermodynamics of Water Splitting

The thermodynamic potential for the electrochemical splitting of water, $2H_2O(l) \rightarrow 2H_2(g) + O_2(g)$, is $E_o = -1.23$ V. This overall reaction is comprised of two half reactions, which differ depending on the pH of the electrolyte in which the reaction is carried out. Scheme 3.1 shows the two half reactions and the associated thermodynamic potentials for the extremes of pH = 0 (acid) and pH = 14 (base). Under acidic conditions, water is oxidized at the anode to produce gaseous O_2, protons, and electrons and protons and electrons are recombined at the cathode to produce gaseous H_2. Under basic conditions, hydroxyl anions are oxidized at the anode to produce O_2, water, and electrons and water and electrons react at the cathode to produce gaseous H_2 and hydroxyl anions. The two half reactions are referred to as the oxygen evolution reaction (OER) and the hydrogen evolution reaction (HER).

In Scheme 3.1, the potential (E) for each reaction at pH = 0 is the same as the standard potential (E^0) which is defined for this pH, assuming that the value of E^0 for the HER is 0.00 V. The potential for each half reaction depends on the Nernst equation, which at unit partial pressure of H_2 and O_2 is given by $E_{HER} = -0.59 \times pH$ V for the HER and $E_{OER} = 1.23 - 0.59 \times pH$ V for

In aqueous acid (pH = 0)

$2\,H_2O(l)$	$\rightarrow O_2(g) + 4\,H^+ + 4\,e^-$	$E = E^0 = 1.23$ V
$4\,H^+ + 4\,e^-$	$\rightarrow 2\,H_2(g)$	$E = E^0 = 0.00$ V
$2\,H_2O(l)$	$\rightarrow O_2(g) + 2\,H_2(g)$	$E = E^0 = 1.23$ V

In aqueous base (pH = 14)

$4\,OH^-$	$\rightarrow O_2(g) + 2\,H_2O(l) + 4\,e^-$	$E = 0.40$ V
$4\,H_2O(l) + 4\,e^-$	$\rightarrow 2\,H_2(g) + 4\,OH^-$	$E = -0.83$ V
$2\,H_2O(l)$	$\rightarrow O_2(g) + 2\,H_2(g)$	$E = E^0 = 1.23$ V

Scheme 3.1 Water-splitting half reactions in aqueous acid and base.

the OER. Since the pH at which experiments are carried out differ from experimenter to experimenter, it is convenient to refer potentials to the reversible hydrogen electrode (RHE). Accordingly, E_{OER} *vs.* the RHE (or simply E_{OER} *vs.* RHE) is always 1.23 V.

Regardless of pH, the actual potential required to split water is higher due to reaction barriers associated with the catalysts used at the anode and cathode. This excess potential, known as the overpotential, is a function of the current density passed through the electrolyzer. Since the overpotential increases with increasing current density, the performance of different catalysts must compared at a fixed current density, which is often taken to be 10 mA cm^{-2} and is based on the superficial area of the anode. The overpotentials for the OER and HER are symbolized by η_{OER} and η_{HER}, respectively. The dependence of η_{OER} and η_{HER} on the current density (j) is given by the Tafel equation, $d\eta_{OER}/d\ln(j) = \beta$, where β is referred to as the Tafel slope. This equation and its role in describing the activity of OER catalysts is discussed in greater detail below.

Extensive research has shown that while the overpotential required to drive the cathode reaction (the HER) at 10 mA cm^{-2} can be as low as 50 mV, the overpotential required to drive the anode reaction (the OER) typically exceeds 300 mV.[3,5] The overpotentials for the OER and HER affect the efficiency of water splitting for a given current density, which is defined as the ratio of the thermodynamic potential for water splitting, E_o, divided by the sum of E_o and the overpotentials for the anode and cathode, η_{OER} and η_{HER}, respectively. For example, if $\eta_{HER} = 50$ mV and $\eta_{OER} = 300$ mV, the efficiency for water splitting at 10 mA cm^{-2} (neglecting ohmic and Nernstian losses) would be 78%. Since η_{OER} is significantly larger than η_{HER}, most research in the field of catalysis for water splitting has focused on the discovery and development of catalysts having a low value of η_{OER}, as well as understanding how the chemical composition and structure affect the magnitude of this parameter. It should be noted, though, that comparison of different OER catalysts solely on the basis of the value of η_{OER} at a fixed anodic current density can be misleading, particularly when catalysts are prepared in radically different ways. This topic and the dependence of η_{OER} on current density are discussed at a later point in this chapter.

3.3 Catalysts for the OER

A number reviews, as well as the Chapter 5 by McCrory within this book, have been published concerning catalysts that can be used to promote the OER under both acidic and basic conditions.[3,5–11] These publications show that with the exception of IrO_2 and RuO_2 all of the materials examined are unstable in acid electrolytes, but that there are a large number of metal oxides that are stable in base electrolytes. To illustrate this latter point, Figure 3.1 shows the results of a recent benchmarking study of OER catalysts carried out in 1 M NaOH and with a current density of 10 mA cm^{-2}.[8] It is evident from this figure that the OER catalysts exhibiting the lowest values of

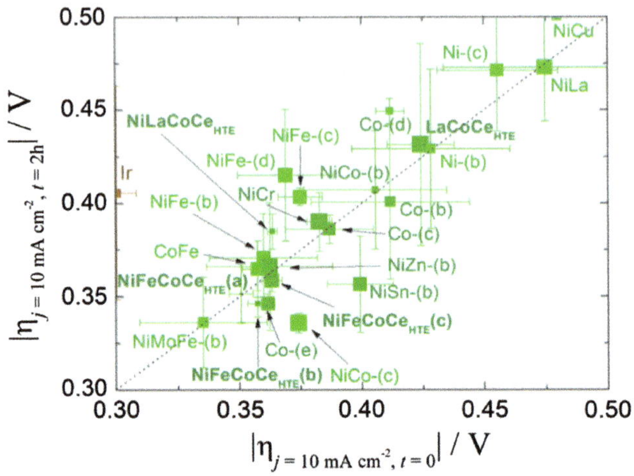

Figure 3.1 Values of η_{OER} determined using a rotating disk electrode in 1 M NaOH. The x-axis is the overpotential required to achieve 10 mA cm^{-2} per geometric area at time $t=0$. The y-axis is the overpotential required to achieve 10 mA cm^{-2} per geometric area at time $t=2$ h. The diagonal dashed line is the expected response for a stable catalyst that does not change in activity during 2 h constant polarization. The color of each point represents the roughness factor of the catalyst with a bin size of 1 order of magnitude with light green representing RF = 1, and dark red representing RF > 104. The size of each point is inversely proportional to the standard deviation in the ECSA measurements. The region of interest for benchmarking is the unshaded white region of the plot where the overpotential required to achieve 10 mA cm^{-2} per geometric area at time $t=0$ and $t=2$ h is less than 0.55 V. There is a break and change in scale in both axes at overpotentials >0.55 V, and the corresponding region of the plot is shown in gray. Catalysts whose activity and stability measurements fall inside this gray area are outside the region of interest for benchmarking, but their activity and stability measurements are included for completeness.
Adapted with permission from C. C. L. McCrory, S. Jung, I. M. Ferrer, S. M. Chatman, J. C. Peters and J. F. Jaramillo, Benchmarking Hydrogen Evolving and Oxygen Evolving Reaction Electrocatalysts for Solar Water Splitting Devices, *J. Am. Chem. Soc.* 2015, **137**, 4347–4357, Copyright 2015 American Chemical Society.[8]

η_{OER} are IrO_2 and mixed oxides based on Ni and Fe. It should be noted, though, that while IrO_2 is stable in an acid electrolyte, it loses activity with time. Further details concerning the performance and stability of metal oxide catalysts that operate in acidic and basic electrolytes can be found in Chapter 5 of this book.

Since Fe and Ni are earth-abundant elements, whereas Ir and Ru are among the least earth-abundant and most expensive elements, a great deal of attention has been devoted to understanding why oxides containing Fe and Ni exhibit a low value of η_{OER} and whether the overpotential of such oxides might be reduced further by addition of other elements. For these

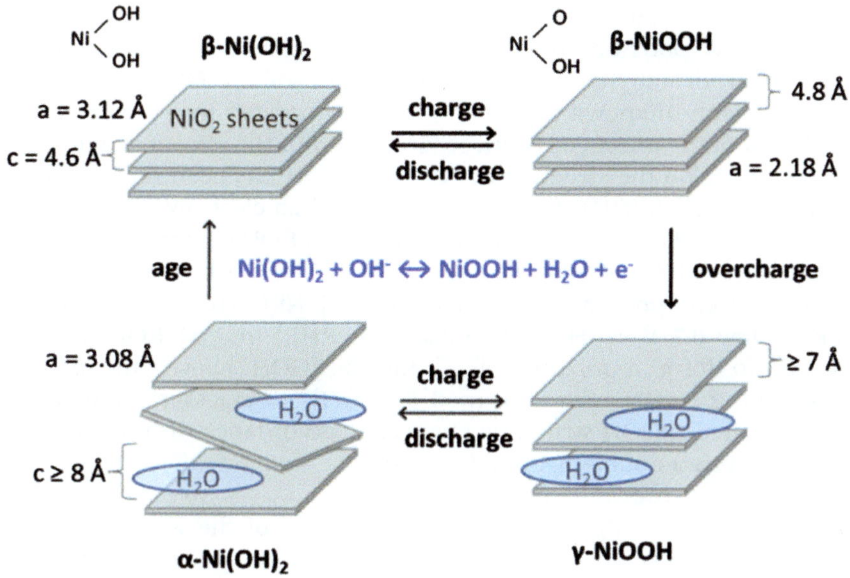

Figure 3.2 Bode scheme for the $Ni(OH)_2$/NiOOH redox transformation. The transformation of β-NIOOH to γ-NiOOH is observed when β-NiOOH is operated at a high OER overpotential for an extended period. This step is referred to as overcharging.

reasons, the balance of this chapter will focus on understanding OER catalysts based on Fe and Ni oxides operated in alkaline media.

An appropriate place to start the discussion of $NiFeO_x$ OER catalysts is by considering what is known about the OER over NiO_x. Identification of Ni oxide as an active catalyst for the OER under alkaline conditions was first reported by Bode in 1966,[12] who noted that the performance of NiO_x depends on the manner in which it is aged. At potentials well below the onset of the OER in KOH solution, the catalyst is present as either α- or β-$Ni(OH)_2$, and prior to the onset of oxygen evolution, this phase oxidizes to form either the β or γ phase of NiOOH *via* the reaction $Ni(OH)_2 + OH^- \rightarrow NiOOH + H_2O + e^-$. The phase of NiOOH depends on the manner of catalyst charging and discharging, as shown in Figure 3.2. Bode also noted that the β phase of NiOOH is more active for the OER than the γ phase. Subsequent studies by various authors have confirmed that the OER activity of NiOOH can be increased by transforming the γ-NiOOH to β-NiOOH phase as a consequence of aging.[13–18] Contemporaneous studies also showed that the OER activity of NiOOH could be enhanced through the addition of Fe.[19–29]

It is important to note, though, that it was not until the work of Boettcher and coworkers,[30] which appeared in 2014, that the increase in OER activity observed with the aging of NiO_x in KOH was recognized to be a consequence of the incorporation of Fe impurities and not due to the transition of γ-NiOOH to β-NiOOH, as had been concluded in earlier studies. These authors also showed that Fe-free NiOOH is quite inactive for the OER, and exhibits an overpotential in excess of 500 mV at 10 mA cm^{-2}. These findings suggest that virtually all of the earlier studies of the OER on NiO_x were performed with unknown amounts of Fe incorporated into the catalyst.

Isolation of the effects of Fe incorporation from the effects of aging in Fe-free electrolyte on the structure-activity of NiOOH has been reported by Bell and coworkers.[31] $Ni(OH)_2$ films aged in unpurified electrolyte were found to incorporate ≥20% Fe after five weeks of aging in KOH containing <1 ppm of Fe and were >10^2 more active than Fe-free films. Films aged in Fe-free electrolyte were predominantly disordered β-$Ni(OH)_2$/β-NiOOH if maintained below 0.7 V *vs.* Hg/HgO (0.84 V *vs.* RHE) in 1 M KOH and 'overcharged' to form a mixture of γ- and β-NiOOH above this potential. Fe-containing $Ni(OH)_2$ films formed β-$Ni(OH)_2$ to a lesser extent and instead exhibited NiOOH structural changes in accordance with the formation of a NiFe-layered double hydroxide phase.

Figure 3.3 illustrates the variation in both the η_{OER} at 10 mA cm^{-2} and the current density for $\eta_{OER} = 300$ mV as a function of the Fe content of the catalyst.[31] As the Fe content increases, the value of η_{OER} decreases from a value of 530 mV in the absence of Fe to a minimum of 280 mV for an Fe content of ~11%. Over the same range of Fe content, the current density at $\eta_{OER} = 300$ mV increases from 0.12 to 20 mA cm^{-2}.

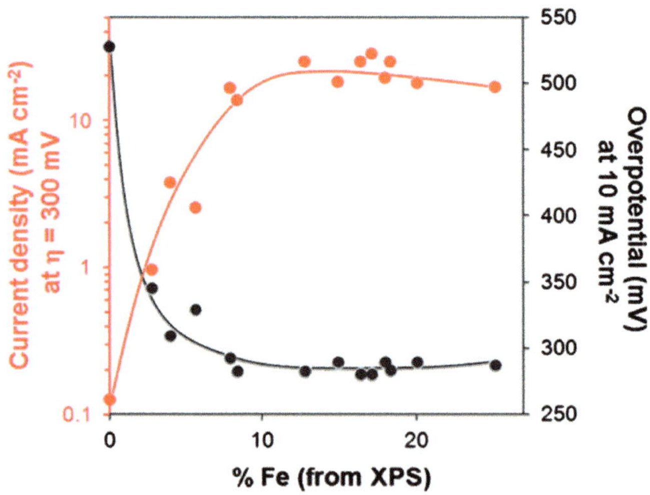

Figure 3.3 Oxygen evolution activity of electrodeposited NiOOH films deposited on Au rotating disk electrodes (RDEs) at 300 mV overpotential and 10 mA cm^{-2} geometric current density in 1 M KOH as a function of Fe content incorporated from the electrolyte. Curves are included to guide the eye.
Reprinted with permission from S. Klaus, Y. Cai, M. W. Louie, L. Trotochaud and A. T. Bell, Effects of Fe Electrolyte Impurities on $Ni(OH)_2$/NiOOH Structure and Oxygen Evolution Activity, *J. Phys. Chem. C*, 2015, **119**, 7243–7254, Copyright 2015 American Chemical Society.[31]

The pattern of activity seen in Figure 3.3 is also observed for electrodeposited and sputtered films of Ni and Fe, as shown in Figure 3.4.[32] In this instance the value of η_{OER} NiOOH is >700 mV and that of FeOOH is ~ 500 mV. The minimum value of η_{OER} is 320 mV and occurs for an Fe content of 20%. It is notable that for the same film thickness, the value of η_{OER} is independent of the manner of film deposition. What is significant, though, is that the values of η_{OER} for NiOOH and FeOOH, as well as the minimum value of η_{OER} and the Fe content at which this occurs are functions of film thickness and extent of Fe and Ni mixing during film deposition. Figure 3.4 also shows that independent of the manner of film deposition, the current density measured at $\eta_{OER} = 300$ mV increases by roughly three orders of magnitude with the addition of even a very small amount of Fe to NiOOH, reaching a maximum current density of ~3 mA cm^{-2} for an Fe content of 20% and then slowly decreasing by nearly three orders of magnitude with the addition of more Fe. The Tafel slope is also a strong function of Fe content. It is 120 mV per dec for pure NiOOH, decreases rapidly to 40 mV per dec with the addition of a small amount of Fe, stays at this value until the Fe content reaches 65%, and then increases to 62 mV per dec for pure FeOOH.

Swierk *et al.* have reported that the Fe content in $Ni_{1-x}Fe_xOOH$ films influences their impedance and apparent activation energy for the OER, in addition to influencing the OER overpotential and Tafel slope.[33] As Fe is incorporated into the NiOOH lattice, the Faradaic resistance associated with the OER decreases and charge relaxation becomes more favorable. These results suggest that the formation of charged surface intermediates is the dominant process for the OER on these films. The activation energy measurements reveal that while isolated Fe-sites in NiOOH lattices have similar energetics regardless of film composition, the overall OER activity is also controlled by structural considerations. Correspondingly, the apparent activation energy for the OER is 76 ± 5 kJ mol^{-1} for pure NiOOH and 66 ± 5 kJ mol^{-1} for pure FeOOH, but decreases to a minimum of ~25 kJ mol^{-1} for $Ni_{0.8}Fe_{0.2}OOH$.

To summarize, recent studies have established that Fe-free NiOOH is not an active OER catalyst but becomes progressively more active with the incorporation of Fe. Therefore, the important questions to address next are what is the structure of $NiFeO_x$ catalysts, what is the nature of the OER-active site (*i.e.*, Fe *vs.* Ni), and what are the effects of preparation on the OER activity of $NiFeO_x$ catalysts?

3.4 The Structure of $FeNiO_x$

Raman and X-ray absorption spectroscopy have been used to characterize the structure of $FeNiO_x$ catalysts as a function of applied potential. Bell and coworkers have reported that a Ni film immersed in 1 M KOH immediately transforms to $Ni(OH)_2$ in the absence of an applied potential.[15,16] Figure 3.5 shows that as the potential is raised from 0 to 1.5 V *vs.* RHE, the spectrum changes from that of $Ni(OH)_2$ (bands at 449 cm^{-1} and 494 cm^{-1}) to that of

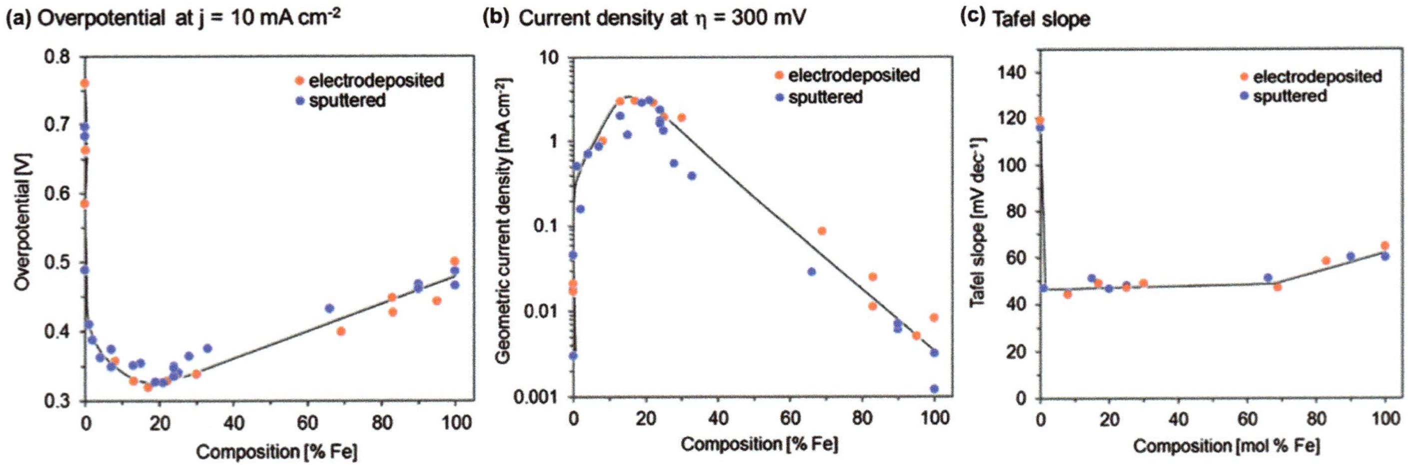

Figure 3.4 Activity plots for sputtered and electrodeposited $Ni_{1-x}Fe_xOOH$ films. Activity values were obtained after holding for 1 h at either constant (a) geometric current density (10 mA cm^{-2}), (b) overpotential (300 mV), or with (c) Tafel slopes obtained from 10 mV s^{-1} CV scans. Measurements were obtained in 0.1 M KOH with a RDE rotation rate of 1600 RPM. Curves are included to guide the eye.
Reprinted with permission from S. Klaus, M. W. Louie, L. Trotochaud and A. T. Bell, Role of Catalyst Preparation on the Electrocatalytic Activity of $Ni_{1-x}Fe_xOOH$ for the Oxygen Evolution Reaction, *J. Phys. Chem. C*, 2015, **119**, 18303–18316, Copyright 2015 American Chemical Society.[32]

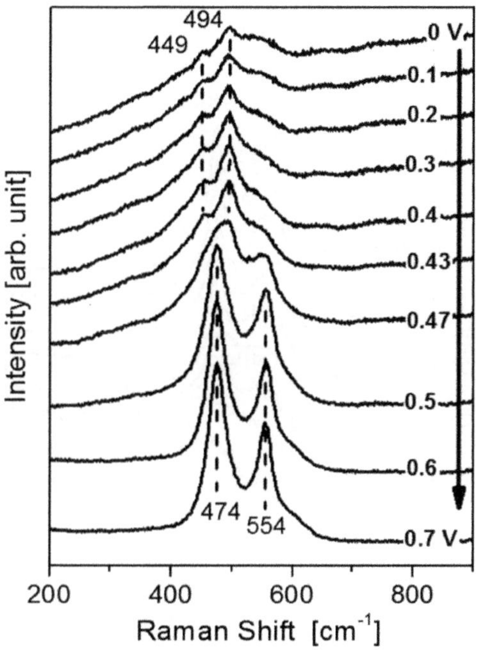

Figure 3.5 *In situ* Raman spectra for Ni films deposited on a roughened Au substrate as a function of potential *vs.* Hg/HgO (1 M KOH) in 0.1 M KOH, for which the equilibrium potential for the OER is 0.365 V.
Reprinted from M. W. Louie and A. T. Bell, An Investigation of Thin-Film Ni-Fe Oxide Catalysts for the Electrochemical Evolution of Oxygen, *J. Am. Chem. Soc.* 2013, **135**, 12329–12337, Copyright 2013 American Chemical Society.[15]

NiOOH (bands at 474 cm^{-1} and 554 cm^{-1}). This change is accompanied by a peak in the plot of current *vs.* voltage due to the process $Ni(OH)_2 + OH^- \rightarrow NiOOH + H_2O + e^-$, which results in the oxidation of Ni^{2+} to Ni^{3+}. It is noted that this process occurs below the potential for the onset of the OER. With progressive addition of Fe to the metal film, the Raman spectrum observed at 1.5 *vs.* RHE changes from that of NiOOH to that of FeOOH, as seen in Figure 3.6. This change is accompanied by a shift in the Ni^{2+} to Ni^{3+} oxidation peak to higher potentials for all samples containing Ni. The interpretation of this shift is discussed below.

Friebel *et al.* have characterized electrochemically deposited $FeNiO_x$ films by *operando* high energy resolution fluorescence detection X-ray adsorption spectroscopy (HERFD XAS).[34] The HERFD XAS spectrum of an Fe-only OER catalyst is shown in Figure 3.7a. The pre-edge region at ~7115 eV exhibits a small discrepancy in the characteristic double-peak structure for octahedrally coordinated Fe^{3+} as observed for γ-FeOOH.[35,36] This suggests that a small amount (10% ± 3%) of tetrahedrally coordinated Fe^{3+} may therefore exist in addition to a majority (90% ± 3%) of octahedrally coordinated Fe^{3+}.

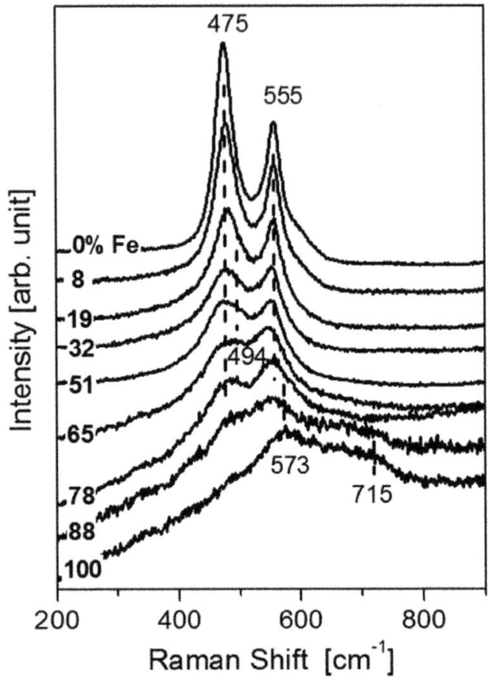

Figure 3.6 *In situ* Raman spectra collected in 0.1 M KOH at an OER overpotential of 0.235 V (or 0.6 V *vs.* Hg/HgO (1 M KOH)) for Ni–Fe films as a function of composition.
Reprinted from M. W. Louie and A. T. Bell, An Investigation of Thin-Film Ni-Fe Oxide Catalysts for the Electrochemical Evolution of Oxygen, *J. Am. Chem. Soc.* 2013, 135, 12329–12337, Copyright 2013 American Chemical Society.[15]

HERFD XAS measurements at potentials of 1.12 V, 1.52 V, 1.62 V, and 1.72 V (*vs.* RHE) show no evidence for potential-induced phase transformation in the pure FeOOH sample. By contrast, the Fe K-edge HERFD XAS for a sample containing 25% Fe and 75% Ni shows strong potential-induced changes (Figure 3.9b) that coincide with changes in the complementary Ni K-edge HERFD XAS spectra presented in Figure 3.7e. The latter are attributed to the spectral signatures of α-$Ni(OH)_2$ at low potentials and γ-NiOOH at high potentials.[37–42] Both α-$Ni(OH)_2$ and γ-NiOOH form layered structures of edge-sharing [NiO_6] octahedra separated by intercalated water molecules and hydrated ions (Figure 3.7d).[44] The Ni–O bond lengths[39–44] differ significantly between 2.05 Å in Ni(II)-containing α-$Ni(OH)_2$, and 1.88 Å in γ-NiOOH which is non-stoichiometric ($NiOOH_{1-x}$) and contains a mixture of Ni^{3+} and Ni^{4+} sites.[43] The significant shift of both the pre-edge peak and the main absorption edge in the Ni K-edge spectra (Figure 3.9e) shows nearly complete oxidation of Ni sites when the potential is increased from 1.12 to 1.52 V *vs.* RHE; the features of the oxidized component then approach saturation with further potential increase. A more detailed analysis confirmed that the observed redox transition is between α-Ni(OH)$_2$ and γ-NiOOH and that the presence of β-Ni(OH)$_2$ or β-NiOOH can be ruled out.

Both the Fe and Ni K-edge spectra exhibit shifts of the structure-sensitive EXAFS dips and peaks to higher energy, indicating significant bond contraction. Using the 'bond length with a ruler' relationship $(E_{peak} - E_0)d^2 = \text{const.}$,[44,45] which is equivalent to an EXAFS analysis with a strongly reduced number of independent fitting parameters, the M–O bond length is estimated to contract by $(7 \pm 1)\%$ for both Fe–O and Ni–O bonds. Complementary *operando* EXAFS measurements over a much larger energy range for an Fe/Ni (25:75) sample in the fully reduced and fully oxidized state are shown in Figure 3.8c. Analysis of these results confirms that the Fe–O and Ni–O bond lengths contract from 2.01 Å and 2.06 Å at 1.12 V to 1.90 Å and 1.89 Å at 1.92 V, respectively. A strong correlation was found not only between Fe–O and Ni–O bond lengths but also between nearest metal–metal distances, indicating that Fe substitutes for Ni in both α-Ni(OH)$_2$ and γ-NiOOH. Moreover, the identical appearance in both Fe and Ni K-edge EXAFS of a peak at approximately twice the nearest Ni–Ni and Fe–Ni distance, due predominantly to multiple-scattering in collinear Fe–Ni–Ni, Ni–Fe–Ni and Ni–Ni–Ni arrangements, clearly shows that Fe is not intercalated between the hexagonal [NiO$_2$] sheets but instead substitutes for Ni within the sheets.

The HERFD XAS also reveals information about the oxidation states of Fe and Ni in Fe-substituted α-Ni(OH)$_2$ and γ-NiOOH. Contracted bond distances are commonly associated with an increased oxidation state, and such oxidation state increase can be clearly seen in the Ni K-edge spectra where the pre-edge centroid and the photoionization threshold both shift to significantly higher energy. By contrast, an intriguing discrepancy can be seen in the Fe K-edge spectra. Comparison of the short Fe–O distance of (1.90 ± 0.01) Å at 1.62 V with bond lengths for different Fe oxides reported in the literature suggests that all Fe sites should have increased their oxidation state from +3 to +4. However, if such a significant oxidation state increase occurred at all Fe sites, a more noticeable shift to higher energies would have been expected for both the pre-edge centroid as well as the photoionization threshold. Further analysis shows that the features characteristic of Fe^{4+}, *e.g.*, a ligand-to-metal charge transfer (LMCT) shakedown feature near the main edge and a strong decrease of the white line intensity, are not observed for Fe-containing γ-NiOOH and is, therefore, concluded that the concentration of Fe^{4+} cation is small.

Ni–O and Fe–O bond length as a function of Fe content are presented in Figure 3.8a for two potentials, the resting state at 1.12 V as well as OER operating conditions at 1.62 V. Irrespective of the presence and concentration of Fe, the Ni–O bond lengths correspond to α-Ni(OH)$_2$ and γ-NiOOH at low and high potentials, respectively. Conversely, the average Fe–O bond length completely follows the Ni–O bond contraction only at low Fe content (10% and 25%). With increasing Fe content, the Fe–O distance at 1.62 V gradually increases towards that of pure γ-FeOOH. The results suggest that two different Fe^{3+} species are present, *i.e.*, Fe^{3+} dopants in α-$Ni(OH)_2$/γ-NiOOH and Fe^{3+} sites within a separate γ-FeOOH phase. While a single Ni/Fe phase exists at low Fe content, FeOOH increasingly contributes to the

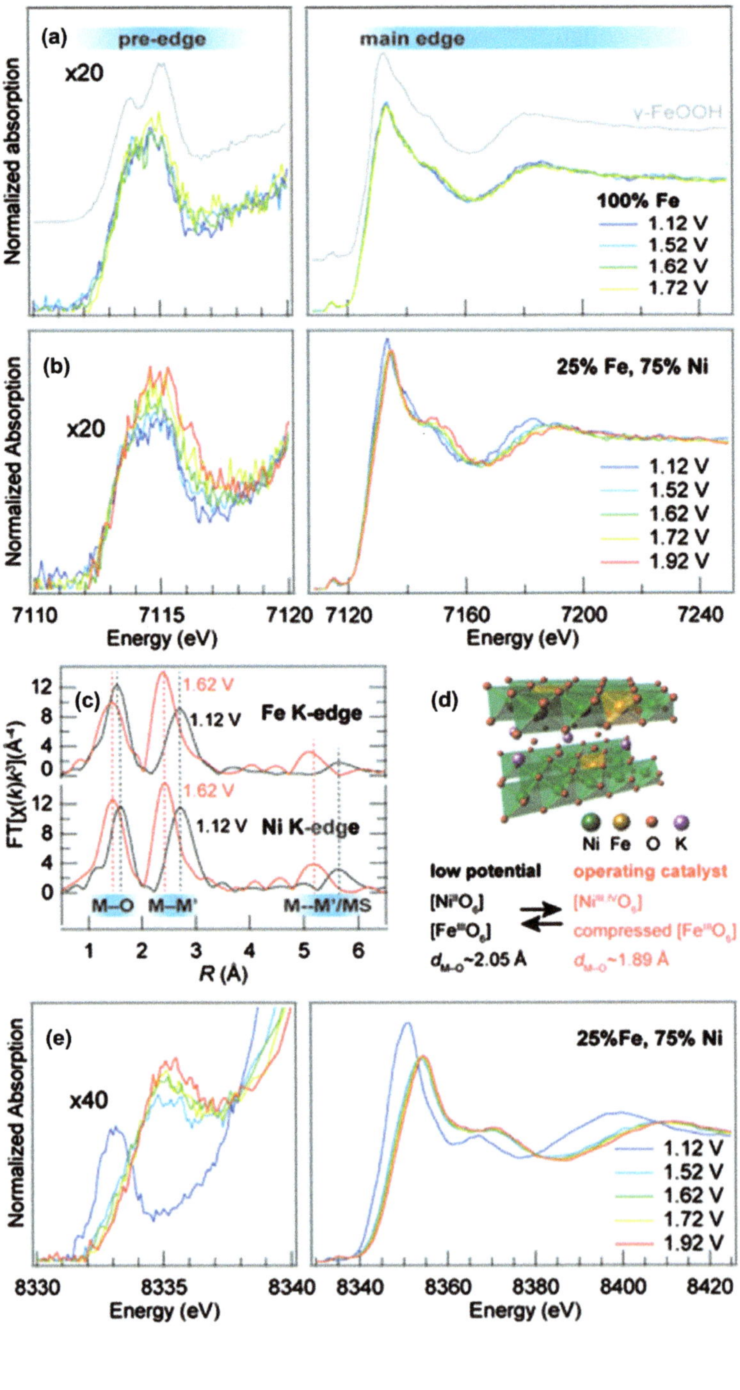

Fe K-edge spectra with increasing overall Fe content, due to limited solubility of Fe^{3+} in α-$Ni(OH)_2$. These results are consistent with the study of Fe uptake into $Ni(OH)_2$ from Fe-containing KOH solution presented above, which indicates a solubility limit near 25% Fe.[31] Since a Ni–O bond expansion even at 75% Fe content is not observed, it is estimated based on experimental uncertainty in the Ni–O bond distance, that Ni doping into γ-FeOOH, if any, does not exceed 3%. The model structures shown in Figure 3.8b capture the known oxidation states and local binding environment of metal sites. A comparison of theoretically predicted metal–oxygen bond lengths (Figure 3.8c) with the experimental values from our measurements (Figure 3.7a) and literature supports the hypothesis that only Fe-doped γ-NiOOH and pure FeOOH, but very little, if any, Ni-doped γ-FeOOH, exist under OER conditions.

The deductions concerning the charges of Fe and Ni cations and Fe–O and Ni–O bond distance determined by Friebel *et al.*[34] are supported by DFT + U calculations (see below for details). Löwdin charges of Ni, Fe, O, and H relative to the free atoms provide a measure of the apparent oxidation states of these elements. Figure 3.8d shows that the Löwdin charges for both Fe and Ni relative to the free atoms remain approximately constant for all model structures investigated. This finding agrees with the experimental observation that oxidation states of Fe and Ni under OER conditions are +3 and +3.6 (average), respectively, independent of the Ni/Fe ratio. The Löwdin charge for Fe is calculated to be significantly higher than for Ni despite the opposite order of formal oxidation states of both cations, which reflects the more ionic and less covalent nature of Fe–O than Ni–O bonds. Likewise, the linear increase in negative charge on the O atoms with increasing Fe content is attributable to the replacement of the more covalent Ni–O bonds with the more ionic Fe–O bonds.

Figure 3.7 Comparison of 100% Fe-containing sample with OER catalyst containing 25% Fe and 75% Ni using *operando* HERFD XAS. (a) Catalyst containing 100% Fe. The spectrum of γ-FeOOH is also shown for comparison. Plots of both pre-edge (enlarged) and the full spectra are shown. (b)-(e) Catalyst containing 25% Fe and 75% Ni. (b) Fe K-edge. While the potential increase does not influence the oxidation-state-sensitive energy of the main absorption threshold (7125 eV), significant Fe–O bond contraction with increasing potential is clearly indicated by the changes of the photoelectron scattering features (energy range above 7140 eV). (c) Complementary *operando* EXAFS measurement confirming the potential-induced bond contraction at both Fe and Ni sites. (d) Structure model of Fe-doped γ-NiOOH. (e) Ni K-edge XAS showing shifts in both oxidation-state-sensitive and structure-sensitive features due to oxidation of Ni^{2+} sites.
Reprinted from D. Friebel, M. W. Louie, M. Abjdich, K. E. Sewald, Y. Cai, A. M. Wise, M.-J. Cheng, D. Sokaras, T.-C. Weng, R. Alonso, R. C. Davis, J. R. Bargar, J. K. Nørskov, A. Nilsson and A. T. Bell, Identification of Highly active Fe Sites in (Ni,Fe)OOH for Electrocatalytic Splitting of Water, *J. Am. Chem. Soc.* 2015, **137**, 1305–1313, Copyright 2015 American Chemical Society.[34]

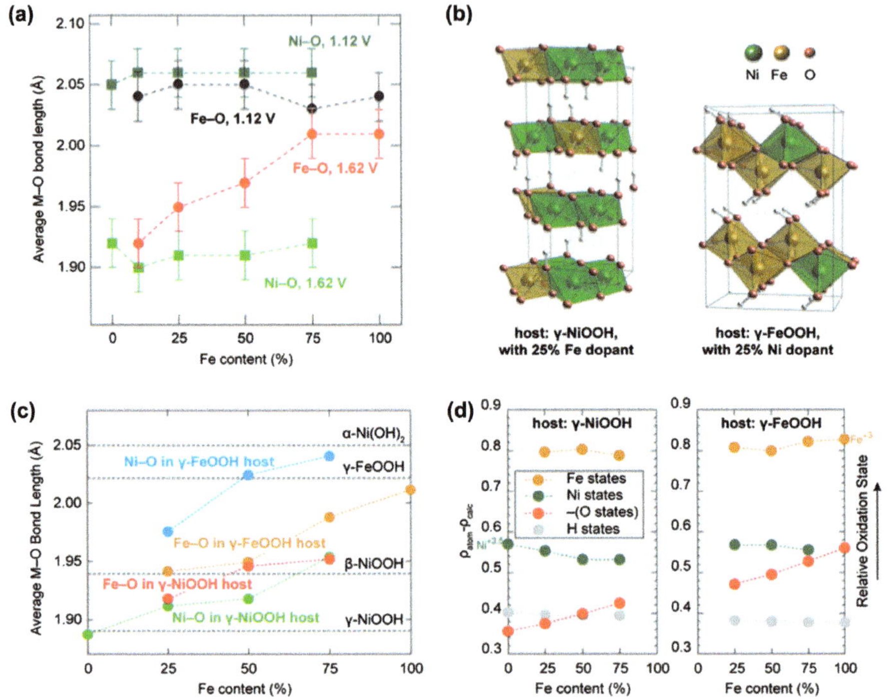

Figure 3.8 Identification of the most likely structural motif for mixed Ni,Fe catalysts by comparison of experimentally obtained metal–oxygen bond lengths with optimized theoretical model structures. (a) Bond lengths extracted from HERFD XAS measurements at 1.12 V and 1.62 V, plotted as a function of Fe content. (b) Examples of unit cells for Fe-substituted γ-NiOOH and Ni-substituted γ-FeOOH model structures. (c) Theoretically predicted bond lengths, corrected by a factor of 0.97 for comparison with experimental data. The dashed lines represent experimental bond lengths from literature. (d) Löwdin charges in Fe-substituted γ-NiOOH and Ni-substituted γ-FeOOH model structures, plotted as a function of Fe content. All dashed lines are guides to the eye.
Reprinted from D. Friebel, M. W. Louie, M. Abjdich, K. E. Sewald, Y. Cai, A. M. Wise, M.-J. Cheng, D. Sokaras, T.-C. Weng, R. Alonso, R. C. Davis, J. R. Bargar, J. K. Nørskov, A. Nilsson and A. T. Bell, Identification of Highly active Fe Sites in (Ni,Fe)OOH for Electrocatalytic Splitting of Water, *J. Am. Chem. Soc.* 2015, **137**, 1305–1313, Copyright 2015 American Chemical Society.[34]

The DFT + U calculations reported by Friebel *et al.*[34] also confirm that an Fe–O bond distance similar to that for Ni–O should be observed when Fe^{3+} cations substituted for Ni^{3+} cations in γ-$Ni_{1-x}Fe_xOOH$. The absence of any experimental evidence for Ni–O bond expansion in materials prepared with Fe contents above 50% suggests that for Fe contents in excess of ∼25%, a γ-FeOOH phase is nucleated that does not contain a large amount of Ni in it. The differences in solubility limits for Fe in $Ni(OH)_2$/NiOOH and Ni in FeOOH are attributed to differences in the capability of the host structures to compensate the charge difference between Ni^{2+} and Fe^{3+} which are present at low potential and, presumably, during catalyst electrodeposition. The α-$Ni(OH)_2$ structure contains intercalated H_2O between the hexagonal [NiO_2] sheets, and can accommodate electrolyte anions. In the resulting charge-neutral $Ni(II)_{1-x}Fe(III)_x(OH)_{2+x}(H_2O)_y$ layered double hydroxide structure,[46] Fe uptake is enabled by closely matching Fe^{3+}–O and Ni^{2+}–O distances but limited by steric hindrance and repulsion between partially hydrated anions. Conversely, there is no obvious pathway for the γ-FeOOH structure to allow intercalation of cations (*e.g.*, K^+) or protonation of O or OH ligands without a significant distortion of linkages between [MO_6] octahedra. Correspondingly, the absence of intercalation in β-$Ni(OH)_2$ could explain why the presence of Fe^{3+} appears to prevent the transformation ('aging') of α-$Ni(OH)_2$ into β-$Ni(OH)_2$.[30]

It is important to compare and contrast the findings of Friebel *et al.*[34] with those recently reported by Strasser and coworkers,[47] who estimated the oxidation state of Ni and Fe from quantitative deconvolution of the charge injected into $Ni_{1-x}Fe_xOOH$ during cyclic voltammetry (CV) complemented by data from differential electrochemical mass spectrometry (DEMS) and from analysis of *ex situ* X-ray absorption measurements (XAS). These authors found that in the absence of intentional Fe addition, the CV data lead to the conclusion that all of the Ni^{2+} undergoes oxidation to Ni^{3+}, whereas the DEMS data suggest that up to 25% of the Ni reaches the Ni^{4+} oxidation state. The addition of Fe stabilized low-valent Ni and suppressed its oxidation to higher oxidation states. These findings are qualitatively consistent with the findings of Friebel *et al.* determined from *operando* XAS.[34]

The XAS data reported by Strasser and coworkers[47] were obtained by freeze-quenching catalysts after conditioning them at 1.63 V *vs.* RHE. The Ni K-edge position observed in the XANES spectrum of freeze quenched NiO_x was consistent with an oxidation state of +3.7 but the oxidation state was reported to decrease to +2 with the addition of more than 4% Fe. It is notable that the Ni–O distance determined EXAFS data for freeze-quenched NiO_x was 1.88 Å but increased to 2.05 Å with the addition of more than 4% Fe. The Fe K-edge position observed in the XANES spectrum of freeze-quenched FeO_x was consistent with an Fe oxidation state of +3.4, and little change in oxidation state was observed when Fe was present together with Ni. Likewise, the Fe–O distance determined from analysis of the EXAFS data gave a value of 2.0 Å independent of whether Fe was present alone or together with Ni.

The oxidation states of Ni and Fe and the corresponding Fe–O and Ni–O bond distances reported by Strasser and coworkers[47] for NiOOH and FeOOH agree reasonably well with those reported by Friebel *et al.*;[34] however, significant differences occur for all samples of $Ni_{1-x}Fe_xOOH$. Whereas the *operando* EXAFS data of Friebel *et al.*[34] show that the Ni–O bond distance retains its value in NiOOH upon the addition of Fe, the freeze-quenched samples examined *ex situ* by Strasser and coworkers[47] show Ni–O bond distances that are more characteristic of $Ni(OH)_2$, a conclusion that would be fully consistent with their observation that Ni is in the 2+ state for these samples. Likewise, whereas Friebel *et al.* find that the Fe–O bond distance in $Ni_{1-x}Fe_xOOH$ undergoes a contraction from 2.01 Å to 1.90 Å for low values of x and then monotonically increases as x increases (see Figure 3.8), Strasser and coworkers[47] find that the Fe–O bond distance remains at about 2.00 Å independent of the value of x. What these findings suggest is that freeze quenching of $Ni_{1-x}Fe_xOOH$ samples after their use for electrochemistry cannot give reliable information about the environments of Ni and Fe cations.

3.5 Identity of the Active Site in $FeNiO_x$

Not addressed to this point is why the substitution of Fe^{3+} into the lattice of γ-NiOOH increases the OER activity. Two options exist—one is that the substituted Fe^{3+} sites become more active when hosted in the lattice of γ-NiOOH due to a change in their electronic environment and the other is that the activity of Ni^{3+} sites becomes more active as a consequence of their electronic properties being altered by the substitution of Fe^{3+} cations into the γ-NiOOH lattice. Another question to ask is whether the small amount of Ni^{3+} that can substitute into γ-FeOOH has an effect on the OER activity of this phase, and if so, how so? As discussed next, both sets of questions can be addressed by DFT + U calculations of the overpotential for the OER.

Friebel *et al.*[34] have reported DFT + U calculations aimed at developing a deeper understanding of the effects of Fe substitution into γ-NiOOH and Ni substitution into γ-FeOOH. Their approach builds on procedures developed by Nørskov and coworkers, which have been demonstrated to give a solid basis for interpreting the relationship between catalyst composition and OER overpotential.[48–50] In acidic conditions, the OER is taken to occur *via* four elementary steps:

$$H_2O + * \rightarrow OH* + e^- + H^+ \quad (3.1)$$

$$OH* \rightarrow O* + e^- + H^+ \quad (3.2)$$

$$O* + H_2O \rightarrow OOH* + e^- + H^+ \quad (3.3)$$

$$OOH* \rightarrow O_2 + e^- + H^+ \quad (3.4)$$

where * indicates an oxygen vacancy site at the surface. It should be noted that this sequence of steps is based on chemical intuition developed from studies of the oxygen reduction reaction on metals, rather than direct observation of the postulated intermediates involved in the OER occurring over NiFeOOH. While several attempts to identify intermediates involved in the OER by *operando* Raman spectroscopy have been reported, these studies have only observed weak evidence for peroxo/superoxo species. For this reason, theoretical analyses have utilized the scheme shown above.

The chemical potentials of protons and electrons at any given pH and applied potential E were described by the computational hydrogen electrode.[48] As a consequence, the theoretical overpotential η_{OER} obtained from Gibbs free energy differences ΔG_i $(i=1...4)$ at each step is independent of pH and is therefore applicable to alkaline conditions.

$$\eta_{OER} = \max[\Delta G_1, \Delta G_2, \Delta G_3, \Delta G_4]/e - 1.23 \text{ [V]} \tag{3.5}$$

Figure 3.9 compares predicted overpotentials for the OER occurring at Ni and Fe surface sites in pure and doped γ-NiOOH and γ-FeOOH. The results shown in Figure 3.9 can be rationalized in terms of the overall affinity of surface sites for adsorbed intermediates and the relative stability of O* with respect to OH* and OOH*. Generally, much weaker adsorption is found for the on-top position of a single 5-fold coordinated metal atom than for the bridge site between two 5-fold coordinated metal atoms, and Fe sites have significantly higher OH affinity than Ni sites. While the difference between adsorption energies of OH* and OOH* is nearly constant, $\Delta E_{OOH} = 0.8\Delta E_{OH} + 3.3$ eV, the O* binding energy increases as function of ΔE_{OH}. Under the optimum condition $\Delta E_O - \Delta E_{OH} = 0.5(\Delta E_{OOH} - \Delta E_{OH})$,[50] η_{OER} reaches a minimum value of 0.4 V. In the absence of cation substitution, all OER intermediates adsorb too strongly on pure γ-FeOOH and too weakly on pure γ-NiOOH. The calculated overpotentials for Fe-free γ-NiOOH are larger than that for pure γ-FeOOH.

Compared to pure γ-FeOOH, Fe sites surrounded by Ni next-nearest neighbors in either γ-NiOOH or γ-FeOOH host structures exhibit decreased affinity for OER intermediates, resulting in a shift in their binding energies towards optimal values. The origin of these trends can be rationalized based on the results in Figure 3.8d, which show that $Ni^{3+/4+}$ cations, due to their higher electron affinity compared to Fe^{3+}, withdraw electron density from oxygen sites. During OER, the formation of HO* and O* requires an oxidation state increase at the active Fe site, which becomes evident, for example, in the notably short bond distance of 1.62 Å between O* and a highly charged Fe site in Fe-doped γ-NiOOH $(01\bar{1}2)$ (Figure 3.9a). The oxidation of the Fe surface site will be less favorable energetically in γ-NiOOH than in γ-FeOOH, because neighboring Ni sites induce lower negative charge density on adsorbed O and OH. We propose that this effect is mostly determined by the local arrangement of neighboring Fe and Ni sites. Computational results for structurally similar Ni-doped hematite indicate the same mechanism for the catalytic enhancement, further supporting our hypothesis.[51]

Further calculations comparing the overpotentials for the OER occurring over Ni or Fe sites confirm that Fe rather than Ni constitutes the active site for the OER at mixed Fe–Ni oxyhydroxides. Subsurface Fe sites in γ-NiOOH (Figure 3.9b) increase the OER overpotential at Ni surface sites because the already too weak oxygen affinity of Ni further decreases. This finding is consistent with the experimentally observed shift of the α-$Ni(OH)_2$/γ-NiOOH redox potential to higher values with increasing Fe content.[15]

The results presented in Figure 3.9 are consistent with those of Li and Selloni,[52] who have reported DFT + U calculations of the OER overpotentials for β- and γ-NiOOH and for Fe substituted into the surface of both NiOOH phases. While the catalyst structure used by these authors differ

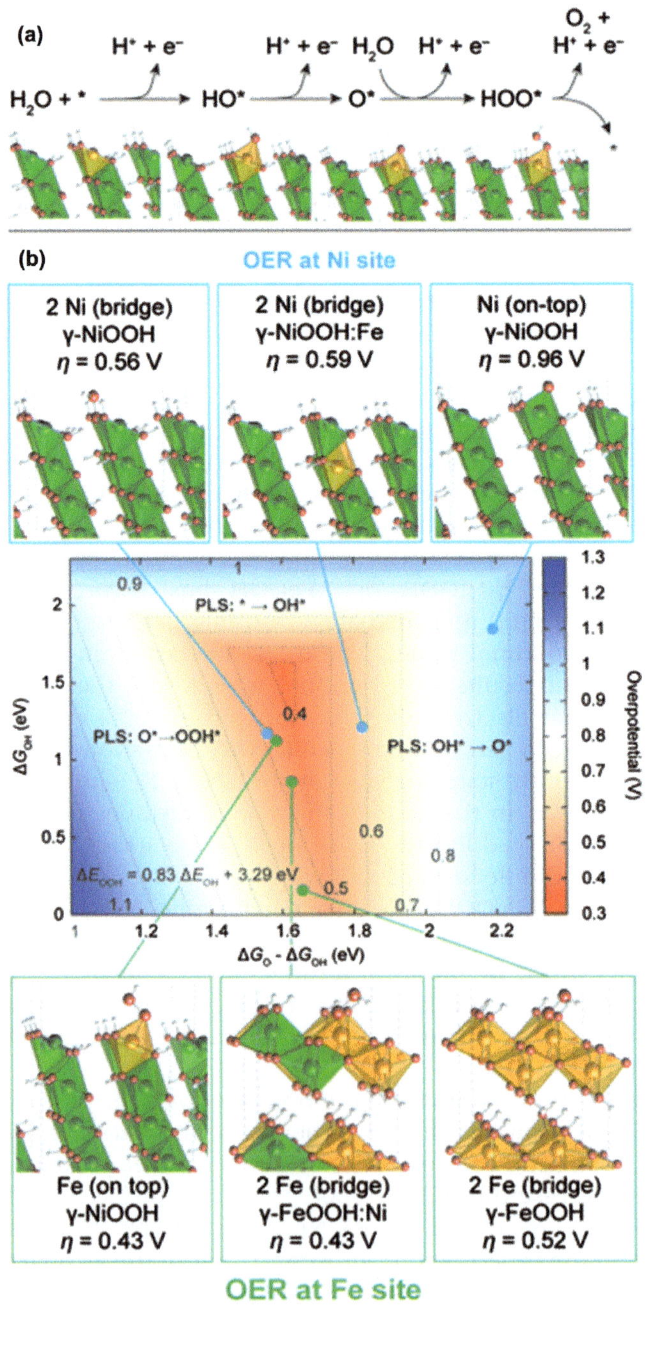

substantially from that used by Friebel *et al.*, both studies agree that the OER overpotential for Fe-doped γ-NiOOH is lower than that for pure γ-NiOOH. Thus, Li and Selloni[54] find that $\eta = 0.48$ V for Fe-doped γ-NiOOH and $\eta = 0.52$ V for pure γ-NiOOH, whereas Friebel *et al.*[34] find that $\eta = 0.43$ V for Fe-doped γ-NiOOH and $\eta = 0.56$ V for pure γ-NiOOH. Li and Selloni[54] also predict that an even lower overpotential for the OER could be achieved on Fe-doped β-NiOOH ($\eta = 0.26$ V) and that the overpotential for pure β-NiOOH should be lower than that for γ-NiOOH ($\eta = 0.46$ V for β-NiOOH *versus* $\eta = 0.52$ V for γ-NiOOH). While these are interesting findings, it should be noted that Friebel *et al.*[34] did not find any experimental evidence for the β phase of NiOOH, and work by Trotochaud *et al.* has shown that the presence of Fe in the active phase of Ni–Fe oxides inhibits the formation of this phase.[30]

Theoretical analysis of the effects of Fe^{3+} substitution for Ni^{3+} in NiOOH has also been reported by Koper and coworkers.[10] In these studies NiOOH was represented by NiO terminated on the (001) surface by *O and *OH. The extent of Fe^{3+} substitution was chosen to be 50%. Based on this model, the potential-limiting value of η_{OER} was found to be 0.37 V for Fe^{3+} cations and 0.57 V for Ni^{3+} cations in qualitative agreement with the findings of Li and Selloni[54] and of Friebel *et al.*[34]

The conclusions of the theoretical analyses reported above have recently been confirmed by experiments reported by Ahn and Bard,[53] using scanning electrochemical microscopy to determine the kinetics of the OER on Ni and Fe cations in NiOOH, FeOOH, and $Ni_{1-x}Fe_xOOH$ ($0 < x < 0.27$). The active centers were assumed to be in the 4+ state. Table 3.1 presents the rate coefficients for the dynamics of oxygen evolution on 'fast' and 'slow' sites.

Figure 3.9 Theoretical OER overpotentials at Ni and Fe surface sites in pure and doped γ-NiOOH and γ-FeOOH model structures. (a) Proposed OER pathway with intermediates HO*, O* and HOO*, illustrated using the example of the on-top site at a substituted Fe surface atom in γ-NiOOH ($01\bar{1}2$). The binding energies of these species are used to estimate the OER overpotential. (b) OER activity volcano plot showing the overpotential as a function of Gibbs free energies of the reaction intermediates. Computed overpotentials are shown for the OER at Ni–Ni bridge and Fe on-top sites located in pure γ-NiOOH ($01\bar{1}2$) and in γ-NiOOH ($01\bar{1}2$) with Fe surface and subsurface doping, at a Ni on-top site in pure γ-NiOOH ($01\bar{1}2$), and at Fe–Fe bridge sites in pure and Ni-doped γ-FeOOH (010) (25% Ni in bulk unit cells). All corresponding model structures are shown with the intermediate whose formation is the potential limiting step (PLS) (HOO* in all cases except for the on-top Ni site in γ-NiOOH ($01\bar{1}2$) and γ-NiOOH ($01\bar{1}2$) with subsurface Fe, where formation of O* determines the overpotential).

Table 3.1 Pseudo-first-order OER rate constants.[53]

Catalyst	Fe Content (%)	'Fast' site rate constant k' (s^{-1})	'Slow' site rate constant k' (s^{-1})	'Fast' site fraction (%)
NiOOH	<0.8	0.04 ± 0.03		
$Ni_{0.91}Fe_{0.09}OOH$	8.97	0.41 ± 0.03	0.01 ± 0.02	6.4
$Ni_{0.82}Fe_{0.18}OOH$	18.97	1.70 ± 0.07	0.04 ± 0.02	17.6
$Ni_{0.73}Fe_{0.27}OOH$	27.31	0.34 ± 0.17		
FeOOH	>98.3	0.18 ± 0.05	0.02 ± 0.02	8.6

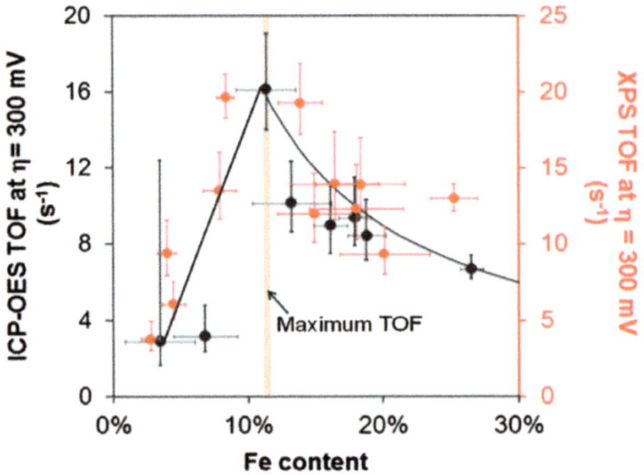

Figure 3.10 Turnover frequency (calculated on a per-Fe site basis) from total Fe content (ICP-OES: black, XPS: red) for $Ni(OH)_2$ films deposited on polished Au-RDEs as a function of percent Fe. The solid line represents the anticipated TOF assuming Fe content exceeding 11% is not OER-active. Error bars are determined from the inherent uncertainty in the Fe content as determined by ICP-OES and XPS for each sample. Reprinted with permission from S. Klaus, Y. Cai, M. W. Louie, L. Trotochaud and A. T. Bell, Effects of Fe Electrolyte Impurities on $Ni(OH)_2$/NiOOH Structure and Oxygen Evolution Activity, *J. Phys. Chem. C* 2015, **119**, 7243–7254, Copyright 2015 American Chemical Society.[31]

The rate constant for oxygen evolution on 'fast' sites in NiOOH is comparable to that measured for 'slow' sites in $Ni_{1-x}Fe_xOOH$ and a factor of 4.5 lower than that for 'fast' sites in FeOOH. With increasing substitution of Fe for Ni in $Ni_{1-x}Fe_xOOH$, the magnitude of the rate coefficient for 'fast' sites increases up to $x = 0.18$ and then decreases for $x = 0.27$, leading to the conclusion that the 'fast' sites are attributable to Fe cations, whereas the 'slow' sites are attributable to Ni cations.

The increase in the apparent rate coefficient for oxygen evolution on 'fast' sites with increasing Fe content reported by Ahn and Bard[53] (see Table 3.1) is fully consistent with the turnover frequency for OER reported by Klaus *et al.*[31] and shown in Figure 3.10. Most notably, both studies demonstrate that with increasing addition of Fe to NiOOH, the apparent rate coefficient for the OER increases up to a maximum value and then declines. As noted earlier the Fe content at which the maximum occurs depends on the manner of sample preparation. These findings suggest that up to the solubility limit of Fe^{3+} cations in NiOOH, the activity of each Fe cation increases due to subtle changes in its environment, a subject not yet fully understood and requiring further investigation.

It is notable that while Ahn and Bard[53] suggest that the oxidation state of the 'fast' Fe sites in $Ni_{1-x}Fe_xOOH$ is +4, based on the observation of this state in XAS studies, no direct evidence for the involvement of Fe^{4+} in OER was presented. In fact, recent work by Stahl *et al.*[54] suggests that, to the contrary, Fe^{3+} cations are the active centers. This conclusion was reached on the basis of *operando* Mössbauer spectroscopy. While direct evidence for Fe^{4+} cations in $Ni_{1-x}Fe_xOOH$ was observed, the persistence of these cations upon lowering the potential during *operando* OER studies, suggests they are not directly responsible for oxygen evolution. This conclusion supports the theoretical findings that indicate the OER occurs preferentially on Fe^{3+} cations.

3.6 Factors Affecting the OER Activity of NiFeOOH

A very large number of papers that have appeared in the recent literature describing different methods for preparing NiFeOOH OER catalysts.[55–78] These include electrodeposition and sputter deposition of thin metal films that upon immersion in alkaline electrolyte convert to the hydroxide phase and to the oxy-hydroxide upon application of a sufficiently high anodic potential, preparation of NiFeOOH nanoparticles by solvothermal or hydrothermal co-precipitation of mixed metal precursors, and pulsed-laser ablation of metals in liquids (PLAL). Comparison of the OER performance of NiFeOOH catalysts prepared by different methods is challenging, since many factors affect their activity.

To understand what factors influence the performance of NiFeOOH catalysts, we begin by noting that in the absence of mass transfer limitation the dependence of the current density based on the geometric area of the anode is described by the Butler–Volmer equation, which for large values of η_{OER}, is given by:[79]

$$j = j_0 \exp\left(\frac{\alpha z F}{RT}\eta_{OER}\right) \tag{3.6}$$

where j_0 is the exchange current density, R is the gas constant, T is the temperature, α is the transfer coefficient, z is the number of electrons transferred in the rate-limiting step, and F is Faraday's constant. The value of j_0 is proportional to the number of electrochemically active centers per unit weight of catalyst placed on the anode, which is, in turn, proportional to the weight of catalyst deposited per unit anode area. If all of the active centers are assumed to be located at the external surface of the catalyst, then the number of electrochemically active centers per unit weight of catalyst should be related to the BET surface area (surface area per unit weight of catalyst). However, the underlying assumption for this relation has been questioned, since active centers can be located between the layers of the NiFe double hydroxide, as well as on the internal surface.[7,22] Further complicating matters is the recognition that the BET area of nanoparticles is rarely reported in experimental studies of OER catalysts and can be difficult to measure if the quantity of catalyst prepared is small. Similar complications arise with the alternative proposal that the surface area of the catalyst relative to surface of the anode can be described by the electrochemically determined roughness factor, since this quantity is also difficult to measure accurately. Yet further complications arise from the fact that the value of j_0 also depends on the intrinsic activity of the catalyst, which is dependent on the content of Fe and its distribution between $Ni_{1-x}Fe_xOOH$ and FeOOH.[34] Another factor not taken into account is the dependence of j_0 on the concentration of base. For KOH, j_0 is proportional to the concentration of OH^- anions for pH > 11.[15] Recent evidence also indicates that the composition of the electrode on which either thin films or nanoparticles of NiFeOOH are deposited can have a significant effect on j_0.[7,16] The preceding discussion shows that many factors in addition to the intrinsic activity of the catalyst affect the value of j_0 and consequently the observation of a lower value of η_{OER} for a given current density does not necessarily mean that the intrinsic activity of the material is superior to another one with a higher value of η_{OER}.

The dependence of η_{OER} on j is characterized by the Tafel slope, β, which is defined as the gradient in the ohmic-drop-corrected value of η_{OER} with respect to the logarithm of the current density (eqn (3.7)).[79]

$$\frac{d\eta_{OER}}{d\log(j)} = \frac{2.3\,RT}{\alpha zF} = \beta \tag{3.7}$$

Eqn (3.7) shows that the lower the Tafel slope, the more rapid the rise in current density. While the Tafel slope of OER catalysts is often reported, the range of current densities for which it is valid is usually not given.

It should be noted that inherent in the use of the Butler–Volmer equation and the determination of β from the Tafel slope is the assumption that a single charge transfer step dictates the overall kinetics of the OER and that all other steps are equilibrated. As discussed below, this assumption greatly simplifies the expression for the dependence of j on η_{OER}, but is overly constraining. Nevertheless, eqn (3.6) and (3.7) continue to be used because experimental data can often be described by these equations over a reasonable range of measured overpotentials.

Until very recently, attempts to interpret the magnitude of measured Tafel slopes have focused on postulating a reaction mechanism, assuming a rate-limiting step, and then determining the Tafel slope from a microkinetic model.[80,81] Agreement between the Tafel slope derived by this means and that measured experimentally was then used to rationalize the nature of the rate-limiting step. This approach has very limited value, since dependence of the rate coefficients for elementary processes involving charge transfer on the potential applied to the cathode are assumed but are not based on a first-principles analysis of the relevant physics. Very recent work by Goddard and coworkers has illustrated a sounder, more physically correct approach.[82] In this study, rate coefficients for the elementary reactions involved in the OER on IrO_2 under acidic conditions are predicted from first principles using quantum mechanics. The authors find that for the reaction conditions used in experimental investigations of the OER, the IrO_2 (110) surface is fully O terminated and the rate-limiting step for O_2 formation involves the attack of aqueous water on a surface O atom leading to the dissociation of water and the formation of OOH*. Two possible pathways for this reaction are found, one for low (<1.5 V) and another for high (>1.5 V) applied potentials. A microkinetic model is then used to develop the relationship between current density and η_{OER}. The Tafel plot of η_{OER} *versus* ln(j) determined by this means exhibits two linear regimes. Above $\eta_{OER} = 0.35$ V, the Tafel slope is 73 meV per dec and below $\eta_{OER} = 0.3$ V, the Tafel slope is 22 meV per dec, in good accord with experimental observation. This work demonstrates that in order to understand the dependence of the current density on the OER overpotential, it is essential to carry out a detailed analysis of the free energies of activation for the OER processes and their dependence on the applied voltage.

The preceding discussion demonstrates why it is so difficult to compare the intrinsic activity of NiFeOOH catalysts prepared by different techniques. For this reason, most authors simply report the values of η_{OER} and β for a specified current density (usually 10 mA cm^{-2}) measured in a KOH solution of specified molarity (usually 0.1 M). From such measurements, it is possible to say that the catalyst with the lowest values of η_{OER} and β is the most active one.

3.7 Effects of Additives Other Than Fe on the OER Activity of $NiMO_x$

Numerous authors have investigated the extent to which elements other than Fe could be added to NiOOH in order to enhance its OER activity.[6,8,10,83–87] It should be noted, though, that all of these studies were conducted using KOH as the electrolyte and many were undertaken prior to the recognition that solutions of this base contain impurity Fe^{3+} cations that can incorporate into NiOOH and thereby increase its OER activity. Thus, the degree to which earlier studies are corrupted by Fe is not clear and for this reason, reliable conclusions about the effects of additive metals should be drawn from studies reported after 2014.

Koper and coworkers have examined the incorporation of Co, Fe, Mn, Cr, Cu, and Zn into NiOOH.[10] Based on DFT analysis of the potential limiting value of η_{OER}, they concluded that the incorporation of Mn should be equivalent to the incorporation of Fe, and that of the remaining elements, only Cr enhances the OER of NiOOH. The order of decreasing activity was predicted to be NiFeOOH > NiMnOOH > NiCrOOH, and this sequence was confirmed by experimental observation. It is also notable that DFT analysis revealed that Fe and Mn are the active sites in NiFeOOH and NiMnOOH but Ni is the active site in NiCrOOH.

In a recent study, Boettcher and coworkers have investigated the effects of adding ~10% and 30% Fe, Mn, La, Ti, or Ce on the OER activity of NiOOH.[88] Catalyst films were prepared by spin-casting from solutions of 0.1 M $Ni(NO_3)_2 \cdot 6H_2O$ and the nitrate salt of the additive metal dissolved in a 1 : 1 mixture of ethanol and water onto Au/Ti or Pt/Ti coated quartz substrates. The coated substrates were then transferred to a hot plate maintained at 150 °C and annealed for 10 min. Following preparation, samples were characterized electrochemically by cycling in Fe-free 1 M KOH at 20 mV s^{-1}. Figure 3.11 illustrates the TOF_{tm} based on total metal content of each sample, taken after 5 and 50 voltammogram cycles. It is observed that after 5 cycles, the activities of the $Ni_{0.85}Ce_{0.15}O_xH_y$ and $Ni_{0.65}Ce_{0.35}O_xH_y$ films are about 8 and 4 times higher that of NiO_xH_y, respectively. However, after 50 cycles, the activity of $Ni_{0.85}Ce_{0.15}O_xH_y$ is only 2-fold higher than that of NiO_xH_y and the activity of $Ni_{0.65}Ce_{0.35}O_xH_y$ is equivalent to that of NiO_xH_y.

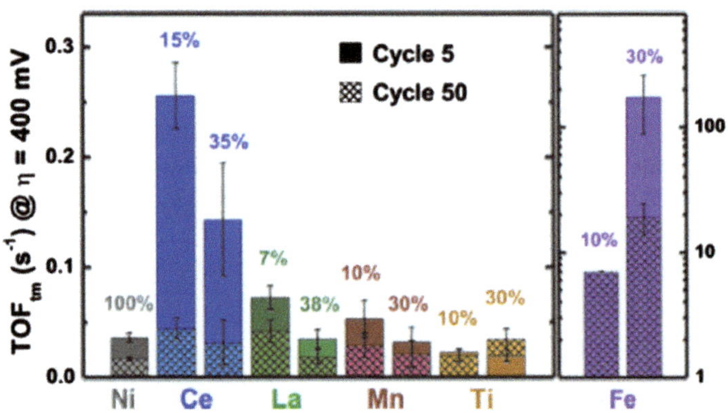

Figure 3.11 OER turnover frequency of spin-cast $Ni_{1-z}M_zO_xH_y$ films at 400 mV overpotential at cycle 5 (solid) and cycle 50 (pattern) in Fe-free 1 M KOH from voltammetry data collected at 20 mV s^{-1}. Values reported are the average, and error bars are the standard deviation of three samples. In the case of $Ni_{0.9}Fe_{0.1}O_xH_y$, there is no activity decrease over 50 cycles, and the error bars are small and not visible on this scale. TOFs are calculated assuming all metal cations are active (and thus are lower limits), and the OER current is taken as the average of the forward and reverse scans. Note the different scale on the *y*-axis for $Ni_{1-z}M_zO_xH_y$ films.
Reprinted from L. J. Enman, M. L. Burke, A. S. Batchellor and S. W. Boettcher, Effects of Intentionally Incorporated Metal Cations on the Oxygen Evolution Electrocatalytic Activity of Nickel (Oxy)hydroxide in Alkaline Media, *ACS Catal.* 2015, **6**, 2416–2423, Copyright 2015 American Chemical Society.[88]

The authors suggest that slow dissolution of Ce from the $Ni(Ce)O_xH_y$ films may be the cause for the decrease in activity with cycling or alternatively that cerium oxide phase segregation occurs during electrochemical cycling. On the other hand, catalysts containing Ti, Mn, or La were found to have lower OER activities than NiO_xH_y. In contrast, $Ni(Fe)O_xH_y$ films maintained their activity over the 50 cycles and were more than two orders of magnitude more active than NiO_xH_y. It was also noted that there was no significant correlation between the potential at which Ni^{2+}/Ni^{3+} redox occurred, suggesting that Ni is not the primary site for the OER. This finding is consistent with Fe being the active site in $Ni(Fe)O_xH_y$.

3.8 Effects of Additive on the OER Activity of $NiFeO_x$

A number of recent studies have examined the possibility of improving the OER activity of $NiFeO_x$ catalysts through the addition of a third metallic element. These efforts have utilized high-throughput methods to produce very large arrays of catalyst compositions, as described in greater detail in Chapter 9 by Gregoire, as well as targeted syntheses of specific catalyst compositions. Both approaches will be reviewed.

Stahl and coworkers have used a fluorescence-based combinatorial array to detect oxygen evolution from arrays of metal oxide catalysts.[89,90] Twenty-one ternary metal oxides were screened, each ternary oxide comprising twenty-one incremental compositions. Comparison of different ternary oxides containing Ni and Fe revealed that catalysts with compositions $Ni_{0.2}Fe_{0.4}Al_{0.4}O_x$, $Ni_{0.2}Fe_{0.4}Cr_{0.4}O_x$, and $Ni_{0.2}Fe_{0.2}Ga_{60}O_x$ were significantly more active at a fixed overpotential than one with the composition $Ni_{60}Fe_{0.4}O_x$. The Tafel slopes for $Ni_{0.2}Fe_{0.4}Al_{0.4}O_x$, $Ni_{0.2}Fe_{0.4}Cr_{0.4}O_x$, and $Ni_{0.2}Fe_{0.2}Ga_{60}O_x$ were in the range of 30–38 mV per dec.

As detailed in Chapter 9 in this book, an array of 5456 quaternary metal oxides containing Ni, Fe, Co, and Ce were explored by Haber *et al.*[91,92] at 3.3% atomic percent intervals. These samples were prepared by ink-jet printing on an FTO plate and each composition was then characterized by using a scanning drop three-electrode system. Of this set, 665 compositions were found to have superior activities than other combinations. Two samples were found to be exceptionally active and exhibited the same OER overpotential at a current density of 10 mA cm^{-2} – $Ni_{0.5}Fe_{0.3}Co_{0.17}Ce_{0.03}O_x$ and $Ni_{0.3}Fe_{0.07}Co_{0.2}Ce_{0.43}O_x$, which are referred to as low-Ce and high-Ce catalysts, respectively. The Tafel slopes for the low-Ce and high-Ce catalysts are approximately 40 mV per dec and 70 mV per dec, respectively. This illustration demonstrates that neither η_{OER} nor the Tafel slope can be used individually to identify catalyst activity, since the high-Ce catalyst is more active at low OER overpotentials but the low-Ce catalyst is more active at high OER overpotentials. Given the unexpected high activity of the high-Ce catalyst at low OER overpotentials, further investigation of this material was undertaken.[93] *Ex situ* synchrotron X-ray absorption spectroscopy (XAS) and low-dose, high-resolution transmission electron microscopy (TEM) revealed that the catalyst is biphasic—consisting of distinct nano-domains of crystalline CeO_2 and particles having a cubic, rock salt structure consistent with NiO containing Co and Fe in a highly disordered state, possibly located at grain boundaries but likely alloyed into the NiO phase. While definitive conclusions about the relationship between the two phases could not be drawn, the authors proposed that superior activity of the high-Ce catalyst could be due to synergistic relationships between the two phases comprising the catalyst.

The effects of incorporating Co into $NiFeO_x$ has been reported in two very recent studies. Zhu *et al.* have investigated the performance of NiCoFe layered double hydroxides (LDH) by hydrothermal deposition of the metals onto a 3D conductive Ni foam (NF).[94] TEM analysis revealed that the NiCoFe LDH sheets have a lateral size of about 100 nm and are ultrathin. The OER activities of $Ni_{2.5}Co_{0.5}Fe$/NF and Ni_2CoFe/NF were compared to that of Ni_3Fe/NF by *iR*-corrected linear sweep voltammetry. For $Ni_{2.5}Co_{0.5}Fe$/NF the value of η_{OER} at a current density of 10 mA cm^{-2} is 275 mV and the Tafel slope is 99 mV per dec. By comparison, for Ni_3Fe/NF the value of η_{OER} is 305 mV and the Tafel slope is 92 mV per dec. The authors also found that the activation energy for the OER for $Ni_{2.5}Co_{0.5}Fe$/NF was 21 kJ mol^{-1}, whereas that for Ni_3Fe/NF was 52.3 kJ mol^{-1}. Based on their results, the authors proposed that Co improves the electrical conductivity of the catalyst and thereby contributes to the superior performance of $Ni_{2.5}Co_{0.5}Fe$/NF compared to Ni_3Fe/NF.

The structure and performance of NiCoFe layered triple hydroxides (LTHs) supported on carbon fiber cloth (CFC) for the electrochemical splitting of water have been investigated by Wang *et al.*[95] NiCoFe LTHs were electrodeposited onto the surface of CFC from a solution containing a mixture of Ni, Co, and Fe nitrates and well as ammonium nitrate. The HRTEM images show that the catalyst consists of thin, leaf-like sheets of LTHs organized on the surface of the CFC to form a very porous open structure. The OER activity of NiCoFe LTHs/CFC (Ni : Co : Fe = 8.7 : 10.5 : 1.0) was investigated in 1 M KOH by scanning voltammetry and compared with that of NiFe double hydroxides nano-sheet arrays (NiFe DHNAs/CFC) (Ni : Fe = 7.3 : 1). At a current density of 10 mA cm^{-2}, $\eta_{OER} = 240$ mV and the Tafel slope is 32 mV per dec for NiCoFe LTHs/CFC, whereas $\eta_{OER} = 260$ mV and the Tafel slope is 65 mV per dec for NiFe DHNAs/CFC. Another remarkable property of NiCoFe LTHs/CFC is its excellent stability. Only a 1% loss in the activity of this catalyst was observed when it was tested for 12 h in 1 M KOH at a current density of 20 mA cm^{-2}.

Improvements in the performance of $NiFeO_x$ catalysts by the incorporation of Mn have been reported by Lu *et al.*[96] Ternary NiFeMn layered double hydroxides were synthesized by dropwise addition of NaOH or Na_2CO_3 to an aqueous solution of Ni, Fe, and Mn nitrates. HRTEM images show that the catalyst consists of rag-like layers that are typically 50 nm wide and 3.7 nm thick. For an optimized composition of Ni : Fe : Mn = 3 : 1 : 0.5, the OER activity at 20 mA cm^{-2} is characterized by $\eta_{OER} = 289$ mV and a Tafel slope of

47 mV per dec. These characteristics compare favorably with those for a NiFe layered double hydroxide, for which $\eta_{OER} = 401$ mV and the Tafel slope is 65 mV per dec.

While most of the efforts aimed at improving the OER activity of $NiFeO_x$ have focused on the role of added cations, Hunter *et al.* have recently reported that the composition of the charge-compensating anions present in NiFe-layered double hydroxides ($[Ni_{1-x}Fe_x(OH)_2](A^{m-})_{x/m}$) also affect the OER activity.[97] $[Ni_3Fe(OH)_2](NO_3)$ was prepared by PLAL[58] and then ion-exchanged in a 1 M solution of BF_4^-, Cl^-, ClO_4^-, CO_3^{2-}, $C_2O_4^{2-}$, F^-, PO_4^{3-} or SO_4^{2-}. Electrochemical characterization of the $[Ni_3Fe(OH)_2](A^{m-})_{1/m}$ catalysts was carried out in a virtually CO_2-free 1 M solution of KOH, to avoid back exchange of the intercalated anion by CO_3^{2-}. Figure 3.12 illustrates the value of η_{OER} measured for a current density of 1 mA cm^{-2} as a function of the anion basicity characterized by the pK_a of the conjugate acid. A strong correlation is observed between the OER activity and the pK_a of the conjugate acid. The effect of the intercalated anion was stable over a period of 3 h in 1M KOH, suggesting that di- and tri-valent anions, in particular, outcompete OH^- anions for charge compensation between the LDH galleries. It was noted though, that if the KOH solution was exposed to air and allowed to absorb CO_2, carbonate anions formed, which rapidly displaced any anions having a lower conjugate acid pK_a. The authors suggest that the observed effect of anions may be due to a base-assisted deprotonation step during the OER. It is notable that no reference is made to the transformation of $[Ni_3Fe(OH)_2](A^{m-})_{1/m}$ to Ni_3FeOOH at a potential lower than that required for the onset of the OER. Therefore, it is conceivable that only a part of the $[Ni_3Fe(OH)_2](A^{m-})_{1/m}$ undergoes oxidation to Ni_3FeOOH and that, somehow, the unconverted part of the $[Ni_3Fe(OH)_2](A^{m-})_{1/m}$ influences the OER activity of Ni_3FeOOH.

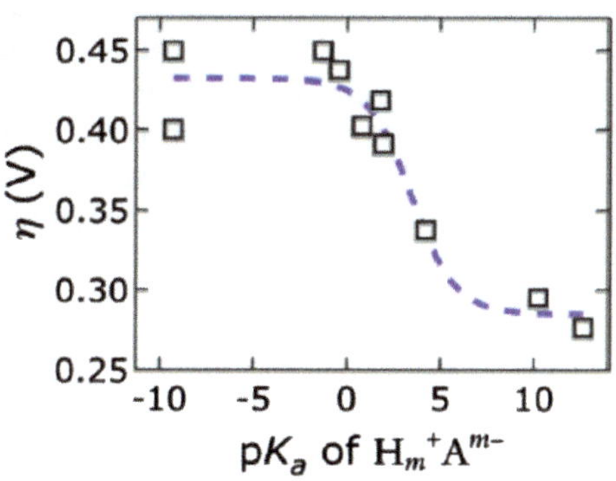

Figure 3.12 OER overpotential for $[Ni_3Fe(OH)_2](A^{m-})_{1/m}$ *versus* the pK_a of the conjugate acid of the anion A^{m-}. Experiment carried out at a current density of 1 mA cm^{-2} in virtually CO_2-free 1 M KOH.
Reproduced from ref. 97 with permission from the Royal Society of Chemistry.

3.9 Conclusions

Extensive research has established nickel iron oxyhydroxide, $Ni_{1-x}Fe_xOOH$, to be one of the most active catalysts for the OER. This material is attractive because it is based on earth-abundant elements and can operate without loss of activity under alkaline conditions. The OER activity of $Ni_{1-x}Fe_xOOH$ increases rapidly with x up to an Fe content of 12–25%, where it reaches a plateau, and then decreases with further addition of Fe. The highest OER activity is characterized by the lowest values for η_{OER} and β, and the lowest activation energy for the OER. Operando XAS studies of $Ni_{1-x}Fe_xOOH$ have shown that the initial increase in OER activity with added Fe is attributable to the substitution of Fe^{3+} cations for Ni^{3+} cations into the lattice of NiOOH. The substituted Fe^{3+} cations have an Fe–O bond length similar to that of Ni–O in NiOOH rather than the Fe–O bond distance in FeOOH, causing the electronic properties of such Fe^{3+} cations to be different from those in FeOOH. DFT calculations of the overpotential for the OER occurring on such Fe^{3+} cations show that it is significantly lower than that for Ni^{3+} sites in NiOOH, Fe^{3+} sites in FeOOH or Ni^{3+} sites in $Ni_{1-x}Fe_xOOH$, strongly suggesting that the superior OER activity of $Ni_{1-x}Fe_xOOH$ over NiOOH is due to the substituted Fe^{3+} cations. This conclusion is supported by scanning electrochemical microscopy measurement of the kinetics of the OER on Ni^{3+} and Fe^{3+} cations in NiOOH, FeOOH, and $Ni_{1-x}Fe_xOOH$. The decline in OER activity observed for Fe contents greater than the optimum is attributed to the formation of FeOOH, which has a much lower OER activity than $Ni_{1-x}Fe_xOOH$. While additives to NiOOH other than Fe have been explored, none of the bimetallic catalysts exhibit higher activity and stability than Fe. While even lower OER overpotentials and Tafel slopes than those observed for $Ni_{1-x}Fe_xOOH$ can be attained by the incorporation of Co or Mn, the means by which these additives have their effect is not understood.

A number of issues concerning the OER activity of $Ni_{1-x}Fe_xOOH$ remain outstanding and should be pursued. The first is the reason for the observed increase in the specific activity (turnover frequency) of Fe^{3+} cations with increasing Fe content. Closely related to this issue is the question of how does the local environment around Fe^{3+} cations affect the Tafel slope. The next issue would be to clearly establish the maximum extent to which Fe^{3+} cations can be substituted into the lattice of NiOOH and whether this level can be controlled through the synthetic strategy used to produce $Ni_{1-x}Fe_xOOH$. A third issue is the extent to which synthetic procedures can be devised to produce 3-D, highly porous $Ni_{1-x}Fe_xOOH$ layers that would enhance the superficial current density achievable for a given anode surface area.

Acknowledgements

The author acknowledges support from the Joint Center for Artificial Photosynthesis, a DOE Energy Innovation Hub, supported through the Office of Science of the U.S. Department of Energy under Award number DE-SC0004993.

References

1. N. S. Lewis and D. G. Nocera, Powering the planet: Chemical challenges in solar energy utilization, *Proc. Natl. Acad. Sci. U. S. A.*, 2006, **103**, 15729–15735.
2. H. B. Gray, Powering the plane with solar fuel, *Nat. Chem.*, 2009, **1**, 7.
3. A. Kudo and Y. Miseki, Heterogeneous photocatlayst materials for water splitting, *Chem. Soc. Rev.*, 2009, **38**, 253–278.
4. T. R. Cook, D. K. Dogutan, S. Y. Reece, Y. Surendranth, T. S. Teets and D. G. Nocera, Solar energy supply and storage for the legacy and non-legacy worlds, *Chem. Rev.*, 2010, **110**, 6474–6502.
5. B. M. Hunter, H. B. Gray and A. M. Müller, Earth-Abundant Heterogeneous Water Oxidation Catalysts, *Chem. Rev.*, 2016, DOI: 10.1021/acs.chem.rev.6b00398.
6. C. C. L. McCrory, S. Jung, J. C. Peters and T. F. Jaramillo, Benchmarking heterogeneous electrocatalysts for the oxygen evolution reaction, *J. Am. Chem. Soc.*, 2013, **135**, 16077–16987.
7. E. Fabbri, A. Habrederer, K. Waltar, R. Kotz and T. J. Schmidt, Development and perspective of oxide-based catalysts for the oxygen evolution reaction, *Catal. Sci. Technol.*, 2014, **4**, 3800–3821.
8. C. C. L. McCrory, S. Jung, I. M. Ferrer, S. M. Chatman, J. C. Peters and J. F. Jaramillo, Benchmarking Hydrogen Evolving and Oxygen Evolving Reaction Electocatalysts for Solar Water Splitting Devices, *J. Am. Chem. Soc.*, 2015, **137**, 4347–4357.
9. M. Gong and H. Dai, A mini review of NiFe-based materials as highly active oxygen evolution catalaysts, *Nano Res.*, 2015, **8**, 23–39.
10. O. Diaz-Morales, I. Ledezema-Yanez, M. T. M. Koper and F. Calle-Vallejo, Guidelines for the Rational Design of Ni-Based Double Hydroxide Electrocatalysts for the Oxygen Evolution Reaction, *ACS Catal.*, 2015, **5**, 5380–5387.
11. S. Jung, C. C. L. McCrory, I. M. Ferrer, J. C. Peters and T. F. Jaramillo, Benchmarking Nanoparticulate Metal Oxide Electrocatalysts for the Alkaline Water Oxidation Reaction, *J. Mater. Chem. A*, 2016, **4**, 3068–3076.
12. H. Bode, K. Dehmelt and J. Witte, Zur Kenntnis der Nickelhydroxidelektrode—I.Über Das Nickel (II)-Hydroxidhydrat, *Electrochim. Acta*, 1966, **11**, 1079–1087.
13. M. Cappadonia, J. Divisek, T. Vonderheyden and U. Stimming, Oxygen Evolution at Nickel Anodes in Concentrated Alkaline-Solution, *Electrochim. Acta*, 1994, **39**, 1559–1564.
14. P. W. T. Lu and S. Srinivasan, Electrochemical-Ellipsometric Studies of Oxide Film Formed on Nickel during Oxygen Evolution, *J. Electrochem. Soc.*, 1978, **125**, 1416–1422.
15. M. W. Louie and A. T. Bell, An Investigation of Thin-Film Ni-Fe Oxide Catalysts for the Electrochemical Evolution of Oxygen, *J. Am. Chem. Soc.*, 2013, **135**, 12329–12337.

16. B. S. Yeo and A. T. Bell, In Situ Raman Study of Nickel Oxide and Gold-Supported Nickel Oxide Catalysts for the Electrochemical Evolution of Oxygen, *J. Phys. Chem. C*, 2012, **116**, 8394–8400.
17. I. J. Godwin and M. E. G. Lyons, Enhanced Oxygen Evolution at Hydrous Nickel Oxide Electrodes via Electrochemical Ageing in Alkaline Solution, *Electrochem. Commun.*, 2013, **32**, 39–42.
18. M. Wehrens-Dijksma and P. H. L. Notten, Electrochemical Quartz Microbalance Characterization of Ni(OH)(2)-Based Thin Film Electrodes, *Electrochim. Acta*, 2006, **51**, 3609–3621.
19. D. A. Corrigan, The Catalysis of the Oxygen Evolution Reaction by Iron Impurities in Thin-Film Nickel-Oxide Electrodes, *J. Electrochem. Soc.*, 1987, **134**, 377–384.
20. X. H. Li, F. C. Walsh and D. Pletcher, Nickel Based Electrocatalysts for Oxygen Evolution in High Current Density, Alkaline Water Electrolysers, *Phys. Chem. Chem. Phys.*, 2011, **13**, 1162–1167.
21. R. N. Singh, J. P. Pandey and K. L. Anitha, Preparation of Electrodeposited Thin-Films of Nickel Iron-Alloys on Mild-Steel for Alkaline Water Electrolysis. 1. Studies on Oxygen Evolution, *Int. J. Hydrogen Energy*, 1993, **18**, 467–473.
22. L. Trotochaud, J. K. Ranney, K. N. Williams and S. W. Boettcher, Solution-Cast Metal Oxide Thin Film Electrocatalysts for Oxygen Evolution, *J. Am. Chem. Soc.*, 2012, **134**, 17253–17261.
23. M. Gong, Y. Li, H. Wang, Y. Liang, J. Z. Wu, J. Zhou, J. Wang, T. Regier, F. Wei and H. Dai, An Advanced Ni-Fe Layered Double Hydroxide Electrocatalyst for Water Oxidation, *J. Am. Chem. Soc.*, 2013, **135**, 8452–8455.
24. J. Landon, E. Demeter, N. Inoglu, C. Keturakis, I. E. Wachs, R. Vasic, A. I. Frenkel and J. R. Kitchin, Spectroscopic Characterization of Mixed Fe-Ni Oxide Electrocatalysts for the Oxygen Evolution Reaction in Alkaline Electrolytes, *ACS Catal.*, 2012, **2**, 1793–1801.
25. E. L. Miller and R. E. Rocheleau, Electrochemical Behavior of Reactively Sputtered Iron-Doped Nickel Oxide, *J. Electrochem. Soc.*, 1997, **144**, 3072–3077.
26. C. C. Hu and Y. R. Wu, Bipolar Performance of the Electroplated Iron-Nickel Deposits for Water Electrolysis, *Mater. Chem. Phys.*, 2003, **82**, 588–596.
27. R. D. L. Smith, M. S. Prevot, R. D. Fagan, Z. P. Zhang, P. A. Sedach, M. K. J. Siu, S. Trudel and C. P. Berlinguette, Photochemical Route for Accessing Amorphous Metal Oxide Materials for Water Oxidation Catalysis, *Science*, 2013, **340**, 60–63.
28. R. D. L. Smith, M. S. Prévot, R. D. Fagan, S. Trudel and C. P. Berlinguette, Water Oxidation Catalysis: Electrocatalytic Response to Metal Stoichiometry in Amorphous Metal Oxide Films Containing Iron, Cobalt, and Nickel, *J. Am. Chem. Soc.*, 2013, **135**, 11580–11586.
29. M. D. Merrill and R. C. Dougherty, Metal Oxide Catalysts for the Evolution of O_2 from H_2O, *J. Phys. Chem. C*, 2008, **112**, 3655–3666.

30. L. Trotochaud, S. L. Young, J. K. Ranney and S. W. Boettcher, Nickel-Iron Oxyhydroxide Oxygen-Evolution Electrocatalysts: The Role of Intentional and Incidental Iron Incorporation, *J. Am. Chem. Soc.*, 2014, **136**, 6744–6753.
31. S. Klaus, Y. Cai, M. W. Louie, L. Trotochaud and A. T. Bell, Effects of Fe Electrolyte Impurities on $Ni(OH)_2$/NiOOH Structure and Oxygen Evolution Activity, *J. Phys. Chem. C*, 2015, **119**, 7243–7254.
32. S. Klaus, M. W. Louie, L. Trotochaud and A. T. Bell, Role of Catalyst Preparation on the Electrocatalytic Activity of $Ni_{1-x}Fe_xOOH$ for the Oxygen Evolution Reaction, *J. Phys. Chem. C*, 2015, **119**, 18303–18316.
33. J. Swierk, S. Klaus, L. Trotochaud, A. T. Bell and T. D. Tilley, An Electrochemical Study of the Energetics of the Oxygen Evolution Reaction at Nickel Iron (oxy)hydroxide Catalysts, *J. Phys. Chem. C*, 2015, **119**, 19022–19029.
34. D. Friebel, M. W. Louie, M. Abjdich, K. E. Sewald, Y. Cai, A. M. Wise, M.-J. Cheng, D. Sokaras, T.-C. Weng, R. Alonso, R. C. Davis, J. R. Bargar, J. K. Norskov, A. Nilsson and A. T. Bell, Identification of Highly active Fe Sites in (Ni,Fe)OOH for Electrocatalytic Splitting of Water, *J. Am. Chem. Soc.*, 2015, **137**, 1305–1313.
35. W. M. Heijboer, P. Glatzel, K. R. Sawant, R. F. Lobo, U. Bergmann, R. A. Barrea, D. C. Koningsberger, B. M. Weckhuysen and F. M. F. de Groot, Kβ-Detected XANES of Framework-Substituted FeZSM-5 Zeolites, *J. Phys. Chem. B*, 2004, **108**, 10002–10011.
36. T. E. Westre, P. Kennepohl, J. G. DeWitt, B. Hedman, K. O. Hodgson and E. I. Solomon, A Multiplet Analysis of Fe K-Edge 1s→3d Pre-Edge Features of Iron Complexes, *J. Am. Chem. Soc.*, 1997, **119**, 6297–6314.
37. D. K. Bediako, B. Lassalle-Kaiser, Y. Surendranath, J. Yano, V. K. Yachandra and D. G. Nocera, Structure–Activity Correlations in a Nickel–Borate Oxygen Evolution Catalyst, *J. Am. Chem. Soc.*, 2012, **134**, 6801–6809.
38. M. Balasubramanian, C. A. Melendres and S. Mini, X-ray Absorption Spectroscopy Studies of the Local Atomic and Electronic Structure of Iron Incorporated into Electrodeposited Hydrous Nickel Oxide Films, *J. Phys. Chem. B*, 2000, **104**, 4300–4306.
39. T. W. Capehart, D. A. Corrigan, R. S. Conell, K. I. Pandya and R. W. Hoffman, In situ extended x-ray absorption fine structure spectroscopy of thin-film nickel hydroxide electrodes, *Appl. Phys. Lett.*, 1991, **58**, 865–867.
40. A. N. Mansour and C. A. Melendres, XAFS investigation of the structure and valency of nickel in some oxycompounds, *Phys. B Condens. Matter*, 1995, **208–209**, 583–584.
41. Y. Hu, I. T. Bae, Y. Mo, D. A. Scherson and M. R. Antonio, In situ X-ray absorption fine structure and optical reflectance studies of electrodeposited nickel hydrous oxide films in alkaline electrolytes, *Can. J. Chem.*, 1997, **75**, 1721–1729.

42. M. Morishita, S. Ochiai, T. Kakeya, T. Ozaki, Y. Kawabe, M. Watada, S. Tanase and T. Sakai, Phase Transformation in the Charge-Discharge Process and the Structural Analysis by Synchrotron XAFS and XRD for Nickel Hydroxide Electrode, *Electrochem*, 2008, **76**, 802.
43. X. Yang, K. Takada, M. Itose, Y. Ebina, R. Ma, K. Fukuda and T. Sasaki, Highly Swollen Layered Nickel Oxide with a Trilayer Hydrate Structure, *Chem. Mater.*, 2008, **20**, 479–485.
44. A. Bianconi; J. Garcia and M. Benfatto, XANES in condensed systems, in *Synchrotron Radiation in Chemistry and Biology I; Topics in Current Chemistry*, Springer, Berlin Heidelberg, 1988, pp. 29–67.
45. J. Stöhr, F. Sette and A. L. Johnson, Near-Edge X-Ray-Absorption Fine-Structure Studies of Chemisorbed Hydrocarbons: Bond Lengths with a Ruler, *Phys. Rev. Lett.*, 1984, **53**, 1684–1687.
46. L. Guerlou-Demourgues, L. Fournès and C. Delmas, In Situ ^{57}Fe Mössbauer Spectroscopy Study of the Electrochemical Behavior of an Iron-Substituted Nickel Hydroxide Electrode, *J. Electrochem. Soc.*, 1996, **143**, 3083–3088.
47. M. Görlin, P. Chernev, J. Ferreira de Araújo, T. Reier, S. Dresp, B. Paul, R. Kränhert, H. Dau and P. Strasser, Oxygen Evolution Reaction Dynamics, Faradic Charge Efficiency, and the Active Metal Redox States of Ni-Fe Oxide Water Splitting Electorcatalysts, *J. Am. Chem. Soc.*, 2016, **138**, 5603–5614.
48. J. K. Nørskov, J. Rossmeisl, A. Logadottir, L. Lindqvist, J. R. Kitchin, T. Bligaard and H. Jónsson, Origin of the Overpotential for Oxygen Reduction at a Fuel-Cell Cathode, *J Phys Chem B*, 2004, **108**, 17886–17892.
49. I. C. Man, H. Su, F. Calle-Vallejo, H. A. Hansen, J. I. Martínez, N. G. Inoglu, J. Kitchin, T. F. Jaramillo, J. K. Nørskov and J. Rossmeisl, Universality in Oxygen Evolution Electrocatalysis on Oxide Surfaces, *ChemCatChem*, 2011, **3**, 1159–1165.
50. M. T. M. Koper, Thermodynamic theory of multi-electron transfer reactions: Implications for electrocatalysis, *J. Electroanal. Chem.*, 2011, **660**, 254–260.
51. P. Liao, J. A. Keith and E. A. Carter, Water oxidation on pure and doped hematite (0001) surfaces: prediction of Co and Ni as effective dopants for electrocatalysis, *J. Am. Chem. Soc.*, 2012, **134**, 13296–13309.
52. Y.-F. Li and A. Selloni, Mechanism and Activity of Water Oxidation on Selected Surfaces of Pure and Fe-Doped NiO_x, *ACS Catal.*, 2014, 1148–1153.
53. H. S. Ahn and A. J. Bard, Surface Interrogation Scanning Electrochemical Microscopy of $Ni_{1-x}Fe_xOOH$ ($0<x<0.27$) Oxygen-Evolving Catalyst: Kinetics of the "Fast" Iron Sites, *J. Am. Chem. Soc.*, 2016, **119**, 8147–8154.
54. J. Y. C. Chen, L. Dang, H. Liang, W. Bi, J. B. Gerken, S. Jin, A. A. Alp and S. S. Stahl, Operando Analysis of NiFe and Fe Oxyhydroxide Electrocatalysts for Water Oxidation: Detection of Fe^{4+} by Mossbauer Spectroscopy, *J. Am. Chem. Soc.*, 2015, **137**, 15090–15093.

55. J. Landon, E. Demeter, N. Đnoğlu, C. Keturakis, I. E. Wachs, R. Vasić, A. I. Frenkel and J. R. Kitchin, Spectroscopic Characterization of Mixed Fe–Ni Oxide Electrocatalysts for the Oxygen Evolution Reaction in Alkaline Electrolytes, *ACS Catal.*, 2012, **2**, 1793–1801.
56. M. Gong, Y. Li, H. Wang, Y. Liang, J. Z. Wu, J. Zhou, J. Wang, T. Regier, F. Wei and H. Dai, An advanced Ni–Fe layered double hydroxide electrocatalyst for water oxidation, *J. Am. Chem. Soc.*, 2013, **135**, 8452–8455.
57. J. B. Gerken, S. E. Shaner, R. C. Massé, N. J. Porubsky and S. S. Stahl, A survey of diverse earth abundant oxygen evolution electrocatalysts showing enhanced activity from Ni–Fe oxides containing a third metal, *Energy Environ. Sci.*, 2014, **7**, 2376–2382.
58. B. M. Hunter, J. D. Blakemore, M. Deimund, H. B. Gray, J. R. Winkler and A. M. Muller, Highly active mixed-metal nanosheet water oxidation catalysts made by pulsed-laser ablation in liquids, *J. Am. Chem. Soc.*, 2014, **136**, 13118–13121.
59. K. L. Nardi, N. Yang, C. F. Dickens, A. L. Strickler and S. F. Bent, Creating Highly Active Atomic Layer Deposited NiO Electrocatalysts for the Oxygen Evolution Reaction, *Adv. Energy Mater.*, 2015, **5**, 1500412.
60. J. A. Bau, E. J. Luber and J. M. Buriak, Oxygen Evolution Catalyzed by Nickel-Iron Oxide Nanocrystals with a Nonequilibrium Phase, *ACS Appl. Mater. Interfaces*, 2015, **7**, 19755–19763.
61. J. Qi, W. Zhang, R. Xiang, K. Liu, H.-Y. Wang, M. Chen, Y. Han and R. Cao, Porous Nickel-Iron Oxide as a Highly Efficient Electrocatalyst for Oxygen Evolution Reaction, *Adv. Sci.*, 2015, **2**, 1500199.
62. X. Lu and C. Zhao, Electrodeposition of hierarchically structured three-dimensional nickel iron electrodes for efficient oxygen evolution at high current densities, *Nat. Commun.*, 2015, **6**, 6616.
63. Y. Feng, H. Zhang, Y. Zhang, X. Li and Y. Wang, Ultrathin Two-Dimensional Free-Standing Sandwiched NiFe/C for High-Efficiency Oxygen Evolution Reaction, *ACS Appl. Mater. Interfaces*, 2015, **7**, 9203–9210.
64. L. Zhiyi, L. Yingjie, L. Xiaodong, J. Liu and X. Sun, Nanoarray based "superaerophobic" surfaces for gas evolution reaction electrodes, *Mater. Horiz.*, 2015, **2**, 294.
65. X. Yan, L. Tian, K. Li, S. Atkins, H. Zhao, J. Murowchick, L. Liu and X. Chen, $FeNi_3$/NiFeOxNanohybrids as Highly Efficient Bifunctional Electrocatalysts for Overall Water Splitting, *Adv. Mater. Interfaces*, 2016, 1600368.
66. Z. Li, M. Shao, H. An, Z. Wang, S. Xu, M. Wei, D. G. Evans and X. Duan, Fast electrosynthesis of Fe-containing layered double hydroxide arrays toward highly efficient electrocatalytic oxidation reactions, *Chem. Sci.*, 2015, **6**, 6624–6631.
67. X. Cui, P. Ren, D. Deng, J. Deng and X. Bao, Single layer graphene encapsulating nonprecious metals as high-performance electrocatalysts for water oxidation, *Energy Environ. Sci.*, 2016, **9**, 123–129.

68. B. M. Hunter, W. Hieringer, J. R. Winkler, H. B. Gray and A. M. Müller, Effect of interlayer anions on [NiFe]-LDH nanosheet water oxidation activity, *Energy Environ. Sci.*, 2016, **9**, 1734–1743.
69. Y. Feng, H. Zhang, L. Fang, Y. Mu and Y. Wang, Uniquely Monodispersing NiFe Alloyed Nanoparticles in Three-Dimensional Strongly Linked Sandwiched Graphitized Carbon Sheets for High-Efficiency Oxygen Evolution Reaction, *ACS Catal.*, 2016, **6**, 4477–4485.
70. A. Sivanantham, P. Ganesan and S. Shanmugam, Hierarchical $NiCo_2S_4$ Nanowire Arrays Supported on Ni Foam: An Efficient and Durable Bifunctional Electrocatalyst for Oxygen and Hydrogen Evolution Reactions, *Adv. Funct. Mater.*, 2016, **26**, 4661–4672.
71. C. Xiao, Y. Li, X. Lu and C. Zhao, Bifunctional Porous NiFe/$NiCo_2O_4$/Ni Foam Electrodes with Triple Hierarchy and Double Synergies for Efficient Whole Cell Water Splitting, *Adv. Funct. Mater.*, 2016, **26**, 3515–3523.
72. F. Rong, J. Zhao, Q. Yang and C. Li, Nanostructured hybrid NiFeOOH/CNT electrocatalysts for oxygen evolution reaction with low overpotential, *RSC Adv.*, 2016, **6**, 74536–74544.
73. W. Xu, Z. Lu, P. Wan, Y. Kuang and X. Sun, High-Performance Water Electrolysis System with Double Nanostructured Superaerophobic Electrodes, *Small*, 2016, **12**, 2492–2498.
74. F. Dionigi and P. Strasser, NiFe-Based (Oxy)hydroxide Catalysts for Oxygen Evolution Reaction in Non-Acidic Electrolytes, *Adv. Energy Mater.*, 2016, 1600621.
75. T. T. H. Hoang and A. A. Gewirth, High Activity Oxygen Evolution Reaction Catalysts from Additive-Controlled Electrodeposited Ni and NiFe Films, *ACS Catal.*, 2016, **6**, 1159–1164.
76. C. G. Morales-Guio, L. Liardet and X. Hu, Oxidatively Electrodeposited Thin-Film Transition Metal (Oxy)hydroxides as Oxygen Evolution Catalysts, *J. Am. Chem. Soc.*, 2016, **138**, 8946–8957.
77. Y. Q. Gao, X. Y. Liu and G. W. Yang, Amorphous mixed-metal hydroxide nanostructures for advanced water oxidation catalysts, *Nanoscale*, 2016, **8**, 5015–5023.
78. K. Zhu, M. Li, X. Li, X. Zhu, J. Wang and W. Yang, Enhancement of oxygen evolution performance through synergetic action between NiFe metal core and NiFeOx shell, *Chem. Commun.*, 2016, DOI: 10.1039/c6cc04951f.
79. J. Newman and K. E. Thomas-Alyea, *Electrochemical Systems*, Wiley, New York, 3rd edn, 2004.
80. (a) M. E. G. Lyons and M. P. Bradon, The oxygen evolution reaction on passive oxide covered transition metal electrodes in aqueous alkaline solution. Part 1-Nickel, *Int. J. Electrochem Soc.*, 2008, **3**, 1386–1424; (b) M. E. G. Lyons and M. P. Bradon, The oxygen evolution reaction on passive oxide covered transition metal electrodes in aqueous alkaline solution. Part 1-Nickel. The Oxygen Evolution Reaction on Passive Oxide Covered Transition Metal Electrodes in Alkaline Solution. Part III-Iron, *Int. J. Electrochem Soc.*, 2008, **3**, 1463–1503.

81. Y. Matsumoto and E. Sato, Electrocatalytic properties of transition-metal oxides for oxygen evolution reaction, *Mater. Chem. Phys.*, 1986, **14**, 397–426.
82. Y. Ping, R. J. Nielsen and W. A. Goddard, The Reaction Mechanism with Free Energy Barrriers at Constant Potential for the Oxygen Evolution Reaction at the IrO_2 (110) surface, *J. Am. Chem. Soc.*, 2016, DOI: 10.1021/jacs.6b07557.
83. D. A. Corrigan and R. M. Bendert, Effect of coprecipitated metal ions on the electrochemistry of nickel hydroxide thin films: cyclic voltammetry in 1M KOH, *J. Electrochem. Soc.*, 1989, **136**, 723–728.
84. S. I. Cordoba, R. E. Carbonia and M. L. Teijelo, The electrochemical response of binary mixtures of hydrous transition metal hydroxides co-precipitated on conducting substrates with reference to the oxygen evolution reaction, *Electrochim. Acta*, 1986, **31**, 1321–1332.
85. P. W. T. Lu and S. Srinivasan, Nickel-Based Alloys as Electrocatalysts for Oxygen Evolution from Alkaline Solutions, *J. Electrochem. Soc.*, 1978, **125**, 265–270.
86. L. Qian, Z. Lu, T. Xu, X. Wu, Y. Tian, Y. Li, Z. Huo, X. Sun and X. Duan, Trinary Double Hydroxides as High-Performance Bifunctional Materials for Oxgen Electrocatalysis, *Adv. Energy Mater.*, 2015, **5**, 1500245.
87. K.-Y. Niu, F. Lin, S. Jung, L. Fang, D. Nordlund, C. C. L. McCrory, T.-C. Weng, P. Ercius, M. M. Doeff and H. Zheng, Tuning Complex Transition Metal Hydroxide Nanostructures as Active Catalysts for Water Oxidation by a Laser–Chemical Route, *Nano Lett.*, 2015, **15**, 2498–2503.
88. L. J. Enman, M. L. Burke, A. S. Batchellor and S. W. Boettcher, Effects of Intentionally Incorporated Metal Cations on the Oxygen Evolution Electorcatalytic Activity of Nickel (Oxy)hydroxide in Alkaline Media, *ACS Catal.*, 2015, **6**, 2416–2423.
89. J. B. Gerken, J. Y. C. Chen, R. C. Masse, A. B. Powell and S. S. Stahl, Development of O_2-senstitive fuorescence-qunenching assay for the combinatorial discovery of electrocatalysts for water oxidation, *Angew. Chem. Inter. Ed.*, 2012, **51**, 6676–6680.
90. J. B. Gerken, S. E. Shanner, R. C. Masse, N. J. Porbusky and S. S. Stahl, A survey of diverse earth abundant oxygen evolution electrocatalysts showing enhanced activity from Ni-Fe oxides containing a third element, Energy Environ. *Sci.*, 2014, 7, 2376–2382.
91. J. A. Haber, Y. Cai, S. Jung, C. Xiang, S. Mitrovic, J. Jin, A. T. Bell and J. M. Gregoire, Discovering Ce-rich oxygen evolution catalysts from high throughput screening to water electrolysis, *Energy Environ. Sci.*, 2014, 7, 682–688.
92. J. A. Haber, C. C. Xiang, D. Gueverra, S. H. Jung, J. Jin and J. M. Gregoire, High-throughput mapping of the electrochemical properties of (Ni-Fe-Co-Ce)O_x oxygen-evolving catalysts, *ChemElectroChem*, 2014, **1**, 524–528.
93. J. A. Haber, E. Anzenburg, J. Yano, C. Kisielowski and J. M. Gregoire, Multiphase Nanostructure of a Quinary Metal Oxide Electrocatalyst

Reveals a New Direction for OER Electrocatalyst Design, *Adv. Energy Mater.*, 2015, **5**, 1402307.
94. X. Zhu, C. Tang, H.-F. Wang, B.-Q. Li, Q. Zhang, C. Li, C. Yang and F. Wei, Monolithic-structured ternary hydroxides as freestanding bifunctional electrocatalysts for overall water splitting, *J. Mater. Chem. A.*, 2016, **4**, 7245–7250.
95. A.-L. Wang, H. Xu and G.-R. Li, NiCoFe Layered Triple Hydroxides with Porous Structures as High-Performance Electrocatalysts for Overall Water Splitting, *ACS Energy Lett.*, 2016, **1**, 445–453.
96. Z. Lu, L. Qian, Y. Tian, Y. Li, X. Sun and X. Duan, Ternary NiFeMn layered double hydroxides as highly-efficient oxygen evolution catalysts, *Chem. Commun.*, 2016, **52**, 908–911.
97. B. M. Hunter, W. Hieringer, J. R. Winkler, H. B. Gray and A. M. Muller, Effect of interlayer anions on [NiFe]-LDH nonsheet water oxidation activity, *Energy Environ. Sci.*, 2016, **9**, 1734–1743.

CHAPTER 4

Surface Science, X-ray and Electron Spectroscopy Studies of Electrocatalysis

DANIEL FRIEBEL

SLAC National Accelerator Laboratory, Menlo Park, CA, USA
Email: dfriebel@slac.stanford.edu

4.1 Introduction

The Joint Center for Artificial Photosynthesis (JCAP) was designed to cover the breadth of research from basic science to prototyping and scale-up, and to benefit from the synergies between basic and applied research. This chapter provides a review of the experimental research on electronic structure effects in electrocatalysis carried out in the Heterogeneous Catalysis group.

All oxygen and fuel forming reactions in water splitting and carbon dioxide reduction have in common that they require electrodes that not only provide for electron transfer to or from the reactants, but that also provide catalytic activity. It is not sufficient to align photoelectrode band edges with the redox potentials of a multi-electron reaction. Without a catalyst, single-electron transfers would result in highly unstable intermediates, *e.g.*, OH, whose generation requires much higher electrochemical potentials than the thermodynamic potential for the final products. The role of the catalyst is to allow for the formation of more stable intermediates that are chemically bonded to its surface. It is the chemical bonding between catalyst and intermediates that partially compensates the energy required to dissociate

Energy and Environment Series No. 22
Integrated Solar Fuel Generators
Edited by Ian D. Sharp, Harry A. Atwater and Hans-Joachim Lewerenz

Published by the Royal Society of Chemistry, www.rsc.org

the starting molecules, H_2O or CO_2. The energy of each such catalyst-intermediate bond is the key criterion that needs to be optimized in order to accelerate the reaction rate. It must be noted here that optimization does not mean merely increasing, but instead actually fine-tuning bond strengths in between too weak and too strong interactions, since every catalytic process requires a sequence of both bond-breaking as well as bond-making steps. Too weak adsorption of intermediates fails to promote the bond-breaking steps at a sufficient rate, while too strong adsorption prevents the reaction from proceeding through bond-making steps, which compete energetically with the catalyst-intermediate bonds. A breakthrough towards understanding existing and predicting new catalyst materials was made with the d-band model, which describes the surface reactivity of transition metals and their alloys as a result of their valence band structures and their interaction with reactant molecular levels.[1,2] Knowledge of adsorption energies, *e.g.* thermodynamic parameters of stable intermediates, which in many cases can be measured experimentally, is sufficient for a prediction of trends in catalytic reaction rates, due to the Brønsted–Evans–Polanyi relation between activation energy and reaction energy.[3] For electrocatalytic reactions, the influence of applied electrochemical potential can be added by means of the "computational hydrogen electrode".[4–7]

Furthermore, the adsorption energies of various adsorbates correlate with each other in linear scaling relationships. This finding allows for drastic simplifications in computational screening for new catalyst materials, since a "volcano plot" can be drawn, using just one adsorbate as a descriptor for catalytic activity in reactions comprising multiple elementary steps.[8] The validity of such descriptor-based models for catalyst screening can be experimentally validated by combining activity measurements with surface science experiments, in particular X-ray spectroscopy, which provide information about the energy and structure of the d-band, and about adsorption energies and metal–adsorbate interactions.

The methodology outlined above has been applied with great success to the electrocatalytic reactions in PEM fuel cells, and good agreement has been achieved between theoretical predictions and experimental studies.[4,9–17] Such good agreement was owed, on the experimental side, to the development of electrochemical surface science as a discipline that uses well-defined model electrodes under narrowly defined conditions and monitors physical parameters, in particular electronic structure, that can be directly compared with theory. It is the synergy between computational screening using density functional theory, and carefully designed surface science experiments, which has brought forth new catalyst materials for the oxygen reduction reaction that minimize the use of platinum.

The key challenges to designing catalyst materials for a solar water splitting device are the same as in fuel cell catalyst design, and this motivates the use—and further improvement—of the same methodology that has already been applied with great success in fuel cell catalysis, in particular experimental electrochemical surface science and theoretical modeling of catalyst-intermediate interactions. Catalysts for photoelectrochemical fuel generation must meet the following requirements:

1. High catalytic activity, achieved by optimized chemical bonding to reaction intermediates for the oxygen evolution reaction (OER) at the photoanode, and for the hydrogen evolution reaction (HER) at the photocathode.
2. Stability against corrosive electrolytes. In order to avoid excessive internal resistance of the photo-electrolyzer cell, the choice of electrolyte is restricted to strong acids or bases. OER catalysts must be stable at strongly oxidizing potentials typically exceeding +1.5 V *vs.* RHE (reversible hydrogen electrode).
3. Minimized use of rare elements such as platinum-group metals. To this date, acid-stable OER catalysts that are free of platinum-group metals have not been discovered. In addition to the search for earth-abundant and acid-stable OER catalysts, which might not exist at all, it is important to follow the alternative approach that maximizes the catalytic efficiency of highly dilute Ir or Ru sites, similar to the successful efforts of reducing the Pt content in fuel cell catalysts.

Traditional surface science methods are best suited to study chemical bonding at surfaces. The use of single-crystal surfaces under ultra-high vacuum (UHV) conditions allows for rigid control of all material-specific parameters that control chemical bonding of adsorbates, and UHV based electron spectroscopy provides highly surface sensitive electronic structure probing. To some extent, even the influence of electrochemical potential can be simulated under UHV conditions with the addition of adsorbates that induce a work function change. However, clear limitations of UHV experiments arise since true solid/electrolyte interfaces can accommodate work function changes of ± 1.5 V or, in non-aqueous electrolytes, far beyond, which cannot be achieved by adsorption of, *e.g.*, oxygen within the UHV pressure range. Furthermore, catalytically active surfaces will convert catalytic intermediates rapidly; therefore, in order to access catalytic conditions at the surface, providing the desired work function change alone is not sufficient, but needs to be accompanied with high electrochemical current density. The following two questions cannot be answered with data from UHV experiments: 1. What is the nature of adsorbates under reaction conditions, *i.e.* at pressures ~9–12 orders of magnitude higher than UHV? 2. What is the surface structure of the catalyst under reaction conditions? These questions can be addressed if we employ surface characterization methods in the presence of condensed electrolyte at the surface (*in situ*). The most desired outcome is to probe such solid–electrolyte interfaces during reaction conditions with technologically relevant electrochemical current densities (*operando*). Section 4.3 will showcase how *in situ*, and in some cases even *operando* conditions can be achieved using ambient pressure X-ray photoemission spectroscopy (APXPS). The conflict between surface sensitivity and *operando* probing also becomes evident in that section, and a workaround using hard X-ray spectroscopy will be presented.

Additional challenges that are not present in fuel cell catalysis, as far as precious metal catalysts are concerned, arise from new classes of materials where the applicability of the traditional d-band model either does not exist or at least is unclear. This is likely the case with some HER catalysts such as MoS_x, and we certainly need to develop new ways to correlate between electronic structure and catalytic activity when we consider OER catalysts, which are all present in the form of metal oxides. Moreover, most of the earth-abundant catalyst materials are much less stable under both HER and OER conditions than most precious metals and their alloys and exhibit redox reactions not only at or near the surface, but in some cases throughout their bulk. The catalytic activity at the surface cannot be understood at all without detailed knowledge of the subsurface and bulk structure of the catalyst at relevant electrochemical potentials. Therefore, this chapter is not limited to the discussion of surface characterization but will also address *operando* studies of the bulk structure of catalyst layers.

4.2 Laboratory Based Methods for Surface Characterization

4.2.1 UHV-based Surface Science

The thorough understanding of the relationships between surface structure, electrocatalytic activity, and catalyst stability in fuel cells is owed to the efforts that have been made in the past to prepare clean catalyst surfaces with well-defined structure using single-crystals, and to characterize these model electrodes both before and after electrochemical reactions using surface analysis techniques. Crucial to the combination of electrochemical and surface science experiments was the design of electrochemistry-surface science apparatus (EC-SSA) that allow fast transfers of samples between preparation, electrochemical treatments, and surface characterization that minimizes surface contamination during the transfer steps. There can be no doubt that, in order to gain the same quality of understanding in the field of solar fuels generation, we need to utilize the same methods that have been successfully applied to fuel cell catalysts and expand their scope from studying precious metal surfaces to earth-abundant materials, which, in addition to transition metal alloys for HER and CO2RR, comprise two new classes of materials, metal sulfides, phosphides, carbides for the HER or CO2RR, and transition metal oxides for the OER. To serve this purpose, a new custom-designed EC-SSA was installed at JCAP's laboratory at the California Institute of Technology. This instrument combines techniques for sample preparation (reactive sputtering and physical vapor deposition), surface and adsorbate characterization in UHV (LEED, AES, XPS, UPS, HREELS, LEISS, TPD), and an electrochemistry chamber that includes the capability to perform differential electrochemical mass spectroscopy (DEMS).

In particular the combination of high-resolution XPS measurements with other characterization methods, including Raman spectroscopy and electron microscopy techniques has furthered the understanding of the syntheses of new earth-abundant and acid-stable HER catalysts. Among the most promising candidates here are Ni–Mo alloys,[18] CoP,[19] and $MoSe_2$,[20] but challenges remain. Ni–Mo alloys exhibit a very low overpotential (~50 mV) for the HER at a current density of 10 mA cm^{-2}, but are unstable in strongly acidic electrolytes.[21,22] The mechanism of the HER activity enhancement on Ni–Mo alloys has not been fully understood. Electrochemical double layer capacitance measurements indicated at least 100 times larger electrochemically active surface areas for Ni–Mo catalysts than for Pt;[22] therefore, it is likely that a significant portion if not all of the catalytic enhancement is an effect of increased surface area. Two possible mechanisms of an additional intrinsic enhancement were proposed: more favorable bonding of the H_{ads} intermediate at Ni sites due to the d-band shift in Ni–Mo alloys,[23–25] or a hydrogen "spillover" effect in which initial H adsorption and subsequent H_2 formation steps occur at different surface sites, and overall high reaction rates then result from the diffusion of H_{ads} from one surface metal site to the other.[26] To gain further insights into the HER reaction mechanism at Ni–Mo electrodes, and their degradation in strongly acidic electrolytes, better knowledge of the surface composition during HER, and of electrochemically induced changes such as metal dissolution, oxide formation, and segregation is required. Initial steps towards such examinations were taken with an electrodeposited Ni–Mo film.[18] Figure 4.1 shows XPS measurements in the Ni 2p and Mo 3d regions, in which clear differences of the surface composition before and after electrochemical testing can be seen. It is quite remarkable that the as-deposited catalyst precursor consists predominantly of highly oxidized Ni^{3+} and Mo^{6+} species with relatively minor contributions from metallic Ni and Mo, despite the reducing conditions that were chosen for the electrodeposition. After constant-current electrolysis at 10 mA cm^{-2} for 3h in H_2-saturated 1M H_2SO_4 solution, only Ni in the metallic state, together with Mo^0 and Mo^{4+}, was detected. The XPS data, which were measured with an Al Kα X-ray source, provide the composition of the near-surface region with electron escape depths on the order of ~1–2 nm. Ongoing efforts are aimed towards obtaining complementary low-energy ion scattering spectroscopy (LEISS) measurements, which provide the composition of the outermost atomic layer.[18]

Another example of an earth-abundant HER catalyst that undergoes drastic structural and chemical changes during its operation was demonstrated with a molybdenum selenide precipitate that was made by mixing ammonium heptamolybdate and sodium selenide solutions.[20] An isopropanol suspension of the precipitate was then coated onto a glassy carbon disk for electrochemical testing and characterization with SEM, XPS and Raman spectroscopy. High-resolution XPS measurements (Figure 4.2) reveal

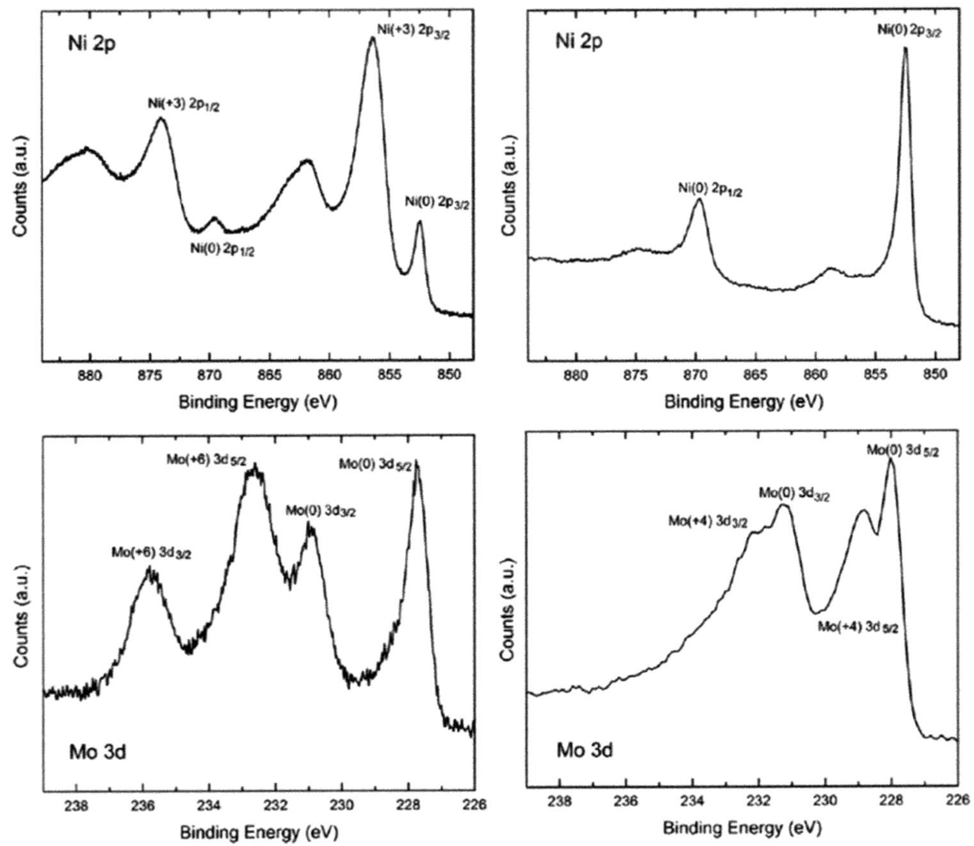

Figure 4.1 XPS measurements of electrodeposited Ni–Mo catalyst on glassy carbon showing the Ni 2p and Mo 3d regions before (left) and after (right) catalytic HER testing.[18]
Reprinted from *Surf. Sci.*, **631**, M. P. Soriaga, J. H. Baricuatro, K. D. Cummins, Y.-G. Kim, F. H. Saadi, G. Sun, C. C. L. McCrory, J. R. McKone, J. M. Velazquez, I. M. Ferrer, A. I. Carim, A. Javier, B. Chmielowiec, D. C. Lacy, J. M. Gregoire, J. Sanabria-Chinchilla, X. Amashukeli, W. J. Royea, B. S. Brunschwig, J. C. Hemminger, N. S. Lewis and J. L. Stickney, Electrochemical surface science twenty years later: Expeditions into the electrocatalysis of reactions at the core of artificial photosynthesis, 285–294, Copyright 2015, with permission from Elsevier.

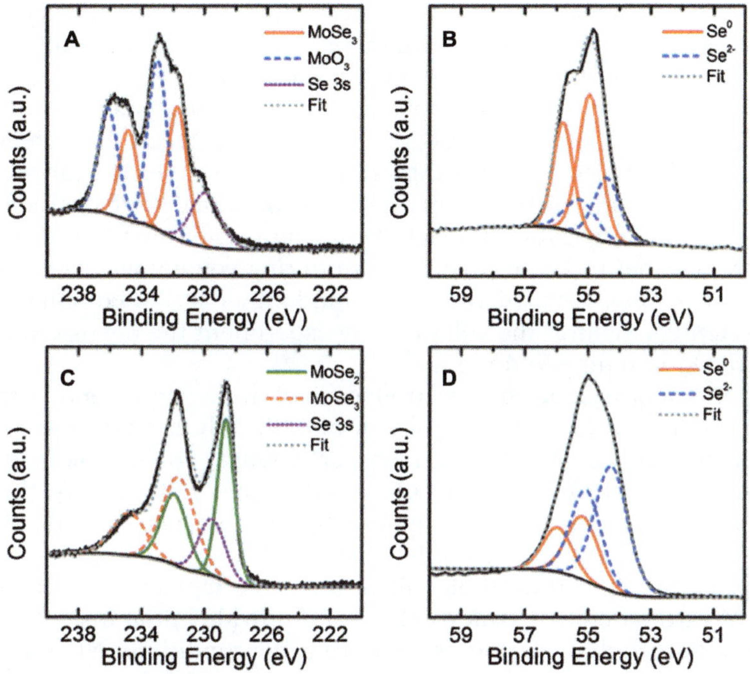

Figure 4.2 XPS analysis of molybdenum selenide films. The Mo 3d and Se 3d regions are shown for the as-deposited film (A, B) and after 10 voltammetric cycles (C, D).
Reprinted with permission from F. H. Saadi, A. I. Carim, J. M. Velazquez, J. H. Baricuatro, C. C. L. McCrory, M. P. Soriaga and N. S. Lewis, *ACS Catal.*, 2014, **4**, 2866–2873, Copyright 2014 American Chemical Society.[20]

the composition of the as-prepared catalyst precursor film with a mixture of two different types of Mo^{6+} species, Se^{2-} and Se^{0}, indicating the formation of $MoSe_3$, MoO_3, and elemental Se. The latter two compounds could result from reactions with air during or after the precipitation of $MoSe_3$.

XPS measurements after the electrode was subjected to 10 potential cycles between 0 and −0.4 V *vs.* RHE showed that all of the MoO_3 and substantial parts of the $MoSe_3$ and Se^0 components were removed during the electrochemical test, and the resulting film contained mostly $MoSe_2$. Since the HER overpotential decreased significantly during the potential cycling experiment, it was proposed that $MoSe_2$ represents the catalytically active species for the HER. In addition, significant porosity developed during the conversion, however, the increase of surface area alone was found to be insufficient to account for the increased HER current density.[20]

While some uncertainty remains about the long-term stability of earth-abundant HER catalysts over the expected lifetime of an industrial-scale photoelectrolyzer with a proton-conducting membrane, the progress from Ni–Mo to the much more stable $MoSe_2$[20] and MoS_2[27–29] compounds is notable, although a tradeoff must be made with the HER overpotential, which is approximately 200 mV higher with the Mo chalcogenides than with Ni–Mo. In this regard, recent evaluations of the performance of transition metal phosphides under acidic HER conditions are very promising, since here the results indicate that both the required acid stability, and HER overpotentials of less than 0.1 V at 10 mA cm^{-2} can be achieved together.[20,30,31] An electrochemical procedure was used for the fabrication of CoP films; the synthesis route consisted of cathodic electrodeposition from a solution containing $CoCl_2$ and NaH_2PO_2. Again, the formation of the active catalyst appeared to be occurring *in situ* during voltammetric testing. SEM images of the as-deposited film show randomly distributed spherical particles (Figure 4.3); after the voltammetry experiment the deposit appears to have eroded into μm-sized mesas.[19]

The XPS analysis of the electrodeposited film before and after HER (Figure 4.4) indicates CoP to be present initially only as a minor component in a mixture with Co_3O_4. Furthermore, an intense, broad peak in the P 2p region was assigned to orthophosphate. After cyclic voltammetry, the Co_3O_4 components in the Co 2p spectrum have disappeared, and the ratio of CoP to phosphate in the P 2p spectrum has increased.

The aforementioned examples illustrate important progress in the synthesis of earth-abundant HER catalysts and provide initial insights into the chemical state of these materials under catalytic conditions. Not understood so far, however, are the mechanisms that enhance the corrosion stability of CoP and NiP over the pure Co and Ni metals, and that reduce the HER overpotential. It is currently not known how the H_{ads} intermediate binds to

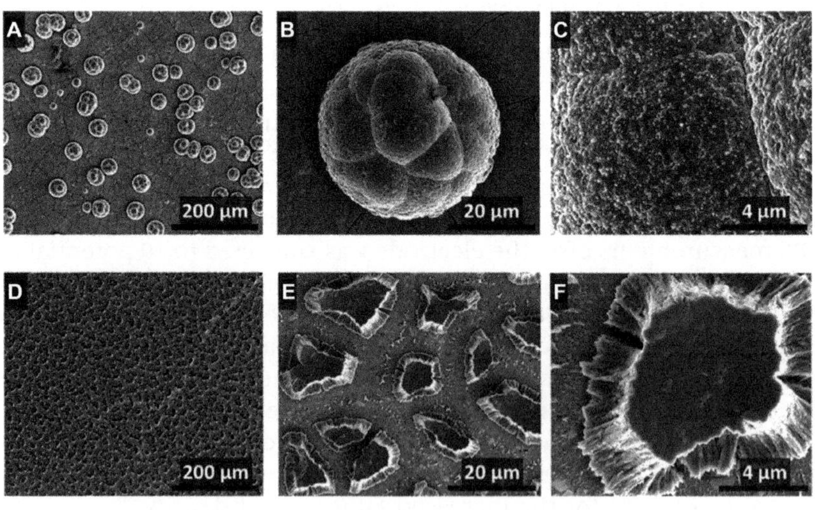

Figure 4.3 Scanning electron micrographs showing drastic morphology changes in electrodeposited CoP catalysts induced by electrochemical treatment. Films before (A–C) and after (D–F) voltammetry.
Reprinted with permission from F. H. Saadi, A. I. Carim, E. Verlage, J. C. Hemminger, N. S. Lewis, and M. P. Soriaga, *J. Phys. Chem. C*, 2014, **118**, 29294–29300, Copyright 2014 American Chemical Society.[19]

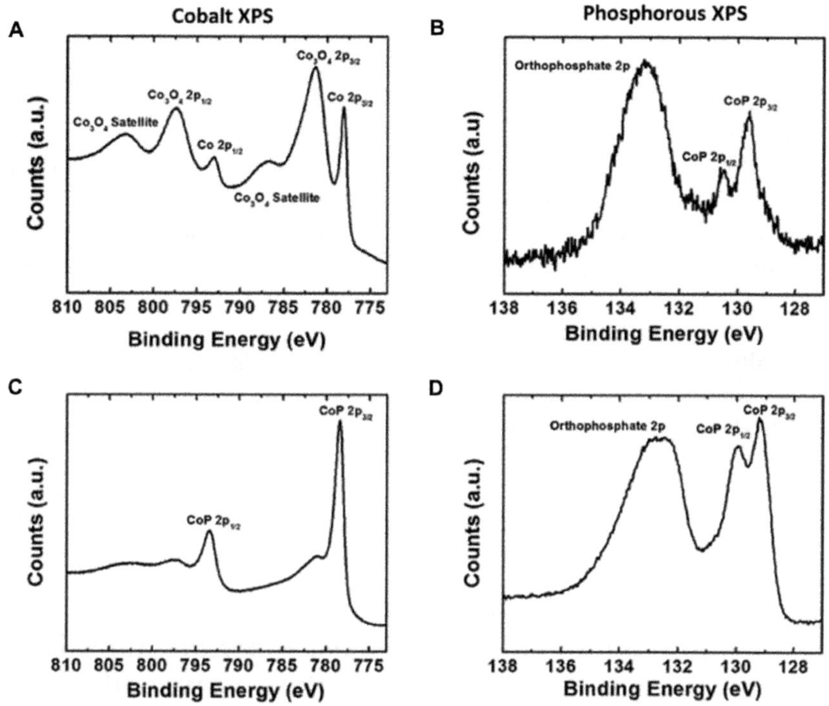

Figure 4.4 XPS measurements showing surface compositional changes in CoP catalysts induced by voltammetric testing. Co 2p and P 2p spectra before (A,B) and after voltammetry (C,D).
Reprinted with permission from F. H. Saadi, A. I. Carim, E. Verlage, J. C. Hemminger, N. S. Lewis, and M. P. Soriaga, *J. Phys. Chem. C*, 2014, **118**, 29294–29300, Copyright 2014 American Chemical Society.[19]

active Co or P surface sites, and how mixing with P influences the electronic structure of Co and Ni. These questions could be answered in the near future with a combination of surface science experiments on transition metal phosphide single-crystals, or epitaxial films, which should include temperature-programmed desorption (TPD), HREELS, and valence band photoemission spectroscopy measurements. Furthermore, synchrotron-based *in situ* X-ray spectroscopy could give important insights into corrosion mechanisms and their inhibition.

4.3 Synchrotron-based *in situ* and *operando* Spectroscopy

The availability of third generation synchrotron light sources provides a suite of powerful material characterization techniques using tunable X-rays, in particular X-ray and electron spectroscopy. A particular advantage of X-ray spectroscopy is its ability to probe occupied and unoccupied states in such a way that occupied and unoccupied densities of states can be obtained separately, and core-level excitations of different chemical elements are accessed selectively. X-ray absorption spectra provide the absorber-atom-projected unoccupied partial density of states (pDOS); in a complimentary way, the occupied pDOS can be obtained from valence-to-core X-ray emission spectra.[32–35]

Depending on the X-ray/electron energies used, realistic catalytic conditions can be approached to various degrees, ranging from "*in situ*" conditions with at least some of the reactants and physical conditions (temperature, pressure, mass transport) present towards "*operando*" experiments, in which all of the parameters of a technologically relevant catalytic process are realized.

Generally, to what extent *in situ* or even *operando* X-ray probing can be achieved depends on (i) the energy and material dependent attenuation length of X-rays or photoelectrons in the reaction environment, and the ability of experimenters to design (ii) reaction environments where X-ray or electron pathways through strongly attenuating materials are minimized, and (iii) the availability of X-ray sources with high photon flux and spectrometers with enhanced signal/noise and signal/background ratios.

4.3.1 Photon-in/photon-out Methods: Experimental Setup for *operando* Spectroscopy, X-ray Absorption, and High Resolution X-ray Spectroscopy

Hard X-ray photon-in/photon-out techniques have the advantage that liquid electrolyte layers of 1 mm thickness can be penetrated with only little signal loss. The attenuation length in the electrolyte inevitably scales with the probing depth in solid electrodes, which in this case is in the μm range. Therefore, hard X-ray absorption spectroscopy is inherently bulk sensitive and work within the scope of JCAP was carried out in order to obtain bulk structure information; however, procedures that can provide enhanced surface sensitivity will be briefly discussed at the end of this section.

4.3.1.1 *Experimental Setup for* operando *Photon-in/photon-out Spectroscopy*

In order to design an *operando* (photo-)electrochemical cell that can be employed successfully, careful considerations must be made of materials that are compatible with exposure to X-rays and corrosive electrolytes and do not cause contamination, and of expected XAS signal intensities and their attenuation in electrolyte and other components of the cell. In the tender to hard X-ray energy range, fluorescence detection is the most appropriate mode for XAS measurements with thin catalyst samples. The fluorescence yield can be further enhanced if a catalyst layer is supported on a sufficiently planar substrate such as a single crystal[46–48] or a polished glassy carbon disk.[49] The X-ray incident angle can be adjusted near the critical angle of total external reflection in order to generate an X-ray standing wave. Using this technique, which is known as grazing incidence X-ray absorption spectroscopy (GI-XAS), the fluorescence yield can be enhanced up to fourfold.[50]

In considering the most efficient geometry for an *operando* (photo)electrochemical XAS cell, the signal enhancements gained from GI-XAS must be weighed against signal losses due to required relatively long paths of X-rays through the electrolyte. At lower X-ray energies, such losses can outweigh the advantages of GI-XAS and the better alternative becomes a cell whose working electrode coated onto an X-ray transparent window; in this setup, incoming and fluorescent X-rays enter and exit the back of the working electrode without needing to pass through the electrolyte at all. At JCAP, electrochemical cell designs of both types were successfully employed. Special consideration was given to the need for high current densities in GI-XAS measurements; therefore, the traditional design of GI-XAS cells, in which a thin electrolyte layer (~20–30 μm) is enclosed between the working electrode and a thin polymer film in order to minimize the X-ray path length through electrolyte, was replaced with a hanging-meniscus cell (Figure 4.5), adopting a design concept originally developed for surface X-ray diffraction measurements.[51,52] At high X-ray energies (>10 keV), the resulting additional loss of fluorescence intensity is acceptable, owing to the high photon fluxes provided at third generation synchrotron facilities. At X-ray energies within the range of 3d transition metal K absorption edges, the X-ray attenuation length in water becomes shorter than 1 mm, whereas typical hard X-ray sources (*e.g.*, SSRL BL11-2) provide a beam profile that results, under grazing

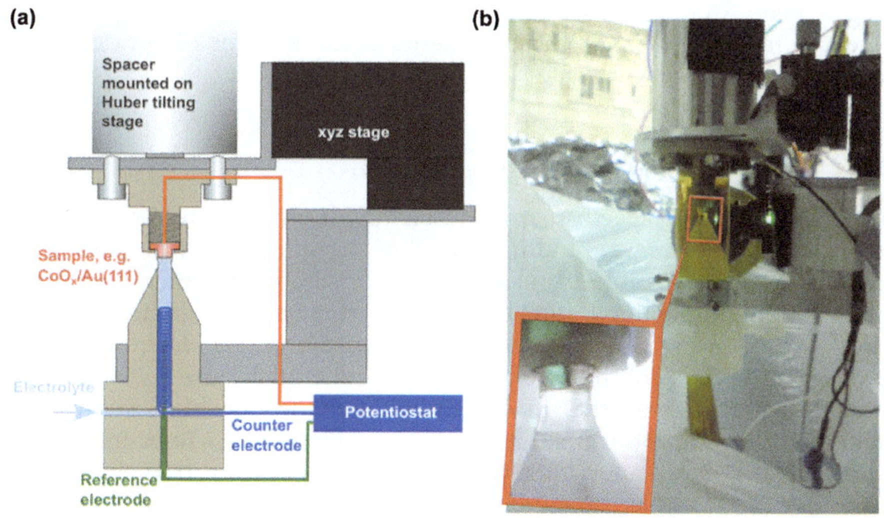

Figure 4.5 (a) Schematic drawing of the hanging meniscus cell for *operando* GI-XAS measurements. (b) Photograph of cell taken during operation at SSRL BL 6-2.

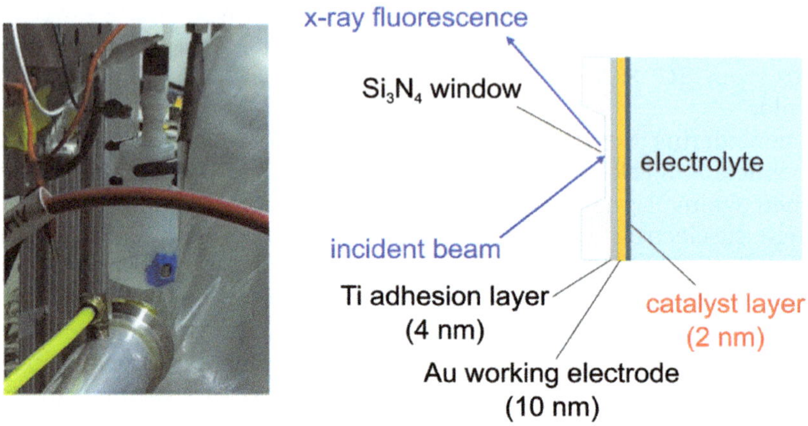

Figure 4.6 Photograph of X-ray back-illumination cell for *operando* XAS. An opening is cut into the side wall of a polypropylene bottle and sealed with a 100 nm thick Si_3N_4 window, which on the electrolyte-facing side is coated with the thin-film working electrode as shown in the schematic drawing. Counter (Pt wire) and reference electrodes (Hg/HgO) are inserted through openings in the bottle cap.

incidence, in a projection on the sample of at least 5 mm length. While improvements can be made, such as shifting the sample position in order to minimize the path of the fluorescent X-rays through the electrolyte meniscus, it is clear that the back-illumination cell (Figure 4.6) becomes more efficient than GI-XAS at lower X-ray energies. In particular, the use of sufficiently thin silicon nitride windows can extend the accessible energy range for *operando* XAS all the way into the soft X-ray regime with the exciting potential for *operando* XAS measurements of the K-edges of light elements such as C and O.[53]

4.3.1.2 X-ray Absorption Spectroscopy

In X-ray absorption spectroscopy, the energy of an incident X-ray beam is swept across a core level binding energy of the sample and the absorption coefficient of the sample is measured either directly *via* measurement of incident and transmitted intensities using the Beer–Lambert law, or indirectly by counting either fluorescence photons or Auger electrons that are emitted in the decay of the core-hole. While quite thick samples ($\sim$1 μm) are required for measurements in transmission mode, the high fluorescence yield of $L\alpha$ lines for 5d transition metal $L_{2,3}$ edges and $K\alpha$ lines for 3d transition metal K edges allows for XAS measurements with extremely dilute or extremely thin samples. For example, successful *operando* EXAFS measurements were made at SSRL BL4-1 with Fe–Ni catalysts made from 2 nm thin sputtered metal films with 25% Fe, and in HERFD XAS measurements even $\sim 10\times$ thinner Fe loadings gave sufficient signal intensity at SSRL BL6-2.[36]

X-ray absorption spectroscopy can be used to probe the nature of chemical bonding of the absorbing atom, as well as the local short-range structure. As depicted schematically in Figure 4.7, the electron that is removed from a core level in the absorption process is excited into excited states that represent a local projection of the unoccupied density of states upon the absorber atom in the presence of a core-hole.[32–35] At relatively low photon energy, the excited electron can remain localized at the absorber atom. At higher energies the resulting photoelectron, while no longer being localized at the absorber atom, is still subject to multiple-scattering as it interacts with surrounding atoms. Therefore, it is convenient for the theoretical understanding of spectral features to divide the XAS energy range into two regions that are most commonly referred to as X-ray absorption near edge structure (XANES) or near-edge X-ray absorption fine structure (NEXAFS) and extended X-ray absorption fine structure (EXAFS). It is important to keep in mind that both XANES and EXAFS region originate from the same physical process, which can be described either in terms of the partial unoccupied density of states (DOS) in the presence of a core-hole, or in terms of

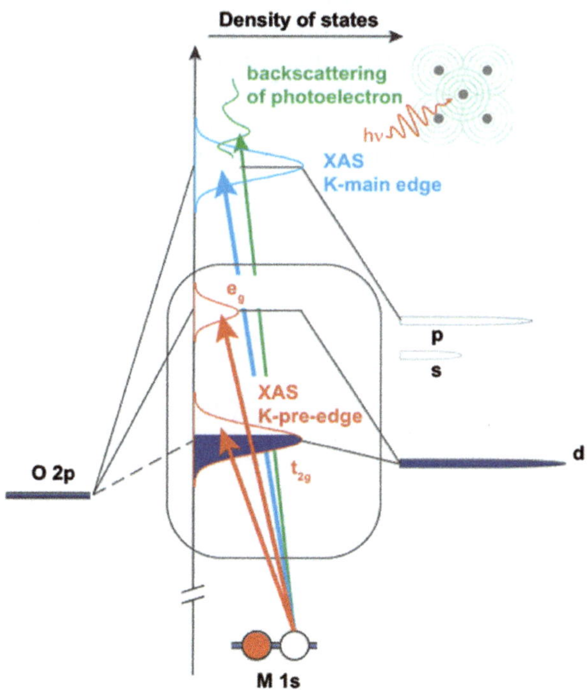

Figure 4.7 Schematic diagram of electronic excitations in XAS at 3d transition metal K-edges. Within the XANES energy range, 1s→3d transitions in the K pre-edge and 1s→4p transitions near the main edge provide local symmetry and oxidation state information. Backscattering of the photoelectron in the EXAFS energy range provides local structure information (coordination numbers and distances of neighbor atoms within ~5 Å).

multiple-scattering of a photoelectron (MS).[37–45] It is the XANES region where spectral features predominantly provide information on chemical bonding such as local symmetry, oxidation states and bond order. This is because in the DOS picture, the excited states have strong contributions from atomic orbitals of the absorber atom and are very sensitive to hybridization. In the MS picture of XANES, scattering paths that involve more than one neighbor atom contribute strongly and are highly sensitive to local symmetry.

While a qualitative understanding of XANES may seem quite straightforward, quantitative *ab-initio* modeling of spectra remains computationally demanding and the agreement between calculated and experimental spectra can be limited. The difficulty of accurate XANES calculations becomes apparent in the multiple-scattering picture, where the number of scattering paths becomes very large. At the same time, the relatively mean free path of photoelectrons at low energy (~0–20 eV) requires large cluster sizes.[39] Therefore, a more empirical approach to the data analysis is often needed, such as fitting linear combinations of spectra obtained with well-known standard compounds. In the EXAFS region, the number of strongly contributing scattering paths is drastically reduced to mostly single-scattering contributions from neighbor atoms within ~5 Å of the absorber atom, and strong multiple-scattering contributions occur only for bond angles near 180°. The development of reliable *ab initio* calculations for use in EXAFS analysis[43] is mature, and calculated single-scattering paths can be used to fit interatomic distances, coordination numbers and interatomic displacement parameters quantitatively with high accuracy of ±0.01 Å and ±1 for interatomic distances and coordination numbers, respectively. Due to the element-specific energy dependence of phase shifts and scattering amplitudes, atomic numbers of neighbors can be determined within $Z \pm 2$. Furthermore, multiple-scattering paths can be analyzed to identify collinear arrangements of atoms or to estimate intramolecular bond angles near 180°. *Operando* EXAFS measurements have been essential to the structure determination of Ni–Fe catalysts during OER conditions;[36] the reader is referred to Chapter 3 by Bell in this book for detailed results.

4.3.1.3 High Resolution X-ray Spectroscopy

The performance of electrochemical *operando* XAS is affected by two challenges that arise in conventional fluorescence detection mode. First, the energy resolution of state-of-the-art solid-state detectors is typically on the order of ~200 eV. The X-ray spectrum that is emitted from a highly dilute electrocatalyst in an electrochemical cell with several other X-ray illuminated components (X-ray windows, electrolyte *etc.*) is dominated by strong elastic and Compton scattering peaks, which can be several orders of magnitude more intense than the desired X-ray fluorescence. During a XAS scan, the energy position of the elastic and Compton peaks shifts and their tails, at 200 eV instrumental broadening, inevitably overlap with the fluorescence signal, resulting in XAS scans with extremely high background. These contributions can be so strong that the use of Soller slits with filters containing a Z-1-element becomes ineffective. Second, the amount of information that can be extracted from XANES at high X-ray energies is limited due to the strong lifetime broadening of deep core holes. The effect of lifetime broadening is small in the soft X-ray region, *e.g.* ~0.2 eV for O 1s, and becomes much more severe for high-energy excitations in heavier elements such as 3d transition metal 1s (>1.5 eV) and 5d transition metal 2p (>5 eV). A key objective in XAS measurements of transition metal-based catalysts is to probe the unoccupied d states, which are particularly affected by lifetime broadening. In 3d transition metal K-edge XAS, the low-energy tail of the strong $1s \rightarrow 4p$ main absorption edge overlaps with the weak $1s \rightarrow 3d$ pre-edge peaks. While the ligand field induced splitting into t_{2g} and e_g features can usually still be resolved, but not the additional multiplet splitting due to interactions within partially filled d-states.

The development of large bent-crystal X-ray analyzers with high energy resolution, together with the construction of intense third generation synchrotron X-ray sources, opened the field for advanced X-ray spectroscopy techniques that dramatically reduce both of the two aforementioned challenges in hard X-ray spectroscopy, *i.e.* the background/signal ratio with dilute samples in *operando* spectroscopy and the effect of core-hole lifetime broadening. Here, the advantages of RIXS and HERFD XAS will be briefly introduced; details of various hard X-ray techniques that are enabled by bent-crystal analyzers can be found elsewhere.[54–57]

In resonant inelastic X-ray scattering (RIXS), the XAS experiment is expanded by adding a second energy axis of energy transfer, *i.e.* the difference between the XAS excitation energy and emission energy. In the resulting RIXS plane, core-hole excited intermediate states generated by the X-ray absorption process are separated not only by their excitation energies but also by the energies of their decay channels that result in different final states, and the lifetime broadenings of intermediate and final states appear on the excitation energy and energy transfer axes, respectively. As an example, the RIXS plane of CoOOH is shown in Figure 4.8.

In high energy resolution fluorescence detection X-ray absorption spectroscopy (HERFD XAS), a narrow bandpass around an X-ray fluorescence peak is recorded. This is accomplished using the same setup as in RIXS, but scanning only the monochromator, while the analyzer remains at a fixed energy. The differences between conventional XAS and HERFD XAS can be understood when we consider again the RIXS plane in Figure 4.8b. In a conventional XAS measurement, one would move through the RIXS plane along the diagonal where the emission energy is constant (yellow dashed line in Figure 4.8b), and all intensities within the energy resolution of the detector (±100 eV perpendicular to the yellow dashed line) are integrated to give the XAS signal. The main-edge features (upper right portion of Figure 4.8b) are much broader in emission energy than the pre-edge features and give rise to much higher integrated intensities than the pre-edge

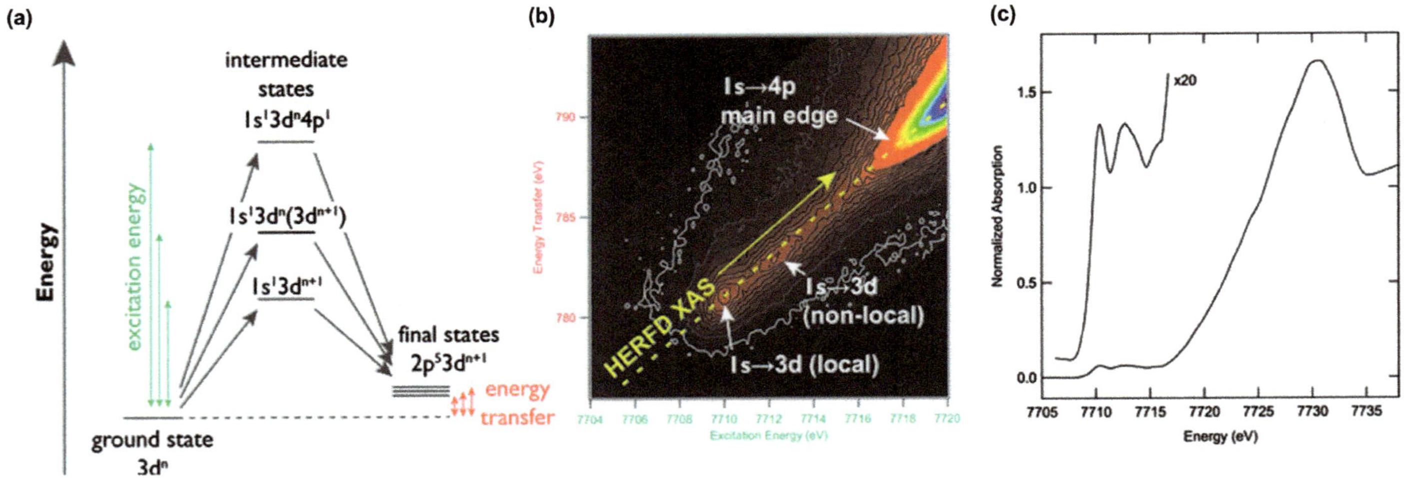

Figure 4.8 (a) Energy diagram for the RIXS process. (b) RIXS plane of CoOOH showing pre-edge and main-edge regions. A narrow slice of the RIXS plane along its diagonal is integrated to obtain HERFD XAS. (c) HERFD XAS corresponding to the full RIXS plane in (b).

features. By contrast, the pre-edge peaks are enhanced dramatically when only a narrow band-pass of emission energies is integrated. The resulting HERFD XAS scan is shown in Figure 4.8c. The RIXS technique is well suited for separating the pre-edge peaks from the main absorption edge in XAS. Furthermore, 2p–3d multiplet interactions in the final state result in pre-edge resonances that appear at off-diagonal energy positions.

If the condition is met that none of the resonances in a RIXS measurement is affected by final state interactions, HERFD XAS scans can be considered as X-ray absorption spectra with reduced effective lifetime broadening. The contribution of lifetime broadening to the spectral resolution in HERFD XAS is reduced according to

$$\frac{1}{\Gamma^2} = \frac{1}{\Gamma_{\text{int}}^2} + \frac{1}{\Gamma_{\text{fin}}^2}$$

where instead of lifetime of the intermediate state, which limits the resolution in conventional XAS, the predominant term is the much narrower lifetime broadening of the final state. The overall resolution is further limited by the instrumental resolution of the incident energy monochromator and the emission spectrometer.

Operando RIXS and HERFD XAS measurements were used in order to characterize a Co oxyhydroxide electrocatalyst with high OER activity.[48] The catalyst was made by anodic electrodeposition from a Co^{2+}-containing precursor solution on a Au(111) electrode, and the chemical nature of the active catalyst during OER was not precisely known. Furthermore, an enhancement of OER activity with decreasing layer thickness had been reported, and *in situ* surface enhanced Raman spectroscopy (SERS) measurements indicated the presence of Co_3O_4 in thicker catalyst layers, which was not detected for very thin films.[58] Since the surface enhancement mechanism in SERS can represent minority species disproportionately, it was not clear whether the reported presence of Co_3O_4 could affect OER activity and if so, how. In RIXS and HERFD XAS, the spectra can be described quantitatively by linear combinations of the spectra of each unique Co-containing component. Furthermore, subtle changes of oxidation states in a non-stoichiometric catalyst can be detected by monitoring the Co 1s→3d transitions in the XAS K pre-edge region.

RIXS measurements of well-defined standard compounds CoO, Co_3O_4 and CoOOH show very distinct pre-edge peaks (Figure 4.9). In particular, the presence of any Co^{2+} ions can be clearly detected since it gives rise to off-diagonal resonances. In RIXS measurements of the electrodeposited Co electrocatalyst, the presence of Co^{2+} was ruled out. Instead, both the pre-edge features as well as the XAS main edge strongly resemble CoOOH. The 1s→3d pre-edge in CoOOH shows two features, a peak at the absorption energy of 7710 eV, and a much broader resonance along the RIXS diagonal between 7711 eV and 7714 eV. A quite similar pattern was reported for the structurally related $LiCoO_2$, and the two resonances can be interpreted in

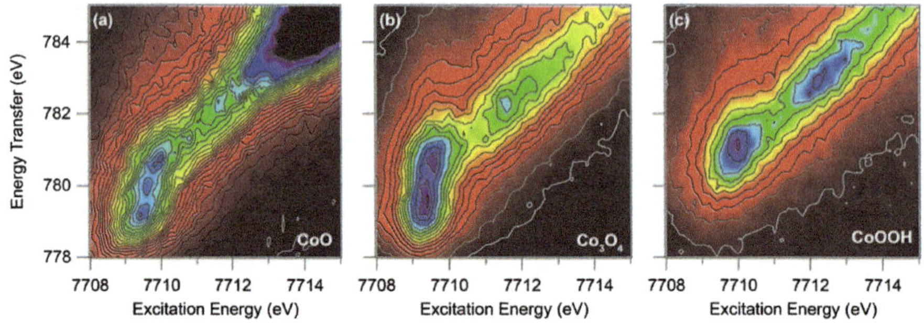

Figure 4.9 RIXS planes showing the Co K pre-edge resonances in reference compounds (a) CoO, (b) Co_3O_4, and (c) CoOOH. Intensities are shown with a color scale ranging from deep red (low) to deep purple (high).[48] Reproduced from ref. 48 with permission from the PCCP Owner Societies.

such a way that the sharp resonance at lower energy corresponds to a transition into 3d states that are strongly localized at the Co absorber atom, while the broader resonance arises from non-local transitions into hybridized states that have 3d contributions from neighboring Co atoms.[59–61] In the CoOOH powder reference, the non-local transition is more pronounced than in the electrodeposited catalyst. We anticipate that the electrodeposited CoOOH film on Au(111) has a tendency to grow epitaxially, due to the small (1%) mismatch between Au(111) and CoOOH(0001) in-plane lattice constants, and that the difference in intensity of the non-local dipole transition, which is strongly polarization dependent, is due to the more anisotropic character of the electrodeposited film.

The absence of off-diagonal features in RIXS was confirmed in several measurements at potentials throughout the stability window of the 0.1 M NaOH working electrolyte and also during OER, and a series of HERFD XAS scans was recorded within the double layer region and up to +1.77 V *vs.* RHE, *i.e.* at OER overpotentials up to 0.54 V. The HERFD spectra are shown in Figure 4.10. Spectra at high potentials continue to agree qualitatively with that of CoOOH powder, but we detected subtle changes that indicate a non-stoichiometric composition $CoOOH_{1\text{-}x}$ of the catalyst, which contains an increased fraction of Co^{4+} under OER conditions. An increased average Co oxidation state can be inferred from the subtle shift of the main absorption edge to higher energies. In addition, the pre-edge region was extracted by subtracting out the main-edge and was analyzed using three pseudo-Voigt profiles (Figure 4.11). The first two components sufficiently describe the local and non-local $1s \rightarrow 3d$ transitions in CoOOH; the third component was introduced in order to account for an intensity increase at high potentials from 0.97 to 1.77 V, and was proposed to represent a transition occurring at Co^{4+} sites. This assignment was supported not only by qualitative consideration of energy shifts induced by oxidation state changes, but also with theoretical XAS calculations using DFT + U.

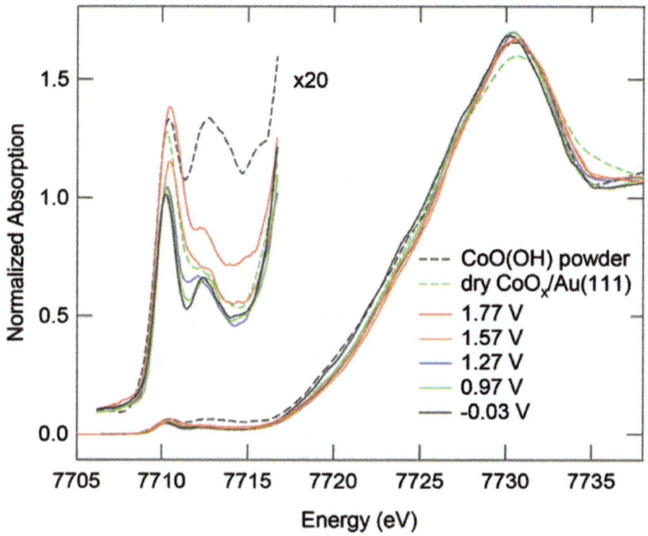

Figure 4.10 *Operando* HERFD XAS measurements of a CoO_x/Au(111) catalyst at potentials ranging from −0.09 V to +1.77 V *vs.* RHE. For comparison, the measurements of the dry, as-prepared Co/Au(111) sample and of the CoOOH powder reference are also shown.[48]
Reproduced from ref. 48 with permission from the PCCP Owner Societies.

The detection of metal sites with increased oxidation states in working OER catalysts has important implications in catalyst design. The variability of surface oxidation states needs to be taken into account, since it will significantly alter the energetics of adsorbed reaction intermediates. Related theoretical work in JCAP on the OER activity of Co oxides indicated that the presence of Co^{4+} surface sites decreases OER activity, and that tuning the stability of metal oxidation states can significantly improve catalytic performance. Synthetic efforts aimed at controlling oxidation states cannot be decoupled from potential-induced redox transitions that can occur during OER, and HERFD XAS and RIXS are essential in order to monitor these changes under realistic operating conditions.

4.3.1.4 *Feasibility of High-energy XAS as* operando *Surface Analysis Tool*

A further advantage of GI-XAS, in addition to enhancing the fluorescence yield from very dilute absorbers, is the reduction of the probing depth to ~2 nm at perfectly flat surfaces. It is therefore worthwhile considering to what extent the GI-XAS technique, and its combination with the HERFD detection mode, could replace XPS as a means of detecting adsorbates under *operando* conditions. Clearly, it is much less difficult to establish *operando* conditions with the hard X-ray photon-in/photon-out XAS measurement

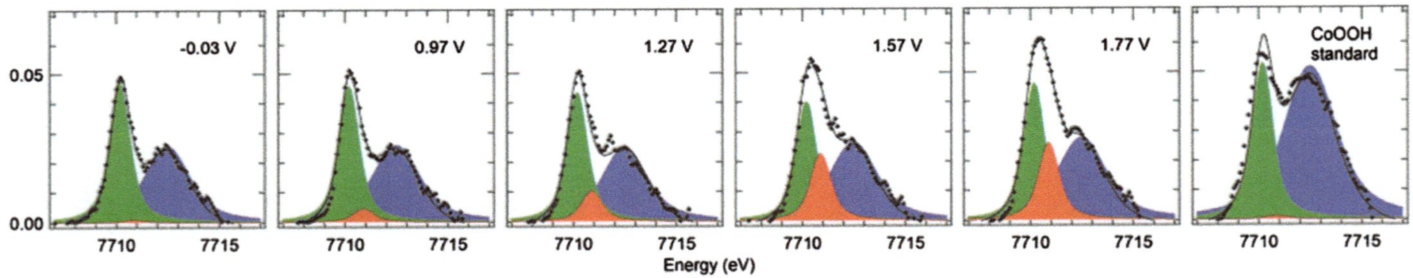

Figure 4.11 HERFD XAS pre-edge region of CoO_x/Au(111) catalyst at potentials ranging from −0.09 V to +1.77 V *vs.* RHE. The main absorption edge was subtracted in order to isolate the pre-edge peaks. Black markers and solid lines represent measured data and fitted curves, respectively. Colored areas show the peak components used in the fitting procedure.[48] Reproduced from ref. 48 with permission from the PCCP Owner Societies.

than with XPS. On the other hand, XPS can probe direct excitations from adsorbate core levels, while hard X-ray XAS only indirectly probes for the presence of adsorbates through the adsorbate-induced changes of the pDOS at a metal atom, which can be very subtle. For a GI-XAS experiment aimed at adsorbate detection, electrocatalysts must be prepared in such a way that at least ~50% of the X-ray absorbing metal atoms are surface atoms. Adsorbate-induced spectral changes are subtle, and they are limited to nearest-neighbors, therefore, relying on the probing-depth effect in GI-XAS alone is not sufficient. Furthermore, the reduction of sample thickness at which saturation effects become significant in fluorescence-detected XAS follows the same laws as the reduction of probing depth. Sufficiently thin samples can be made by growing ultrathin layers (1–2 ML) of the catalyst material on a suitable support, or by supporting sufficiently small nanoparticles (~2–3 nm) with a narrow size distribution. This approach was applied successfully in XAS studies of fuel cell catalysts and for the detection of adsorbed H and O/OH at Pt surfaces, and might be suitable for the detection of C-containing adsorbates during CO_2RR, or for probing the oxidation state of surface metal sites if sufficiently thin model catalyst layers can be prepared.[46,47,49]

4.3.2 Ambient Pressure XPS

A clear advantage of XPS over high-energy XAS for the detection of adsorbed species is that XPS can directly excite adsorbate core levels (*e.g.* O 1s), and that reference measurements of precisely defined model adsorbates in UHV conditions can be made. The development of ambient pressure XPS (APXPS) has revolutionized the capabilities of electron spectroscopy where measurements at atmospheric pressure have become possible with spectroscopic techniques that were traditionally restricted to UHV. Conventional XPS requires UHV conditions because photoelectrons have very short mean free paths in gases, and high-voltage components of electron energy analyzers and detectors need to be protected against discharges. APXPS measurements are made possible by using a differentially pumped electron lens system with very small aperture that is brought into close proximity to the sample. The size of the aperture and its distance from the sample are optimized to minimize the flow of gases into the pumping system, to minimize the local pressure drop at the sample induced by pumping through the aperture, to minimize the distance that photoelectrons must travel through the gas atmosphere and maximize the solid angle for photoelectrons that can enter the analyzer.

Figure 4.12 shows the APXPS system that was developed at SSRL beamline 13-2 for heterogeneous catalysis and electrochemistry studies.[62] Major improvements over earlier designs include the use of a narrowly focused beam under grazing incidence, in order to further reduce the size of the aperture to 50 μm, and a short and adjustable distance between the X-ray window and the sample to minimize attenuation of the incident beam in the gas

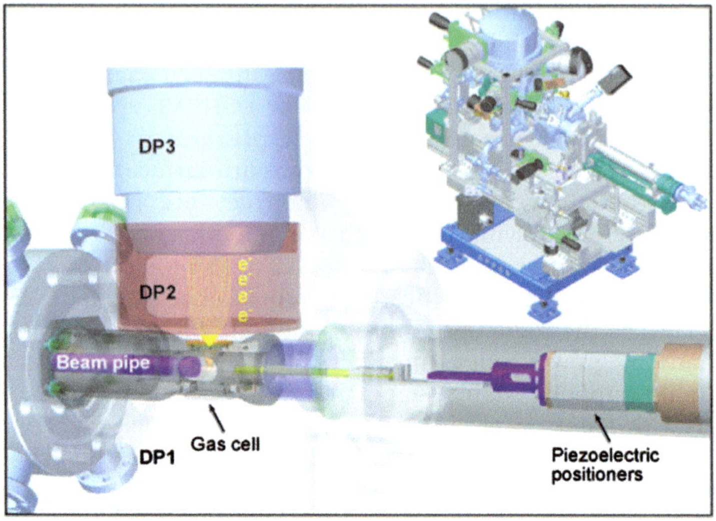

Figure 4.12 The APXPS instrument at SSRL BL 13-2. The enlarged portion shows the gas cell and the arrangement of the first three differential pumping sections DP1, DP2 and DP3. X-rays from the beam pipe enter the gas cell through a Si_3N_4 window. The sample rod forms a gas-tight seal with the high pressure cell upon insertion; subsequently the sample position near the analyzer cone is optimized using piezoelectric positioners.[62] Reprinted from *Catal. Today*, **205**, S. Kaya, H. Ogasawara, L.-Å. Näslund, J-O. Forsell, H.S. Casalongue, D. J. Miller and A. Nilsson, Ambient-pressure photoelectron spectroscopy for heterogeneous catalysis and electrochemistry, 101–105, Copyright 2012, with permission from Elsevier.

atmosphere. This setup was specified for routinely conducting XPS measurements at pressures near the saturation pressures of liquids such as water, but even measurements under pressures as high as 100 Torr were successfully made with sufficiently long data acquisition times.[62] The sample holder is mounted on a transfer rod that is equipped with a Teflon seal, which upon insertion of the rod into the gas cell separates the gas-filled volume from the sample preparation section of the chamber, which always remains under UHV and is equipped for typical cleaning procedures such as sputter-annealing cycles. The gas cell is connected by leak valves to a gas manifold that can provide mixtures of various gases and vapors; their composition can be monitored using a quadrupole mass spectrometer in the differentially pumped section.

In addition to APXPS measurements, the photon energy can also be scanned for XAS measurements; absorption spectra can be acquired by total electron yield (sample current) and, using the XPS analyzer, by Auger electron yield detection.

An *operando* electrochemical cell compatible with the APXPS instrument at SSRL BL13-2 was designed for electrocatalysis studies in JCAP

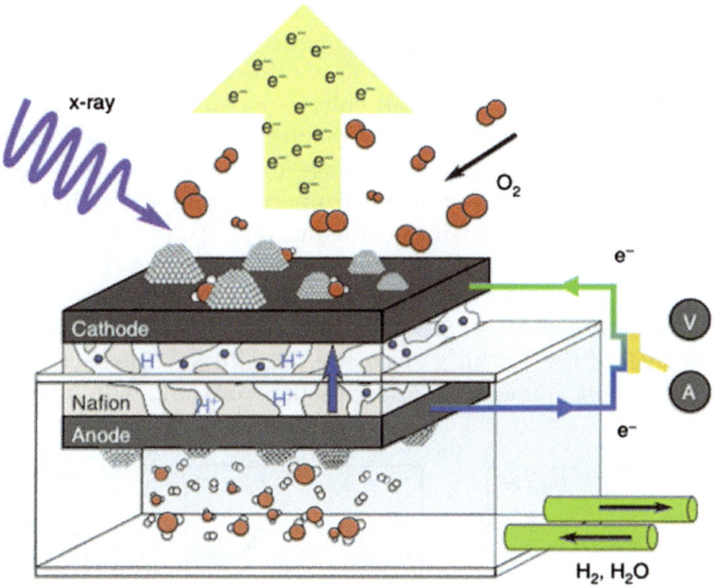

Figure 4.13 Schematic drawing of a membrane-electrode assembly for electrochemical APXPS studies. The working electrode faces the XPS analyzer aperture and is exposed to the gases in the APXPS gas cell. The counter electrode has a separate compartment that can be purged with a different gas mix.[63]

(Figure 4.13).[63] With the short mean free paths of soft X-ray photoelectrons in mind, this experiment was designed to capture photoelectrons that can escape from regions near the triple-phase boundary between polymer electrolyte, catalyst nanoparticles and reactive gas atmosphere. Procedures similar to those used in PEM fuel cell production were followed to make a membrane-electrode assembly (MEA) consisting of a Nafion membrane that was coated on each side with the respective Nafion/electrocatalyst/solvent suspension. The MEA was tightened over the circular opening of a small receptacle made of PEEK. The receptacle provides a separate gas compartment for the counter electrode that is purged with its own gas mixture, while the working electrode is exposed to the gases that are dosed into the gas cell.

The electrochemical APXPS cell was employed for the first time to identify the oxygen-containing adsorbates that form at the Pt surface in PEMFC fuel cell cathodes.[63] The identification of OH had been a major challenge; for example, an adsorbed species was detected with HERFD XAS at the Pt L_3 edge, and could be clearly distinguished from Pt surface oxides, but it was impossible to distinguish between spectral changes caused by OH_{ads} or by O_{ads}.[46,49,64]

A fuel cell MEA was made using carbon-black supported Pt nanoparticles on both electrodes. The gas cell was supplied with oxygen at various pressures, while the counter electrode compartment was purged with humidified forming gas (95% N_2/5% H_2) to provide both hydrogen fuel and to humidify the Nafion membrane. APXPS measurements were made under variation of the O_2 pressure both under open circuit conditions (Figure 4.14) and under a current load; in both cases the cell voltage was also measured. Significant changes in the O1s XP spectra can be seen after the introduction of 0.4 Torr O_2, and during the subsequent increase of the open-circuit voltage; the composition of adsorbates changes from predominantly H_2O (multilayer and adsorbed) with increasing contributions from OH and O species (Figure 4.14a). In Figure 4.14b, the characteristic doublet signature of O_2 gas appears after dosing, and its energy position shifts to lower binding energies (BE) with increasing voltage. Since the BE scale is calibrated with respect to the Fermi level of the working electrode, BE shifts of gaseous species near the sample, whose ionization energies are constant with respect to the

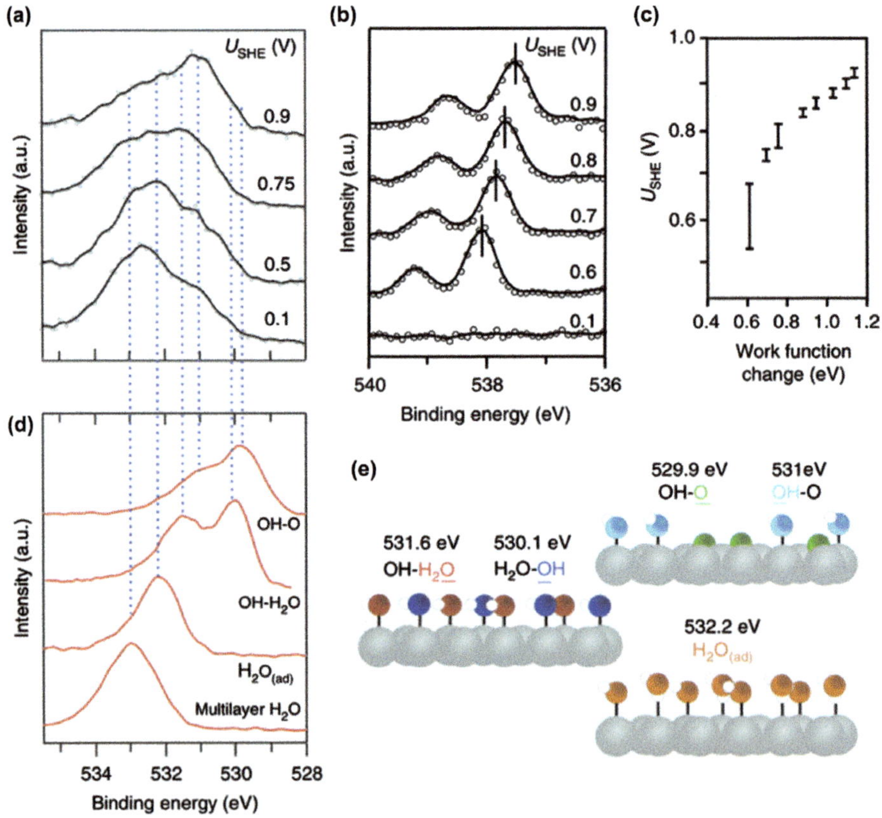

Figure 4.14 APXPS characterization of oxygen-containing adsorbates on a PEMFC cathode. XP spectra covering the energy regions of (a) adsorbed species ranging from H_2O (533 eV) to O (529.9 eV), and (b) O_2 in the gas phase. (c) Variation of the work function change, obtained from the O_2(g) peak shifts, as function of cell voltage. (d) Reference spectra measured on Pt(111) with well-defined model adsorbates; the corresponding adsorbate structure models are shown in (e).[63]

vacuum level, directly represent the work function change;[65–67] the latter is equivalent with the change of the electrochemical potential. The work function change closely follows a (1 : 1) correlation with cell voltage; this indicates that all potential changes occur at the working electrode and that the counter electrode is functioning reliably as a reversible hydrogen electrode.

In order to identify the adsorbed species at the Pt working electrode, the potential-dependent spectra were compared with measurements of well-defined model adsorbates on Pt(111) that included monolayer and multi-layer water, and coadsorbate layers composed of OH/H_2O and of OH/O (Figure 4.14d,e).[68,69] This comparison made it possible not only to distinguish different adsorbates H_2O, OH and O, but also to identify two different OH species with and without hydrogen-bonding to neighboring H_2O adsorbates. Theoretical calculations suggest that both types of OH adsorbate can occur as intermediate in two alternative ORR pathways, those OH adsorbates that are embedded in a hydrogen-bonding network with adsorbed H_2O are less reactive than "non-hydrated" OH, resulting in an higher overall activation barrier for the ORR.[63]

The same type of MEA design was also used to study the OER; in this case the working electrode was made with IrO_2 nanoparticles and the counter electrode with Pt/C; water vapor was supplied at a pressure of 10 Torr and a voltage of 1.75 V was applied to drive water electrolysis.[70] Figure 4.15 shows O1s spectra of the dry working electrode, at open circuit after the addition of water vapor, and during OER at 1.75 V. Peak fitting results reveal the surface composition, which initially in the dry state consists entirely of oxide anions. The surface reacts with water vapor at open circuit voltage, resulting in a surface composition of 1 : 2 between hydroxide and oxide. The hydroxide/oxide ratio decreases to 1 : 3 during OER. The mechanisms that change the OH/O ratio are different for humidification and OER, respectively: the rise of the hydroxide component during humidification occurs without an oxidation state change of the Ir cations,[70] which indicates a hydration reaction of the type $Ir^{IV}–O + H_2O \rightarrow Ir^{IV}(OH)_2$. By contrast, when the PEM electrolyzer is switched from open circuit to OER at 1.75 V, the decrease of the OH/O ratio is accompanied by the appearance of a new component in the Ir 4f signal at higher binding energy, which was assigned to Ir^V sites (Figure 4.16).[70] Therefore, the loss of OH in this case is an oxidation reaction, $Ir^{IV}–OH \rightarrow Ir^V–O + H^+ + e^-$.

Furthermore, the comparison between Ir 4f spectra taken at different photon energies (Figure 4.16) reveals that the oxidation of Ir sites during OER is limited to surface sites. When the photon energy is increased from 390 eV to 760 eV, which corresponds to an increase of the photoelectron inelastic mean free path from 0.7 nm to 1.1 nm, the Ir^V component that appears at 1.75 V is diminished. This finding demonstrates an advantage of the high surface sensitivity of soft X-ray photoemission spectroscopy that allows for the detection of surface redox reactions, which did not become evident with bulk-sensitive XAS measurements.[71,72] Knowledge of the Ir

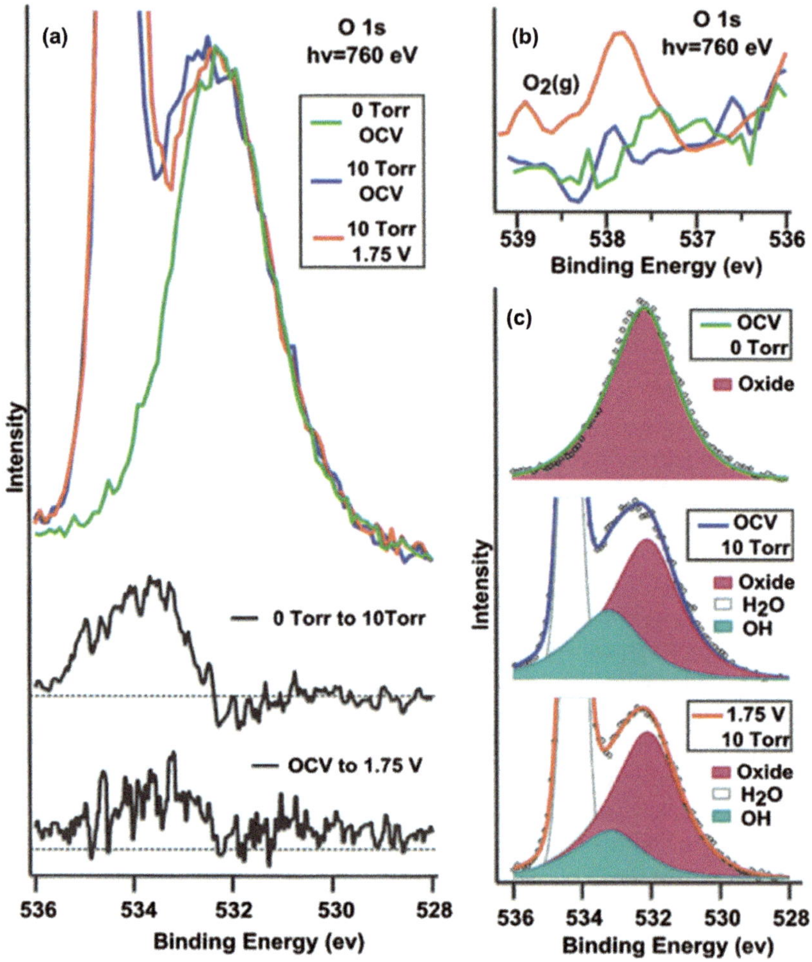

Figure 4.15 (a) Comparison of O1s spectra of an IrO_x electrocatalyst under dry conditions (green line), at open circuit after introducing H_2O at 10 Torr (blue line), and during OER at 10 Torr and an applied voltage of 1.75 V. (b) Weak signal from O_2(g) emerging during OER. (c) O1s peaks with fitted components for oxide, surface hydroxide and liquid water. Reproduced from ref. 70 with permission from John Wiley and Sons, © 2014 Wiley-VCH Verlag GmbH & Co. KGaA, Weinheim.

surface oxidation state under OER conditions is essential for the correct choice of input parameters in the theoretical modeling of the OER mechanism.[6]

A key objective of JCAP has been to develop new catalyst materials that are purely made from earth-abundant materials, and this ultimately requires the discovery of abundant metal oxides that are stable in acidic conditions and, at the same time, catalyze the OER at a reasonable overpotential. Understanding the criteria that make IrO_2 suitable for OER in acidic conditions is

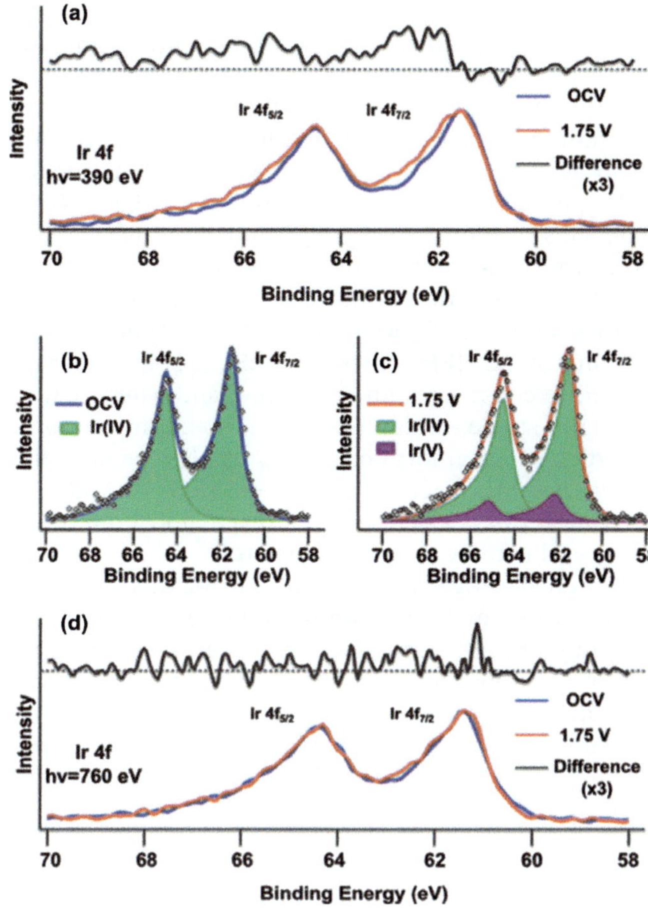

Figure 4.16 (a) Comparison of Ir 4f spectra of an IrO_2 electrocatalyst at open circuit conditions and during OER at 1.75 V. Peak fitting results for the two spectra in (a) with Ir^{IV} and Ir^{V} components are shown in (b) and (c). (d) Ir 4f spectra taken under the same conditions as in (a), but with the photon energy increased from 390 eV (a) to 760 eV, resulting in greater probing depth.
Reproduced from ref. 70 with permission from John Wiley and Sons, © 2014 Wiley-VCH Verlag GmbH & Co. KGaA, Weinheim.

the first step necessary to move towards abundant materials that mimic the performance of IrO_2. Instead of entirely dismissing the use of Ir, it might be necessary to develop inert earth-abundant oxides that are doped with highly dilute amounts of Ir atoms, which would then constitute the catalytically active sites. On the other hand, an important first step towards stabilizing MnO_2 by incorporating small amounts of TiO_2 was achieved more recently by Frydendal *et al.*[73] In a similar spirit to the progress described in this chapter, the combination of computational and experimental tools, in particular X-ray and electron spectroscopy, was essential to this discovery.

These are the tools that enable the design of electrocatalysts based on tailored surface structuring; rather than choosing from a limited number of bulk compounds with high thermodynamic stability, the local coordination of surface sites is altered in order to kinetically hinder dissolution reactions, or to expose certain surface sites with enhanced catalytic activity. The latter motive has also been pursued in the search for earth-abundant HER catalysts, where the pioneering work of Jaramillo *et al.* that revealed the HER activity of edge sites in monolayer MoS_2 islands on Au(111) led the way towards scalable synthesis routes for highly active nanostructured MoS_x catalysts.[74] Multiple synthetic efforts have since been made with the aim of maximizing the amount of active sites.[27,28,75–80] While success has been made towards improving HER activity of MoS_x, it is not clear exactly which local Mo/S structure constitutes an active site, and whether these configurations will remain stable under long-term operating conditions. Initial *ex situ* XPS studies of an amorphous MoS_x catalyst were indicative of the reduction of the as-synthesized catalyst from MoS_3 to MoS_2, but the conditions under which this reduction takes place, and how this may impact catalyst stability and activity, were not known.[28]

The electrochemical APXPS setup at SSRL BL13-2 was used to elucidate the electrochemical transformation of amorphous MoS_x nanoparticles during HER.[29] An electrolyzer MEA was made with a Nafion membrane, which was loaded with MoS_x nanoparticles on the working electrode side, while the counter electrode was coated with a Pt catalyst. S 2p spectra were taken under various conditions with increasing HER current density. Peak fitting results (Figure 4.17) show that the catalyst composition, as seen in the ratio

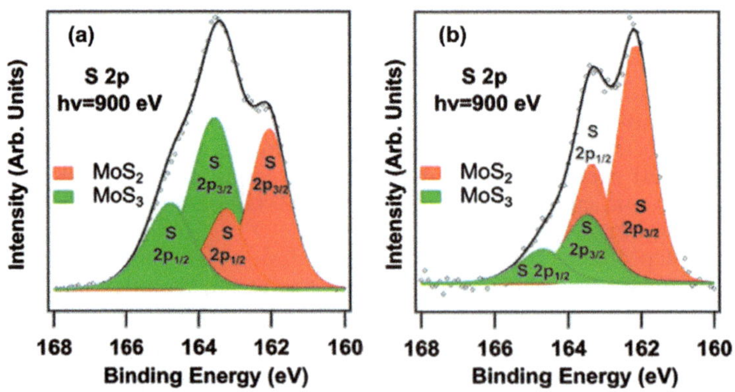

Figure 4.17 APXPS measurement of amorphous MoS_x catalyst, S 2p spectra (a) at open circuit, $p(H_2O) = 1$ Torr, (b) during hydrogen evolution, the HER current density is 1.5 mA cm^{-2} under $p(H_2O) = 5$ Torr.
Reprinted with permission from H. G. S. Casalongue, J. D. Benck, C. Tsai, R. K. B. Karlsson, S. Kaya, M. L. Ng, L. G. M. Pettersson, F. Abild-Pedersen, J. K. Nørskov, H. Ogasawara, T. F. Jaramillo and A. Nilsson, *J. Phys. Chem. C*, 2014, **118**, 29252–29259, Copyright 2014 American Chemical Society.[29]

of components assigned to disulfide sites in MoS_3 and monosulfide sites in MoS_2, changes from predominantly MoS_3 (56.5% at OCV) towards MoS_2 (79% during HER at $-1.5\ mA\,cm^{-2}$). Due to the apparent correlation between HER current density and the increased relative amount of monosulfide sites, it was proposed that the monosulfide sites constitute the most active sites for the HER. This model was further corroborated with DFT calculations, in which hydrogen adsorption energies were found to be $\Delta G_H = 0.06$ eV at monosulfide sites, which is close to the optimum (0.0 eV) for HER, whereas disulfide sites were predicted to be catalytically inactive with $\Delta G_H > 1$ eV.[29]

4.3.2.1 Methods: Tender X-ray APXPS

A limitation of the soft X-ray APXPS technique is the short attenuation length of the photoelectrons, which on one hand provides superior surface sensitivity, but on the other hand precludes studies of electrochemical systems which, instead of a solid–gas interface[65] or a solid–gas–electrolyte triple-phase boundary,[29,63,70,81] involve a solid–liquid interface with a thicker electrolyte layer. The probing depth and, correspondingly, the attenuation length of photoelectrons in liquid electrolyte, increased with the excitation energy in XPS. An APXPS end station using a "tender" X-ray source with an energy range between 2.3 and 5.2 keV has been developed at BL 9.3.1 at the Advanced Light Source.[82] For electrochemical APXPS measurements, a glass beaker with a conventional three-electrode arrangement is installed inside the ambient-pressure section of the measurement chamber, which is held near the water vapor pressure of ~18 Torr during operation. The working electrode is initially fully immersed into the electrolyte, and subsequently raised in such a manner that a thin electrolyte film is preserved on the immersed portion of the electrode, which remains in electric contact through the bulk electrolyte with the counter and reference electrodes (Figure 4.18).

The vertical position of the working electrode is thereby adjusted in order to minimize attenuation of photoelectrons from the working electrode by the electrolyte film, but without losing electric contact to the bulk electrolyte. In a typical experiment, the wetting of the partially immersed working electrode is enhanced by electrochemical conditioning with extended potential cycles. This technique was employed successfully in the electrochemical oxidation of a Pt electrode in 6M KF electrolyte.[82] Evidence that the applied electrochemical potential was maintained in the region of the Pt electrode that is probed with XPS was provided with the binding energy shift of the photoemission peaks from electrolyte species H_2O and K^+. The formation of oxidized Pt^{2+} and Pt^{4+} species was found at the applied potential of +1.2 V *vs.* Ag/AgCl, in agreement with other *in situ* experiments using APXPS[81] and HERFD XAS.[49,84] The energy-band relations in Si/TiO_2/electrolyte and Si/TiO_2/Ni/electrolyte junctions were explored using an APXPS measurement, in which the binding energy scale was referenced to the reference electrode

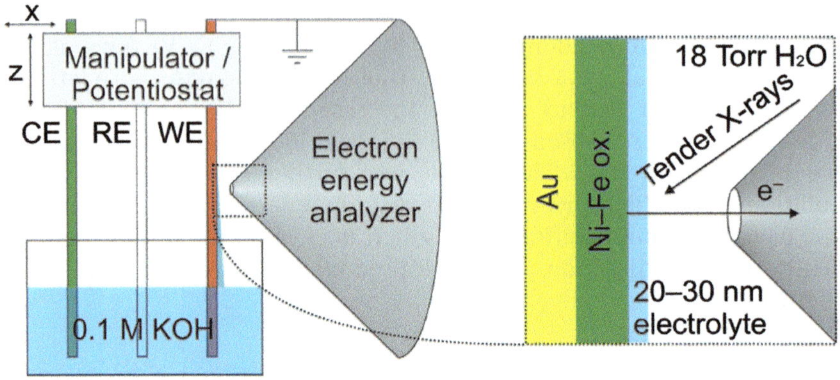

Figure 4.18 Schematic illustration of the setup used for tender X-ray APXPS measurements using the "dip & pull" technique.
Adapted with permission from H. Ali-Löytty, M. W. Louie, M. R. Singh, L. Li, H. G. Sanchez Casalongue, H. Ogasawara, E. J. Crumlin, Z. Liu, A. T. Bell, A. Nilsson, and D. Friebel, *J. Phys. Chem. C*, 2016, **120**, 2247–2253, Copyright 2016 American Chemical Society.[83]

instead of the XPS analyzer. The resulting energy shifts of the Ti 2p and Ni 2p signals were analyzed to determine the band-edge shifts as a function of applied potential, and it was shown that the addition of the Ni layer removes the rectifying behavior of the TiO_2.[85]

These APXPS experiments were carried out under conditions where the electrochemical current densities that are required to sustain the applied electrochemical potential are very low. Limitations of the "dip & pull" technique can be noticed when the goal is to characterize a highly active electrocatalyst under technologically relevant operating conditions, *i.e.* for the case of photoelectrochemical water splitting, at a current density of 10 mA cm^{-2}. In APXPS measurements of an electrodeposited Fe–Ni electrocatalyst on an Au working electrode,[83] it was observed that the O1s binding energy shifts of H_2O in the vapor phase and in the liquid electrolyte followed the expected 1 : 1 relationship with the applied potential until the onset of the OER was reached. A small deviation of the peak shift appeared at +0.55 V *vs.* Ag/AgCl, and the correlation was lost at +0.65 V (Figure 4.19). These two potentials corresponded to current densities in the cyclic voltammogram of 0.07 and 0.13 mA cm^{-2}, respectively. The lack of correlation between peak shifts and potential at high current density can be explained with the small thickness of the electrolyte film in the XPS probing region, which introduces a significant ohmic drop that limits the effective local OER overpotential. A numerical simulation of the ohmic drop as a function of the vertical position in the thin-film electrolyte was carried out. The resulting effective potentials at the XPS probing position were converted into binding energy shifts for $H_2O(l)$ and $H_2O(g)$, which were found to be in good agreement with the measured peak positions. A relative increase of the oxide/hydroxide ratio is remarkable because it was not yet accompanied by an oxidation of Ni

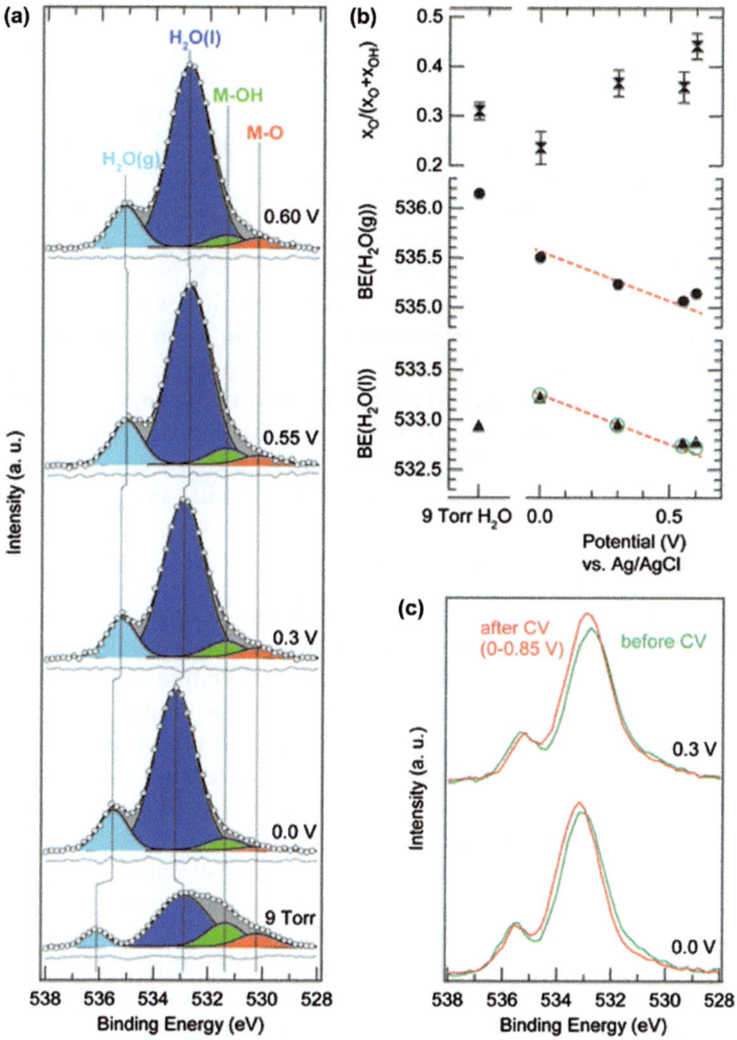

Figure 4.19 Tender X-ray APXPS measurement ($h\nu = 4020$ eV) of a 7 nm thick Ni–Fe electrocatalyst layer electrodeposited on Au. (a) Bottom to top: O1s spectra of the as-prepared catalyst under 9 Torr H_2O, and potential-dependent measurements under a ~30 nm thick 0.1 M KOH electrolyte film. The electrode was conditioned with electrochemical potential cycles prior to the XPS measurements under applied potentials. (b) Fit results for the spectra in (a): binding energy shifts of H_2O(g) and H_2O(l), and relative intensity of the oxide component in (Fe,Ni)O_xOH_y, as function of potential. Peak positions from theoretical modeling of the ohmic potential drop are indicated with green open circles. (c) Effect of different electrochemical conditioning protocols: 6 cycles, 0–0.65 V (green) *vs.* 6 cycles, 0–0.65 V and then 6 cycles, 0–0.85 V (red). The potential-dependent spectra in (a) were acquired after applying the second conditioning protocol.
Reprinted with permission from H. Ali-Löytty, M. W. Louie, M. R. Singh, L. Li, H. G. Sanchez Casalongue, H. Ogasawara, E. J. Crumlin, Z. Liu, A. T. Bell, A. Nilsson, and D. Friebel, *J. Phys. Chem. C*, 2016, **120**, 2247–2253, Copyright 2016 American Chemical Society.[83]

sites; it was proposed that a partial deprotonation of $Ni(OH)_2$ precedes the oxidation of Ni sites.[83] Further evidence that the high-current density OER regime could not be accessed in the APXPS measurement stems from the absence of the formation of highly oxidized Ni^{3+} and Ni^{4+} species. The oxidation of $(Fe^{3+},Ni^{2+})O_xH_y$ to $(Fe^{3+}, Ni^{3+/4+})OOH_{1\text{-}z}$ during OER is evident from *operando* HERFD XAS measurements (see Chapter 3 by Bell in this book).[36] Tender-X-ray APXPS provides a powerful tool to investigate the potential distribution across the electrolytic double layer. An advantage over other hard X-ray spectroscopies is the ability to probe low-Z elements such as oxygen directly. This can be combined with capabilities to probe at different depths with respect to the electrode surface, on both electrode and electrolyte sides, if APXPS is performed under X-ray standing wave conditions.[86] Strategies to achieve higher current density in the thin liquid electrolyte film could involve a substantial reduction of the distance between working and counter electrodes. Maybe this could be realized by coating both electrodes onto one insulating substrate, with a narrow gap in between.

4.4 Summary and Outlook

Significant progress has been made towards the understanding of structure–activity–stability relationships in electrocatalysis for solar water splitting. Some of the important and well-established concepts of electrochemical surface science have been successfully adapted to explore the field of earth-abundant electrocatalyst materials, where several promising candidates have been identified for HER catalysis under acidic conditions and for alkaline OER. What is still missing is an acid-stable, earth abundant OER catalyst, which cannot be found without profound understanding of the electrochemical surface science of metal oxides. This marks the beginning of a steep learning curve, where non-noble metals, chalcogenides, phosphides and oxides are much more difficult to examine under well-controlled conditions than the traditionally studied and much less reactive noble metals. We have yet to see experiments with single-crystal model catalysts and surface alloys made with physical methods. The majority of work performed in JCAP has identified the chemical state of catalysts under working conditions but not the reaction intermediates of the HER or OER. Even with APXPS, which was successful in detecting O and OH adsorbates on fuel cell electrodes,[63] the identification of OER intermediates could be difficult, due to the very similar chemical shifts of lattice oxide and hydroxide in a metal oxide-based catalyst. This problem could be overcome with time-resolved *operando* spectroscopy. Noteworthy in this regard is the detection of an OOH species during photoinduced water oxidation at an IrO_2 nanoparticle catalyst with rapid-scan FT-IR.[87] A variety of time-resolved measurement techniques have become available at X-ray synchrotrons[88,89] and free-electron lasers[90] that cover a wide range of time scales and various possibilities for pump-probe schemes, which could open up a new era of electrocatalysis research.

References

1. B. Hammer and J. K. Nørskov, *Nature*, 1995, **376**, 238–240.
2. B. Hammer and J. K. Nørskov, *Surf. Sci.*, 1995, **343**, 211–220.
3. T. Bligaard, J. K. Nørskov, S. Dahl, J. Matthiesen, C. H. Christensen and J. Sehested, *J. Catal.*, 2004, **224**, 206–217.
4. J. K. Nørskov, J. Rossmeisl, A. Logadottir, L. Lindqvist, J. R. Kitchin, T. Bligaard and H. Jónsson, *J. Phys. Chem. B*, 2004, **108**, 17886–17892.
5. J. Rossmeisl, A. Logadottir and J. K. Nørskov, *Chem. Phys.*, 2005, **319**, 178–184.
6. J. Rossmeisl, Z.-W. Qu, H. Zhu, G.-J. Kroes and J. K. Nørskov, *J. Electroanal. Chem.*, 2007, **607**, 83–89.
7. J. Rossmeisl, J. K. Nørskov, C. D. Taylor, M. J. Janik and M. Neurock, *J. Phys. Chem. B*, 2011, **110**, 21833–21839.
8. F. Abild-Pedersen, J. Greeley, F. Studt, J. Rossmeisl, T. R. Munter, P. G. Moses, E. Skúlason, T. Bligaard and J. K. Nørskov, *Phys. Rev. Lett.*, 2007, **99**, 016105.
9. V. Stamenkovic, B. S. Mun, K. J. J. Mayrhofer, P. N. Ross, N. M. Markovic, J. Rossmeisl, J. Greeley and J. K. Nørskov, *Angew. Chem., Int. Ed.*, 2006, **45**, 2897–2901.
10. V. R. Stamenkovic, B. Fowler, B. S. Mun, G. F. Wang, P. N. Ross, C. A. Lucas and N. M. Markovic, *Science*, 2007, **315**, 493–497.
11. V. R. Stamenkovic, B. S. Mun, M. Arenz, K. J. J. Mayrhofer, C. A. Lucas, G. F. Wang, P. N. Ross and N. M. Markovic, *Nat. Mater.*, 2007, **6**, 241–247.
12. J. Greeley, I. E. L. Stephens, A. S. Bondarenko, T. P. Johansson, H. A. Hansen, T. F. Jaramillo, J. Rossmeisl, I. Chorkendorff and J. K. Nørskov, *Nat. Chem.*, 2009, **1**, 552–556.
13. P. Strasser, S. Koh, T. Anniyev, J. Greeley, K. More, C. Yu, Z. Liu, S. Kaya, D. Nordlund, H. Ogasawara, M. F. Toney and A. Nilsson, *Nat. Chem.*, 2010, **2**, 454–460.
14. J. Zhang, M. B. Vukmirovic, Y. Xu, M. Mavrikakis and R. R. Adzic, *Angew. Chem., Int. Ed.*, 2005, **44**, 2132–2135.
15. J. L. Zhang, M. B. Vukmirovic, K. Sasaki, A. U. Nilekar, M. Mavrikakis and R. R. Adzic, *J. Am. Chem. Soc.*, 2005, **127**, 12480–12481.
16. C. Chen, Y. Kang, Z. Huo, Z. Zhu, W. Huang, H. L. Xin, J. D. Snyder, D. Li, J. A. Herron, M. Mavrikakis, M. Chi, K. L. More, Y. Li, N. M. Markovic, G. A. Somorjai, P. Yang and V. R. Stamenkovic, *Science*, 2014, **343**, 1339–1343.
17. I. E. L. Stephens, A. S. Bondarenko, U. Grønbjerg, J. Rossmeisl and I. Chorkendorff, *Energy Environ. Sci.*, 2012, **5**, 6744.
18. M. P. Soriaga, J. H. Baricuatro, K. D. Cummins, Y.-G. Kim, F. H. Saadi, G. Sun, C. C. L. McCrory, J. R. McKone, J. M. Velazquez, I. M. Ferrer, A. I. Carim, A. Javier, B. Chmielowiec, D. C. Lacy, J. M. Gregoire, J. Sanabria-Chinchilla, X. Amashukeli, W. J. Royea, B. S. Brunschwig, J. C. Hemminger, N. S. Lewis and J. L. Stickney, *Surf. Sci.*, 2015, **631**, 285–294.

19. F. H. Saadi, A. I. Carim, E. Verlage, J. C. Hemminger, N. S. Lewis and M. P. Soriaga, *J. Phys. Chem. C*, 2014, **118**, 29294–29300.
20. F. H. Saadi, A. I. Carim, J. M. Velazquez, J. H. Baricuatro, C. C. L. McCrory, M. P. Soriaga and N. S. Lewis, *ACS Catal.*, 2014, **4**, 2866–2873.
21. J. R. McKone, B. F. Sadtler, C. A. Werlang, N. S. Lewis and H. B. Gray, *ACS Catal.*, 2013, **3**, 166–169.
22. C. C. L. McCrory, S. Jung, I. M. Ferrer, S. M. Chatman, J. C. Peters and T. F. Jaramillo, *J. Am. Chem. Soc.*, 2015, **137**, 4347–4357.
23. C. Fan, D. L. Piron, A. Sleb and P. Paradis, *J. Electrochem. Soc.*, 1994, **141**, 382–387.
24. Q. Han, S. Cui, N. Pu, J. Chen, K. Liu and X. Wei, *Int. J. Hydrogen Energy*, 2010, **35**, 5194–5201.
25. E. Navarro-Flores, Z. Chong and S. Omanovic, *J. Mol. Catal. A: Chem.*, 2005, **226**, 179–197.
26. J. G. Highfield, E. Claude and K. Oguro, *Electrochim. Acta*, 1999, **44**, 2805–2814.
27. J. Kibsgaard, Z. Chen, B. N. Reinecke and T. F. Jaramillo, *Nat. Mater.*, 2012, **11**, 963–969.
28. J. D. Benck, Z. Chen, L. Y. Kuritzky, A. J. Forman and T. F. Jaramillo, *ACS Catal.*, 2012, **2**, 1916–1923.
29. H. G. S. Casalongue, J. D. Benck, C. Tsai, R. K. B. Karlsson, S. Kaya, M. L. Ng, L. G. M. Pettersson, F. Abild-Pedersen, J. K. Nørskov, H. Ogasawara, T. F. Jaramillo and A. Nilsson, *J. Phys. Chem. C*, 2014, **118**, 29252–29259.
30. E. J. Popczun, J. R. McKone, C. G. Read, A. J. Biacchi, A. M. Wiltrout, N. S. Lewis and R. E. Schaak, *J. Am. Chem. Soc.*, 2013, **135**, 9267–9270.
31. J. F. Callejas, J. M. McEnaney, C. G. Read, J. C. Crompton, A. J. Biacchi, E. J. Popczun, T. R. Gordon, N. S. Lewis and R. E. Schaak, *ACS Nano*, 2014, **8**, 11101–11107.
32. A. Nilsson and L. G. M. Pettersson, *Surf. Sci. Rep.*, 2004, **55**, 49–167.
33. A. Nilsson, L. G. M. Pettersson, B. Hammer, T. Bligaard, C. H. Christensen and J. K. Nørskov, *Catal. Lett.*, 2005, **100**, 111–114.
34. A. Nilsson and L. G. M. Pettersson, in *Chemical Bonding at Surfaces and Interfaces*, Elsevier, Amsterdam, 2008, pp. 57–142.
35. A. Nilsson, *J. Electron Spectrosc. Relat. Phenom.*, 2002, **126**, 3.
36. D. Friebel, M. W. Louie, M. Bajdich, K. E. Sanwald, Y. Cai, A. M. Wise, M.-J. Cheng, D. Sokaras, T.-C. Weng, R. Alonso-Mori, R. C. Davis, J. R. Bargar, J. K. Nørskov, A. Nilsson and A. T. Bell, *J. Am. Chem. Soc.*, 2015, **137**, 1305–1313.
37. A. Bianconi, M. Dell'Ariccia, P. J. Durham and J. B. Pendry, *Phys. Rev. B*, 1982, **26**, 6502–6508.
38. M. Benfatto, C. R. Natoli, A. Bianconi, J. Garcia, A. Marcelli, M. Fanfoni and I. Davoli, *Phys. Rev. B*, 1986, **34**, 5774–5781.
39. A. Bianconi, J. Garcia, and M. Benfatto, in *Synchrotron Radiation in Chemistry and Biology I*, Springer Berlin Heidelberg, 1988, pp. 29–67.
40. J. J. Rehr, R. C. Albers, C. R. Natoli and E. A. Stern, *Phys. Rev. B*, 1986, **34**, 4350–4353.

41. S. I. Zabinsky, J. J. Rehr, A. Ankudinov, R. C. Albers and M. J. Eller, *Phys. Rev. B*, 1995, **52**, 2995–3009.
42. A. L. Ankudinov, B. Ravel, J. J. Rehr and S. D. Conradson, *Phys. Rev. B*, 1998, **58**, 7565–7576.
43. J. J. Rehr and R. C. Albers, *Rev. Mod. Phys.*, 2000, **72**, 621–654.
44. Y. Joly, D. Cabaret, H. Renevier and C. R. Natoli, *Phys. Rev. Lett.*, 1999, **82**, 2398–2401.
45. Y. Joly, *Phys. Rev. B*, 2001, **63**, 125120.
46. D. Friebel, V. Viswanathan, D. J. Miller, T. Anniyev, H. Ogasawara, A. H. Larsen, C. P. O'Grady, J. K. Nørskov and A. Nilsson, *J. Am. Chem. Soc.*, 2012, **134**, 9664–9671.
47. D. Friebel, F. Mbuga, S. Rajasekaran, D. J. Miller, H. Ogasawara, R. Alonso-Mori, D. Sokaras, D. Nordlund, T.-C. Weng and A. Nilsson, *J. Phys. Chem. C*, 2014, **118**, 7954–7961.
48. D. Friebel, M. Bajdich, B. S. Yeo, M. W. Louie, D. J. Miller, H. S. Casalongue, F. Mbuga, T.-C. Weng, D. Nordlund, D. Sokaras, R. Alonso-Mori, A. T. Bell and A. Nilsson, *Phys. Chem. Chem. Phys.*, 2013, **15**, 17460–17467.
49. L. R. Merte, F. Behafarid, D. J. Miller, D. Friebel, S. Cho, F. Mbuga, D. Sokaras, R. Alonso-Mori, T.-C. Weng, D. Nordlund, A. Nilsson and B. Roldan Cuenya, *ACS Catal.*, 2012, 2371–2376.
50. G. A. Waychunas, *Reviews in Mineralogy and Geochemistry, Mineral. Soc. Am.*, 2002, **49**, 267–315.
51. F. U. Renner, Y. Gründer and J. Zegenhagen, *Rev. Sci. Instrum.*, 2007, **78**, 033903.
52. O. M. Magnussen, K. Krug, A. H. Ayyad and J. Stettner, *Electrochim. Acta*, 2008, **53**, 3449–3458.
53. D. K. Bora, P.-A. Glans, J. Pepper, Y.-S. Liu, C. Du, D. Wang and J. Guo, *Rev. Sci. Instrum.*, 2014, **85**, 043106.
54. F. de Groot, *Chem. Rev.*, 2001, **101**, 1779–1808.
55. P. Glatzel and U. Bergmann, *Coord. Chem. Rev.*, 2005, **249**, 65–95.
56. P. Glatzel, M. Sikora, G. Smolentsev and M. Fernández-García, *Catal. Today*, 2009, **145**, 294–299.
57. P. Glatzel, T.-C. Weng, K. Kvashnina, J. Swarbrick, M. Sikora, E. Gallo, N. Smolentsev and R. A. Mori, *J. Electron Spectrosc. Relat. Phenom.*, 2013, **188**, 17–25.
58. B. S. Yeo and A. T. Bell, *J. Am. Chem. Soc.*, 2011, **133**, 5587–5593.
59. W. M. Heijboer, P. Glatzel, K. R. Sawant, R. F. Lobo, U. Bergmann, R. A. Barrea, D. C. Koningsberger, B. M. Weckhuysen and F. M. F. de Groot, *J. Phys. Chem. B*, 2004, **108**, 10002–10011.
60. F. M. F. de Groot, P. Glatzel, U. Bergmann, P. A. van Aken, R. A. Barrea, S. Klemme, M. Hävecker, A. Knop-Gericke, W. M. Heijboer and B. M. Weckhuysen, *J. Phys. Chem. B*, 2005, **109**, 20751–20762.
61. G. Vankó, F. M. F. de Groot, S. Huotari, R. J. Cava, T. Lorenz, and M. Reuther, arXiv:0802.2744, 2008.
62. S. Kaya, H. Ogasawara, L.-Å. Näslund, J.-O. Forsell, H. S. Casalongue, D. J. Miller and A. Nilsson, *Catal. Today*, 2013, **205**, 101–105.

63. H. S. Casalongue, S. Kaya, V. Viswanathan, D. J. Miller, D. Friebel, H. A. Hansen, J. K. Nørskov, A. Nilsson and H. Ogasawara, *Nat. Commun.*, 2013, **4**, 2817.
64. D. Friebel, D. J. Miller, C. P. O'Grady, T. Anniyev, J. Bargar, U. Bergmann, H. Ogasawara, K. T. Wikfeldt, L. G. M. Pettersson and A. Nilsson, *Phys. Chem. Chem. Phys.*, 2011, **13**, 262–266.
65. C. Zhang, M. E. Grass, A. H. McDaniel, S. C. DeCaluwe, F. E. Gabaly, Z. Liu, K. F. McCarty, R. L. Farrow, M. A. Linne, Z. Hussain, G. S. Jackson, H. Bluhm and B. W. Eichhorn, *Nat. Mater.*, 2010, **9**, 944–949.
66. S. Axnanda, M. Scheele, E. Crumlin, B. Mao, R. Chang, S. Rani, M. Faiz, S. Wang, A. P. Alivisatos and Z. Liu, *Nano Lett.*, 2013, **13**, 6176–6182.
67. E. J. Crumlin, H. Bluhm and Z. Liu, *J. Electron Spectrosc. Relat. Phenom.*, 2013, **190**, 84–92, part A.
68. T. Schiros, O. Takahashi, K. J. Andersson, H. Öström, L. G. M. Pettersson, A. Nilsson and H. Ogasawara, *J. Chem. Phys.*, 2010, **132**, 094701.
69. T. Schiros, L. Å. Näslund, K. Andersson, J. Gyllenpalm, G. S. Karlberg, M. Odelius, H. Ogasawara, L. G. M. Pettersson and A. Nilsson, *J. Phys. Chem. C*, 2007, **111**, 15003–15012.
70. H. G. S. Casalongue, M. L. Ng, S. Kaya, D. Friebel, H. Ogasawara, and A. Nilsson, *Angew. Chem. Int. Ed.*, 2014, **53**, 7169–7172.
71. Y. Mo, I. C. Stefan, W.-B. Cai, J. Dong, P. Carey and D. A. Scherson, *J. Phys. Chem. B*, 2002, **106**, 3681–3686.
72. A. R. Hillman, M. A. Skopek and S. J. Gurman, *Phys. Chem. Chem. Phys.*, 2011, **13**, 5252–5263.
73. R. Frydendal, E. A. Paoli, I. Chorkendorff, J. Rossmeisl and I. E. L. Stephens, *Adv. Energy Mater.*, 2015, **5**, 1500991.
74. T. F. Jaramillo, K. P. Jørgensen, J. Bonde, J. H. Nielsen, S. Horch and I. Chorkendorff, *Science*, 2007, **317**, 100–102.
75. H. Vrubel, D. Merki and X. Hu, *Energy Environ. Sci.*, 2012, **5**, 6136–6144.
76. D. Kong, H. Wang, J. J. Cha, M. Pasta, K. J. Koski, J. Yao and Y. Cui, *Nano Lett.*, 2013, **13**, 1341–1347.
77. Z. Chen, D. Cummins, B. N. Reinecke, E. Clark, M. K. Sunkara and T. F. Jaramillo, *Nano Lett.*, 2011, **11**, 4168–4175.
78. A. B. Laursen, S. Kegnæs, S. Dahl and I. Chorkendorff, *Energy Environ. Sci.*, 2012, **5**, 5577–5591.
79. D. Merki and X. Hu, *Energy Environ. Sci.*, 2011, **4**, 3878–3888.
80. B. Hinnemann, P. G. Moses, J. Bonde, K. P. Jørgensen, J. H. Nielsen, S. Horch, I. Chorkendorff and J. K. Nørskov, *J. Am. Chem. Soc.*, 2005, **127**, 5308–5309.
81. R. Arrigo, M. Hävecker, M. E. Schuster, C. Ranjan, E. Stotz, A. Knop-Gericke and R. Schlögl, *Angew. Chem., Int. Ed.*, 2013, **52**, 11660–11664.
82. S. Axnanda, E. J. Crumlin, B. Mao, S. Rani, R. Chang, P. G. Karlsson, M. O. M. Edwards, M. Lundqvist, R. Moberg, P. Ross, Z. Hussain and Z. Liu, *Sci. Rep.*, 2015, **5**, 9788.

83. H. Ali-Löytty, M. W. Louie, M. R. Singh, L. Li, H. G. Sanchez Casalongue, H. Ogasawara, E. J. Crumlin, Z. Liu, A. T. Bell, A. Nilsson and D. Friebel, *J. Phys. Chem. C*, 2016, **120**, 2247–2253.
84. D. Friebel, D. J. Miller, D. Nordlund, H. Ogasawara and A. Nilsson, *Angew. Chem., Int. Ed.*, 2011, **50**, 10190–10192.
85. M. F. Lichterman, M. H. Richter, S. Hu, E. J. Crumlin, S. Axnanda, M. Favaro, W. Drisdell, Z. Hussain, B. S. Brunschwig, N. S. Lewis, Z. Liu and H.-J. Lewerenz, *J. Electrochem. Soc.*, 2016, **163**, H139–H146.
86. S. Nemšák, A. Shavorskiy, O. Karslioglu, I. Zegkinoglou, A. Rattanachata, C. S. Conlon, A. Keqi, P. K. Greene, E. C. Burks, F. Salmassi, E. M. Gullikson, S.-H. Yang, K. Liu, H. Bluhm and C. S. Fadley, *Nat. Commun.*, 2014, **5**, 5441.
87. N. Sivasankar, W. W. Weare and H. Frei, *J. Am. Chem. Soc.*, 2011, **133**, 12976–12979.
88. D. Ferri, M. S. Kumar, R. Wirz, A. Eyssler, O. Korsak, P. Hug, A. Weidenkaff and M. A. Newton, *Phys. Chem. Chem. Phys.*, 2010, **12**, 5634.
89. D. Ferri, M. A. Newton and M. Nachtegaal, *Top. Catal.*, 2011, **54**, 1070–1078.
90. H. Öström, H. Öberg, H. Xin, J. LaRue, M. Beye, M. Dell'Angela, J. Gladh, M. L. Ng, J. A. Sellberg, S. Kaya, G. Mercurio, D. Nordlund, M. Hantschmann, F. Hieke, D. Kühn, W. F. Schlotter, G. L. Dakovski, J. J. Turner, M. P. Minitti, A. Mitra, S. P. Moeller, A. Föhlisch, M. Wolf, W. Wurth, M. Persson, J. K. Nørskov, F. Abild-Pedersen, H. Ogasawara, L. G. M. Pettersson and A. Nilsson, *Science*, 2015, **347**, 978–982.

CHAPTER 5

Evaluating Electrocatalysts for Solar Water-splitting Reactions

CHARLES C. L. McCRORY,*[a] SUHO JUNG[b] AND JEREMY KALLICK[a]

[a] Department of Chemistry, University of Michigan, Ann Arbor, MI 48109, USA; [b] Joint Center for Artificial Photosynthesis, California Institute of Technology, Pasadena, CA 91125, USA

*Email: cmccrory@umich.edu

5.1 Introduction

The efficient electrochemical conversion of H_2O to $\frac{1}{2}$ O_2 and H_2 is crucial for the development of overall water-splitting technologies. However, an important challenge in the implementation of new integrated solar-water splitting devices is the discovery of efficient electrocatalysts for the oxygen evolution reaction (OER) and the hydrogen evolution reaction (HER), the two component reactions of water splitting (Scheme 5.1). In order to address this challenge, electrocatalysis researchers have reported thousands of materials for the OER and the HER that differ in their compositions, structures, morphologies, and preparations. Standardization in the evaluation and reporting of the performance of these materials as water-splitting catalysts is necessary to assess the utility of these catalysts for integration into solar fuels devices.

In this chapter, we will provide an overview of the methods and results of the studies by the Benchmarking project at the Joint Center for Artificial Photosynthesis (JCAP) evaluating solid-state, heterogeneous electrocatalysts for the OER and the HER.[1–3] These studies were conducted under conditions

Energy and Environment Series No. 22
Integrated Solar Fuel Generators
Edited by Ian D. Sharp, Harry A. Atwater and Hans-Joachim Lewerenz

Published by the Royal Society of Chemistry, www.rsc.org

$2\,H^+(aq) + 2\,e^- \rightleftarrows H_2(g)$	HER
$2\,H_2O(l) \rightleftarrows O_2(g) + 4\,e^- + 4\,H^+$	OER
$2\,H_2O(l) \rightleftarrows O_2(g) + 2\,H_2(g)$	Overall Water Splitting

Scheme 5.1

considered relevant to an integrated solar-water splitting device under 1 sun illumination, strongly acidic or alkaline electrolyte (*e.g.* 1 M H_2SO_4 or 1 M NaOH/KOH) under 1 atm of O_2 (in the case of OER) or 1 atm H_2 (in the case of HER). The chapter will specifically overview the choice of appropriate electrochemical apparatus and electrodes for conducting electrocatalytic studies, provide details regarding the use and limitations of the benchmarking protocols used in our studies to evaluate catalyst activity and stability, and summarize the general results of our investigations of catalyst performance.

5.2 Experimental Considerations

5.2.1 Cell Design

The typical electrochemical cells we use to evaluate electrocatalytic water splitting by heterogeneous catalysts are two-chamber U-cells (or H-cells) in which the auxiliary electrode is sequestered in a separate chamber from the working and reference electrodes.[2,4] Note that using glass electrochemical cells with alkaline NaOH solutions can lead to the possible introduction of trace transition metal impurities due to glass corrosion.[5–7] The two chambers are separated by a semi-permeable barrier which inhibits the mass transport of gases and solvated ions between the working and auxiliary electrodes.[8] This separation is important in electrochemical water splitting measurements because O_2 and H_2 are generated concurrently at the working and auxiliary electrodes. Without a semi-permeable barrier separating the two chambers, O_2 (or H_2) generated at the auxiliary electrode during HER (or OER) could be reduced (or oxidized) at the working electrode, leading to inaccurate measurements of catalytic current.[4] This is primarily an issue in long-term electrolyses where large amounts of both gases are produced. Moreover, O_2 and H_2 generated concurrently in a single-chamber cell could, in theory, create an explosive or flammable mixture of gases,[9] although the amount of gas produced in typical benchmarking measurements is sufficiently small that this risk is minimal.

Single or double fine porosity glass frits sufficiently attenuate the transport of gas bubbles from one cell compartment to the other, and are a frequently used and acceptable semipermeable barrier for typical short-term electrochemical measurements.[10,11] However, because glass frits do not prevent the transport of ions or dissolved gases between the chambers, they are not as effective for long-term measurements. This limitation is especially important when conducting Faradaic efficiency measurements by controlled-current or controlled-potential electrolysis in a sealed cell, where the pressure gradients generated during electrolysis make gaseous crossover through the frit more likely.[4]

Gas-impermeable ion-exchange membranes, as described in greater depth in Chapter 10 by Miller and Houle within this book, are a viable alternative to porous glass frits as barriers to gas transport between the working and auxiliary electrode in two-chamber electrochemical cells.[4,12,13] Cation exchange membranes such as Nafion® facilitates the transport of H^+ in acidic solutions while inhibiting the transport of anions and dissolved gases,[4,12] whereas anion-exchange membranes such as Selemion® AMV and NEOSEPTA® AMX promote the transport of OH^- in alkaline solutions and inhibit the transport of cations and dissolved gases.[13] Careful selection of the membrane material is necessary to prevent contaminating the working electrode by specific species produced at the auxiliary electrode.

5.2.2 Auxiliary Electrode

Generally, the selection of the auxiliary electrode should be centered on minimizing potential interactions between species produced at the auxiliary electrode and the working electrode during the electrochemical measurements.[14] Platinum wires, meshes, and gauzes are a popular choice for auxiliary electrodes because of their conductivity and high resistance to corrosion. However, platinum has a propensity for dissolution into platinum (IV) and (VI) under anodic currents.[15,16] The rate of dissolution is enhanced in coordinating acid media (*e.g.* H_2SO_4) and in the presence of the chloride ion. Studies have shown that during long experiments, the dissolved platinum cations can diffuse through the glass frit in a dual compartment electrolysis cell and electrodeposit onto the cathode.[17] This presents an immediate problem when studying the HER, where the working and auxiliary electrodes are held at potentials that facilitate the oxidation of platinum at the auxiliary electrode and subsequent electrodeposition onto the working electrode. Because of platinum's high activity for the HER, seemingly small amounts of platinum electrodeposited on the working electrode can account for large measured HER activity. An anion-exchange membrane should be used to separate the auxiliary and working electrodes if one is using a Pt auxiliary electrode during HER measurements.

Carbon electrodes, including multiple forms of glassy carbon and graphite, represent an attractive alternative to platinum for use as auxiliary electrodes because of their comparative chemical inertness, low cost, high conductivity, and electrochemical inactivity over a wide potential range.[18] The largest concern about using a carbon rod as an auxiliary electrode is the possible oxidation of graphitic carbon to carbon dioxide and carbon monoxide.[19] If carbon monoxide crossed over into the working electrode chamber it could poison or otherwise affect the catalyst being tested. While a glass frit would not be effective at preventing gas diffusion, an ion-exchange membrane should prevent CO crossover. Note that the electrochemical corrosion of carbon is kinetically slow under normal HER and OER conditions, making this concern somewhat minor when evaluating water splitting catalysts.[19] We have found that commercial graphite rods (available as >99.99% graphitic carbon rods from several commercial sources including Strem Chemicals, Inc., Goodfellow USA, Alfa Aesar, among many others) are a relatively inexpensive and effective alternative to platinum wires as auxiliary electrodes in typical benchmarking experiments.

5.2.3 Reference Electrodes

The most convenient standard for comparing catalytic activity for the OER and the HER is the measured potential *versus* the reversible hydrogen electrode (RHE), which is defined by the half reaction shown in eqn (5.1):

$$2\ H_3O^+(aq) + 2\ e^- \rightleftarrows H_2(g) + H_2O(l) \tag{5.1}$$

The RHE typically consists of a platinized platinum wire or foil immersed in an aqueous solution with known activities of H^+ and H_2. In order to maintain a constant concentration of $H_2(aq)$, H_2 is bubbled continuously across the Pt surface. Because the RHE is immersed in the same electrolyte solution in which the electrochemical water-splitting measurements are occurring, the thermodynamic potential for the HER is 0 V *vs.* RHE, independent of solution pH. Likewise, the thermodynamic potential for the OER is 1.23 V *vs.* RHE.

Although the RHE provides a convenient standard reference potential for water-splitting potentials, practical considerations limit its utility as an actual reference electrode for activity measurements. For instance, common impurities in the H_2 gas used to maintain the RHE, including CO and O_2, can poison the immersed platinum foil and cause potential drift.[20,21] The requirement for an oxygen-free environment and continuously bubbling high purity H_2 make the RHE an inconvenient reference electrode for measuring electrochemical water-splitting reactions. Instead, researchers typically use an alternative commercial reference electrode for actual measurements, and then convert their measured potentials to an RHE scale.

One of the most commonly used commercially available reference electrodes is the saturated calomel electrode (SCE). The SCE interface consists of a paste of Hg_2Cl_2 and KCl solution layered on a small amount of Hg(*l*). SCEs are stable in acidic and neutral media,[22] showing negligible potential drift (~10 μV) over the course of several hours of immersion in electrolyte. However, during longer measurements (*ca.* 24 h) in highly acidic conditions (>1 M) and at elevated temperatures (>45 °C), the acid-promoted disproportionation of $HgCl_2$ to species such as Hg_2Cl^+ and $HgCl^+$ can result in positive potential drift.[23] In most applications these disproportion effects will be negligible because the reference is separated from strongly acidic solutions by a frit, thus slowing the diffusion of H^+ into the reference electrode. SCE electrodes can also suffer from mercury leaking into the electrolyte solution through the formation of nanometer scale colloidal droplets,[24] which can deposit onto the working electrode surface and decrease electrocatalytic activity. The formation of these Hg droplets is due to the spontaneous emulsification of mercury at the mercury–water interface.[25] While commercial SCE references are constructed to prevent Hg(*l*) migration outside of the electrode, when using SCE reference some leakage may still occur and so the electrochemical cell solution should be changed regularly.

When conducting experiments in 1 M sulfuric acid using an SCE reference, the formation of small amounts of Hg_2SO_4 in the reference electrode may occur, causing potential drift. A possible alternative to the SCE for use in sulfate solutions, or where the Cl^- ions in the SCE could contaminate the electrolyte solution, is the mercury sulfate reference electrode (MSRE).[26] The MSRE consists of metallic mercury coated with a mercury sulfate slurry in a saturated solution of K_2SO_4 or H_2SO_4. However, use of the MSRE has some slight drawbacks due its use of Hg_2SO_4, which has tendency to hydrolyze and is relatively soluble in aqueous solutions compared to other mercury salts.[27]

Both SCE and MRSE electrodes can exhibit noticeable drift in the reference potential when exposed to alkaline solutions for long times due to the formation of HgO(*s*). Although we have found that the reference potential for an SCE does not appreciably drift in 1 M NaOH during the course of our activity and short-term stability measurements, an alternative reference electrode that may be more appropriate for use in alkaline solutions is the mercury–mercuric oxide (Hg/HgO) reference electrode. The reference potential of the Hg/HgO reference electrode depends on the concentration of OH^- in the filling solution. Commercial HgO electrodes are available with known concentrations of NaOH. Saturated solutions of $Ba(OH)_2$ and $Ca(OH)_2$ are also sometimes used. The mercury–mercuric oxide reference has been shown to be particularly stable and well suited for use in alkaline media.[7,28,29]

The silver–silver chloride reference is one of the most used reference electrodes due its low cost and ease of construction.[21] A standard silver–silver chloride electrode consists of a silver wire coated with silver chloride immersed in a solution of KCl. In order to obtain reproducible results when using a silver–silver chloride electrode, the chloride ion concentration must be constant. The concentration is typically controlled by using 3.5 M KCl or saturated (~4.77 M) KCl filling solution. The reference electrode solution should also be saturated with AgCl in order to prevent the stripping of AgCl from the silver wire by from formation of $AgCl_{2-}$ (*aq*).[30] Although silver–silver chloride electrodes are generally stable, some drift may occur in reducing[31] or highly basic solutions.[32]

Because of the possibility of any reference electrode to undergo potential drift over time it is important to regularly compare the reference to an external standard. Ferrocenecarboxylic acid in 0.2 M pH 7 phosphate buffer is a convenient external standard with a reversible redox couple of 0.284 V *vs.* SCE at pH 7 (0.525 V *vs.* SHE).[33] The ferro/ferricyanide redox couple can also be used as an external reference. It has a reversible redox couple at of 0.112 V *vs.* SCE (0.356 V *vs.* SHE). However, the potential of the ferro/ferricyanide redox couple shifts based on the ratio of the salts, the total ionic strength, and the presence of any other electrolytic compounds.[34,35] Therefore, if employing the ferro/ferricyanide redox couple as an external reference it is essential to note the concentration and ratio and compare it to previously tabulated values. An alternative to using the redox couples of dissolved species is to instead directly compare the reference electrode to a reversible hydrogen electrode. This is done by measuring the open circuit potential of a Pt wire in the H_2-saturated electrolyte solution to be used in the electrochemical measurements.[30]

The ideal reference electrode for a specific set of experimental conditions may differ, so the measured potentials must be normalized to a single reference for comparison. As mentioned previously, the most convenient reference standard for water electrolysis is the RHE. The potentials determined by the aforementioned reference electrodes can all be converted to RHE using the relationship in eqn (5.2)

$$E_{\mathrm{RHE}} = E_{\exp} + E_{\mathrm{ref}} + 0.059V * \mathrm{pH} \tag{5.2}$$

Here, E_{RHE} is the potential *vs.* RHE, $E_{\exp}$ is the measured potential, E_{ref} is the potential of the reference electrode *vs.* SHE, and pH is the pH of the electrochemical cell solution.

5.2.4 Working Electrode Material

In order to effectively evaluate and compare the performance of water-splitting catalysts, they first must be adhered to working electrodes that do not appreciably contribute to the measured catalytic activity. An ideal working electrode material has metallic conductivity but shows no activity for HER or OER, thus ensuring that the measured activity from an electrocatalyst deposited on the working electrode comes entirely from the deposited electrocatalyst. We have found that glassy carbon is an appropriate support material for HER and OER under most conditions. It has very high conductivity[18] and demonstrates relatively low activity towards OER and HER in both acid and alkaline solution at moderate potentials.[36] Glassy carbon is resistant to chemical attack under conditions relevant to electrochemical water splitting,[18,37] and degradation of the material typically occurs only after exposure to oxygen at high temperatures (800 °C),[38] highly corrosive substances such as HF,[39] or potentials much more positive than those studied in typical OER experiments.[40]

Electrocatalysts are also sometimes deposited onto metal supports including nickel,[41,42] copper,[43,44] and titanium.[45,46] However, using these metal supports can lead to difficulties in measuring electrocatalytic HER/OER performance because the underlying support can demonstrate innate HER/OER activity.[47–50] When using metal supports, careful controls must be taken in order to ensure that measured activity comes solely from the deposited catalyst. In addition, corrosion of the metallic supports is possible under anodic conditions in acidic solutions.[49,51–53]

For many electrocatalytic materials that have been investigated for the OER and the HER, the reported electrocatalytic activity is similar regardless of whether the material is deposited onto glassy carbon or a metallic support as long as the deposition conditions remain constant. An important exception is $NiCeO_x$, which shows higher activity for the OER when deposited onto a Ni surface compared to when deposited on glassy carbon.[3] The reason for the enhanced activity of this catalyst on Ni has not been determined. In addition it has been reported that thin films of cobalt,[54] nickel,[55] and manganese oxides,[56] show enhanced activity for the OER when deposited onto a Au electrode compared to other supports, although there is discussion within the community as to the primary cause of this enhancement.[57] Regardless of the cause of the enhancement, its existence underscores the importance of selecting working electrode support materials that do not affect catalytic measurements when attempting to evaluate electrocatalytic activity.

5.2.5 Catalyst Deposition and Characterization

Multiple strategies exist for depositing electrocatalytic films onto the working electrode including electrodeposition, reactive sputtering, atomic-layer deposition, chemical vapor deposition, adhering nanoparticles in thin polymer films, ink-jet printing onto electrode supports (see Chapter 9 by Gregoire *et al.*), *etc.* Electrocatalytic films evaluated in our benchmarking studies were deposited primarily *via* electrodeposition,[2,3] or the adherence of nanoparticles to the surface with a thin Nafion® film.[1]

Electrodeposition involves immersing a polished electrode support in a deposition solution consisting of soluble metal precursors and any additional necessary additives. The electrode is then polarized at a specific potential to facilitate the formation of an electrocatalyst film. Electrodeposition can be used to synthesize a wide variety of single and mixed metal catalysts,[2,49,58,59] and the electrocatalytic activity of the catalyst material depends on the electrodeposition conditions.[49,59] Electrodeposited catalysts are often amorphous films,[49] making it difficult to characterize the composition and structure of electrodeposited catalysts. In our benchmarking studies, the materials were classified instead by their deposition conditions and catalytic activity.

In addition to electrodeposited films, nanoparticulate catalysts are among the most commonly-studied materials for the OER and the HER. Nanoparticulate catalysts offer a greater degree of control of both surface structure and shape compared to electrodeposited films,[60,61] and their well-defined material format makes it easier to characterize their composition and structure. Typically, nanoparticles are adhered to the catalyst surface by first mixing them with a polymer such as Nafion® which acts as a binder and a solvent that helps disperse the ionomer.[1,62–66] The resulting nanoparticle ink is then drop-casted onto the electrode surface and dried, resulting in an adhered nanoparticle catalyst film.

To increase the overall conductivity of deposited nanoparticle films, the nanoparticles are often supported on carbon black, a form of graphitized carbon that consists of high surface area particles of <50 nm diameter.[67–69] Supporting the nanoparticles on the carbon black prior to deposition onto the working electrode has the effect of increasing the overall conductivity between the nanoparticle catalyst film and the electrode, and can help prevent agglomeration of the catalyst nanoparticles thereby increasing their catalytic surface areas with smaller loadings.[70,71] This can result in an increase in the observed electrocatalytic performance.[68] However, using a carbon support complicates measurements of catalyst surface area for non-Pt metals because the measured surface area will include that of the high-surface area carbon support. Moreover, carbon black is sensitive to oxidative corrosion at potentials more positive than 0.6–1.2 V *vs.* RHE in both acidic[68,72,73] and alkaline conditions,[74] and the rate of the corrosion is significantly enhanced by the presence of OER electrocatalysts.[74,75] It is important to note that different carbon supports, such as graphene and carbon nanotubes, may have enhanced oxidative stability compared to carbon black.[68] Regardless, in our benchmarking studies we chose to measure the activity of unsupported nanoparticles films deposited onto the glassy carbon working electrode due to concerns over the possible corrosion of carbon black supports, and to avoid the difficulty in comparing the specific activities of carbon-black supported materials.

5.3 Catalyst Performance

We have previously proposed a general protocol for the screening of electrocatalysts for the HER and the OER designed to meet two specific goals: (1) primarily using experiments that are readily accessible to a typical researcher in electrocatalysis, and (2) to minimize the time and number of experiments required to make meaningful comparisons of catalytic performance to enable a throughput of ~2 catalysts evaluated per week.[1,58] An outline for the procedure can be seen in Figure 5.1. We suggest that every water-splitting electrocatalyst reported be studied—*at minimum*—for its elemental composition, catalytic activity, and short-term stability. Note that all activity and stability measurements should be conducted under 1 atm of H_2 or O_2 in order to establish the equilibrium conditions necessary to define the thermodynamic potential for the HER or the OER, respectively. For particularly active catalysts, additional experiments should be conducted to confirm the Faradaic efficiency for oxygen or hydrogen production, and the longer-term stability under device-relevant conditions. Surface area measurements can be used to estimate the specific activity per catalyst site, although difficulties in accurately determining the electrochemically-active surface area (ECSA) can complicate such comparisons.

We suggest that the primary figure of merit for an electrocatalyst for water-splitting under conditions relevant to an integrated solar-fuels device is the overpotential, or extra potential beyond the thermodynamic potential for the

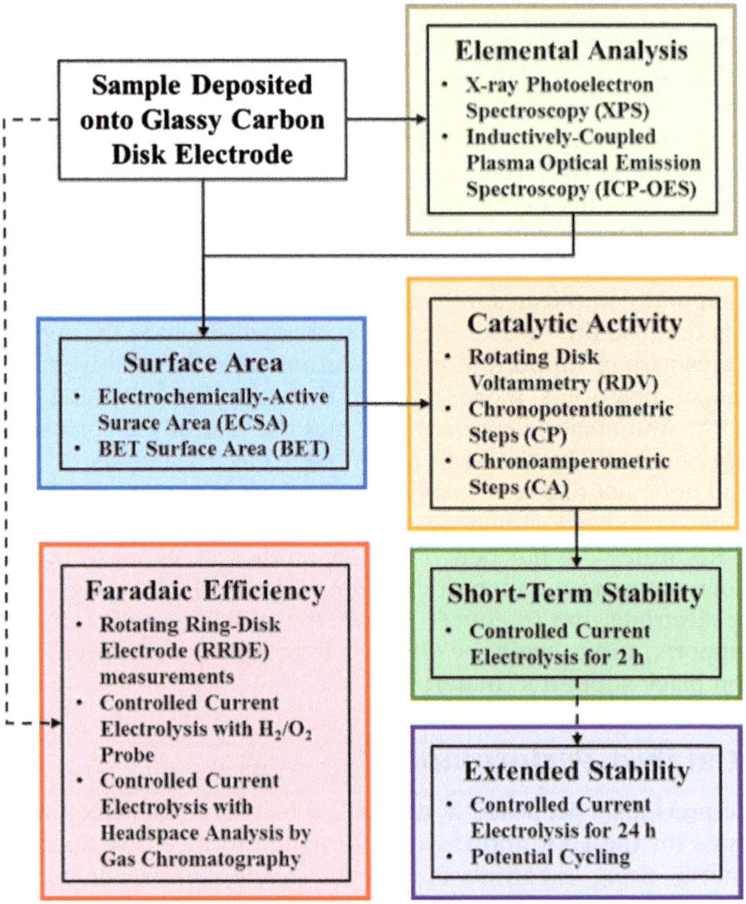

Figure 5.1 General protocol for evaluating the performance and stability of heterogeneous electrocatalysts for the HER and the OER.
Adapted with permission from C. C. L. McCrory, S. Jung, I. M. Ferrer, S. M. Chatman, J. C. Peters, and T. F. Jaramillo, *J. Am. Chem. Soc.*, 2015, **137**, 4347, Copyright 2015 American Chemical Society.

OER or HER, required to drive the electrocatalytic reaction at a magnitude current density of 10 $mA\,cm^{-2}$ geometric surface area, or $\eta_{j=10\,mA\,cm_{geo}^{-2}}$.[1–3] This current density (10 $mA\,cm_{geo}^{-2}$) is roughly the current density with which a 10% efficient integrated solar-water splitting device is expected to operate under 1 sun illumination.[76–78] Note that using the $\eta_{j=10\,mA\,cm_{geo}^{-2}}$ metric as an effective figure of merit for electrocatalytic water splitting performance requires that the catalyst material be deposited onto a planar working electrode surface, such as a 5 mm diameter disk electrode. Whenever possible, the specific activity normalized per geometric area or mass loading should be included to provide information regarding the intrinsic activity of the catalyst material.

5.3.1 Elemental Analysis

The extent to which a catalyst's composition and structure are determined will depend on the type of catalyst being investigated, and the purpose of the investigation. We suggest at minimum that after deposition of the electrocatalyst on the working electrode, the resulting surface be analyzed using X-ray photoelectron spectroscopy (XPS) survey scans from 0 to 1200 eV to confirm the qualitative elemental composition of the catalyst surface, and hi-resolution scans between 270 and 340 eV to test for the presence of noble metal contaminants including Pt, Ir, and Ru.[2,3] When reporting new catalyst materials, more in-depth elemental analysis of the deposited catalysts should be conducted using XPS hi-resolution scans of metal regions to quantify elemental ratios and oxidation states. However, given the limited sensitivity of XPS for minority elemental species, complimentary approaches to composition analysis, such as dissolution of the catalyst in acid followed by use of inductively-coupled plasma optical emission spectroscopy (ICP-OES) to quantify metal composition or other techniques, should be performed. For crystalline materials, XRD should be used to confirm the phase of the bulk material. Because the surface composition of the material may change during the course of electrocatalytic measurements, it may be useful to measure the elemental composition using XPS post-electrolysis to determine the extent of these changes.[79]

5.3.2 Catalytic Activity

In order to measure the activity of an electrocatalyst for the HER or the OER under steady-state conditions, a combination of rotating disk voltammetry (RDV), chronoamperometric steps, and chronopotentiometric steps are used at a constant rotation rate of 1600 rpm. For HER and OER measurements, the concentration of substrate is sufficiently large such that the rotation rate does not appreciably influence the delivery of the substrate to the catalyst surface. Instead, the rotation rate of 1600 rpm is used to sweep product bubbles from the catalyst surface. A scan rate of 0.01 $V\,s^{-1}$ is used to ensure steady-state behavior at the electrode surface, and this is confirmed by performing controlled-current chronopotentiometric (CP) steps and controlled-potential chronoamperometric (CA) steps under identical conditions.[58] In the CP (and CA) measurements, the current (or potential) is held constant for 30 s which allows the measured potential (or current) to achieve steady-state, typically after ~2 s.[80,81] The resulting steady-state potentials and currents are overlaid with the RDV measurements, and close agreement in the overlays confirms steady-state behavior. An example of the overlay of RDV, CA, and CP measurements for an electrodeposited Ni oxide catalyst confirming steady-state behavior under the reaction conditions is shown in Figure 5.2. The magnitude overpotential necessary to achieve 10 $mA\,cm_{geo}^{-2}$ can be determined from these activity measurements.

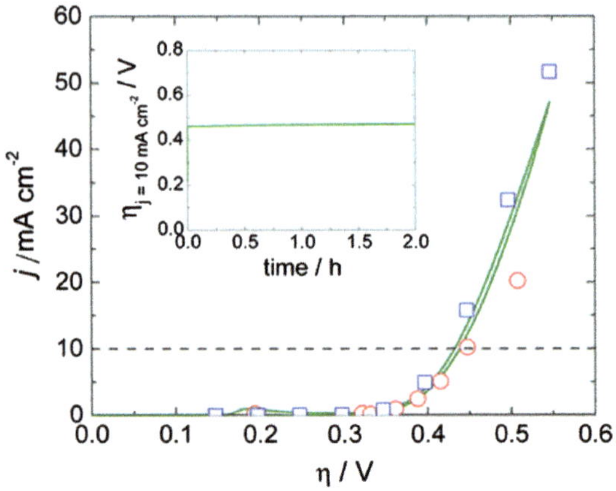

Figure 5.2 Representative activity and short-term stability measurements for the OER at an electrodeposited Ni oxide catalyst on a glassy carbon rotating disk electrode in O_2-saturated 1 M NaOH at 1600 rpm and 0.01 V s^{-1} scan rate. The green line is and RDV measurement, and the blue squares and red circles overlaid are results from 30 s^{-1} CP and CA measurements. The close overlay of the data near 10 mA cm^{-2} confirms steady-state conditions. The inset is a 2 h controlled-current electrolysis at 10 mA cm^{-2} that shows stable catalysis over the course of the electrolysis. Reprinted with permission from C. C. L. McCrory, S. Jung, J. C. Peters, and T. F. Jaramillo, *J. Am. Chem. Soc.*, 2013, **135**, 16977, Copyright 2013 American Chemical Society.

5.3.3 Short-term Stability

The short-term stability of an electrocatalytic material is determined from 2 h controlled-current electrolysis at a constant current density of 10 mA cm_{geo}^{-2} at 1600 rpm rotation rate. The potential required to maintain the steady-state current density is then recorded as a function of time. By comparing $\eta_{j=10\,mA\,cm_{geo}^{-2}}$ at time zero to that after two hours of constant polarization, one can ascertain the stability of the catalyst performance. No change in the measured $\eta_{j=10\,mA\,cm_{geo}^{-2}}$ after 2 h of polarization from the value at time zero is evidence of short-term catalyst stability, whereas an increase in the magnitude of $\eta_{j=10\,mA\,cm_{geo}^{-2}}$ after 2 h of polarization is evidence of a loss of catalytic activity over time. Representative data from a 2 h short-term stability measurement for the OER with an electrodeposited Ni oxide catalyst is shown in the inset of Figure 5.2. The method of catalyst deactivation is not evident from the experiment and could be due to corrosion, material degradation, surface passivation, catalyst film delamination, or other mechanisms. Conversely, a decrease in the magnitude of $\eta_{j=10\,mA\,cm_{geo}^{-2}}$ after 2 h of polarization is evidence of an increase in catalytic activity over time, which may be due to catalyst conditioning or surface activation.

An additional consideration when interpreting catalyst stability is the effect film thickness can have on the apparent stability of a catalyst. For example, Ir oxide films have been reported to be unstable in alkaline solutions under anodic conditions,[82–85] possibly due to the dissolution of oxidized Ir(VI) ions.[85–87] However, thick Ir oxide films may exhibit enhanced stability over the course of short-term stability measurements in 1 M NaOH compared to thinner films,[58] possibly because comparatively more material remains in the thicker film during the short-term stability measurement. These results highlight an important point that the stability referred to here is an operational stability—a measure of a catalyst's ability to maintain a certain activity level over a fixed amount of time—and is not a measure of the catalyst's resistance to corrosion or degradation.

5.3.4 Extended Stability

For catalysts that show promising HER or OER activity and 2 h stability, additional experiments to assess the extended stabilities of the catalysts should be conducted. One such extended stability measurement is a 24 h stability test in which the electrocatalyst is held at 10 $\text{mA cm}_{\text{geo}}^{-2}$ for 24 h during which the operating potential is measured. An increase in the magnitude of $\eta_{j=10\,\text{mA cm}_{\text{geo}}^{-2}}$ during the 24 h electrolysis at 10 $\text{mA cm}_{\text{geo}}^{-2}$ is indicative of a loss of catalyst activity. Although 24 h stability tests are useful to determine the catalyst stability under singular catalytic conditions, they do not show the stability of the catalyst when exposed to the transient conditions expected in water-splitting devices attached to intermittent solar-energy sources.[88] In order to test for stability under transient conditions, potential cycling experiments are conducted where the potential is continually cycled from near the thermodynamic potential to the potential at which ~100 $\text{mA cm}_{\text{geo}}^{-2}$ is initially reached.[2,89,90] Examples of 24 h stability measurements and potential-cycling stability measurements for a NiMo HER catalysts in 1 M H_2SO_4 are shown in Figure 5.3.

The importance of conducting appropriate extended stability measurements is illustrated by recent investigations of the stability of electrodeposited NiMo catalysts during HER measurements in acidic solution. Although previous studies have reported that NiMo nanoparticles are moderately unstable under HER conditions in acidic solution,[91] we found that the electrodeposited NiMo catalysts show good 2 h and 24 h stability under HER conditions in 1 M H_2SO_4.[2] However, during potential-cycling measurements between −0.1 V to 0.1 V *vs.* RHE, there is a modest increase in the magnitude of the overpotential needed to achieve 10 mA cm^{-2} of roughly 0.05 V from $\eta_{j=10\,\text{mA cm}_{\text{geo}}^{-2}} = 0.4$ V to $\eta_{j=10\,\text{mA cm}_{\text{geo}}^{-2}} = 0.87 \pm 0.046$ V, as shown in Figure 5.3. Longer-term cycling experiments resulted in more appreciable catalyst deactivation, with an increase in the magnitude overpotential at 10 mA cm^{-2} of ~0.15 V after 20 000 cycles, and ~0.4 V when cycled more than 30 000 times.[2] The electrodeposited NiMo system we investigated shows comparable stability with numerous other reported acid-stable HER catalysts subjected to 16–24 constant polarization or ≤10 000 potential cycles,[92–96]

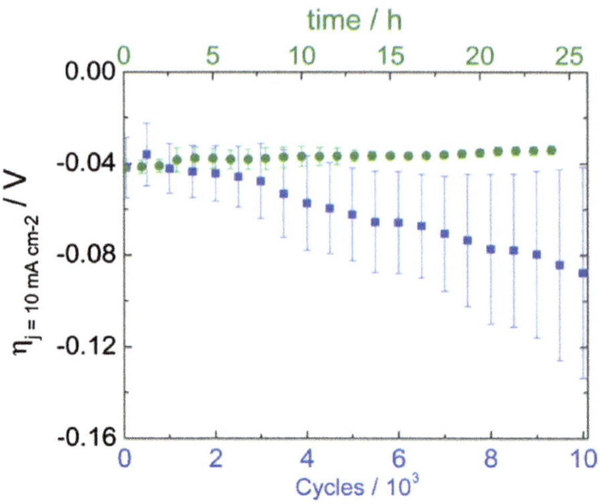

Figure 5.3 Extended stability measurements for HER with an electrodeposited NiMo catalyst on a glassy carbon surface. The blue squares are average measured overpotentials at 10 mA cm_{geo}^{-2} from 30 s CP steps following every 500 potential cycles in a 25 h, 10 000 potential cycle experiment plotted as a function of the number of cycles on the bottom *x*-axis. The green circles are the average measured overpotentials at 10 mA cm_{geo}^{-2} after 1 h intervals during a 24 h controlled-current electrolysis measurement. The small change in the operating overpotential suggests that the catalyst remains active over the course of both extended stability experiments.
Reprinted with permission from C. C. L. McCrory, S. Jung, I. M. Ferrer, S. M. Chatman, J. C. Peters, and T. F. Jaramillo, *J. Am. Chem. Soc.*, 2015, **137**, 4347, Copyright 2015 American Chemical Society.

and only shows substantial loss of catalytic activity after an aggressive 20 000–30 000 cycle potential-cycling test. These results highlight the need for the development and adoption of accelerated lifetime testing protocols for OER and HER electrocatalysts.

One possibility for accelerated lifetime stability measurements is through the incorporation of an electrochemical quartz crystal microbalance (EQCM) in order to measure catalyst dissolution under relevant conditions.[97–99] Using an EQCM as a working electrode allows for the observation of minute mass changes on the order of 1 ng cm^{-2} during the electrolysis experiment, allowing for the real-time detection of catalyst dissolution even for slow corrosion rates. EQCM dissolution measurements can be supplemented with inductively coupled plasma mass spectroscopy (ICP-MS) analysis of the electrolyte solution post-electrolysis to look for the presence of dissolved metal ions which would be indicative of catalyst dissolution.[100] In addition to EQCM, comparisons of pre- and post-electrolysis compositional analyses can provide information regarding possible catalyst deactivation or corrosion.

5.3.5 Faradaic Efficiency Measurements

For electrocatalysts that show promising electrochemical activity for the HER or the OER, it is important to confirm that materials catalyze the production of H_2 or O_2 with 100% Faradaic efficiency. Here, the Faradaic efficiency is the percentage of the charge from an electrochemical measurement that goes towards the production of a specific product. This ensures that the measured activity results come from the formation of useful products, rather than from the formation of undesired side products, or non-productive side reactions such as the oxidation/reduction of surface sites or slow corrosion of the catalyst material. Two methods that are often used to assess Faradaic efficiency are to quantify product formation post-electrolysis with gas chromatography (GC) or O_2/H_2 probes, or to use rotating-ring disk electrode measurements to measure real time product evolution.

Rotating ring-desk electrode (RRDE) measurements are sometimes used to determine the Faradaic efficiency of electrocatalysts for water-splitting reactions.[3,101–103] An RRDE consists of a central disk working electrode with a separate ring electrode a fixed distance from the center. The potential of each electrode is controlled separately. Species produced at the central disk during a voltammetric sweep or short CP or CA step are collected (oxidized or reduced) at the outer ring.[104] The benefit of this method is that it allows for relatively rapid screening of Faradaic efficiency compared to electrolysis measurements, and provides real-time analysis of product formation. However, because only solution-phase species can be detected at the ring, the catalyst must be operated at sufficiently low current densities, typically $\leq$1 mA cm_{geo}^{-2} rather than the standard 10 mA cm_{geo}^{-2},[101] to prevent local product saturation and bubble formation at the disk. Moreover, RRDE electrodes typically have low (15–20%) collection efficiencies, where the collection efficiency is determined from measuring the percentage of a probe species reduced at the electrode disk that is then reoxidized at the ring. The low collection efficiencies in RRDE measurements suggest that it is not sensitive to small differences in Faradaic efficiency, and small deviations in the ring current or measured collection efficiency can result in large changes in the measured Faradaic efficiency.

A more widely-used alternative to RRDE measurements for Faradaic efficiency determination is to electrocatalytically generate H_2 or O_2 from a controlled-current electrolysis experiment, and then quantify the product using either O_2 or H_2 probes, or gas chromatography (GC). Electrolyses are conducted for sufficiently long times to produce an appreciable amount of product, and are conducted in a two-compartment sealed electrolysis cell in which the auxiliary electrode is separated from the working electrode by a gas-impermeable membrane. When using O_2 or H_2 probes to quantify products, a sealed electrolysis cell is completely filled with electrolyte solution so that there is no appreciable gaseous headspace, and the concentration of the dissolved gases is continuously monitored throughout a controlled-potential electrolysis at 10 mA cm_{geo}^{-2}. When using GC to detect products, the sealed electrolysis cell is filled with electrolyte leaving a small headspace of known volume. A sample of this headspace is collected post-electrolysis, and the amount of H_2 or O_2 is then quantified using GC. Some benefits of quantifying the O_2 and H_2 produced post-electrolysis compared to RRDE measurements are (1) the electrolyses can be run at the standard 10 mA cm^{-2} current density relevant to device conditions, and (2) the quantification is more sensitive to small variations in Faradaic efficiency due to the comparatively large amounts of gas evolved. However, the electrolysis experiments typically take significantly longer to set up and conduct compared to RRDE measurements, and are cumulative measurements of total gas evolved that do not show real-time analysis of the product evolution.

5.3.6 Measuring Catalyst Surface Area

In order to evaluate the *intrinsic* activity of catalysts with different surface areas or loadings, one must normalize the measured activity by the number of active sites. Because the number of active sites on a catalyst surface is related to the electrochemically-active surface area (ECSA) of the material, the specific activity per ECSA is an effective measurement to compare the intrinsic activity of catalyst materials. However, accurately determining the ECSA for electrocatalytic materials is non-trivial, and general methods used for measuring ECSA cannot be applied universally to all materals.[100] In this section, we will limit our discussion to the approach we used in our benchmarking studies to determine order-of-magnitude approximations of the ECSA of catalyst based on double-layer capacitance estimates, and we will highlight the significant limitations of this approach. For more comprehensive discussion of the various methods that have been used to estimate ECSA for specific materials, we refer the reader to the extensive literature on this topic (a small subset is cited here).[7,105–109]

In our benchmarking studies, the ECSA of the catalysts were approximated by estimating the double-layer capacitance of the surface from the capacitative current in cyclic voltammograms (CVs).[1–3] This is done by first choosing a potential range where there is minimal contribution from Faradaic processes and all measured current can be assumed to be due to double layer charging. The double-layer charging current is related to the double-layer capacitance according to eqn (5.3)

$$i_c = \nu C_{DL} \tag{5.3}$$

Here, i_c is the measured double-layer charging current, ν is the scan rate of the CV, and C_{DL} is the electrochemical double-layer capacitance. CVs are then collected at different scan rates in the non-Faradaic region, and the plot of i_c *versus* ν results in a straight line with slope equal to C_{DL}. The ECSA is calculated by dividing C_{DL} by the specific capacitance of the material, C_S, according to eqn (5.4)

$$\text{ECSA} = \frac{C_{DL}}{C_S} \tag{5.4}$$

Based on typical values reported for metal electrodes in acidic and alkaline solutions, general values of of $C_S = 0.035$ mF cm^{-2} in 1 M H_2SO_4 and $C_S = 0.040$ mF cm^{-2} in 1 M NaOH are used to approximate ECSA.[3] The ECSA values reported are suggested as order-of-magnitude approximations of the surface area, and are intended only as approximate guides for comparisons of catalyst area, and are not meant to be interpreted as accurate measurements of ECSA.[3]

It is important to note the significant limitations of this approach to determine ECSA. First, because a general value for C_S was used in these studies, we estimate that errors as large as a factor of seven can exist in the ECSA approximations due to the large range of C_S values reported for metal electrodes.[3] Moreover, in the case of electrodeposited oxy(hydroxide) catalyst films, electrolyte ions can freely permeate into the catalyst film, and as a result the surface area of the film may not contibute to the electrochemical double-layer.[110–113] Thus, the measured C_{DL} may not be reflective of the electrochemical surface areas of oxy(hydroxide) catalyst films. Furthermore, the C_{DL} values can be heavily influenced by the conductivity of the film as determined by the material and the potential range of the C_{DL} measurement. For example, it has been recently reported that the measured C_{DL} of a NiFe oxy(hydroxide) film increases linearly with film deposition when measured at a potenetial at which the film is conductive, whereas the C_{DL} measured at a potential at which the film is insulating shows a constant C_{DL} similar to that of the working electrode substrate.[7,114] This suggests that the benchmarking approach to approximate ECSA may be valid for only those materials that are conductive in the potential range of the C_{DL} measurement.[114]

As a check of the validity of our ECSA approach, we compared the *total* surface area of several metal–oxide particle catalysts as determined by nitrogen gas physisorption analysis (BET) to the ECSA estimated from the measured C_{DL} of a non-Faradaic region (Figure 5.4). Here, the measured ECSA is converted into units of m^2 g^{-1} to facilitate direct comparison with the measured BET surface areas. For most metal–oxide nanoparticles, there is no obvious correlation between the BET surface area and the ECSA. In particular, the BET surface areas of the metal oxide particles range over four orders of magnitude between 0.1 and 1000 m^2 g^{-1}, but the measured ECSA values for most of the metal oxide particles vary only from 0.1 to 1 m^2 g^{-1}.[1] Some important exceptions include the conductive oxides IrO_2, RuO_2, $LaCoO_3$, and $LaNiO_3$. For each of these materials, the ECSA and BET surface area measurements yield similar results.[1] This further supports that ECSA measurements may be more valid for conductive materials compared to more insulating metal oxides.

While empirical measurements of surface area using techniques such as BET or electron microscopy can be used to describe the *total* surface area of the material, they provide little insight into what fraction of the total area is electrocatalytically active. Because the ECSA is a subset of the total surface area, using the total surface area determined from BET or electron microscopy to normalize activity measurements may underestimate the specific

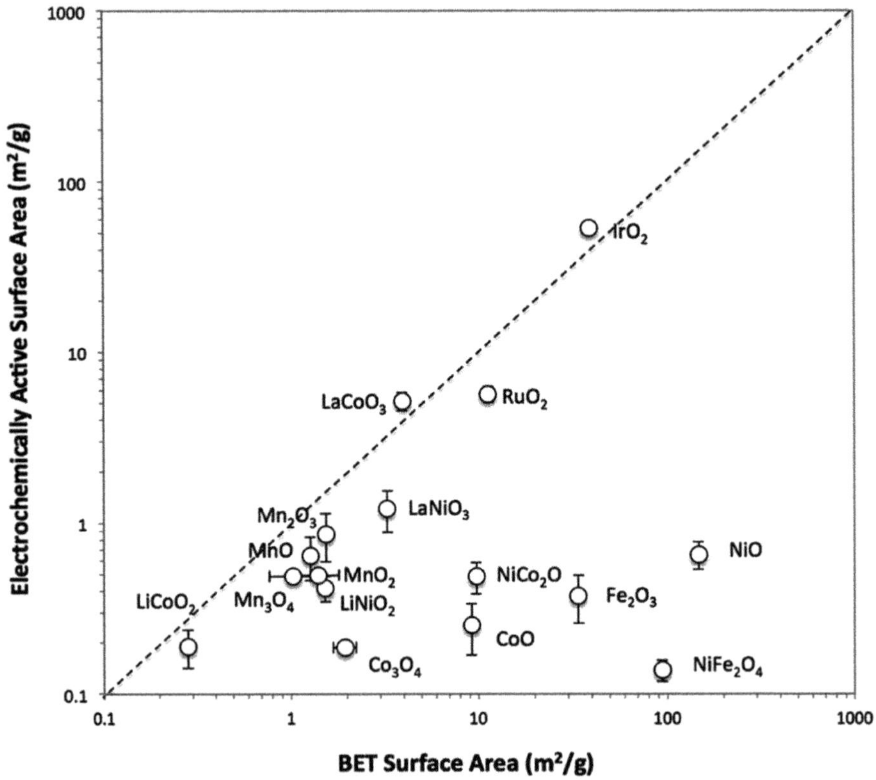

Figure 5.4 Comparisons of surface area measured using BET gas adsorption with the electrochemically-active surface area estimated from double-layer capacitance for various metal oxide particles. The BET surface area measurements were conducted on the powder samples, and the ECSA measurements were conducted on particles immobilized within a Nafion® film on the electrode surface.
Reproduced from ref. 1 with permission from the Royal Society of Chemistry.

activity of a catalyst. However, the lack of a universal method to accurately estimate the ECSA of a catalyst limits the utility of this descriptor for estimating specific activity. In general, we suggest measuring and reporting both ECSA and total surface area measurements to provide a limiting range of specific activities to facilitate comparisons of catalysts.

5.4 Benchmarking Catalyst Performance

5.4.1 Primary Figure of Merit

There are three activity metrics that are most relevant for comparing catalyst performance for solar water splitting devices. The first is the activity normalized by the geometric area of the working electrode substrate. This activity metric is the measured current–voltage response of the deposited catalyst film, and does not take into account the loading or active surface area. The second is the specific activity normalized for the electroactive surface area of the material. For two materials that operate with the same activity per mass or geometric area, the material with the higher specific activity operates with a higher turnover frequency per active site, and thus has a higher *intrinsic* activity. The third is the activity normalized for the catalyst loading, which is sometimes called the *mass* activity of the catalyst material. This metric is useful when considering the cost of device construction and operation, especially for noble metal catalysts. For adhered particle films, the mass loading is calculated directly from the mass of the particle catalysts in the dropcasted solution. For thin film catalysts, the mass loading can be measured during deposition using ECQM, or can be estimated by dissolving the catalyst from the surface and using ICP-OES to determine the mass of the dissolved metals in the resulting solutions.

Choice of the primary figure of merit depends on the purpose of the catalyst performance evaluation. The specific activity per ECSA is a measurements of catalyst performance per active site, and thus is most useful when comparing the intrinsic activity of catalyst materials independent of surface area and loading effects. When considering an integrated solar fuels device in which electrocatalysts are immobilized onto the photoelectrode, electrocatalyst layers can be deposited onto the photoelectrode either as thin continuous films (Figure 5.5a), or as an array of electrocatalyst islands (Figure 5.5b,c). Both device designs have similar optimal solar-to-fuel

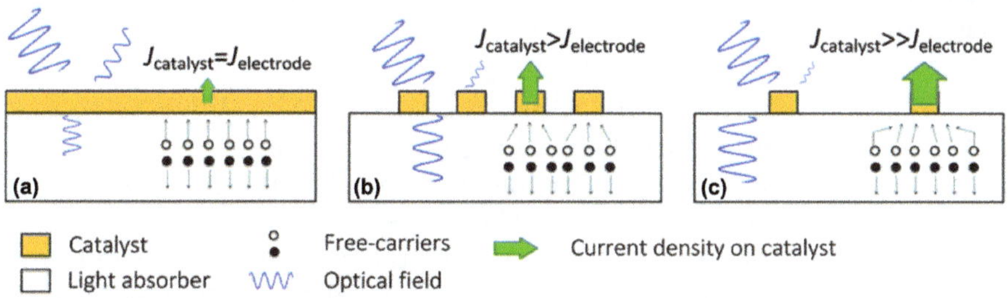

Figure 5.5 Schematic illustration of photoabsorbers (a) coated with continuous electrocatalyst films, (b) coated with patterned electrocatalyst films with high (1–10% photoelectrode geometric area) coverage of the surface, and (c) coated with patterned electrocatalyst films with low (<1% photoelectrode geometric area) coverage of the surface. In the case of (a), parasitic absorption by the electrocatalyst is dictated by the thickness and opacity of the electrocatalyst film. In the case of (b) and (c), the coverage of the patterned electrocatalyst film dictates the extent of parasitic adsorption. Low coverage patterned films have less adsorption by the electrocatalyst but require higher operating current densities at the catalyst sites compared to higher coverage films.
Reproduced from ref. 115 with permission from the Royal Society of Chemistry.

efficiencies under ideal conditions.[115] In the first case, the continuous electrocatalyst film must be sufficiently thin (<2 nm) to minimize the parasitic light absorption by the electrocatalyst.[115–117] In such a design, the optimal catalyst will achieve high activity at low catalyst loading, thus making the mass activity the most relevant figure-of-merit.

When considering the second case, where the photoelectrode is coated with electrocatalyst islands (Figure 5.5b), the parasitic absorption by the electrocatalyst is a function of the fractional coverage of the electrocatalyst on the geometric surface are of the photoelectrode regardless of the opacity of the individual electrocatalyst islands.[115,118] In this design, the optimal catalyst will achieve high activity with minimal geometric area, thus making the activity per geometric area the most relevant figure-of-merit.

The goal of the benchmarking project is to evaluate the performance of materials for eventual consideration for incorporation into an integrated solar-fuels device. Given the widespread use of the patterned-catalyst design in recent devices, we reiterate our suggestion that the primary figure of merit for the purpose of benchmarking catalyst activity under conditions relevant to integrated solar fuels devices is the overpotential necessary to achieve 10 mA cm^{-2} geometric area, $\eta_{j=10\,\mathrm{mA\,cm}_{\mathrm{geo}}^{-2}}$. Note that we encourage authors to report electrocatalysts' specific activities per surface area and mass activities per mass loading in addition to their activities per geometric area, as all three metrics are useful for comparing electrocatalytic activity.

5.4.2 Comparing Electrocatalytic Performance

The results of the benchmarking measurements of electrocatalytic HER and OER activity and short-term stability are summarized in Figure 5.6. When comparing catalytic activity, we propose that promising HER and OER electrocatalysts should operate with $\eta_{j=10\,\mathrm{mA\,cm}_{\mathrm{geo}}^{-2}} \leq 0.1$ V and $\eta_{j=10\,\mathrm{mA\,cm}_{\mathrm{geo}}^{-2}} \leq$ 0.35 V, respectively. These values come from device models that suggest the target combined overpotential for the OER and the HER in a 10% efficient integrated solar-to-hydrogen device should be ≤0.45 V.[78,119–121]

Of the 16 electrodeposited non-noble metal HER catalysts tested during our benchmarking studies, four electrodeposited materials that exhibited the highest activity were various combinations of Mo with Ni, Co, and Fe, including CoMo,[122] NiMoCo,[123] and two different preparations of NiMo.[124,125] Each of these materials operated with $\eta_{j=10\,\mathrm{mA\,cm}_{\mathrm{geo}}^{-2}} \leq 0.1$ V in both 1 M H_2SO_4 and 1 M NaOH, and showed good operational stability over the course of 24 h of constant polarization in both solutions.[2] Of particular promise was NiMo-(a),[124] which catalyzed the HER with $\eta_{j=10\,\mathrm{mA\,cm}_{\mathrm{geo}}^{-2}} \leq 0.05$ V in both acidic and alkaline solutions, and showed little loss of catalyst activity in 1 M NaOH over the course of the 24 h constant polarization stability measurements or over the course of 40 000 potential cycles in potential-cycling stability studies.[2] In 1 M H_2SO_4, NiMo also showed good stability

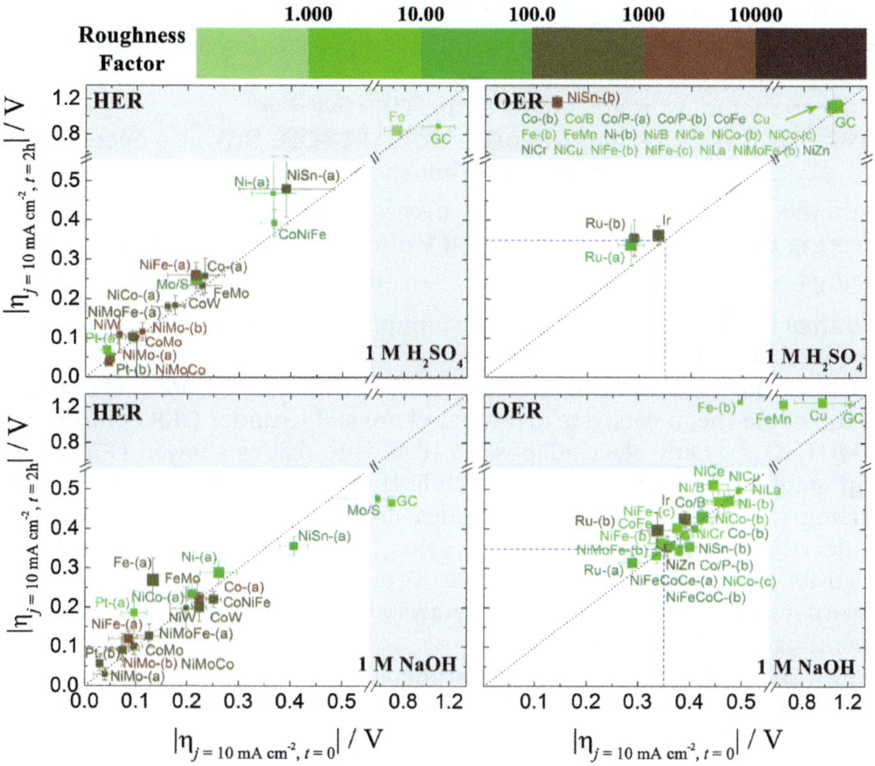

Figure 5.6 Comprehensive plots summarizing the electrocatalytic activity, short-term operational stability, and ECSA for HER (left) and OER (right) electrocatalysts in acidic (top) and alkaline (bottom) solutions. The x-axis is $\eta_{j=10\,\text{mA}\,\text{cm}_{\text{geo}}^{-2}}$ at time = 0 and is a measure of the catalyst activity. The y-axis is $\eta_{j=10\,\text{mA}\,\text{cm}_{\text{geo}}^{-2}}$ after 2 h of constant polarization at 10 mA cm^{-2} and is related to the catalyst's operational stability. Catalysts that have high operational stability will sit on the diagonal dashed line, which is the expected response for a catalyst that does not change in activity during 2 h of constant polarization. The color of each point represents the roughness factor of the catalyst, calculated from ECSA estimates, with a bin size of one order of magnitude. The size of each point is inversely proportional to the standard deviation in the ECSA measurements, and is an indication of the precision (not the accuracy) of the ECSA measurements. There is a break and change in scale in both axes at $\eta > 0.55$ V, and the corresponding region of the plot is shown in gray. Catalysts in this region have low activity and/or operational stability, but are included for completeness. The blue boxed regions are the target regions for HER and OER electrocatalysis based on device models.
Reprinted with permission from C. C. L. McCrory, S. Jung, I. M. Ferrer, S. M. Chatman, J. C. Peters, and T. F. Jaramillo, *J. Am. Chem. Soc.*, 2015, **137**, 4347, Copyright 2015 American Chemical Society.

over the course of 24 h constant polarization, but exhibited a moderate loss of catalyst activity over the course of 10 000 potential cycles and significant loss of activity after 30 000 cycles.[2]

Other materials that catalyzed the HER with $\eta_{j=10\text{ mA cm}_{\text{geo}}^{-2}} \leq 0.1$ V are NiW[126] in 1 M H_2SO_4 and NiMoFe[127] in 1 M NaOH. NiW also operated with $\eta_{j=10\text{ mA cm}_{\text{geo}}^{-2}} \leq 0.06$ V in 1 M H_2SO_4, but showed comparatively poor stability with the magnitude overpotential necessary to achieve 10 mA cm^{-2} increasing to $\eta_{j=10\text{ mA cm}_{\text{geo}}^{-2}} = 0.33 \pm 0.06$ V after 24 h of constant polarization.[2] NiMoFe operated with $\eta_{j=10\text{ mA cm}_{\text{geo}}^{-2}} = 0.09$ V in 1 M NaOH and showed minimal loss of activity over 2 h constant polarization, but was not tested for extended stability.[2]

One of the more notable results of the benchmarking studies is that none of the noble-metal catalysts investigated are stable under OER conditions in 1 M H_2SO_4.[2,3] Only electrodeposited Ir and Ru oxides showed OER activity and stability in acidic solution.[1] While these results are perhaps not surprising given that many metal–oxides are not thermodynamically stable under oxidative conditions in strong acids,[87,128] it nonetheless highlights the need for the development of new, acid-stable OER materials to enable the incorporation of non-noble metal catalysts into solar-water splitting devices operating in acidic solution.

In 1 M NaOH, most of the electrodeposited OER catalysts investigated operate with roughly equivalent activity of 0.35 V $\leq \eta_{j=10\text{ mA cm}_{\text{geo}}^{-2}} \leq 0.5$ V.[2,3] Similar results were observed for catalyst films of adhered particles in a thin Nafion® binder at catalyst loadings of 0.8 mg cm^{-2},[1,65] the standardized loading for this system based on the loading-dependent performance of IrO_2.[1] The fact that nearly all of the OER materials investigated operate with similar activity may suggest they operate *via* a common mechanism. Such a common mechanism has been proposed, and a key step is the oxidation of a surface hydroxide *OH by two electrons to form a surface hydroxyperoxy *OOH species.[129–132] Theoretical studies based on this common mechanism for the OER at planar metal electrodes suggest the existence of a "thermodynamic overpotential" of ~0.4 V.[131,133] The similarity of the measured activities in our studies is consistent with this hypothesis. The existence of a thermodynamic overpotential for planar metal oxide catalyst would suggest that a key strategy to develop new OER electrocatalysts that operate with high OER activity per geometric area is to focus on increasing the number of accessible active sites in the material.[134,135]

5.5 Conclusions

In this chapter, we presented an overview of the protocol used in, and results from, our studies for the moderate-throughput evaluation of thin-film and nanoparticulate catalysts under conditions relevant to an integrated solar-water splitting device under 1 sun illumination in acidic or alkaline solution. Because this chapter was a broad overview of our studies, there were several topics of which the reader should be aware that were not covered in detail here including: the measurement of and correction for uncompensated resistance in electrocatalytic measurements,[7] the role of electrolyte impurities in electrocatalytic activity measurements,[136–138] the possible introduction of such impurities from the corrosion of glass in alkaline solutions of NaOH,[5–7] and the evaluation of a catalyst's corrosion stability (rather than performance stability),[99,139,140] among others.

Note that in addition to the work presented in this chapter, there has been significant discussion in the literature regarding the appropriate choice of standard protocols, benchmarks, and best practices for measuring and reporting the HER and OER activity and stability of electrocatalysts.[7,99,108,141–145] In particular, we refer the reader to a recent report that suggests a detailed complementary set of techniques and benchmarks for evaluating thin-film OER catalyst performance to those presented here.[7] We emphasize that the studies summarized in this chapter reflect the work conducted at JCAP to evaluate catalysts for incorporation into an integrated, low-current density solar-water splitting device. Devices that operate at significantly higher current densities (such as water electrolyzes or photoelectrochemical devices under multiple-sun illumination) will require different testing protocols to accommodate for their different operating parameters and figures of merit.[141,146,147]

References

1. S. Jung, C. C. L. McCrory, I. M. Ferrer, J. C. Peters and T. F. Jaramillo, *J. Mater. Chem. A*, 2016, **4**, 3068–3076.
2. C. C. L. McCrory, S. Jung, I. M. Ferrer, S. M. Chatman, J. C. Peters and T. F. Jaramillo, *J. Am. Chem. Soc.*, 2015, **137**, 4347–4357.
3. C. C. L. McCrory, S. Jung, J. C. Peters and T. F. Jaramillo, *J. Am. Chem. Soc.*, 2013, **135**, 16977–16987.
4. C. Xiang, K. M. Papadantonakis and N. S. Lewis, *Mater. Horiz.*, 2016, **3**, 169–173.
5. K. J. J. Mayrhofer, A. S. Crampton, G. K. H. Wiberg and M. Arenz, *J. Electrochem. Soc.*, 2008, **155**, P78–P81.
6. K. J. J. Mayrhofer, G. K. H. Wiberg and M. Arenz, *J. Electrochem. Soc.*, 2008, **155**, P1–P5.
7. M. B. Stevens, L. J. Enman, A. S. Batchellor, M. R. Cosby, A. E. Vise, C. D. M. Trang and S. W. Boettcher, *Chem. Mater.*, 2017, **29**, 120–140.
8. F. M. Hawkridge, in *Laboratory Techniques in Electroanalytical Chemistry*, ed. P. T. Kissinger and W. R. Heineman, Marcel Dekker, Inc., New York, 1996, p. 270.
9. A. Ursua, L. M. Gandia and P. Sanchis, *Proc. IEEE*, 2012, **100**, 410–426.
10. D. T. Sawyer, A. Sobkowiak and J. L. Roberts Jr., *Electrochemistry for Chemists*, John Wiley & Sons, Inc., New York, 2nd edn, 1995, p. 250.

11. A. J. Bard and L. R. Faulkner, *Electrochemical Methods: Fundamentals and Applications,* John Wiley & Sons, Inc., Hoboken, NJ, 2nd edn, 2001, pp. 26–27.
12. A. Z. Weber and J. Newman, *J. Electrochem. Soc.*, 2004, **151**, A326–A339.
13. E. Verlage, S. Hu, R. Liu, R. J. R. Jones, K. Sun, C. Xiang, N. S. Lewis and H. A. Atwater, *Energy Environ. Sci.*, 2015, **8**, 3166–3172.
14. A. J. Bard and L. R. Faulkner, *Electrochemical Methods: Fundamentals and Applications*, Wiley, New York, 2001, p. 127.
15. D. A. J. Rand and R. Woods, *J. Electroanal. Chem. Interfacial Electrochem*, 1972, **35**, 209–218.
16. J. Llopis and A. Sancho, *J. Electrochem. Soc.*, 1961, **108**, 720–726.
17. R. P. Frankenthal and H. W. Pickering, *J. Electrochem. Soc.*, 1965, **112**, 514–517.
18. R. L. McCreery, *Chem. Rev.*, 2008, **108**, 2646–2687.
19. S.-E. Jang and H. Kim, *J. Am. Chem. Soc.*, 2010, **132**, 14700–14701.
20. P. Spitzer and S. Wunderli, in *Handbook of Reference Electrodes*, ed. G. Inzelt, A. Lewenstam and F. Scholz, Springer Berlin Heidelberg, Berlin, Heidelberg, 2013, p. 84.
21. D. T. Sawyer, A. Sobkowiak and J. L. Roberts, *Electrochemistry for Chemists*, Wiley, New York, 1995, p. 188.
22. P. Spitzer and S. Wunderli, in *Handbook of Reference Electrodes*, ed. G. Inzelt, A. Lewenstam and F. Scholz, Springer Berlin Heidelberg, Berlin, Heidelberg, 2013, p. 108.
23. D. J. G. Ives and J. G. Hills, in *Reference Electrodes*, ed. D. J. G. Ives and G. J. Janz, Academic Press, New York, 1961, p. 142.
24. J. Chen, K. Iwama, K. J. Aoki and T. Nishiumi, *Int. J. Electrochem. Sci.*, 2013, **8**, 7394–7400.
25. K. Aoki, C. Li, T. Nishiumi and J. Chen, *J. Electroanal. Chem.*, 2012, **682**, 66–71.
26. D. T. Sawyer, A. Sobkowiak and J. L. Roberts, *Electrochemistry for Chemists*, Wiley, New York, 1995, p. 197.
27. P. Spitzer and S. Wunderli, in *Handbook of Reference Electrodes*, ed. G. Inzelt, A. Lewenstam and F. Scholz, Springer Berlin Heidelberg, Berlin, Heidelberg, 2013, p. 115.
28. D. J. G. Ives, in *Reference Electrodes*, ed. D. J. G. Ives and G. J. Janz, Academic Press, New York, 1961, p. 336.
29. P. Spitzer and S. Wunderli, in *Handbook of Reference Electrodes*, ed. G. Inzelt, A. Lewenstam and F. Scholz, Springer Berlin Heidelberg, Berlin, Heidelberg, 2013, p. 127.
30. D. T. Sawyer, A. Sobkowiak and J. L. Roberts, in *Electrochemistry for Chemists*, Wiley, New York, 1995, p. 191.
31. P. Spitzer and S. Wunderli, in *Handbook of Reference Electrodes*, ed. G. Inzelt, A. Lewenstam and F. Scholz, Springer Berlin Heidelberg, Berlin, Heidelberg, 2013, p. 92.
32. Y. S. Femenias, U. Angst, F. Caruso and B. Elsener, *Mater. Struct.*, 2016, **49**, 2637–2651.

33. E. Liaudet, F. Battaglini and E. J. Calvo, *J. Electroanal. Chem. Interfacial Electrochem.*, 1990, **293**, 55–68.
34. I. M. Kolthoff and W. J. Tomsicek, *J. Phys. Chem.*, 1934, **39**, 945–954.
35. G. I. H. Hanania, D. H. Irvine, W. A. Eaton and P. George, *J. Phys. Chem.*, 1967, **71**, 2022–2030.
36. M. Noel and P. N. Anantharaman, *Surf. Coat. Technol.*, 1986, **28**, 161–179.
37. D. T. Sawyer, A. Sobkowiak and J. L. Roberts, *Electrochemistry for Chemists*, Wiley, New York, 1995, p. 188.
38. S. Yamada and H. Sato, *Nature*, 1962, **193**, 261–262.
39. G. M. Swain, *J. Electrochem. Soc.*, 1994, **141**, 3382–3393.
40. A. Dekanski, J. Stevanović, R. Stevanović, B. Ž. Nikolić and V. M. Jovanović, *Carbon*, 2001, **39**, 1195–1205.
41. X. Li, F. C. Walsh and D. Pletcher, *Phys. Chem. Chem. Phys.*, 2011, **13**, 1162–1167.
42. N. Krstajic and S. Trasatti, *J. Appl. Electrochem.*, 1998, **28**, 1291–1297.
43. J. C. K. Ho and D. L. Piron, *J. Appl. Electrochem.*, 1996, **26**, 515–521.
44. M. Manazoğlu, G. Hapçı and G. Orhan, *J. Appl. Electrochem.*, 2016, **46**, 191–204.
45. D. Santos, A. Lopes, M. J. Pacheco, A. Gomes and L. Ciríaco, *J. Electrochem. Soc.*, 2014, **161**, H564–H572.
46. X. Yi, P. He, Y. Chen, W. Wang, D. Yang and F. Dong, *ECS Trans.*, 2010, **28**, 13–20.
47. W. Sheng, M. Myint, J. G. Chen and Y. Yan, *Energy Environ. Sci.*, 2013, **6**, 1509–1512.
48. S. Marini, P. Salvi, P. Nelli, R. Pesenti, M. Villa and Y. Kiros, *Int. J. Hydrogen Energy*, 2013, **38**, 11484–11495.
49. D. Kang, T. W. Kim, S. R. Kubota, A. C. Cardiel, H. G. Cha and K.-S. Choi, *Chem. Rev.*, 2015, **115**, 12839–12887.
50. J. van Drunen, B. K. Pilapil, Y. Makonnen, D. Beauchemin, B. D. Gates and G. Jerkiewicz, *ACS Appl. Mater. Interfaces*, 2014, **6**, 12046–12061.
51. B. Beverskog and I. Puigdomenech, *J. Electrochem. Soc.*, 1997, **144**, 3476–3483.
52. E. Deltombe, N. Zoubov and M. Pourbaix, in *Atlas d'équilibres électrochemiques.English*, ed. M. Pourbaix, Pergamon Press, Oxford, New York, 1966, p. 333.
53. S. M. Bhola and B. Mishra, *Int. J. Electrochem. Sci.*, 2013, **8**, 7075–7087.
54. B. S. Yeo and A. T. Bell, *J. Am. Chem. Soc.*, 2011, **133**, 5587–5593.
55. B. S. Yeo and A. T. Bell, *J. Phys. Chem. C*, 2012, **116**, 8394–8400.
56. M. S. El-Deab, M. I. Awad, A. M. Mohammad and T. Ohsaka, *Electrochem. Commun.*, 2007, **9**, 2082–2087.
57. Y. Gorlin, C.-J. Chung, J. D. Benck, D. Nordlund, L. Seitz, T.-C. Weng, D. Sokaras, B. M. Clemens and T. F. Jaramillo, *J. Am. Chem. Soc.*, 2014, **136**, 4920–4926.
58. C. C. McCrory, S. Jung, J. C. Peters and T. F. Jaramillo, *J. Am. Chem. Soc.*, 2013, **135**, 16977–16987.

59. I. Roger and M. D. Symes, *J. Mater. Chem. A*, 2016, **4**, 6724–6741.
60. Y. Xia, Y. Xiong, B. Lim and S. E. Skrabalak, *Angew. Chem. Int. Ed.*, 2009, **48**, 60–103.
61. C. M. Sanchez-Sanchez, J. Solla-Gullon and V. Montiel, in *Electrochemistry: Volume 11 – Nanosystems Electrochemistry*, The Royal Society of Chemistry, 2013, vol. 11, pp. 34–70.
62. T. Reier, M. Oezaslan and P. Strasser, *ACS Catal.*, 2012, **2**, 1765–1772.
63. J. Suntivich, H. A. Gasteiger, N. Yabuuchi and Y. Shao-Horn, *J. Electrochem. Soc.*, 2010, **157**, B1263–B1268.
64. T. J. Schmidt, H. A. Gasteiger, G. D. Stäb, P. M. Urban, D. M. Kolb and R. J. Behm, *J. Electrochem. Soc.*, 1998, **145**, 2354–2358.
65. K.-Y. Niu, F. Lin, S. Jung, L. Fang, D. Nordlund, C. C. L. McCrory, T.-C. Weng, P. Ercius, M. M. Doeff and H. Zheng, *Nano Lett.*, 2015, **15**, 2498–2503.
66. D.-C. Huang, P.-J. Yu, F.-J. Liu, S.-L. Huang, K.-L. Hsueh, Y.-C. Chen, C.-H. Wu, W.-C. Chang and F.-H. Tsau, *Int. J. Electrochem. Sci.*, 2011, **6**, 2551–2565.
67. A. L. Dicks, *J. Power Sources*, 2006, **156**, 128–141.
68. P. Trogadas, T. F. Fuller and P. Strasser, *Carbon*, 2014, **75**, 5–42.
69. T. Ralph, *Platinum Met. Rev.*, 1997, **41**, 102–113.
70. E. Gracia-Espino, X. Jia and T. Wågberg, *J. Phys. Chem. C*, 2014, **118**, 2804–2811.
71. T. Ralph and M. Hogarth, *Platinum Met. Rev.*, 2002, **46**, 117–135.
72. K. H. Kangasniemi, D. A. Condit and T. D. Jarvi, *J. Electrochem. Soc.*, 2004, **151**, E125–E132.
73. J. Wang, G. Yin, Y. Shao, S. Zhang, Z. Wang and Y. Gao, *J. Power Sources*, 2007, **171**, 331–339.
74. P. N. Ross and H. Sokol, *J. Electrochem. Soc.*, 1984, **131**, 1742–1750.
75. N. Staud and P. N. Ross, *J. Electrochem. Soc.*, 1986, **133**, 1079–1084.
76. M. G. Walter, E. L. Warren, J. R. McKone, S. W. Boettcher, Q. Mi, E. A. Santori and N. S. Lewis, *Chem. Rev.*, 2010, **110**, 6446–6473.
77. Y. Gorlin and T. F. Jaramillo, *J. Am. Chem. Soc.*, 2010, **132**, 13612–13614.
78. M. F. Weber and M. J. Dignam, *J. Electrochem. Soc.*, 1984, **131**, 1258–1265.
79. M. P. Soriaga, J. H. Baricuatro, K. D. Cummins, Y.-G. Kim, F. H. Saadi, G. Sun, C. C. L. McCrory, J. R. McKone, J. M. Velazquez, I. M. Ferrer, A. I. Carim, A. Javier, B. Chmielowiec, D. C. Lacy, J. C. Hemminger, N. S. Lewis and J. L. Stickney, *Surf. Sci.*, 2015, **631**, 285–294.
80. A. J. Bard and L. R. Faulkner, *Electrochemical Methods: Fundamentals and Applications*, Wiley, New York, 2001, pp. 353–354.
81. K. B. Prater and A. J. Bard, *J. Electrochem. Soc.*, 1970, **117**, 207–213.
82. E. Guerrini, H. Chen and S. Trasatti, *J. Solid State Electrochem.*, 2007, **11**, 939–945.
83. M. Morimitsu, C. Murakami, K. Kawaguchi, R. Otogawa and M. Matsunaga, *J. New Mater. Electrochem. Syst.*, 2004, **7**, 323–327.

84. S. Cherevko, S. Geiger, O. Kasian, N. Kulyk, J.-P. Grote, A. Savan, B. R. Shrestha, S. Merzlikin, B. Breitbach, A. Ludwig and K. J. J. Mayrhofer, *Catal. Today*, 2016, **262**, 170–180.
85. J. Juodkazytė, B. Šebeka, G. Stalnionis and K. Juodkazis, *Electroanalysis*, 2005, **17**, 1734–1739.
86. M. Pourbaix, *Atlas of Electrochemical Equilibria in Aqueous Solutions*, Pergamon Press, Oxford, New York, 1966.
87. A. Minguzzi, F.-R. F. Fan, A. Vertova, S. Rondinini and A. J. Bard, *Chem. Sci.*, 2012, **3**, 217–229.
88. S. Song, H. Zhang, X. Ma, Z. Shao, R. T. Baker and B. Yi, *Int. J. Hydrogen Energy*, 2008, **33**, 4955–4961.
89. Y. Li, H. Wang, L. Xie, Y. Liang, G. Hong and H. Dai, *J. Am. Chem. Soc.*, 2011, **133**, 7296–7299.
90. W. F. Chen, K. Sasaki, C. Ma, A. I. Frenkel, N. Marinkovic, J. T. Muckerman, Y. Zhu and R. R. Adzic, *Angew. Chem. Int. Ed.*, 2012, **51**, 6131–6135.
91. J. R. McKone, B. F. Sadtler, C. A. Werlang, N. S. Lewis and H. B. Gray, *ACS Catal.*, 2013, **3**, 166–169.
92. D. Kong, H. Wang, Z. Lu and Y. Cui, *J. Am. Chem. Soc.*, 2014, **136**, 4897–4900.
93. W. F. Chen, C. H. Wang, K. Sasaki, N. Marinkovic, W. Xu, J. T. Muckerman, Y. Zhu and R. R. Adzic, *Energy Environ. Sci.*, 2013, **6**, 943–951.
94. J. D. Benck, Z. Chen, L. Y. Kuritzky, A. J. Forman and T. F. Jaramillo, *ACS Catal.*, 2012, **2**, 1916–1923.
95. E. J. Popczun, C. G. Read, C. W. Roske, N. S. Lewis and R. E. Schaak, *Angew. Chem. Int. Ed.*, 2014, **53**, 5427–5430.
96. E. J. Popczun, J. R. McKone, C. G. Read, A. J. Biacchi, A. M. Wiltrout, N. S. Lewis and R. E. Schaak, *J. Am. Chem. Soc.*, 2013, **135**, 9267–9270.
97. B. Wickman, H. Grönbeck, P. Hanarp and B. Kasemo, *J. Electrochem. Soc.*, 2010, **157**, B592–B598.
98. D. A. Buttry and M. D. Ward, *Chem. Rev.*, 1992, **92**, 1355–1379.
99. R. Frydendal, E. A. Paoli, B. P. Knudsen, B. Wickman, P. Malacrida, I. E. L. Stephens and I. Chorkendorff, *ChemElectroChem*, 2014, **1**, 2075–2081.
100. S. Cherevko, A. A. Topalov, A. R. Zeradjanin, I. Katsounaros and K. J. J. Mayrhofer, *RSC Adv.*, 2013, **3**, 16516–16527.
101. A. Le Goff, V. Artero, B. Jousselme, P. D. Tran, N. Guillet, R. Métayé, A. Fihri, S. Palacin and M. Fontecave, *Science*, 2009, **326**, 1384–1387.
102. J. Bernard, M. Chatenet and F. Dalard, *Electrochim. Acta*, 2006, **52**, 86–93.
103. M. Vukovic, *J. Chem. Soc., Faraday Trans.*, 1990, **86**, 3743–3746.
104. A. J. Bard and L. R. Faulkner, *Electrochemical Methods: Fundamentals and Applications*, Wiley, New York, 2001, p. 348.
105. S. Trasatti and O. A. Petrii, *Pure Appl. Chem.*, 1991, **63**, 711–734.
106. S. E. F. Kleijn, S. C. S. Lai, M. T. M. Koper and P. R. Unwin, *Angew. Chem. Int. Ed.*, 2014, **53**, 3558–3586.

107. M. Grdeń, M. Alsabet and G. Jerkiewicz, *ACS Appl. Mater. Interfaces*, 2012, **4**, 3012–3021.
108. V. Čolić, J. Tymoczko, A. Maljusch, A. Ganassin, W. Schuhmann and A. S. Bandarenka, *ChemElectroChem*, 2015, **2**, 143–149.
109. S. Zhao, H. Yu, R. Maric, N. Danilovic, C. B. Capuano, K. E. Ayers and W. E. Mustain, *J. Electrochem. Soc.*, 2015, **162**, F1292–F1298.
110. L. D. Burke and E. J. M. O'Sullivan, *J. Electroanal. Chem.*, 1981, **117**, 155–160.
111. R. L. Doyle, I. J. Godwin, M. P. Brandon and M. E. G. Lyons, *Phys. Chem. Chem. Phys.*, 2013, **15**, 13737–13783.
112. K. Klingan, F. Ringleb, I. Zaharieva, J. Heidkamp, P. Chernev, D. Gonzalez-Flores, M. Risch, A. Fischer and H. Dau, *ChemSusChem*, 2014, **7**, 1301–1310.
113. F. Lin and S. W. Boettcher, *Nat. Mater.*, 2014, **13**, 81–86.
114. A. S. Batchellor and S. W. Boettcher, *ACS Catal.*, 2015, **5**, 6680–6689.
115. Y. Chen, K. Sun, H. Audesirk, C. Xiang and N. S. Lewis, *Energy Environ. Sci.*, 2015, **8**, 1736–1747.
116. L. Trotochaud, T. J. Mills and S. W. Boettcher, *J. Phys. Chem. Lett.*, 2013, **4**, 931–935.
117. A. Shinde, D. Guevarra, J. A. Haber, J. Jin and J. M. Gregoire, *J. Mater. Res.*, 2014, **30**, 442–450.
118. C. Xiang, A. Z. Weber, S. Ardo, A. Berger, Y. Chen, R. Coridan, K. T. Fountaine, S. Haussener, S. Hu, R. Liu, N. S. Lewis, M. A. Modestino, M. M. Shaner, M. R. Singh, J. C. Stevens, K. Sun and K. Walczak, *Angew. Chem. Int. Ed.*, 2016, **55**, 12974–12988.
119. M. G. Walter, E. L. Warren, J. R. McKone, S. W. Boettcher, Q. X. Mi, E. A. Santori and N. S. Lewis, *Chem. Rev.*, 2010, **110**, 6446–6473.
120. M. F. Weber and M. J. Dignam, *Int. J. Hydrogen Energy*, 1986, **11**, 225–232.
121. L. C. Seitz, Z. Chen, A. J. Forman, B. A. Pinaud, J. D. Benck and T. F. Jaramillo, *ChemSusChem*, 2014, **7**, 1372–1385.
122. C. Fan, D. L. Piron, A. Sleb and P. Paradis, *J. Electrochem. Soc.*, 1994, **141**, 382–387.
123. C. Fan, D. L. Piron and P. Paradis, *Electrochim. Acta*, 1994, **39**, 2715–2722.
124. E. Navarro-Flores, Z. Chong and S. Omanovic, *J. Mol. Catal. A: Chem.*, 2005, **226**, 179–197.
125. N. V. Krstajić, V. D. Jović, L. Gajić-Krstajić, B. M. Jović, A. L. Antozzi and G. N. Martelli, *Int. J. Hydrogen Energy*, 2008, **33**, 3676–3687.
126. I. A. Raj and K. I. Vasu, *J. Appl. Electrochem.*, 1990, **20**, 32–38.
127. I. A. Raj and K. I. Vasu, *J. Appl. Electrochem.*, 1992, **22**, 471–477.
128. M. Pourbaix, *Atlas of Electrochemical Equilibria in Aqueous Solution*, Pergamon Press, Oxford, 1966.
129. I. C. Man, H.-Y. Su, F. Calle-Vallejo, H. A. Hansen, J. I. Martínez, N. G. Inoglu, J. Kitchin, T. F. Jaramillo, J. K. Nørskov and J. Rossmeisl, *ChemCatChem*, 2011, **3**, 1159–1165.

130. J. Rossmeisl, Z. W. Qu, H. Zhu, G. J. Kroes and J. K. Norskov, *J. Electroanal. Chem.*, 2007, **607**, 83.
131. M. T. M. Koper, *J. Electroanal. Chem.*, 2011, **660**, 254–260.
132. J. Rossmeisl, A. Logadottir and J. K. Nørskov, *Chem. Phys.*, 2005, **319**, 178–184.
133. I. C. Man, H. Y. Su, F. Calle-Vallejo, H. A. Hansen, J. I. Martinez, N. G. Inoglu, J. Kitchin, T. F. Jaramillo, J. K. Norskov and J. Rossmeisl, *ChemCatChem*, 2011, **3**, 1159–1165.
134. K. L. Pickrahn, Y. Gorlin, L. C. Seitz, A. Garg, D. Nordlund, T. F. Jaramillo and S. F. Bent, *Phys. Chem. Chem. Phys.*, 2015, **17**, 14003–14011.
135. F. Song and X. Hu, *Nat. Commun.*, 2014, **5**, 4477.
136. D. A. Corrigan and R. M. Bendert, *J. Electrochem. Soc.*, 1989, **136**, 723–728.
137. L. Trotochaud, S. L. Young, J. K. Ranney and S. W. Boettcher, *J. Am. Chem. Soc.*, 2014, **136**, 6744–6753.
138. S. Klaus, Y. Cai, M. W. Louie, L. Trotochaud and A. T. Bell, *J. Phys. Chem. C*, 2015, **119**, 7243–7254.
139. S. Zou, M. S. Burke, M. G. Kast, J. Fan, N. Danilovic and S. W. Boettcher, *Chem. Mater.*, 2015, **27**, 8011–8020.
140. N. Danilovic, R. Subbaraman, K.-C. Chang, S. H. Chang, Y. J. Kang, J. Snyder, A. P. Paulikas, D. Strmcnik, Y.-T. Kim, D. Myers, V. R. Stamenkovic and N. M. Markovic, *J. Phys. Chem. Lett.*, 2014, **5**, 2474–2478.
141. A. Ganassin, A. Maljusch, V. Colic, L. Spanier, K. Brandl, W. Schuhmann and A. Bandarenka, *ACS Catal.*, 2016, **6**, 3017–3024.
142. R. A. Rincón, E. Ventosa, F. Tietz, J. Masa, S. Seisel, V. Kuznetsov and W. Schuhmann, *ChemPhysChem*, 2014, **15**, 2810–2816.
143. T. Bligaard, R. M. Bullock, C. T. Campbell, J. G. Chen, B. C. Gates, R. J. Gorte, C. W. Jones, W. D. Jones, J. R. Kitchin and S. L. Scott, *ACS Catal.*, 2016, **6**, 2590–2602.
144. C. Costentin, G. Passard and J.-M. Savéant, *J. Am. Chem. Soc.*, 2015, **137**, 5461–5467.
145. C. Costentin and J.-M. Savéant, *ChemElectroChem*, 2014, **1**, 1226–1236.
146. J. Turner, G. Sverdrup, M. K. Mann, P.-C. Maness, B. Kroposki, M. Ghirardi, R. J. Evans and D. Blake, *Int. J. Energy Res.*, 2008, **32**, 379–407.
147. J. D. Holladay, J. Hu, D. L. King and Y. Wang, *Catal. Today*, 2009, **139**, 244–260.

Semiconductor Light Absorbers

CHAPTER 6

Heterojunction Approaches for Stable and Efficient Photoelectrodes

JOEL W. AGER[a,b,c]

[a] Joint Center for Artificial Photosynthesis, Lawrence Berkeley National Laboratory, Berkeley, CA 94720, USA; [b] Materials Sciences Division, Lawrence Berkeley National Laboratory, Berkeley, CA 94720, USA; [c] Department of Materials Science and Engineering, University of California Berkeley, Berkeley, CA 94720, USA
Email: jwager@lbl.gov

6.1 Introduction

There has been continuous scientific interest in the prospect of direct photoelectrochemical (PEC) conversion of sunlight to hydrogen since the pioneering demonstration of water splitting with a TiO_2 photo-electrode reported by Fujishima and Honda in 1972.[1] In the 2010s there has been an increase in attention, some of which is driven by the prospect of grid-parity pricing for solar photovoltaic power and the resulting need, at high penetrations of intermittent sources into the electrical power grid, of large scale energy storage.[2–5] Hydrogen has been identified as a potential energy carrier for this purpose.[6] Also, the release of fuel cell vehicles into the commercial market creates a potential market for renewable hydrogen in the transportation sector.[7–9]

Of the various routes to sustainable hydrogen production,[10] production *via* PEC water splitting has received significant attention.[11–17] In this

Energy and Environment Series No. 22
Integrated Solar Fuel Generators
Edited by Ian D. Sharp, Harry A. Atwater and Hans-Joachim Lewerenz

Published by the Royal Society of Chemistry, www.rsc.org

approach, generation of photovoltaic current is coupled to the hydrogen evolution reaction (HER, reaction 1) and the oxygen evolution reaction (OER, reaction 2) such that overall water splitting, reaction 3, is performed:

$$2H^+ + 2e^- \rightarrow H_2(g), \quad E^0 = 0 \text{ V } \mathit{vs.} \text{ RHE} \tag{6.1}$$

$$H_2O \rightarrow \tfrac{1}{2}\, O_2(g) + \; 2H^+ + 2e^-, \quad E^0 = 1.23 \text{ V } \mathit{vs.} \text{ RHE} \tag{6.2}$$

$$H_2O \leftrightarrow H_2(g) + \tfrac{1}{2}\, O_2(g), \quad \Delta G^0 = +237 \text{ kJ mol}^{-1}, \quad \Delta E^0 = 1.23 \text{ eV} \tag{6.3}$$

The degree of integration of experimental devices which perform overall water splitting ranges from decoupled PV systems and electrolytes to photoelectrode based systems which couple more directly the HER and OER chemistry to the light absorber; Nielander *et al.* has a full discussion of the taxonomy of such systems.[18] It would be expected that the limiting conversion efficiency would be a convolution of the radiative limit for photovoltaic power conversion (Shockley–Queisser limit,[19] which is 33.7% for unconcentrated AM1.5G illumination) and the kinetics limits for reactions 1 and 2. Indeed, theoretical predictions of the 1 sun efficiency limit are in the range of 20–25% depending on assumptions regarding device design.[20–24]

There have been numerous experimental demonstrations of unassisted, solar-driven overall water splitting using a wide variety of photovoltaic and catalytic components. Separation of the gas products has also been addressed, but to a lesser extent. Rongé *et al.*[25] and Ager *et al.*[26] are recent reviews of these efforts. Notably, there are a number of demonstrations of >10% solar to hydrogen (STH) efficiency using PEC components[27–30] and, recently, demonstrations of >20% STH efficiency using concentrated sunlight, high performance photovoltaics, and electrolyzers.[31–33]

From both a technoeconomic and life cycle assessment point of view, a viable solar to hydrogen conversion technology must have both high efficiency and a long operational lifetime.[34–38] A comparison point may be obtained from the commercial photovoltaic market; manufacturers are required, essentially by market forces, to warranty the power output of their panels for 20 years or more.[39,40] The technological readiness level (TRL) of solar to hydrogen conversion is much lower than that of solar PV, so there is little field data on device operational stability and long term performance.[3] Nevertheless, guidance can be obtained from energy return on investment (EROI) calculations performed in the course of life cycle assessments, which found that a minimum of five years lifetime is required for a positive return on energy investment (at a STH efficiency of 10%) but, given the uncertainties in the estimates, a lifetime of ten years or more is preferable.[37,38] This value contrasts with the reported operational lifetime of demonstration devices which, with only a few exceptions, are one week or less.[26]

Similar to photovoltaic modules, PEC devices are multi-component systems, the failure or degradation of any part of which can compromise the entire device.[41–43] For a PEC device, not only the "active" components

(PV and catalytic elements) but also the separation membrane (if present), windows and encapsulation, and the gas handling and compression for the H_2 product should be considered.[36,37] Membranes and gas handling are common to both solar PEC hydrogen generation and water electrolysis. This latter technology is considerably more advanced and has been reviewed extensively elsewhere.[44–46] This book chapter will therefore concentrate on the photovoltaic and catalysis materials in the system.

Quite obviously, driving reactions (1) and (2) requires transporting electrons across a solid–liquid interface. From an efficiency point of view, the faradaic efficiency of (1) and (2) should be 100% and electrochemical reactions with the device materials, unless they are perfectly reversible,[48,49] should be avoided, as they might be expected to lead to corrosion and, eventually, to device failure. The stability requirements for materials under illuminated conditions, where non-equilibrium populations of electrons and holes can exist, were first analysed by Gerischer in the 1970s and are shown in Figure 6.1.[47,50] Quite simply, the redox potential for decomposition of the materials should lie within with the conduction or valence bands of the materials; otherwise, a dark or light-driven decomposition process is possible. Examination of Pourbaix diagrams can be used to assess materials for stable operation.[51,52] Computational screening can also be employed, as exemplified by the Joint Center for Artificial Photosynthesis (JCAP) study of Chen and Wang,[53] which found that only a very small group of oxides such as TiO_2, $SrTiO_3$, *etc.* was kinetically stable under illuminated HER and OER conditions.

Given the thermodynamic instability of all group IV (Si, Ge), III–V (GaAs, InP), and II–VI (CdSe, CdS, *etc.*) semiconductors under illuminated HER and, especially, OER conditions,[53–55] it is attractive to create a heterostructure of a PV element and a "protection" coating formed of a material which is thermodynamically stable.[56] This concept was first explored by Bard and coworkers in the 1970s utilizing TiO_2 coatings on Si, CdS, GaAs, and GaP.[57]

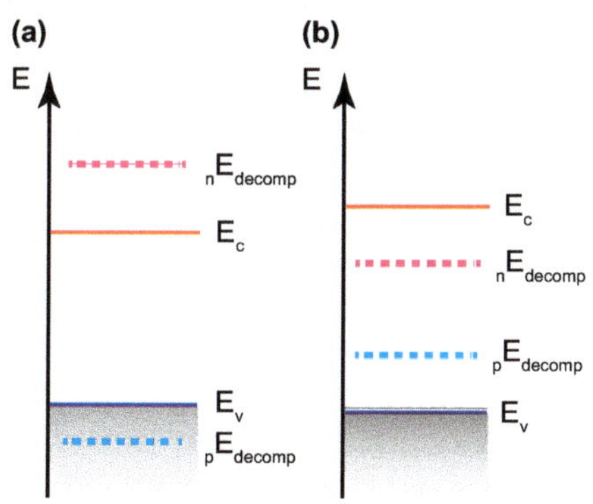

Figure 6.1 (a) Stable and (b) unstable band alignment for semiconductor electrodes where $_nE_{\text{decomp}}$ indicates the cathodic decomposition potential and $_pE_{\text{decomp}}$ indicates the anodic decomposition potential. After Gerischer.[47]

While some promising results were reported, delamination of the sputter-deposited coating and transport of holes in the case of n-type absorbers emerged as key issues. Improvements in conformal thin film deposition, especially by atomic layer deposition,[58] and also improved understanding of the interface between the absorber and the electrolyte, mediated by the protection layer,[59,60] created the opportunity to revisit the protected photoelectrode concept.[61] This chapter will review JCAP efforts in this area in the time period 2011–2016 and also provide contemporary context.

6.2 Semiconductor–Electrolyte Interface in the Context of Chemical Conversion

6.2.1 Overview

For photoelectrical energy conversion to occur, the photo-generated carriers have to pass through a solid–liquid interface. This is common not only to PEC cells but also to dye sensitized solar cells and similar devices that use a redox couple as part of the charge transport chain.[62,63] The distinguishing feature of PEC energy conversion is the number of electrons transferred, which must be at least two to make a chemical bond. The hydrogen evolution reaction can thus be thought of as the simplest PEC energy conversion product as the transfer of just two electrons is required.[64] The oxygen evolution reaction represents a step up in complexity, with the involvement of at least two water molecules and the transfer of 4 electrons.[65–67]

6.2.2 Simple Picture of an Unpinned Semiconductor–Liquid Junction (SLJ)

Before discussing heterojunction design, it is useful to examine the function of a semiconductor–liquid junction (SLJ) from the point of view of charge selectivity. In principle, if this solid–liquid interface could provide efficient collection of minority carriers, suppression of interface recombination, and also selective catalysis of the desired fuel-forming reaction (*i.e.* HER for photocathodes and OER for photoanodes) and hence suppress ion of any corrosion paths, efficient energy conversion could be possible. The simplest picture is the classical view of a one-electron redox couple in contact with a simple unpinned semiconductor surface.[68] This approach is exemplified by the JCAP work of Hu *et al.*[69] which examined the PEC properties of n-type GaAs nanowires in ferrocenium/ferrocene ($FeCp_2^{+/0}$) in acetonitrile. This system achieved near-unity external quantum efficiency due to light trapping in a nanowire array and a photovoltage of nearly 600 mV.

We consider a hypothetical n-type semiconductor with a band gap of 2.0 eV and a flat band potential ϕ_{FB}, referenced to the one-electron redox couple, of 0.7 V, as illustrated in Figure 6.2. The depletion or space charge

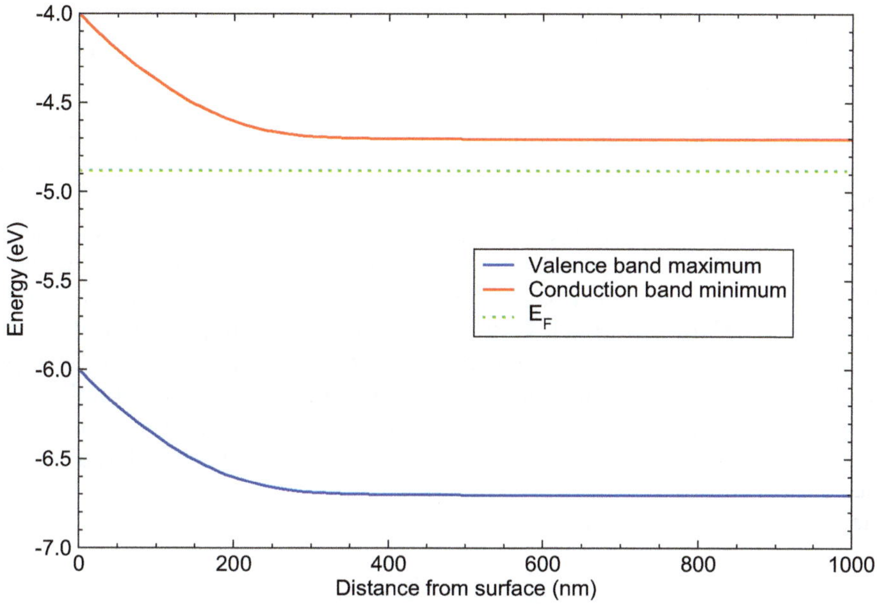

Figure 6.2 Equilibrium band diagram of a hypothetical n-type semiconductor with a band gap of 2 eV in contact with a one-electronic redox couple. The flat band potential is 0.7 V, the donor density, N_D, is 10^{16} cm^{-3}, and the dielectric constant is $12\varepsilon_0$. The density of states of oxidized and reduced states, D_{ox} and D_{red}, are calculated by Marcus theory. The motion of photoexcited holes toward the electrolyte and electrons into the bulk of the semiconductor is depicted.
Adapted from Memming[68] and Reichman.[71]

region (SCR) has width W which depends on the applied potential with respect to the flat band potential ϕ_{FB}:

$$W = \left(\frac{2(\phi_{FB} - V)\varepsilon\,\varepsilon_0}{q\,N_D}\right)^{\frac{1}{2}} \tag{6.4}$$

where ε and N_D are the dielectric constant and donor density of the semiconductor, respectively. The simplest treatment of the photocurrent under illumination is due to Gärtner,[70] who assumed that all carriers generated the SCR and within one minority carrier diffusion length of it contribute to the photocurrent J_g:

$$J_g = J_0 + I_0\left(1 - \frac{e^{-\alpha W}}{1 + \alpha L}\right) \tag{6.5}$$

where J_0 is the current density at the edge of the SCR,

$$J_0 = \frac{q\,n_0 L}{\tau}$$

and n_0 is the electron density in the bulk (p_0 for p-type absorbers), τ is the minority carrier lifetime, I_0 is the light intensity, α is the absorption coefficient and L is the minority carrier diffusion length. In this form, the equation describes monochromatic illumination; the equation can be integrated to take into account illumination over a spectral range. This approach neglects all carrier recombination in the SCR and in this sense represents an ideal limiting case.

This approach was extended by Reichman,[71] who considered explicitly the possibility of recombination in the SCR, which becomes more pronounced if the kinetics of electron transfer from (hole transfer to) the solution is slow. The essence of this approach is the calculation of the majority and minority carrier concentrations at the surface (electron and holes in the example given here) and the assumption that the currents are proportional to the excess carrier density. The current passing through the semiconductor-electrolyte interface due to electrons (J_n) and holes (J_p) is related to surface concentrations of the corresponding charge carriers (n_s, p_s), relative to their values at equilibrium (n_{s0}, p_{s0}):

$$J_n = -J_n^0\left[\frac{n_s}{n_{s0}} - 1\right] \tag{6.5a}$$

$$J_p = -J_p^0\left[\frac{p_s}{p_{s0}} - 1\right], \tag{6.5b}$$

where J_n^0 and J_n^0 are the exchange current densities for electrons and holes, which reflect the kinetics of the cathodic and anodic surface reactions, respectively. Under the low injection conditions produced by 1 sun illumination, for an n-type semiconductor, the surface electron concentration is unaffected by light and it is simply a function of the applied potential, relative to flat band, which changes the value of $\frac{n_s}{n_{s0}}$:

$$J_n = -J_n^0\left[\exp\left[\frac{qV}{k_B T}\right] - 1\right]. \tag{6.6}$$

The surface concentration of holes, p_s, is related to the illumination and also to recombination in the SCR:

$$\frac{p_s}{p_{s0}} = \left(\frac{-K + (K^2 + 4AB)^{\frac{1}{2}}}{2A}\right)^2 \quad \text{where} \tag{6.7a}$$

$$K = \frac{\pi\, k_B T\, n_i\, W}{4\tau(\phi_{FB} - V)}\exp\left[\frac{qV}{2\, k_B T}\right], \tag{6.7b}$$

$$A = J_p^0 + J_0 \exp\left[\frac{qV}{2k_B T}\right], \quad \text{and} \tag{6.7c}$$

$$B = J_p^0 + J_g, \tag{6.7d}$$

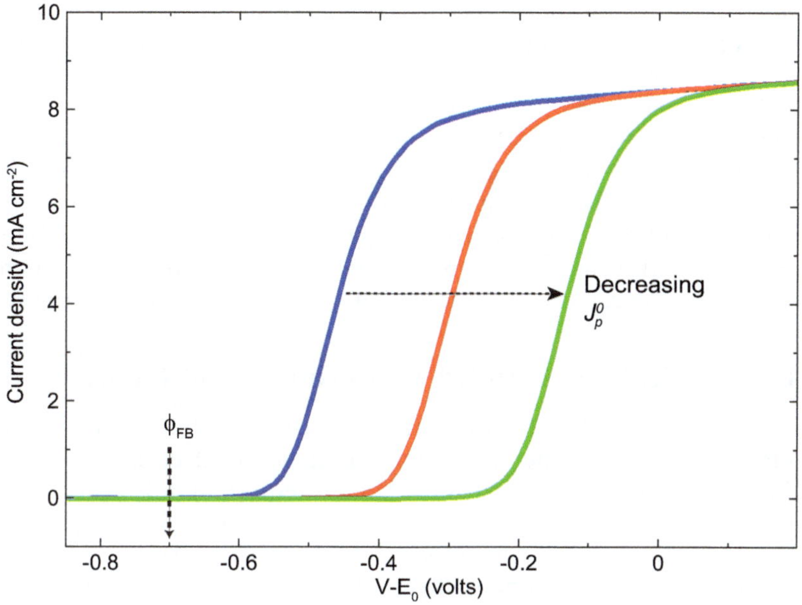

Figure 6.3 Reichman model for illuminated current density. Three cases of hole exchange current J_p^0 are considered: 10^{-12} mA cm^{-2}, 10^{-15} mA cm^{-2}, and 10^{-18} mA cm^{-2}. Other model parameters are I_0 (light intensity) = 10 mA cm^{-2}. $L = 0.5\times10^{-4}$ cm, $\tau = 0.1$ ns, $N_\text{D} = 10^{16}$ cm^{-3}, $\alpha = 3\times10^4$ cm^{-1}, $n_i = 10^7$ cm^{-3}, $\varepsilon = 12\varepsilon_0$, $\phi_\text{FB} = 0.7$ V, $J_\text{n}^0 = 10^{-20}$ mA cm^{-2}.

where n_i is the intrinsic electron concentration. The textbook by Memming has an excellent discussion of the limiting cases predicted by this model;[68] only a summary will be given here for a hypothetical n-type absorber with a band gap of 2.0 eV. The magnitude of the exchange current densities affects the predicted photocurrent, as illustrated in Figure 6.3. The parameters, in particular the low value of the electron compared to the hole exchange current, are chosen to approximate the situation of a water oxidation photoanode.

It is notable that the approach of Reichmann is essentially homologous with the formulae used to specify surface recombination velocities (SRV) in the context of charge collection from solar cells. In this case, the interface is characterized by its SRV for electron and holes, S_n and S_p, respectively. In fact, it is possible to make a direct connection between J_n^0 and J_p^0 in the Reichman model and the surface recombination velocities as follows:

$$S_\text{n} = \frac{J_\text{n}^0}{q\,N_\text{D}\exp\left[\dfrac{-q\phi_\text{FB}}{k_\text{B}T}\right]} \tag{6.8a}$$

$$S_\text{p} = \frac{J_\text{p}^0}{q\,p_0\exp\left[\dfrac{+q\phi_\text{FB}}{k_\text{B}T}\right]} \tag{6.8b}$$

This allows the practitioner to use a 1D simulation program such as PC-1D or AFORS-HET to perform simulations of the semiconductor electrolyte interface.[72,73] Notable, this approach also can be used for both cathodic and anodic bias (where eqn (6.4) would diverge).

Models which consider the effect of surface states have also been developed. The work of Peter and co-workers is exemplary here,[74] and these types of models are used to interpret time modulated experiments such as intensity modulated photocurrent spectroscopy and electrochemical impedance spectroscopy.[75] Full numerical treatments have also been developed and compared to experimental data, such as the work of Anz *et al.*[76]

6.2.3 Electrically Decoupled Photovoltaic and Catalyst

The discussion above has focused on the case where the potential between the semiconductor and the electrolyte manifests itself exclusively by band bending in the semiconductor as described by eqn (6.4). It is also noted that the rates for charge exchange with the solution, J_n^0 and J_p^0 (also S_n and S_p in eqn (6.8)) do not depend on potential. This is the case of the idealized semiconductor–liquid junction. The other extreme is where the semiconductor has a carrier selective interface (*e.g.* a pn junction) which decouples its band bending from the electrolyte. Electrically, this can be thought of as a solar cell in electrical series with an electrocatalyst. Intermediate cases where the potential produces changes in the electrocatalyst are also possible: the reader is referred to the work of Boettcher and coworkers which considered this case and also that of a coupled molecular catalyst.[59,77]

Modelling of a decoupled system can be performed by assuming that the solar cell has the following JV characteristics:[78]

$$J_{PV} = J_L - J_{sat}\left[\exp\left[\frac{-q(V + J_{PV}R_S)}{n\, k_B T}\right] - 1\right] - \frac{V + J_{PV}R_S}{R_{sh}}, \tag{6.9}$$

where R_s and R_{sh} are the specific series and shunt resistances. When R_s is zero and R_{sh} infinite eqn (6.9) reduces to the sum of the ideal diode equation with reverse saturation current density J_{sat}, ideality factor n, and illumination current density of J_L. The catalyst can be modelled with Butler–Volmer kinetics (shown here for single electron kinetics):

$$J_{cat} = J_0\left\{\left[\exp\left[\frac{(1-\alpha)q(E-E_0)}{k_B T}\right] - \exp\left[\frac{(1-\alpha)q(E-E_0)}{k_B T}\right]\right]\right\}, \tag{6.10}$$

where α is the transfer coefficient and the electrochemical nomenclature of using E to designate the potential is used (E_0 is the equilibrium potential). Alternatively, experimental dark electrocatalytic data taken with the oppositely doped semiconductor (*e.g.* heavily p-type doped for anodes) coupled to the catalyst can be used. The analysis then consists of finding the split in voltage between the photovoltaic (V, eqn (6.9)) and the catalyst (E, eqn (6.10)) which allows the current density to match.

An illustration of this type of analysis is shown in Figure 6.4, which is adapted from the work of Ager and co-workers. As discussed in Section 6.3.3, the photoanode had a n–p^+-Si/$NiCo_2O_4$/NiFe structure constructed in such a way to decouple the Si n–p^+ junction from the electrolyte and the NiFe OER catalyst. Control experiments were performed with a p^+-Si/$NiCo_2O_4$/NiFe structure in dark to determine the OER kinetics. Data from the illuminated n–p^+-Si/$NiCo_2O_4$/NiFe photoanodes is fit with eqn (6.9) and the experimental data from the p^+-Si/$NiCo_2O_4$/NiFe dark anode under conditions of current matching. This essentially shifts the dark *J–E* curve anodically by an amount approximately equal to the open circuit voltage of the PV component. One can also observe the light-limited current density at high anodic bias, above 1.6 V *vs.* RHE. The calculated open circuit voltage is 658 mV, which is typical for this type of solar cell structure under 1 sun illumination. The short circuit current density, which also limits the photocurrent at large anodic potentials, is somewhat smaller than in a typical Si cell, due to absorption of light by the $NiCo_2O_4$, which is not completely transparent.

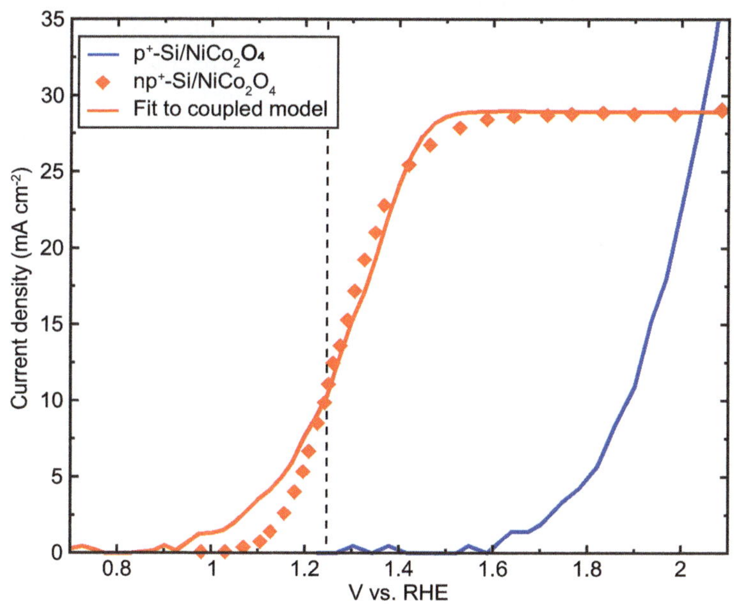

Figure 6.4 Analysis of decoupled PV and electrocatalyst system. Blue curve is for p^+-Si/$NiCo_2O_4$ dark anode. Data points are for n–p^+-Si/$NiCo_2O_4$ illuminated with simulated AM 1.5 light. The solid line is obtained by convolving the purple curve with eqn (6.9) under the constraint of current density matching. The analysis used the following parameters for the Si pn junction: $J_0 = 2.5 \times 10^{-10}$ mA cm^{-2}, $n = 1$, $R_s = 0.1$ Ω-cm^2, $R_{sh} = 50$ MΩ-cm^2, and $J_{SC} = 29$ mA cm^{-2}. All experiments were performed in water at pH 14.
Adapted from Chen *et al.*[79]

6.2.4 Heterojunction Design for Stability and Efficiency

Examination of Figure 6.3 shows that for a directly coupled junction, a high exchange current density (or SRV) for the majority carrier and a low one for the minority carrier is desired. This same principle applies in heterojunction solar cells,[80] and use of this approach of surface passivation and highly carrier-selective contacts has led to recent improvements in the open circuit voltage and overall efficiency of solar cells.[81–83] The same principle applies here in the design of interfaces for PEC water splitting whenever minority carriers are present and are subject to recombination. It should be further mentioned, that all interfaces, including majority carrier interfaces with other elements of the system, would also benefit from passivation, again by analogy with state of the art PV devices. These principles drove a significant amount of work in JCAP in the period 2011–2015.

PEC water splitting has significantly more stringent requirements for fast kinetics at the photoelectrode–electrolyte interface, in order to increase the selectivity for the desired catalyst pathways and mitigate corrosion processes. Recently, considerable scepticism has arisen regarding the idea that a single material could have both the excellent photovoltaic and surface catalytical properties to function as either a photocathode or photoanode in a high efficiency water-splitting device.[15] This again suggests the use of different materials for surface passivation/carrier collection and for catalytic interactions with the electrolyte. Conductivity through any passivation and/or protection layers is another crucial factor, as first outlined by Campet *et al.*[56] However, the required conductivity here is considerably less than is required for photovoltaic cells, as the charge transport paths are short compared to the much larger lateral length scales in PV.

6.3 JCAP Experimental Work

This section describes JCAP work on heterojunction PEC and also provides some contemporary context by discussing leading work which occurred concurrently. Photocathodes based on p-type absorbers and photoanodes based on n-type absorbers are described separately.

6.3.1 Photocathodes

The early 2010s saw two important breakthroughs in heterojunction photocathodes for PEC hydrogen production. Grätzel and co-workers showed that Cu_2O, which is reductively unstable under HER conditions, could be operated as a stable photocathode if protected by a dual layer coating of Al-doped ZnO (AZO) and TiO_2.[84,85] In this system, the AZO forms a charge separating pn heterojuction with the Cu_2O, reducing surface recombination, and the TiO_2 provides corrosion protection, epitomizing the principles discussed above.

The early 2010s saw an increased interest in the use of nanowire-based structures for use in PEC energy conversion. Such a structure can optimize light absorption along the length of the wire while increasing the surface area available for catalysis. At the same time, the increased surface area provides additional challenges for reducing surface recombination. Work by Lewis and co-workers[86] and Chorkendorff and co-workers[87] explored this approach with Si nanowires. However, as with planar Si photocathodes, the overall energy conversion was low due to the small photovoltages, relative to H^+/H_2, available with p-Si photoabsorbers.[61]

Another reference point for JCAP work was the pioneering work of Heller and co-workers in the 1980s who showed that p-InP, when interfaced with thin layers of precious metal catalysts, could produce near-PV quality half-cell efficiencies for HER.[88–90] Intriguingly, it was necessary to anodically poise the electrodes from time to time to arrest a decrease in current density. This, and work performed by Lewerenz and co-workers in the 2000s, showed that the maintenance of very thin surface layers was necessary to retain high photocurrents for HER and also for one-electron reduction reactions.[91–93]

JCAP worked to integrate the advantages of an engineered charge separating and protection interface in a nanostructured III–V platform for highly efficient solar-driven HER. Javey, Ager, and co-workers developed a mask-free nanostructuring method based on reactive ion etching which produces InP nanocone arrays.[94] The nanocone dimensions are on the order of the electron diffusion length in the p-InP substrate. Atomic layer deposition (ALD) was used to deposit TiO_2 as an electron selective contact and protection layer. As shown in Figure 6.5, the passivated and protected photocathodes produce AM 1.5 current densities over 30 mA cm^{-2} with >650 mV of open-circuit potential. Interestingly, even very thick layers of ALD-grown TiO_2 (up to 20 nm) do not significantly reduce the current density or fill factor, suggesting that this approach might be generally applicable to high surface area photocathodes.

As discussed in Section 6.2.2, the onset potential for PEC current depends critically on the surface condition, with passivation against minority carrier recombination being crucial. Javey, Ager, and co-workers carried out a detailed passivation study using the model InP system and ALD-deposited TiO_2.[95] It was found that the TiO_2 deposition conditions had a strong influence on the band alignment with the InP, as measured by the flat band potential. In particular, use of a nitrogen containing ALD precursor led to a smaller barrier for holes at the interface and low onset potentials. Use of an optimized ALD process led to a large barrier for holes at the heterointerface, Figure 6.6, drastically reducing the electron–hole recombination and leading to onset potentials >800 mV *vs.* RHE.

InP is an excellent model system for the exploration of fundamental concepts in PEC hydrogen generation but is considered by many to be too expensive for scale up and manufacturing. However, in 2014 Javey and co-workers developed an inexpensive thin-film vapor liquid solid (TF-VLS) process which yields photovoltaic quality InP thin films and nanostructures

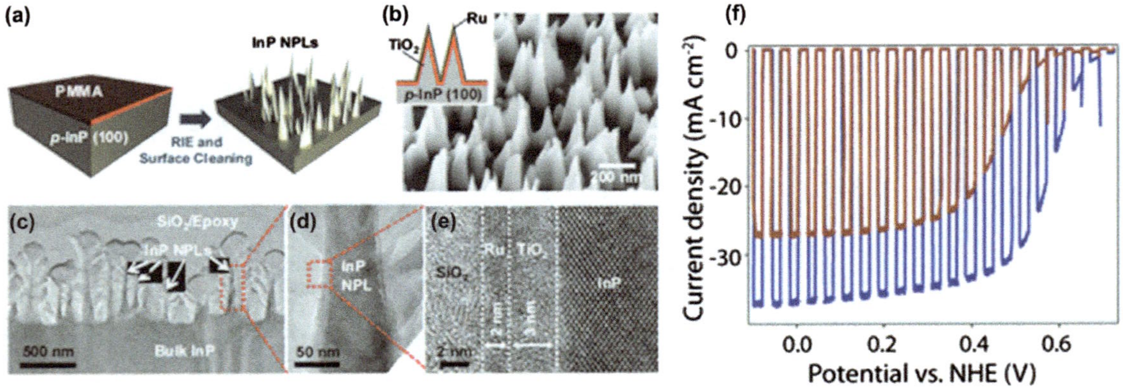

Figure 6.5 (a) Process flow for fabrication of InP nanopillar (NPL) arrays. (b) SEM image of the resulting NPL arrays. Inset shows the schematic of a PEC cell made of p-InP NPLs, TiO_2 passivation layer (thickness, 3–5 nm), and Ru co-catalyst (thickness, ~2 nm). (c–e) TEM images of p-InP NPLs coated with TiO_2/Ru layers. Note that for TEM sample preparation and imaging purposes, the surface of the substrate is coated with a thick layer of evaporated SiO_x. (f) shows the PEC performance under simulated AM1.5 illumination at pH 0.5.
Adapted from ref. 94 with permission from John Wiley and Sons, ©2012 Wiley-VCH Verlag GmbH & Co. KGaA, Weinheim.

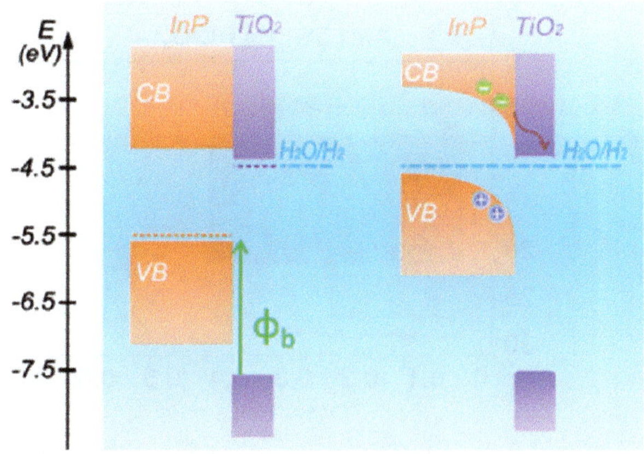

Figure 6.6 Band alignment of p-InP/TiO_2 in solution based on the Mott–Schottky measurement.
Reprinted with permission from *J. Phys. Chem. C*, 2015, **119**, 2308–2313, Copyright (2015) American Chemical Society.

at a greatly decreased prospective cost.[96–98] Javey, Ager, and co-workers used InP material made by this method, combined with the ALD TiO_2 process previously developed, to demonstrate a potentially scalable and efficient (>10% half-cell efficiency) solar to hydrogen conversion, as shown in Figure 6.7.[99]

Having previously demonstrated the efficacy of conformal TiO_2 protection layers on InP, Javey, Ager and co-workers applied the method to textured amorphous Si n-i-p structures.[100] This material corrodes rapidly under HER conditions in the absence of a protective layer. A *ca.* 80 nm TiO_2 layer was deposited on the structure to serve as an electron collector and as a corrosion protection layer, as shown in Figure 6.8(a). With the use of a Pt HER catalyst, a very high onset potential, over 900 mV, was achieved, with a photocurrent in excess of the JCAP goal of 10 mA cm^{-2}, as shown in Figure 6.8(b). Stability under 12 hours of sustained illumination was demonstrated. Similarly, high onset potentials were obtained using an earth-abundant HER catalyst, Ni–Mo, although there was a reduction in fill factor.

Table 6.1 summarizes JCAP work on protected photocathodes used for hydrogen generation.

6.3.2 Photoanodes

The idea of protecting otherwise unstable semiconductor surfaces with metal oxide coatings dates back to the 1970s; photoanodes which perform the oxygen evolution reaction (OER) are particularly challenging in this regard due to the reactive nature of the photo-generated surface holes. Bard and co-workers attempted to protect photoanodes for OER in the 1970s but limitations in the thin film deposition methods led to coating delamination and failure.[57] A significant advance in photoanode protection was reported

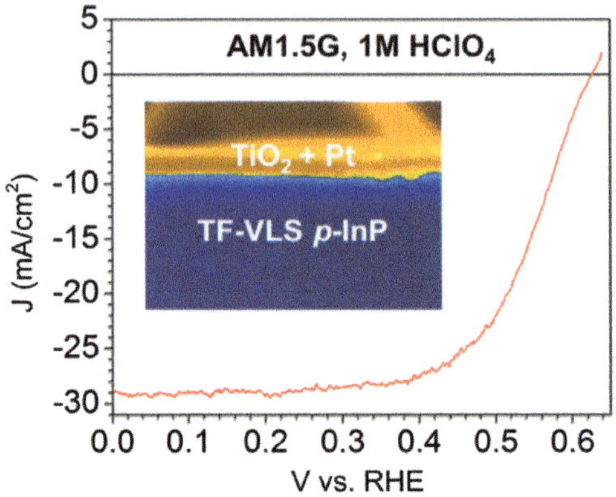

Figure 6.7 Thin-film vapor–liquid–solid (TF-VLS) non-epitaxial growth method combined with an atomic-layer deposition protection process to create thin-film InP photocathodes with large grain size and high performance, in the first reported solar device configuration generated by materials grown with this technique. Current–voltage measurements show a photocurrent (29.4 mA cm^{-2}) and onset potential (630 mV) approaching single-crystalline wafers and an overall power conversion efficiency of 11.6%, making TF-VLS InP a promising photocathode for scalable and efficient solar hydrogen generation.
Reprinted with permission from *J. Phys. Chem. Lett.*, 2015, **6**, 2177–2182, Copyright 2015 American Chemical Society.

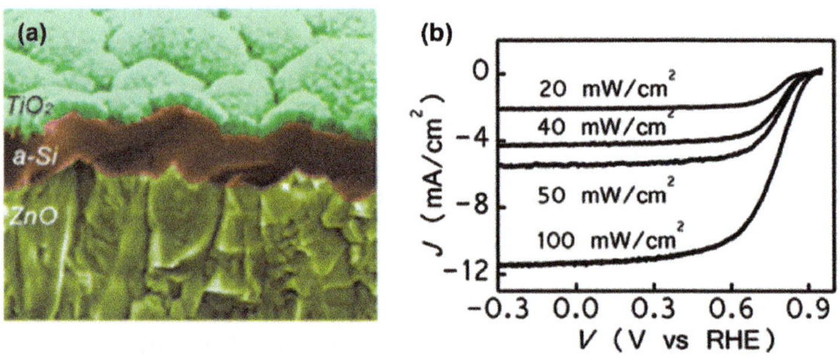

Figure 6.8 (a) Cross sectional scanning electron microscopy image of an amorphous Si photocathode. A sputtered layer of TiO_2 provides a top contact for electron collection from the pin structure, which is back illuminated. (b) *J–E* curves in 0.5 M aqueous potassium hydrogen phthalate solution as a function of illumination intensity using a solar simulator.
Reprinted with permission from *Nano Lett.*, 2013, **13**, 5615–5618, Copyright 2013 American Chemical Society.

by McIntyre and co-workers in 2011.[58] In this work, ALD deposited TiO_2 was shown to provide protection of n-Si photoanodes in both highly acidic and basic conditions. The inferred photovoltage was >500 mV, which suggests some degree of effective interface passivation. The charge transport mechanism was posited to be tunnelling, as TiO_2 layers thicker than 2 nm reduced the photocurrent. Since 2011, McIntyre and co-workers have published a

Table 6.1 Summary of performance characteristics of notable hydrogen evolution photocathodes produced by JCAP researchers in the period 2011–2015.

Study	Absorber(s)	Protection layer, deposition method	OER co-catalyst	Electrolyte and illumination	Photocurrent at 0 V *vs.* RHE; onset potential	Stability notes
Lee *et al.*,[94] 2012	Planar and nanotextured p-InP	3–5 nm TiO_2, ALD	Ru	1 M $HClO_4$, 100 mW cm^{-2}	35 mA cm^{-2}, 0.73 *vs.* RHE	4 hours stable photocurrent at +0.23 *vs.* RHE
Lin *et al.*,[100] 2013	Amorphous Si pin structure	80 nm TiO_2, reactive sputtering	Pt, Ni-Mo	0.5 M KHP, 20–100 mW cm^{-2} AM 1.5	11.6 mA cm^{-2}, 0.93 *vs.* RHE with Pt co-catalyst and 100 mW cm^{-2}	12 hours stable photocurrent at +0.0 *vs.* RHE
Lin *et al.*,[95] 2015	Planar p-InP	10 nm TiO_2, ALD	Pt	1 M $HClO_4$, 100 mW cm^{-2}	25.5 mA cm^{-2}, >0.8 *vs.* RHE	Not reported
Hettick *et al.*,[99] 2015	TF-VLS grown polycrystalline p-InP	30 nm TiO_2, ALD	Pt	1 M $HClO_4$, 100 mW cm^{-2}	29.4 mA cm^{-2}, 0.63 V *vs.* RHE	1.2 hours stable photocurrent at 0 V *vs.* RHE, measurement of H_2 product

number of studies on the fundamentals of this type of metal–insulator–metal (electrolyte) interface.[60,101]

It has been well established that TiO_2 can form an effective corrosion protective layer. However, the relatively low position of its valence band would suggest that it would be ineffective as a hole conductor for photoanodes due to the barrier which would be expected. The idea of using hopping conduction to create a density of states in the gap sufficient to allow hole transport for efficient water oxidation had been proposed before,[56] but the precision and flexibility of atomic layer deposition presents a new opportunity to implement the scheme. In 2014, Lewis and co-workers discovered that thick (>100 nm) layers of TiO_2 could efficiently conduct anodic currents, enabling efficient photoanodes to be made from Si, GaAs, and GaP.[102] Si photoanodes stabilized in this manner and operated in 1 M KOH electrolyte solutions demonstrated the continuous and stable oxidation of water for over 100 hours at photocurrent densities >30 mA cm^{-2} , as shown in Figure 6.9. The synthesis conditions need to be carefully controlled to allow for hole conduction; for example, annealing the TiO_2 layer drastically reduced the anodic photocurrent.

In parallel, Sharp and co-workers worked to develop interfaces based on heterojunctions with a p-type oxide, CoO_x.[103] Cobalt oxides have a relatively high valence band position,[104] compared to TiO_2, and thus would be expected to form heterojunctions with n-type absorbers which would allow for

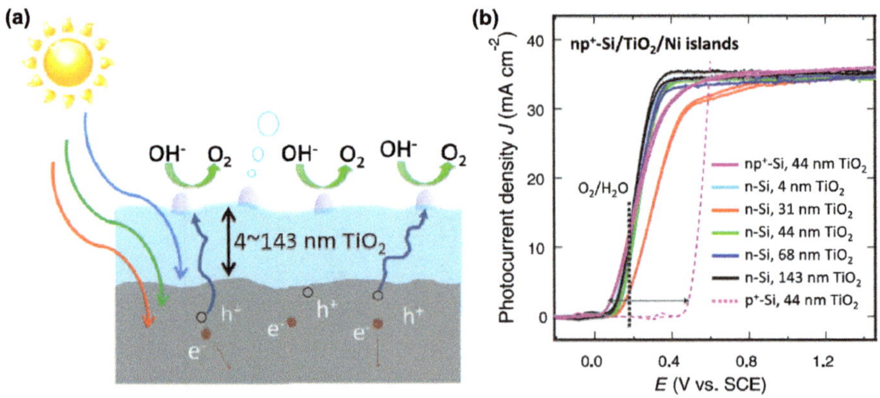

Figure 6.9 (a) Schematic of a photo-absorber with an ALD deposited TiO_2 layer which allows hole conduction. (b) Illuminated *J–E* curves for TiO_2 heterojunction photoanodes with n-Si as a function of TiO_2 thickness; NiO_x is used as the water oxidation co-catalyst. Data for a np^+-Si/TiO_2/NiO_x structure is also shown, along with dark data from a p^+-Si/TiO_2 control structure. Experiments were performed in 1 M aqueous KOH and under 1.25 suns of simulated solar illumination.
Adapted from S. Hu, M. R. Shaner, J. A. Beardslee, M. Lichterman, B. S. Brunschwig and N. S. Lewis, Amorphous TiO2 coatings stabilize Si, GaAs, and GaP photoanodes for efficient water oxidation, *Science*, 2014, **344**, 1005–1009.[102] Reprinted with permission from AAAS.

band conduction of photo-generated holes. Plasma-enhanced atomic layer deposition was used to deposit CoO_x on np^+-Si. It was discovered that the thickness of the SiO_x interlayer which forms between the Si and CoO_x, has a large influence on the performance of the integrated device. Reducing the thickness of this layer by sputter etching achieves a more intimate interfacial contact between the light absorber and catalyst and an improvement in the performance of the photoanodes; 1 sun photocurrents of *ca.* 30 mA cm^{-2} were observed as well as 24 hour stability, as shown Figure 6.10.

An "ideal" protection layer for photoanodes would have a hole selective contact with the underlying photoabsorber and would have a wider band gap to reduce parasitic absorption.[56] Structural stability is also critical; if the material undergoes a phase transition during oxidation/reduction cycling, as is the case with NiO_x,[105,106] the volume change could lead to failure of the coating. Ager and co-workers investigated the p-type conductor $NiCo_2O_4$ as a protection layer that meets many of the above criteria.[79] The bandgap of $NiCo_2O_4$ is smaller than that of NiO but higher than Co_3O_4 and has a hole conductivity (up to 300 S cm^{-1}) at least two orders of magnitude higher than either NiO or Co_3O_4.[107] $NiCo_2O_4$ is relatively transparent in the visible, allowing thick layers (*ca.* 40 nm) to be used as protection layers without large reductions in photocurrent. Moreover, it can be deposited at low temperature by reactive sputtering, which is a more scalable process than atomic layer deposition. In agreement with the Yang *et al.*[103] study the thickness of the SiO_2 interlayer was an important parameter; deposited at higher substrate temperature or annealing the structure increased the thickness of this layer and reduced performance. As shown in Figure 6.11(a), np^+-Si/$NiCo_2O_4$ structures integrated with a NiFe OER co-catalyst achieved 1 sun limiting photocurrents of over 30 mA cm^{-2}, with a current density of 26 mA cm^{-2} at the reversible potential for water oxidation. This work also introduced more

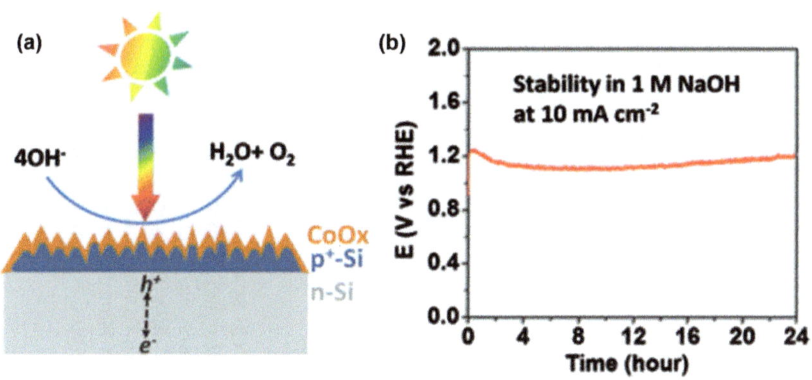

Figure 6.10 (a) Schematic depiction of plasma-enhanced atomic layer deposition of cobalt oxide onto nanotextured p^+n-Si. (b) 24 hour stability test at 10 mA cm^{-2} under 1 sun illumination in 1 M NaOH.
Reprinted with permission from *J. Am. Chem. Soc.*, 2014, **136**, 6191–6194, Copyright 2014 American Chemical Society.

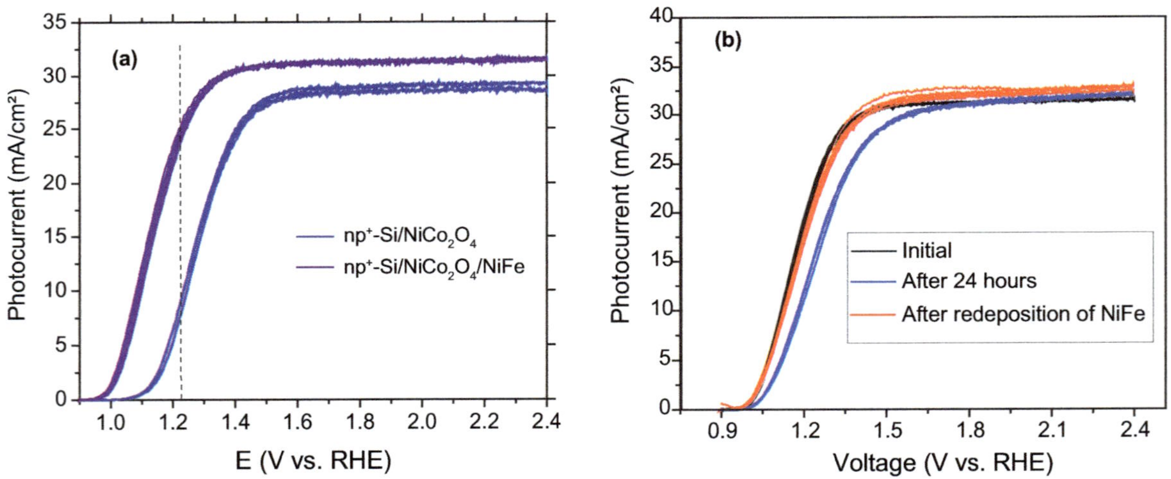

Figure 6.11 (a) CV scans of np$^+$-Si/NiCo$_2$O$_4$ photoanode with (purple) and without (blue) a 2 nm Ni–Fe (50/50) coating serving as an oxygen evolving catalyst. For this device, the onset potential for OER is 0.95 V *vs.* RHE and the current density at the reversible potential for oxygen evolution is 26 mA cm^{-2}. The vertical dashed line is at 1.23 V *vs.* RHE. (b) CV curves from a np$^+$-Si/NiCo$_2$O$_4$/NiFe structure: (1) after NiFe deposition (black), (2) after partial loss of NiFe (blue) leading to decreased performance and (3) after regeneration of NiFe (red). Redeposition of NiFe restores the original performance. Testing was performed in pH 14 KOH electrolyte under simulated 1 sun illumination.
Adapted with permission from *J. Am. Chem. Soc.*, 2015, **137**, 9595–9603, Copyright 2015 American Chemical Society.

stringent accelerated wear diagnostics, such as CV cycling after long term testing, Figure 6.11(b), and also sensitive detection of corrosion products in the electrolyte *via* ICP-MS. This analysis revealed that while the NiFe co-catalyst appear to corrode during operation, the $NiCo_2O_4$ itself was stable with an extrapolated lifetime of several years.

In solar cells, the top contact structure can also function to optimize light trapping in the device, the refractive index and thickness of the layer being critical parameters.[108,109] Transparent protection layers allows this concept to be applied to photoanodes. Lewis and co-workers used NiO_x, deposited under conditions which produce excellent optical transparency ($E_g > 3.5$ eV) to form layers which combine anti-reflection and corrosion resistance.[110] The concept was explored with a heterojunction (HTJ) Si structure with a passivated front contact (n-Si/i-a-Si/p^+-a-Si), a passivated rear contact, and texturing for light trapping. CdTe and amorphous Si absorbers were explored as well. Notably, optimized HTJ Si/75 nm NiO_x structures achieved very high (>90%) incident photon to conversion efficiency (IPCE) when operated under OER conditions, as shown in Figure 6.12. The IPCE values, remarkably, were higher than the solar cell external quantum efficiencies reported for the same devices.[111] Stable operation for >1000 hours was also demonstrated.

As a center, JCAP researchers generated a large body of work on protected photoanodes. Table 6.2 summarizes the more notable studies, focusing on

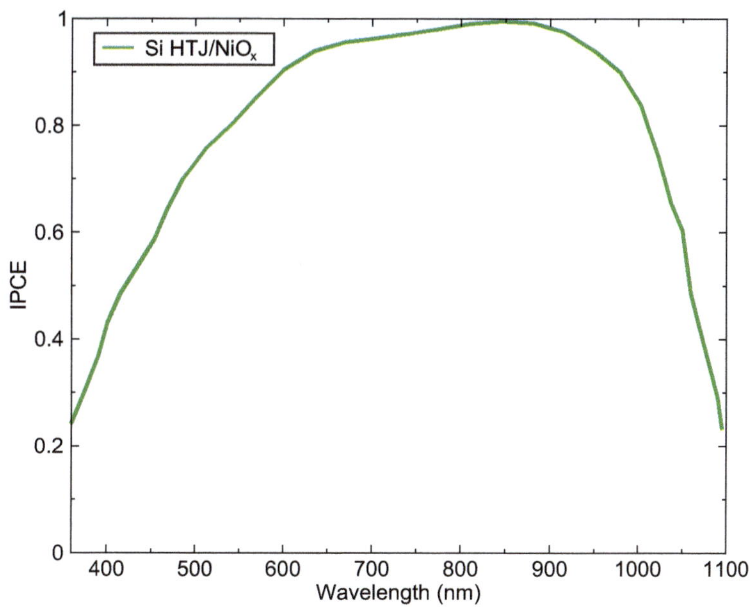

Figure 6.12 IPCE for HTJ-Si/NiO_x photoanodes. Testing was performed at 1.93 V *vs.* RHE in aqueous 1 M KOH.
Adapted from Sun *et al.*[110] Copyright (2015), National Academy of Sciences.

Table 6.2 Summary of performance characteristics of notable water oxidation photoanodes produced by JCAP researchers from 2014–2016.

Study	Absorber(s)	Protection layer, deposition method	OER co-catalyst	Electrolyte and illumination	Photocurrent at 1.23 V *vs.* RHE; onset potential	Stability notes
Hu *et al.*,[102] 2014	n-Si, np^+-Si, np^+-GaAs, n-GaP	4–143 nm TiO_2, ALD	NiO_x	1 M KOH, 1.25 suns	10 mA cm^{-2} for np^+-Si/TiO_2/NiO_x, ~1.1 V *vs.* RHE	>100 hours at 0.93 V *vs.* SCE
Yang *et al.*,[103] 2014	np^+-Si	4–5 nm CoO_x, plasma ALD	None	1 sun AM 1.5G	17 mA cm^{-2}, 1.05 V *vs.* RHE	24 hours held at 10 mA cm^{-2}
Chen *et al.*,[79] 2015	n-Si, np^+-Si, n-InP	40 nm $NiCo_2O_4$, reactive sputtering	NiFe	1 M KOH, 1 sun AM1.5G	26 mA cm^{-2} for np^+-Si/$NiCo_2O_4$/NiFe, 0.95 V *vs.* RHE	>72 hours at 1.4 V *vs* RHE, also CV cycling and analysis of corrosion products
Sun *et al.*,[112] 2015	n-Si, np^+-Si	NiO_x, reactive sputtering	None	1 M KOH, 1 sun AM1.5G	29 mA cm^{-2} for np^+-Si/NiO_x, ~1 V *vs.* RHE	>1000 hours
Sun *et al.*,[110] 2015	HTJ Si, n-CdTe, n-a-Si:H	NiO_x, reactive sputtering	None	1 M KOH, 1 sun AM1.5G	20.7 mA cm^{-2} for HTJ-Si/NiO_x, 0.95 V *vs.* RHE	>100 hours with measurement of O_2 product
Zhou *et al.*,[113] 2016	n-Si	50 nm CoO_x, ALD	None	1 M KOH, 1.1 suns AM1.5G	23.2 mA cm^{-2}, 1.02 V *vs.* RHE	>2500 hours with 10% decline in current density, 1.63 V *vs.* RHE
Yang *et al.*,[114] 2016	np^+-Si	Co_3O_4 spinel, plasma ALD	None	1 M NaOH, 1 sun AM1.5G	30.8 mA cm^{-2}, 0.95 V *vs.* RHE	72 hours constant photocurrent, no Co corrosion products detected by ICP-MS

the performance, as measured by the photocurrent at the reversible potential for OER, and the stability. The stability was usually measured by monitoring the photocurrent as a function of time during continuous illumination at a fixed bias. However, some studies studied stability under more realistic conditions such as oxidation/reduction cycling.

6.4 Summary and Outlook

During the period 2011–2015, there was significant research attention on both fundamental science and also technical performance limits of photoelectrodes used for hydrogen and oxygen evolution. From this work, it has become clear that ultimate performance and operational stability can only be attained by considering carefully the properties of all the heterojunction in the system, both solid–solid and solid–electrolyte.[77] Analogous to solar cells, performance is optimized by reducing minority carrier recombination in the absorber and at all interfaces. In most cases, a solid–solid heterojunction which provides surface passivation and a large barrier for majority carriers produces the best results.[61] Corrosion protection can also be accomplished with a protection layer; here, a conformal coating is crucial if the underlying semiconductor is unstable, as is the case for most non-oxides such as Si, GaAs, InP, *etc.* Significant progress was made with the use of TiO_2, CoO_x, and $NiCo_2O_4$ for this purpose. Even with materials which do not need protection, such as hematite, catalyst integration and the use of additional layers to improve the directionality of the charge transport have become typical in the field.[115]

Acknowledgements

This work was supported by the Joint Center for Artificial Photosynthesis, a DOE Energy Innovation Hub, supported through the Office of Science of the US Department of Energy under Award Number DE-SC0004993.

References

1. A. Fujishima and K. Honda, Electrochemical Photolysis of Water at a Semiconductor Electrode, *Nature*, 1972, **238**, 37.
2. C. Breyer and A. Gerlach, Global overview on grid-parity, *Prog. Photovoltaics*, 2013, **21**, 121–136.
3. K. W. K. Harrison, G. D. G. Martin, T. G. Ramsden, W. E. Kramer and F. J. Novachek, *Wind-to-Hydrogen Project: Operational Experience, Performance Testing, and Systems Integration*, Golden, CO, 2009.
4. H. Zhou, T. Fan and D. Zhang, An Insight into Artificial Leaves for Sustainable Energy Inspired by Natural Photosynthesis, *ChemCatChem*, 2011, **3**, 513–528.

5. Y. Okada and M. Shimura, in *Joint GCC-JAPAN Environment Symposia*, 2013.
6. J. A. Turner, Sustainable hydrogen production, *Science*, 2004, **305**, 972–974.
7. M. Z. Jacobson, W. G. Colella and D. M. Golden, Cleaning the air and improving health with hydrogen fuel-cell vehicles, *Science*, 2005, **308**, 1901–1905.
8. J. J. López Cascales, M. C. Juan-Segovia, J. Ibáñez Molina, J. Sánchez Vera and P. M. Vivo Vivo, Environmental impact associated with the substitution of internal combustion vehicles by fuel cell vehicles refueled with hydrogen generated by electrolysis using the power grid. An estimation focused on the Autonomous Region of Murcia (Spain), *Renewable Energy*, 2015, **77**, 79–85.
9. J. Töpler, Hydrogen as energy-storage-medium and fuel, *Renewable Energy Environ. Sustainable Dev.*, 2016, **1**, 31.
10. C. Graves, S. D. Ebbesen, M. Mogensen and K. S. Lackner, Sustainable hydrocarbon fuels by recycling CO2 and H2O with renewable or nuclear energy, *Renewable Sustainable Energy Rev.*, 2011, **15**, 1–23.
11. A. J. Bard and M. A. Fox, Artificial Photosynthesis: Solar Splitting of Water to Hydrogen and Oxygen, *Acc. Chem. Res.*, 1995, **28**, 141–145.
12. M. G. Walter, E. L. Warren, J. R. McKone, S. W. Boettcher, Q. Mi, E. A. Santori and N. S. Lewis, Solar Water Splitting Cells, *Chem. Rev.*, 2010, **110**, 6446.
13. Y. Tachibana, L. Vayssieres and J. R. Durrant, Artificial photosynthesis for solar water-splitting, *Nat. Photonics*, 2012, **6**, 511–518.
14. M. S. Prévot and K. Sivula, Photoelectrochemical Tandem Cells for Solar Water Splitting, *J. Phys. Chem. C*, 2013, **117**, 17879–17893.
15. T. J. Jacobsson, V. Fjällström, M. Edoff and T. Edvinsson, Sustainable solar hydrogen production: from photoelectrochemical cells to PV-electrolyzers and back again, *Energy Environ. Sci.*, 2014, **7**, 2056.
16. E. L. Miller, Photoelectrochemical water splitting, *Energy Environ. Sci.*, 2015, **8**, 2809–2810.
17. N. S. Lewis, Research opportunities to advance solar energy utilization, *Science*, 2016, **351**, aad1920.
18. A. C. Nielander, M. R. Shaner, K. M. Papadantonakis, S. A. Francis and N. S. Lewis, A taxonomy for solar fuels generators, *Energy Environ. Sci.*, 2015, **8**, 16–25.
19. W. Shockley and H. J. Queisser, Detailed balance limit of efficiency of p-n junction solar cells, *J. Appl. Phys.*, 1961, **32**, 510–519.
20. J. R. Bolton, S. J. Strickler and J. S. Connolly, Limiting and realizable efficiencies of solar photolysis of water, *Nature*, 1985, **316**, 495–500.
21. M. Weber and M. Dignam, *Int. J. Hydrogen Energy*, 1986, **11**, 225–232.
22. L. C. Seitz, Z. Chen, A. J. Forman, B. A. Pinaud, J. D. Benck and T. F. Jaramillo, Modeling Practical Performance Limits of

Photoelectrochemical Water Splitting Based on the Current State of Materials Research, *ChemSusChem*, 2014, **7**, 1372–1385.

23. H. Döscher, J. Geisz, T. Deutsch and J. Turner, Sunlight absorption in water – efficiency and design implications for photoelectrochemical devices, *Energy Environ. Sci.*, 2014, **7**, 2951–2956.
24. S. Haussener, S. Hu, C. Xiang, A. Z. Weber and N. S. Lewis, Simulations of the irradiation and temperature dependence of the efficiency of tandem photoelectrochemical water-splitting systems, *Energy Environ. Sci.*, 2013, **6**, 3605.
25. J. Rongé, T. Bosserez, D. Martel, C. Nervi, L. Boarino, F. Taulelle, G. Decher, S. Bordiga and J. A. Martens, Monolithic cells for solar fuels, *Chem. Soc. Rev.*, 2014, **43**, 7963–7981.
26. J. W. Ager, M. R. Shaner, K. A. Walczak, I. D. Sharp and S. Ardo, Experimental demonstrations of spontaneous, solar-driven photoelectrochemical water splitting, *Energy Environ. Sci.*, 2015, **8**, 2811–2824.
27. O. Khaselev and J. A. Turner, A Monolithic Photovoltaic-Photoelectrochemical Device for Hydrogen Production via Water Splitting, *Science*, 1998, **280**, 425–427.
28. O. Khaselev, High-efficiency integrated multijunction photovoltaic/electrolysis systems for hydrogen production, *Int. J. Hydrogen Energy*, 2001, **26**, 127–132.
29. M. M. May, H.-J. Lewerenz, D. Lackner, F. Dimroth and T. Hannappel, Efficient direct solar-to-hydrogen conversion by in situ interface transformation of a tandem structure, *Nat. Commun.*, 2015, **6**, 8286.
30. E. Verlage, S. Hu, R. Liu, R. J. R. Jones, K. Sun, C. Xiang, N. S. Lewis and H. A. Atwater, A monolithically integrated, intrinsically safe, 10% efficient, solar-driven water-splitting system based on active, stable earth-abundant electrocatalysts in conjunction with tandem III–V light absorbers protected by amorphous TiO2 films, *Energy Environ. Sci.*, 2015, **8**, 3166–3172.
31. A. Nakamura, Y. Ota, K. Koike, Y. Hidaka, K. Nishioka, M. Sugiyama and K. Fujii, A 24.4% solar to hydrogen energy conversion efficiency by combining concentrator photovoltaic modules and electrochemical cells, *Appl. Phys. Express*, 2015, **8**, 107101.
32. S. A. Bonke, M. Wiechen, D. R. MacFarlane and L. Spiccia, Renewable fuels from concentrated solar power: towards practical artificial photosynthesis, *Energy Environ. Sci.*, 2015, **8**, 2791–2796.
33. J. Jia, L. C. Seitz, J. D. Benck, Y. Huo, Y. Chen, J. W. D. Ng, T. Bilir, J. S. Harris and T. F. Jaramillo, Solar water splitting by photovoltaic-electrolysis with a solar-to-hydrogen efficiency over 30%, *Nat. Commun.*, 2016, **7**, 13237.
34. B. A. Pinaud, J. D. Benck, L. C. Seitz, A. J. Forman, Z. Chen, T. G. Deutsch, B. D. James, K. N. Baum, G. N. Baum, S. Ardo, H. Wang, E. Miller and T. F. Jaramillo, Technical and economic feasibility of centralized facilities for solar hydrogen production via photocatalysis and photoelectrochemistry, *Energy Environ. Sci.*, 2013, **6**, 1983–2002.

35. C. A. Rodriguez, M. A. Modestino, D. Psaltis and C. Moser, Design and cost considerations for practical solar-hydrogen generators, *Energy Environ. Sci.*, 2014, **7**, 3828–3835.
36. P. Zhai, S. Haussener, J. Ager, R. Sathre, K. Walczak, J. Greenblatt and T. McKone, Net primary energy balance of a solar-driven photoelectrochemical water-splitting device, *Energy Environ. Sci.*, 2013, **6**, 2380.
37. R. Sathre, C. D. Scown, W. R. Morrow, J. C. Stevens, I. D. Sharp, J. W. Ager, K. Walczak, F. A. Houle and J. B. Greenblatt, Life-cycle net energy assessment of large-scale hydrogen production via photoelectrochemical water splitting, *Energy Environ. Sci.*, 2014, **7**, 3264–3278.
38. R. Sathre, J. B. Greenblatt, K. Walczak, I. D. Sharp, J. C. Stevens, J. W. Ager and F. A. Houle, Opportunities to improve the net energy performance of photoelectrochemical water-splitting technology, *Energy Environ. Sci.*, 2016, **9**, 803–819.
39. M. E. Flowers, M. K. Smith, A. W. Parsekian, D. S. Boyuk, J. K. McGrath and L. Yates, Climate impacts on the cost of solar energy, *Energy Policy*, 2016, **94**, 264–273.
40. A. Limmanee, S. Songtrai, N. Udomdachanut, S. Kaewniyompanit, Y. Sato, M. Nakaishi, S. Kittisontirak, K. Sriprapha and Y. Sakamoto, Degradation analysis of photovoltaic modules under tropical climatic conditions and its impacts on LCOE, *Renewable Energy*, 2017, **102**, 199–204.
41. T. J. Peshek, J. S. Fada, Y. Hu, Y. Xu, M. A. Elsaeiti, E. Schnabel, M. Köhl and R. H. French, Insights into metastability of photovoltaic materials at the mesoscale through massive I–V analytics, *J. Vac. Sci. Technol., B: Nanotechnol. Microelectron.: Mater., Process., Meas., Phenom.*, 2016, **34**, 50801.
42. R. H. French, R. Podgornik, T. J. Peshek, L. S. Bruckman, Y. Xu, N. R. Wheeler, A. Gok, Y. Hu, M. A. Hossain, D. A. Gordon, P. Zhao, J. Sun and G. Q. Zhang, Degradation science: Mesoscopic evolution and temporal analytics of photovoltaic energy materials, *Curr. Opin. Solid State Mater. Sci.*, 2015, **19**, 212–226.
43. N. R. Wheeler, A. Gok, T. J. Peshek, L. S. Bruckman, N. Goel, D. Zabiyaka, C. L. Fagerholm, T. Dang, C. Alcantara, M. L. Terry and R. H. French, A data science approach to understanding photovoltaic module degradation, *SPIE Opt. + Photonics Sustain. Energy*, 2015, **9563**, 95630L.
44. M. Grotheer, R. Alkire and R. Varjian, Industrial Electrolysis and Electrochemical Engineering, *Electrochem. Soc. Interface*, 2006, 53.
45. T. Smolinka, Fuels–hydrogen production, water electrolysis, in *Encyclopedia of Electrochemical Power Sources*, Fraunhofer Institute for Solar Energy Systems (ISE), Freiburg, Germany, 2009, pp. 394–413.
46. A. Ursua, L. M. L. M. Gandia and P. Sanchis, Hydrogen Production From Water Electrolysis: Current Status and Future Trends, *Proc. IEEE*, 2012, **100**, 410–426.

47. H. Gerischer, On the stability of semiconductor electrodes against photodecomposition, *J. Electroanal. Chem. Interfacial Electrochem.*, 1977, **82**, 133–143.
48. M. W. Kanan and D. G. Nocera, In Situ Formation of an Oxygen-Evolving Catalyst in Neutral Water Containing Phosphate and Co2 +, *Science*, 2008, **321**, 1072–1075.
49. M. Dincă, Y. Surendranath and D. G. Nocera, Nickel-borate oxygen-evolving catalyst that functions under benign conditions, *Proc. Natl. Acad. Sci. U. S. A.*, 2010, **107**, 10337–10341.
50. H. Gerischer, The role of semiconductor structure and surface properties in photoelectrochemical processes, *J. Electroanal. Chem. Interfacial Electrochem.*, 1983, **150**, 553–569.
51. M. Pourbaix, R. W. Staehle and S. (Online Service), *Lectures on Electrochemical Corrosion*, 1973.
52. M. Pourbaix, *Atlas of Electrochemical Equilibria in Aqueous Solutions*, National Association of Corrosion Engineers, Houston, Texas, 2d English, 1974.
53. S. Chen and L.-W. Wang, Thermodynamic Oxidation and Reduction Potentials of Photocatalytic Semiconductors in Aqueous Solution, *Chem. Mater.*, 2012, **24**, 3659–3666.
54. T. G. Deutsch, J. L. Head and J. A. Turner, Photoelectrochemical Characterization and Durability Analysis of GaInPN Epilayers, *J. Electrochem. Soc.*, 2008, **155**, B903–B907.
55. H. Wang, T. Deutsch, A. Welch and J. A. Turner, The stability of illuminated p-GaInP2 semiconductor photoelectrode, *Int. J. Hydrogen Energy*, 2012, **37**, 14009–14014.
56. G. Campet, C. Puprichitkun and Z. W. Sun, Protection of photoanodes against photocorrosion by surface deposition of oxide films: Criteria for choosing the protective coating, *J. Electroanal. Chem. Interfacial Electrochem.*, 1989, **269**, 435–445.
57. P. A. Kohl, S. N. Frank and A. J. Bard, Semiconductor Electrodes, *J. Electrochem. Soc.*, 1977, **124**, 225.
58. Y. W. Chen, J. D. Prange, S. Dühnen, Y. Park, M. Gunji, C. E. D. Chidsey and P. C. McIntyre, Atomic layer-deposited tunnel oxide stabilizes silicon photoanodes for water oxidation, *Nat. Mater.*, 2011, **10**, 539–544.
59. T. J. Mills, F. Lin and S. W. Boettcher, Theory and Simulations of Electrocatalyst-Coated Semiconductor Electrodes for Solar Water Splitting, *Phys. Rev. Lett.*, 2014, **112**, 148304.
60. A. G. Scheuermann, J. P. Lawrence, K. W. Kemp, T. Ito, A. Walsh, C. E. D. Chidsey, P. K. Hurley and P. C. McIntyre, Design principles for maximizing photovoltage in metal-oxide-protected water-splitting photoanodes, *Nat. Mater.*, 2015, **15**, 99–105.
61. S. Hu, N. S. Lewis, J. W. Ager, J. Yang, J. R. McKone and N. C. Strandwitz, Thin-Film Materials for the Protection of Semiconducting Photoelectrodes in Solar-Fuel Generators, *J. Phys. Chem. C*, 2015, **119**, 24201–24228.

62. M. Grätzel, Photoelectrochemical cells, *Nature*, 2001, **414**, 338–344.
63. S. Licht, Multiple Band Gap Semiconductor/Electrolyte Solar Energy Conversion, *J. Phys. Chem. B*, 2001, **105**, 6281.
64. A. Lasia, in *Handbook of Fuel Cells*, John Wiley & Sons, Ltd, Chichester, UK, 2010.
65. M. Risch, J. Suntivich and Y. Shao-Horn, in *Encyclopedia of Applied Electrochemistry*, Springer, New York, NY, 2014, pp. 1475–1480.
66. C. C. L. McCrory, S. Jung, J. C. Peters and T. F. Jaramillo, Benchmarking Heterogeneous Electrocatalysts for the Oxygen Evolution Reaction, *J. Am. Chem. Soc.*, 2013, **135**, 16977–16987.
67. C. C. L. McCrory, S. Jung, I. M. Ferrer, S. M. Chatman, J. C. Peters and T. F. Jaramillo, Benchmarking Hydrogen Evolving Reaction and Oxygen Evolving Reaction Electrocatalysts for Solar Water Splitting Devices, *J. Am. Chem. Soc.*, 2015, **137**, 4347–4357.
68. R. Memming, *Semiconductor Electrochemistry*, Wiley, 2000.
69. S. Hu, C.-Y. Chi, K. T. Fountaine, M. Yao, H. A. Atwater, P. D. Dapkus, N. S. Lewis and C. Zhou, Optical, electrical, and solar energy-conversion properties of gallium arsenide nanowire-array photoanodes, *Energy Environ. Sci.*, 2013, **6**, 1879–1890.
70. W. W. Gärtner, Depletion-layer photoeffects in semiconductors, *Phys. Rev.*, 1959, **116**, 84–87.
71. J. Reichman, The current-voltage characteristics of semiconductor-electrolyte junction photovoltaic cells, *Appl. Phys. Lett.*, 1980, **36**, 574–577.
72. P. A. Basore, Numerical modeling of textured silicon solar cells using PC-1D, *IEEE Trans. Electron Devices*, 1990, **37**, 337–343.
73. R. Stangl, M. Kriegel and M. Schmidt, in *Conference Record of the 2006 IEEE 4th World Conference on* Photovoltaic Energy Conversion, 2006, vol. 2, pp. 1350–1353.
74. L. M. Peter, J. Li, R. Peat and L. M. Peter, Surface recombination at semiconductor electrodes, *J. Electroanal. Chem. Interfacial Electrochem.*, 1985, **165**, 29–40.
75. L. M. Peter, K. G. U. Wijayantha and A. A. Tahir, Kinetics of light-driven oxygen evolution at [small alpha]-Fe2O3 electrodes, *Faraday Discuss.*, 2012, **155**, 309–322.
76. S. J. Anz, O. Kru, N. S. Lewis and H. Gajewski, Conditions Under Which Heterogeneous Charge-Transfer Rate Constants Can Be Extracted from Transient Photoluminescence Decay Data of Semiconductor/Liquid Contacts As Determined by Two-Dimensional Transport Modeling, *J. Phys. Chem. B*, 1998, **102**, 5625–5640.
77. M. R. Nellist, F. A. L. Laskowski, F. Lin, T. J. Mills and S. W. Boettcher, Semiconductor–Electrocatalyst Interfaces: Theory, Experiment, and Applications in Photoelectrochemical Water Splitting, *Acc. Chem. Res.*, 2016, **49**, 733–740.
78. S. M. Sze, *Physics of Semiconductor Devices*, 2nd edn., Wiley, 1981.
79. L. Chen, J. Yang, S. Klaus, L. J. Lee, R. Woods-Robinson, J. Ma, Y. Lum, J. K. Cooper, F. M. Toma, L.-W. Wang, I. D. Sharp, A. T. Bell and

J. W. Ager, p-Type Transparent Conducting Oxide/n-Type Semiconductor Heterojunctions for Efficient and Stable Solar Water Oxidation, *J. Am. Chem. Soc.*, 2015, **137**, 9595–9603.

80. U. Wurfel, A. Cuevas and P. Wurfel, Charge Carrier Separation in Solar Cells, *IEEE J. Photovoltaics*, 2015, **5**, 461–469.
81. S. De Wolf, A. Descoeudres, Z. C. Holman and C. Ballif, High-efficiency Silicon Heterojunction Solar Cells: A Review, *Green*, 2012, **2**, 7.
82. C. Battaglia, S. M. De Nicolás, S. De Wolf, X. Yin, M. Zheng, C. Ballif and A. Javey, Silicon heterojunction solar cell with passivated hole selective MoOx contact, *Appl. Phys. Lett.*, 2014, **104**, 113902.
83. J. Bullock, M. Hettick, J. Geissbühler, A. J. Ong, T. Allen, C. M. Sutter-Fella, T. Chen, H. Ota, E. W. Schaler, S. De Wolf, C. Ballif, A. Cuevas and A. Javey, Efficient silicon solar cells with dopant-free asymmetric heterocontacts, *Nat. Energy*, 2016, **1**, 15031.
84. A. Paracchino, V. Laporte, K. Sivula, M. Grätzel and E. Thimsen, Highly active oxide photocathode for photoelectrochemical water reduction, *Nat. Mater.*, 2011, **10**, 456–461.
85. A. Paracchino, N. Mathews, T. Hisatomi, M. Stefik, S. D. Tilley and M. Grätzel, Ultrathin films on copper(I) oxide water splitting photocathodes: a study on performance and stability, *Energy Environ. Sci.*, 2012, **5**, 8673.
86. S. W. Boettcher, E. L. Warren, M. C. Putnam, E. A. Santori, D. Turner-Evans, M. D. Kelzenberg, M. G. Walter, J. R. McKone, B. S. Brunschwig, H. A. Atwater and N. S. Lewis, Photoelectrochemical Hydrogen Evolution Using Si Microwire Arrays, *J. Am. Chem. Soc.*, 2011, **133**, 1216–1219.
87. Y. Hou, B. L. Abrams, P. C. K. Vesborg, M. E. Björketun, K. Herbst, L. Bech, A. M. Setti, C. D. Damsgaard, T. Pedersen, O. Hansen, J. Rossmeisl, S. Dahl, J. K. Nørskov and I. Chorkendorff, Bioinspired molecular co-catalysts bonded to a silicon photocathode for solar hydrogen evolution, *Nat. Mater.*, 2011, **10**, 434–438.
88. A. Heller, Conversion of sunlight into electrical power and photoassisted electrolysis of water in photoelectrochemical cells, *Acc. Chem. Res.*, 1981, **14**, 154–162.
89. A. Heller, E. Aharon-Shalom, W. A. Bonner and B. Miller, Hydrogen-evolving semiconductor photocathodes: nature of the junction and function of the platinum group metal catalyst, *J. Am. Chem. Soc.*, 1982, **104**, 6942–6948.
90. A. Heller, Hydrogen-Evolving Solar Cells, *Science*, 1984, **223**, 1141–1148.
91. H. J. Lewerenz and K. H. Schulte, Combined photoelectrochemical conditioning and surface analysis of InP photocathodes: II. Photoelectron spectroscopy, *Electrochim. Acta*, 2002, **47**, 2639–2651.
92. K. H. Schulte and H. J. Lewerenz, Combined photoelectrochemical conditioning and photoelectron spectroscopy analysis of InP photocathodes. I. The modification procedure, *Electrochim. Acta*, 2002, **47**, 2633–2638.

93. H.-J. Lewerenz, Epitaxial III–V Thin Film Absorbers: Preparation, Efficient InP Photocathodes and Routes to High Efficiency Tandem Structures, *Photoelectrochem. Water Split. Mater. Process. Archit.*, 2013, **9**, 223.
94. M. H. Lee, K. Takei, J. Zhang, R. Kapadia, M. Zheng, Y.-Z. Chen, J. Nah, T. S. Matthews, Y.-L. Chueh, J. W. Ager and A. Javey, p-Type InP Nanopillar Photocathodes for Efficient Solar-Driven Hydrogen Production, *Angew. Chem., Int. Ed.*, 2012, **51**, 10760–10764.
95. Y. Lin, R. Kapadia, J. Yang, M. Zheng, K. Chen, M. Hettick, X. Yin, C. Battaglia, I. D. Sharp, J. W. Ager and A. Javey, Role of TiO_2 Surface Passivation on Improving the Performance of p-InP Photocathodes, *J. Phys. Chem. C*, 2015, **119**, 2308–2313.
96. R. Kapadia, Z. Yu, H.-H. H. Wang, M. Zheng, C. Battaglia, M. Hettick, D. Kiriya, K. Takei, P. Lobaccaro, J. W. Beeman, J. W. Ager, R. Maboudian, D. C. Chrzan and A. Javey, A direct thin-film path towards low-cost large-area III-V photovoltaics, *Sci. Rep.*, 2013, **3**, 2275.
97. M. Zheng, H.-P. Wang, C. M. Sutter-Fella, C. Battaglia, S. Aloni, X. Wang, J. Moore, J. W. Beeman, M. Hettick, M. Amani, W.-T. Hsu, J. W. Ager, P. Bermel, M. Lundstrom, J.-H. He and A. Javey, Thin-Film Solar Cells with InP Absorber Layers Directly Grown on Nonepitaxial Metal Substrates, *Adv. Energy Mater.*, 2015, **5**, 1501337.
98. M. Zheng, K. Horowitz, M. Woodhouse, C. Battaglia, R. Kapadia and A. Javey, III-Vs at scale: a PV manufacturing cost analysis of the thin film vapor-liquid-solid growth mode, *Prog. Photovoltaics Res. Appl.*, 2016, **24**, 871.
99. M. Hettick, M. Zheng, Y. Lin, C. M. Sutter-Fella, J. W. Ager and A. Javey, Nonepitaxial Thin-Film InP for Scalable and Efficient Photocathodes, *J. Phys. Chem. Lett.*, 2015, **6**, 2177–2182.
100. Y. Lin, C. Battaglia, M. Boccard, M. Hettick, Z. Yu, C. Ballif, J. W. Ager and A. Javey, Amorphous Si Thin Film Based Photocathodes with High Photovoltage for Efficient Hydrogen Production, *Nano Lett.*, 2013, **13**, 5615–5618.
101. A. G. Scheuermann, J. D. Prange, M. Gunji, C. E. D. Chidsey and P. C. McIntyre, Effects of catalyst material and atomic layer deposited TiO2 oxide thickness on the water oxidation performance of metal–insulator–silicon anodes, *Energy Environ. Sci.*, 2013, **6**, 2487.
102. S. Hu, M. R. Shaner, J. A. Beardslee, M. Lichterman, B. S. Brunschwig and N. S. Lewis, Amorphous TiO2 coatings stabilize Si, GaAs, and GaP photoanodes for efficient water oxidation, *Science*, 2014, **344**, 1005–1009.
103. J. Yang, K. Walczak, E. Anzenberg, F. M. Toma, G. Yuan, J. Beeman, A. Schwartzberg, Y. Lin, M. Hettick, A. Javey, J. W. Ager, J. Yano, H. Frei and I. D. Sharp, Efficient and Sustained Photoelectrochemical Water Oxidation by Cobalt Oxide/Silicon Photoanodes with Nanotextured Interfaces, *J. Am. Chem. Soc.*, 2014, **136**, 6191–6194.

104. M. T. Greiner, M. G. Helander, W.-M. M. Tang, Z.-B. B. Wang, J. Qiu and Z.-H. H. Lu, Universal energy-level alignment of molecules on metal oxides, *Nat. Mater.*, 2011, **11**, 76–81.
105. D. Singh, Characteristics and Effects of γ-NiOOH on Cell Performance and a Method to Quantify It in Nickel Electrodes, *J. Electrochem. Soc.*, 1998, **145**, 116.
106. F. Lin and S. W. Boettcher, Adaptive semiconductor/electrocatalyst junctions in water-splitting photoanodes, *Nat. Mater.*, 2013, **13**, 81–86.
107. C. F. Windisch Jr., K. F. Ferris, G. J. Exarhos and S. K. Sharma, Conducting spinel oxide films with infrared transparency, *Thin Solid Films*, 2002, **420–421**, 89–99.
108. D. Ginley, H. Hosono and D. C. Paine, *Handbook of Transparent Conductors*, Springer Science & Business Media, 2010.
109. A. E. Delahoy and S. Guo, in *Handbook of Photovoltaic Science and Engineering*, ed. A. Luque and S. Hegedus, John Wiley & Sons, Ltd, 2nd edn, 2011, pp. 716–796.
110. K. Sun, F. H. Saadi, M. F. Lichterman, W. G. Hale, H.-P. Wang, X. Zhou, N. T. Plymale, S. T. Omelchenko, J.-H. He, K. M. Papadantonakis, B. S. Brunschwig and N. S. Lewis, Stable solar-driven oxidation of water by semiconducting photoanodes protected by transparent catalytic nickel oxide films, *Proc. Natl. Acad. Sci.*, 2015, 201423034.
111. H. P. Wang, T. Y. Lin, C. W. Hsu, M. L. Tsai, C. H. Huang, W. R. Wei, M. Y. Huang, Y. J. Chien, P. C. Yang, C. W. Liu, L. J. Chou and J. H. He, Realizing high-efficiency omnidirectional n-type Si solar cells via the hierarchical architecture concept with radial junctions, *ACS Nano*, 2013, **7**, 9325–9335.
112. K. Sun, M. T. McDowell, A. C. Nielander, S. Hu, M. R. Shaner, F. Yang, B. S. Brunschwig and N. S. Lewis, Stable Solar-Driven Water Oxidation to O2 (g) by Ni-Oxide-Coated Silicon Photoanodes, *J. Phys. Chem. Lett.*, 2015, **6**, 592–598.
113. X. Zhou, R. Liu, K. Sun, K. M. Papadantonakis, B. S. Brunschwig and N. S. Lewis, 570 mV photovoltage, stabilized n-Si/CoOx heterojunction photoanodes fabricated using atomic layer deposition, *Energy Environ. Sci.*, 2016, **9**, 892–897.
114. J. Yang, J. K. Cooper, F. M. Toma, K. A. Walczak, M. Favaro, J. W. Beeman, L. H. Hess, C. Wang, C. Zhu, S. Gul, J. Yano, C. Kisielowski, A. Schwartzberg and I. D. Sharp, A multifunctional biphasic water splitting catalyst tailored for integration with high-performance semiconductor photoanodes, *Nat. Mater.*, 2016, **16**, 335–341.
115. Gurudayal, D. Sabba, M. H. Kumar, L. H. Wong, J. Barber, M. Grätzel and N. Mathews, Perovskite–Hematite Tandem Cells for Efficient Overall Solar Driven Water Splitting, *Nano Lett.*, 2015, **15**, 3833–3839.

CHAPTER 7

Artificial Photosynthesis with Inorganic Particles

FRANK E. OSTERLOH

Department of Chemistry, University of California, Davis, CA 95616, USA
Email: fosterloh@ucdavis.edu

7.1 Why Particles?

Solar hydrogen generating devices can be classified into photoelectrochemical cells (PEC), buried photovoltaic devices (PV), photovoltaic-biased photoelectrochemical cells (PV-PEC), photocatalysts (PCs), and hybrids thereof.[1–3] Of these, PV devices achieve the highest efficiencies today, followed by PV-PEC devices, and photocatalytic systems (Figure 7.1).[4–6] Even though costs of PV have come down over the last decade, the technology is still not competitive with the low price of fossil fuels. PV costs are associated with expensive vacuum deposition techniques, high purity light absorbers, and noble metals for the electrolyzers. In PV-PEC devices, cost savings are realized by replacing the electrolyzer and part of the PV with photoelectrodes based on abundant metal compounds. However, efficiencies remain below those of PV and stability is not sufficient for industrial applications. These obstacles make it difficult to implement PEC and PV technology on a scale necessary to meet the global demand (13.5 TW in 2001) for renewable fuels.[7]

Technoeconomic analyses of hydrogen production confirm that particle based water splitting systems (photocatalyst suspensions) can be over one order of magnitude cheaper than photoelectrochemical or photovoltaic cells.[5,8–10] For example, a single compartment photocatalyst reactor

Energy and Environment Series No. 22
Integrated Solar Fuel Generators
Edited by Ian D. Sharp, Harry A. Atwater and Hans-Joachim Lewerenz

Published by the Royal Society of Chemistry, www.rsc.org

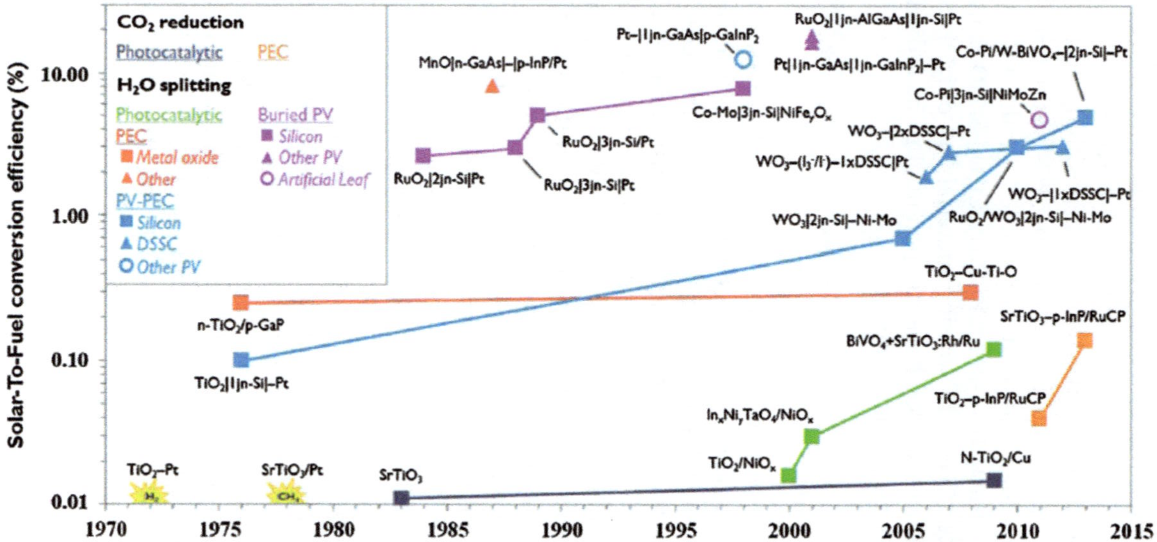

Figure 7.1 Evolution of record solar-to-fuel efficiencies of different approaches, reported in the absence of chemical or electrical bias and under (simulated) solar illumination.
Reproduced from ref. 4 with permission from The Royal Society of Chemistry. For a related analysis see Ager *et al.*[5]

operating at 10% solar-to-hydrogen (STH) efficiency should be able to produce H_2 at a cost of \$1.63 per kg. This would be sufficient to outcompete oil as a fuel and induce a transition to a global solar energy economy. Additionally, production of PCs is more scalable than PV or PEC technology, since particle semiconductors are accessible through many synthetic pathways (solid-state, solution phase, flux, combustion methods),[11–13] and integration with cocatalyst components can be readily achieved through physical mixing, photodeposition, electrochemical deposition, and other methods.[14,15] Furthermore, there are indications that PCs do not require the same high purity starting materials that are needed in the PV industry. For example, a recent RuO_x–$BiVO_4$:Mo/Au/$Ru_{2-y}Cr_yO_4$–$SrTiO_3$:La,Rh tandem system achieves 1.1% STH with starting materials of 'only' 99.5% purity.[16] These arguments show that PCs have the potential to be a disruptive technology for renewable fuel production.

7.1.1 Photoreactors

Half of all photosynthetic activity on Earth is achieved with particles made of phytoplankton.[17] This microorganism consists of 20–100 micrometer particles (Figure 7.2A and B) that drift near the surface of the ocean where they are exposed to essential minerals, sunlight, and carbon dioxide. In this way, the sea serves as the world's largest suspension-reactor for natural photosynthesis. Artificial photosynthesis of hydrogen will need to be conducted on a similar scale to meet the clean energy demands of the future.

The simplest implementation of solar hydrogen production from water uses the type 1 photoreactor shown in Figure 7.2C and D and also in Figure 7.3. Here, the photocatalysts are immobilized on a conductive sheet or suspended in water inside a transparent container. As both H_2 and O_2 are co-evolved in the same sample space, there is a need for downstream separation of the gases. Technology to separate hydrogen from nitrogen, argon, or methane with membranes already exists,[18] but integration with a photoreactor has not yet been demonstrated. Due to its smaller kinetic diameter, hydrogen can be relatively easily separated from gas mixtures using membranes made of polyimide or polycarbonate derivatives, for example.[19] The cost for gas separation is unknown,[8,9] but the minimum theoretical energy expense of such a process has recently been estimated as 8.7% of the energy content of the H_2/O_2 product coming from the photoreactor.[10] Of this loss, 6.7% is due to pressurization losses for the H_2/O_2 mixture, and 2% are due to non-recovered H_2 (Figure 7.3C). This means, with a theoretical 8.7% energy loss, the STH efficiency of a hypothetical reactor operating at 10% would drop to 9.13%. Actual losses will depend on the efficiency of the membrane and the power consumption of the compressor. More research on the integration of membrane separation technology and photocatalytic water splitting reactors is strongly desired. Critics also often point out that the explosion risk of co-evolution of hydrogen and oxygen in the same sample space prohibits implementation of type 1 reactors on a large scale.

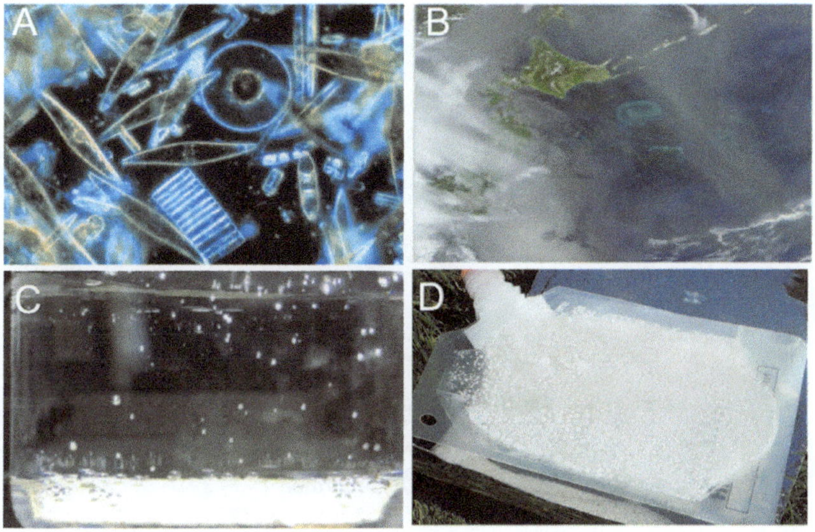

Figure 7.2 Particle based photosynthetic systems. (A) Phytoplankton, individual particles (from NOAA Central Library), (B) phytoplankton bloom offshore of Hokkaido, Japan (from NASA Earth Observatory). (C) H_2/O_2 evolution from a photocatalyst sheet inside of a type 1 photoreactor. From Wang *et al.* Reprinted by permission from Macmillan Publishers Ltd: Nature Materials (Q. Wang, T. Hisatomi, Q. Jia, H. Tokudome, M. Zhong, C. Wang, Z. Pan, T. Takata, M. Nakabayashi, N. Shibata, Y. Li, I. D. Sharp, A. Kudo, T. Yamada and K. Domen, Scalable water splitting on particulate photocatalyst sheets with a solar-to-hydrogen energy conversion efficiency exceeding 1%. *Nat. Mater.* 2016, 15, 611–615) Copyright 2016,[16] (D) H_2/O_2 Evolution from photocatalyst particle suspension in transparent membrane (Osterloh, unpublished results).

Compartmentalization offers a potential solution to this problem. By reducing the size of the photoreactor (for water splitting in microreactors see Dionigi *et al.*[20]), by building it from soft materials (baggies)[8–10] and by encasing critical components inside of an explosion proof shield, operational risks can be minimized (Figure 7.3C).

Gas separation is not required in type 2 photoreactors, which contain separate compartments or water reduction and oxidation (Figure 7.3B). As no reactive gas mixtures are produced, the explosion risk would be lower, allowing for larger reactor sizes. However, because the water oxidation and water reduction half reactions are coupled, some degree of electrolyte cycling needs to be implemented to avoid diffusion overpotentials from the consumption of protons in one compartment and the production of protons in the other.[21] Electric connection between compartments would be achieved with soluble redox shuttles or using solid electrodes. The coupling of photocatalyst suspensions to electrodes was achieved in the 80s by the Bard group.[22,23] However, the diffusion controlled photocurrents in such 'slurry photoreactors' were found to be small. Charge transport limitations can be

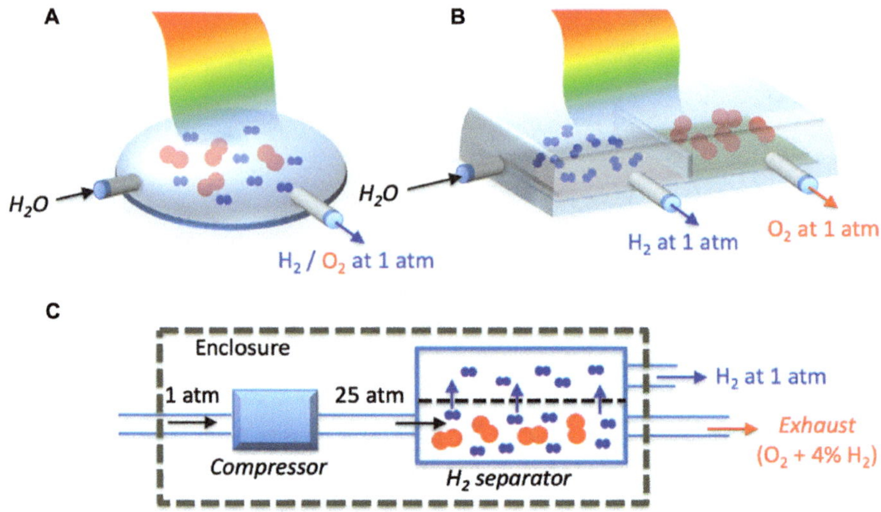

Figure 7.3 (A) Type 1 single compartment particle suspension reactor. (B) Type 2 dual compartment fixed panel reactor. (C) Proposed H_2/O_2 gas separation system in explosion-proof enclosure. Removal of 98% of H_2 from 2 : 1 (vol) H_2/O_2 product stream leaves 4% (vol) in the exhaust. Compare also ref. 5,8–10.

reduced by immobilizing photocatalyst particles directly on the electrodes. Early examples of such photocatalyst sheets were designed to split water *vapor* only,[20,24] while more recent examples (Figure 7.2C) were shown to convert liquid water into H_2/O_2 with an STH efficiency of 1.1%.[16] Particle immobilization would also reduce the mechanical abrasion of the particles that is difficult to avoid in a suspension system.

7.2 Absorber Configurations

First generation PCs operate on the basis of the single absorber configuration (Figure 7.4).[10] Here light absorption occurs at a single semiconductor particle and charge carriers are separated at the *particle–liquid* or particle–cocatalyst interfaces. Two photons are required to split one molecule of water into H_2 and $\frac{1}{2}$ O_2. The first examples of single absorber PCs were publicized in the late 70s by Michael Grätzel and coworkers and consisted of dye-sensitized TiO_2 and RuO_2 nanoparticles.[25,26] Even though complete water splitting was not achieved (O_2 evolution was later attributed to air contamination),[27,28] these photocatalysts exemplified the basic design principles of miniaturized photoelectrochemical cells.[29,30]

Second Generation PCs, or so-called tandem (or *z*-scheme) photocatalysts employ the dual light absorber configuration that is also found in natural photosynthesis.[31–37] In this PC, one absorber, the photoanode, drives the water oxidation reaction and the other, the photocathode, promotes the

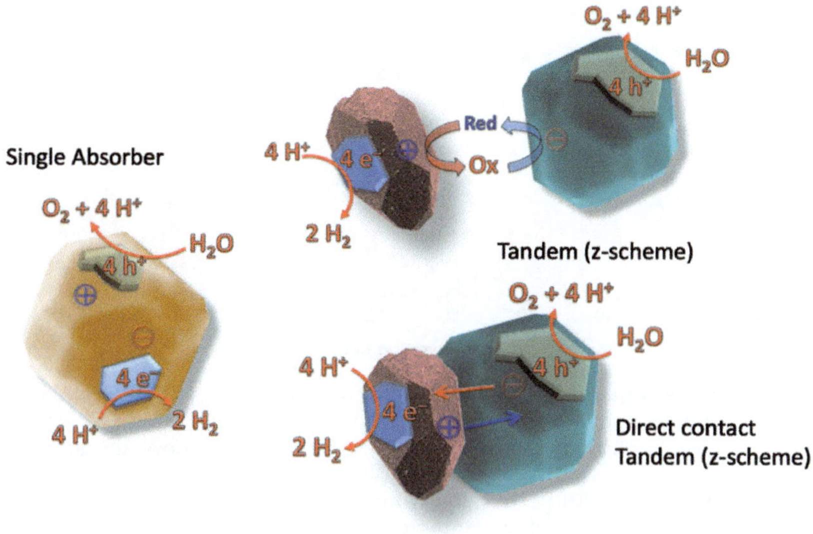

Figure 7.4 Types of microheterogeneous water splitting photocatalysts. See also Maeda *et al.*[36] In tandem devices, electronic contact between the absorbers is established *via* molecular redox shuttles, *e.g.* $Fe^{3+/2+}$ or IO_3^-/I^- or *via* an Ohmic contact.

water reduction reaction. Electrical connection between the sub-systems is established either *via* a solid interface or *via* a liquid interface (Figure 7.4) using redox couples as mediators. For a tandem catalyst, four photons are required to split one molecule of water into H_2 and $\frac{1}{2}$ O_2. The first functional tandem PCs were made by Hironori Arakawa's group in Japan.[38]

7.3 Stability

Recent techno-economic and life cycle assessments of solar water systems have identified STH efficiency and system longevity as the primary factors contributing to positive energy return and cost.[5,39] While long term stability data now is available for a number of electrocatalysts[40,41] and photoelectrodes,[42,43] only a few experimental studies have focused on suspended water splitting photocatalysts. In 2012, Ohno *et al.* examined the GaN:ZnO/$Rh_{2-y}Cr_yO_3$ system with respect to long-term durability and regeneration of photocatalytic activity.[44] Nearly constant and stoichiometric H_2 and O_2 production was observed for three months (2160 h, Figure 7.5). A 50% loss of initial activity occurred during six months of operation which was attributed to detachment of the $Rh_{2-y}Cr_yO_3$ cocatalyst from the light absorber. 80% of the initial activity for H_2 evolution could be regained after $Rh_{2-y}Cr_yO_3$ reloading. However, irreversible decay occurred by photocorrosion of the nitride by valence band holes. Potentially, this could be avoided with all-oxide photoanode materials or with particles protected by

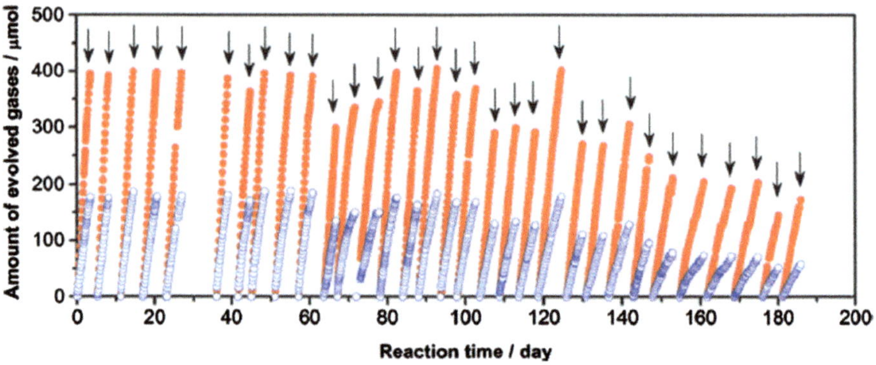

Figure 7.5 Time course of overall water splitting with $Rh_{2-y}Cr_yO_3$/GaN:ZnO under visible light (300 W xenon lamp, $400 < \lambda < 500$ nm). Red circles, H_2; blue circles, O_2. Arrows depict evacuation to remove the accumulated gases.
Reprinted with permission from T. Ohno, L. Bai, T. Hisatomi, K. Maeda and K. Domen, Photocatalytic Water Splitting Using Modified GaN:ZnO Solid Solution under Visible Light: Long-Time Operation and Regeneration of Activity. *J. Am. Chem. Soc.* 2012, 134 (19), 8254–8259, Copyright 2012 American Chemical Society.[44]

passivation layers. While these results showcase the possibility of long term PC performance, it is clear that more studies are needed to evaluate the stability of particle-based water splitting systems.

7.4 Ideal Limiting Solar-to-hydrogen (STH) Efficiency

The ideal limiting solar-to-hydrogen (STH) efficiency η defined in eqn (7.1) provides the maximum possible performance of a water splitting PC. The value depends on the type of PC (tandem or single absorber) and on the band gap of the material. Here we present the original calculation by Bolton *et al.*[45] for a single absorber configuration, although refined calculations for tandem devices have since been published.[46–52] According to Bolton, the efficiency of a single absorber PC is given by a product of the fraction of absorbed photons with $E > E_g$ (η_g), the quantum efficiency of the solar to fuel into chemical energy (η_c):

$$\eta = \eta_g \cdot \eta_c \cdot \phi_{con} \tag{7.1}$$

These parameters are defined in Figure 7.6. Here, η_g is the fraction of absorbed photons *versus* the total solar flux (solar spectral irradiance: air mass 1.5).[53] The larger the band gap E_g of the absorber, the smaller this fraction becomes because only photons of energy $E > E_g$ can be absorbed. For example, a $E_g = 2.0$ eV absorber absorbs 24% of the solar flux.

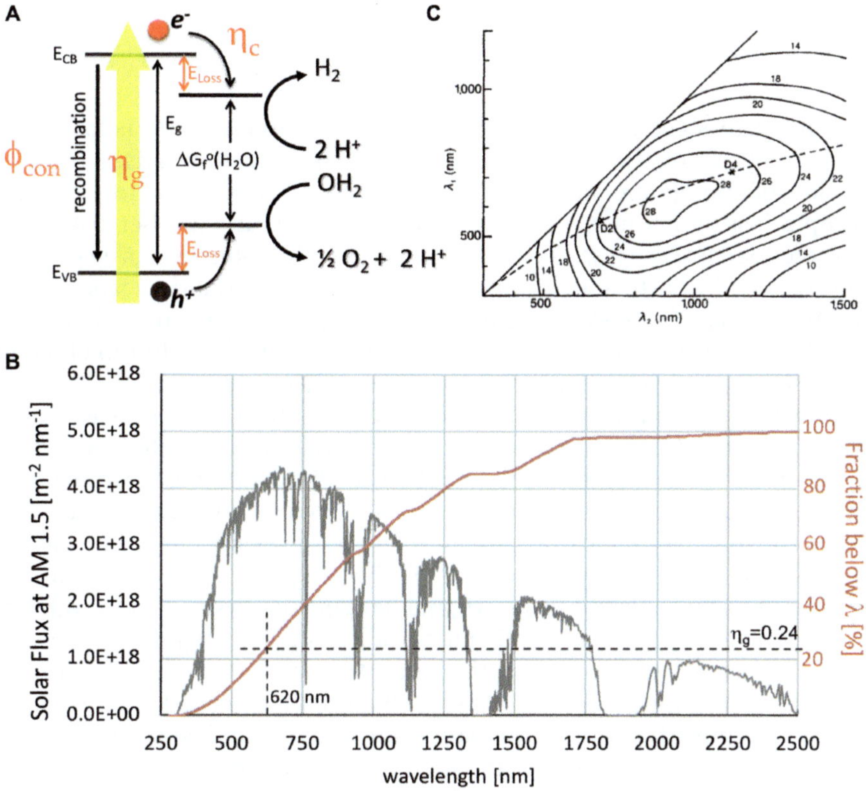

Figure 7.6 (A) Limiting processes for conversion of photon energy into chemical energy. (B) Plot of solar flux *versus* wavelength (AM 1.5). (C) Calculated efficiencies for two light absorbers with E_{g1} and E_{g2} in tandem configuration. Adapted from Bolton.[45] Reprinted by permission from Macmillan Publishers Ltd: Nature (J. R. Bolton, J. S. J. Strickler and J. S. Connolly, Limiting and Realizable Efficiencies of Solar Photolysis of Water, *Nature*, 1985, 316 (6028), 495–500).[45] Copyright 1985.

$\boldsymbol{\phi}_{con}$ is the quantum efficiency of the solar to fuel conversion process, and can be expressed as follows, with ϕ as solar flux:

$$\boldsymbol{\phi}_{con} = \frac{2 \cdot r(H_2)}{\phi} \tag{7.2}$$

The ideal limit for $\boldsymbol{\phi}_{con}$ is 100%, but lower values are generally caused by radiative or non-radioactive electron–hole recombination, which cannot be avoided completely. $\boldsymbol{\eta}_c$ is defined as the fraction of excited state energy converted into chemical energy, according to:

$$\eta_c = \frac{E_g - E_{loss}}{E_g} \tag{7.3}$$

with

$$E_{\text{loss}} = \eta_{\text{pol}} + T\Delta S \tag{7.4}$$

Even for an ideal absorber, some of the band gap energy E_g is lost due to the entropy ΔS of the electron–hole pairs.[54] The entropy arises from the dilution of the charge carriers in the bands and from their motion. This entropic loss is about 0.4 eV per electron–hole pair. Another 0.4 eV are overpotential losses, η_{pol}, associated with the slow kinetics of water electrolysis, and in particular the more complex water oxidation reaction. Assuming a loss of 0.8 eV per electron–hole pair, and a 2.0 eV band gap, $\eta_c = (2.0\ \text{eV} - 0.\ 8\ \text{eV})/2.0\ \text{eV} = 0.6$. Then, the theoretical maximum efficiency of the single absorber system is $\eta = 0.24 \times 1.0 \times 0.6 = 14.4\%$.

Tandem photocatalysts with two separate light absorbers can reach higher STH efficiency. This is because the photovoltage is generated by two absorbers instead of one, which allows the use of smaller band gap materials with better solar light absorbing properties.[55] For example, a combination of absorbers with band gaps of 2.07 and 1.37 eV gives an ideal limiting STH efficiency of 28%, nearly twice that of the single absorber configuration (Figure 7.6c).[45,46] Strictly, this original tandem analysis is only valid for systems in which the large band gap absorber is placed *in front* of the small gap absorber, with minimal optical interference between the two layers (Figure 7.7A). In a particle suspension, when both absorbers are mixed, the relative placement between the large and small band absorbers will be random (Figure 7.7B). This means the large band gap absorber will experience shading from the small band gap absorber in 25% of the cases, as demonstrated in Figure 7.7B, which will reduce the efficiency of this particle arrangement to 0%. In 50% of the cases both absorbers are placed next to each other which doubles the absorption cross section of the tandem and halves the efficiency of the side-by-side configuration. This means that as a

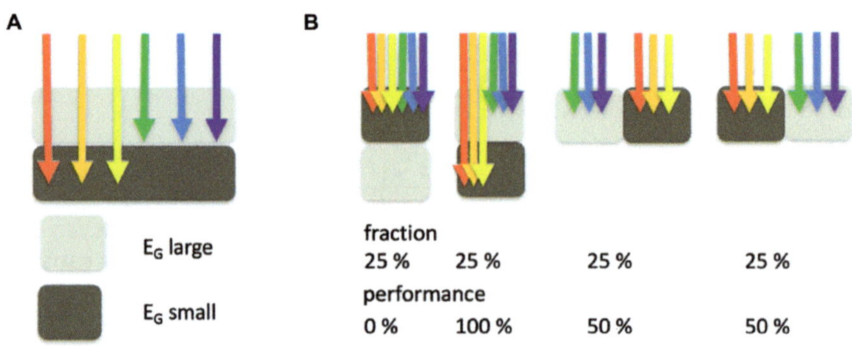

Figure 7.7 Light absorption in bilayer tandem film (A) and in tandem suspension (B). In the suspension, the small gap absorber blocks light for the large band gap absorber 25% of the time, and is next to it 50% of the time, reducing the overall efficiency of the system by 50%.

whole, the suspension tandem only achieves half (50%×28% = 14%) the STH of a layer tandem.

7.5 Experimental Efficiencies

Over the years, a great number of visible light responsive single and dual absorber photocatalytic systems have been developed.[11,33,36,56–58] However, of these only a very small fraction is able to catalyze the overall water splitting reaction (OWS). Figure 7.8 gives an overview of the best visible light active OWS photocatalysts reported to date.[10]

The best single absorber PC is Cr/Rh-modified GaN:ZnO (AQE = 5.9%, 420–440 nm),[59–61] High efficiencies have also been reported for a number of other PCs, including a C_3N_4/C-dot composite (16%, 420 nm),[62] $In_{1-x}Ni_xTaO_4$ (x = 0–0.2),[63,64] CoO,[65] and Cu_2O,[66] but their performance has not yet been reproduced.[67,68] In 2015, Jo *et al.* reported that In/Mo co-doped $BiVO_4$ ($Bi_{1-x}In_xV_{1-x}Mo_xO_4$, x = 0.1) is effective for OWS (AQE = 3.2%) after modification with a RuO_2 cocatalyst. The activity of the material was attributed to a raised conduction band level of the material.[69]

Among tandem structures, several catalysts are notable. The $MgTa_2O_{6-x}N_y$/TaON + WO_3/Pt tandem with a iodate/iodide redox shuttle achieves 6.8% AQE at 420 nm,[37] while the related tandem ZrO_2–TaON/Pt + WO_3/Pt reaches 6.3% AQE (420.5 nm).[34] The direct contact tandem $SrTiO_3$:Rh/$BiVO_4$/Ru has AQE = of 1.7% at 420 nm.[32] This value rises to 3.9–4.2% (420 nm) and 0.1% STH when the components are connected by a soluble $Fe^{2+/3+}$ redox couple,[70] and to 33% (419 nm) and 1.1% STH when the components are co-assembled as a thin film and electrically connected by vacuum-evaporated gold.[16] A derivative using a carbon contact achieves a

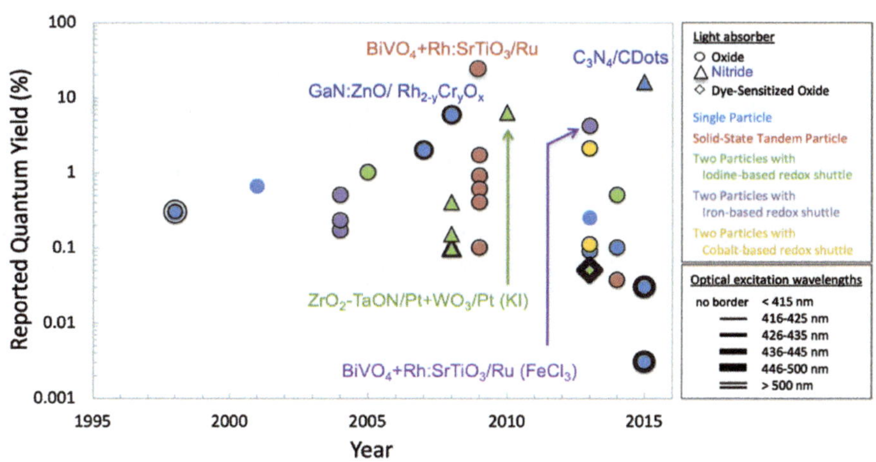

Figure 7.8 Apparent quantum efficiencies (AQE) for selected visible-light-driven water splitting systems.
Reproduced from ref. 10 with permission from The Royal Society of Chemistry.

STH of 1.2% at 331 K and 10 kPa.[71] These performances are noteworthy, but only represent a fraction of the theoretical efficiency limit, as defined above. The discrepancy can be attributed to several factors, the most important ones being intrinsic materials limitations, surface recombination of charge carriers, ineffective charge separation, and the H_2/O_2 backreaction. Understanding these limitations requires a deeper analysis into the mechanism of photocatalytic energy conversion with particles, which will be undertaken in the following.

7.6 Mechanism of Water Splitting Photocatalysis

The basic principles of photoelectrochemical conversion of light into chemical energy are well understood,[27,30,72–80] and include:

1. Absorption of super band gap photons and generation of electron–hole pairs.
2. Separation of electrons and holes.
3. Transport of electrons and holes to the interfaces.
4. Electrochemical water oxidation and reduction.

These processes will be discussed below. Additionally, photocatalysts that drive thermodynamic unfavorable forward reactions, must be able to inhibit the backreaction.[81] For water splitting PCs, this means the H_2/O_2 recombination reaction must be suppressed. Strategies to accomplish this task will be presented.

7.7 Free Energy of Photocatalysts

In general, the electrical power available for water splitting is limited by the free energy of an illuminated semiconductor. This energy is given by the difference of the quasi-Fermi energies for electrons and holes, as shown in Figure 7.9. These quasi-Fermi energies are defined by the potential energy of the charge carriers (the band edges E_{CB} and E_{VB}) and by the excess charge carrier concentrations under illumination.

The free energy and the open circuit potential resulting from the excess electron and hole concentrations Δn and Δp, respectively, in an illuminated semiconductor can be calculated with eqn (7.5):

$$eV_{OC} = E_{F,n} - E_{F,p} = E_{CB} - E_{VB} + kT \ln\left[\frac{n_0 + \Delta n}{N_{CB}} \cdot \frac{p_0 + \Delta p}{N_{VB}}\right] \quad (7.5)$$

Here, N_{CB} and N_{CV} are the effective density of states in the conduction and valence bands, respectively.[54,82–85] The equation predicts a 59 mV open circuit voltage gain for every tenfold increase of Δn and Δp. This explains the dependence of the water splitting rate on the intensity of the absorbed light. Typically, V_{OC} is only about half of the value of E_G, but it can reach over 70%

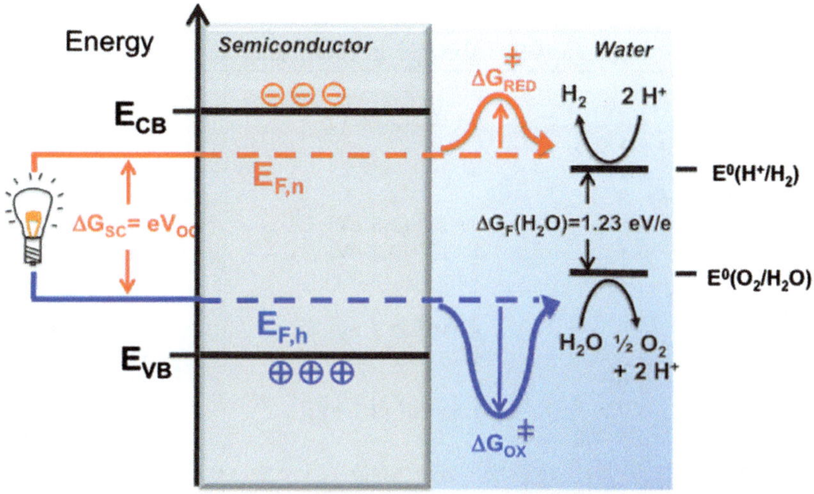

Figure 7.9 Free energy of an illuminated semiconductor. $E_{F,n}$ and $E_{F,p}$ denote the quasi-Fermi levels of electrons and holes. V_{OC}: open circuit potential. $\Delta G^{\#}_{RED}$ and $\Delta G^{\#}_{OX}$ denote the free activation energies for water reduction and oxidation.

for selected semiconductors, *e.g.* GaAs or lead perovskite, when the lifetimes of the carriers are very high.[54] For water splitting to occur at appreciable rate, V_{OC} must exceed the thermodynamic water electrolysis potential of 1.23 V by about 0.3 V to compensate for kinetic overpotentials. This is more easily achieved for large band gap semiconductors because of the greater potential energy of the electrons and holes. This explains why all of the single absorber PCs in Figure 7.8 have band gaps above 2 eV.

7.8 Light Absorption and Exciton Generation

The ability of a semiconductor particle to absorb light and generate excitons is determined by the Lambert–Beer law and the wavelength-dependent absorption coefficient, **α**. Absorption coefficients for many semiconductors and for selected metal oxides are listed in the *Semiconductors Data Handbook*.[86] While optical band gaps are tabulated for many metal oxides,[87] absorption coefficients have been measured only for the most popular absorbers (see Table 7.1).

From the absorption coefficient, the light penetration depth α^{-1} can be calculated. This parameter defines the distance over which the light intensity is reduced to $1/e$ of the incident illumination. For example, for Cu_2O, $\alpha^{-1} = 125$ nm at $\lambda = 500$ nm; for CdTe, $\alpha^{-1} = 106$ nm (550 nm);[88] and for Si, $\alpha^{-1} = 680$ nm (510 nm). To ensure 95% absorption of the incident light, the diameter d of a particle must be three times the value of α^{-1}. Smaller particles only absorb a fraction of the incident light, and consequentially, have a lower number of excitons generated in them. This leads to lower water

Table 7.1 Selected semiconductors for water splitting photocatalysis.

Compound	Band gap/eV	Absorption coefficient/cm^{-1}	References
Photoanodes			
$BiVO_4$	2.4	8×10^4 (2.7 eV)	97, 283
TaO_xN_y	2.1–2.5		282, 284
$LaTiO_2N$	2.1		175, 282
$Ba(Mg)TaO_2N$	1.8		37, 282
Ta_3N_5	2.0	8×10^4 (2.5 eV)	285, 286
$SrTiO_3$	3.1	1×10^4 (3.5 eV)	287
ZnO	3.25	1.8×10^5 (3.3 eV)	288
Photocathodes			
Cu_2O	2.0	8×10^4 (2.5 eV)	97, 282, 289
$CuFeO_2$	1.5	1.45×10^5 (2.48 eV)	241, 289
$CaFe_2O_4$:Ag	1.9		290
$CuBi_2O_4$	1.75	1.5×10^4 (2.1 eV)	282, 291, 292
$Cu(In,Ga)S_2$	1.5–1.8		212, 282, 293, 294
$(Ag,Cu)(In,Ga)Se_2$	1.6–1.7		282
WSe_2	1.2		282
$GaInP_2$	1.8		282, 295
InP	1.28–1.35		296

splitting performance if the energy loss mechanisms (electron–hole recombination, back reaction) are not minimized through additional means. On the other hand, particles larger than $3/\alpha$ will be unevenly illuminated, and thus may not contribute optimally to the photocatalytic reaction.

Separately, the light absorption of particle suspensions can be adjusted with the particle concentration and the optical path length of the reactor. The dependence of the H_2 evolution rate on the PC concentration was described by Qureshi *et al.* (Figure 7.10D).[89] At low particle concentrations, only a fraction of the incident light can be absorbed, and the H_2 evolution rate is found to increase with PC concentration. At high particle concentrations, all light is absorbed in the front of the suspension and the rate becomes independent of PC concentration. Accordingly, photocatalytic activity should not be normalized to catalyst mass, as often found in the literature.

A second aspect of particle suspensions is their ability to randomize the original light direction (Figure 7.10) by scattering (refraction, diffraction, reflection, absorption and re-emission). As a consequence, particles are illuminated from all directions, not just from the front. This reduces reflection losses that would occur on a perfectly flat light absorber. Polman and Atwater have argued that randomizing the light directions in photovoltaic cells affects the entropy balance of the absorber, causing up to 315 mV of open circuit loss at room temperature.[90] This is due to the difference in solid angles between absorbed and emitted light.[83] To date, the effect of light scattering on the STH efficiency of suspended PCs has not yet been investigated systematically. However, for silicon nanorod photoelectrochemical cells, Lewis and coworkers noted a distinct performance dependence on the illumination direction, which they attributed to scattering and the appearance of 'dark spots' in the photoelectrodes.[91]

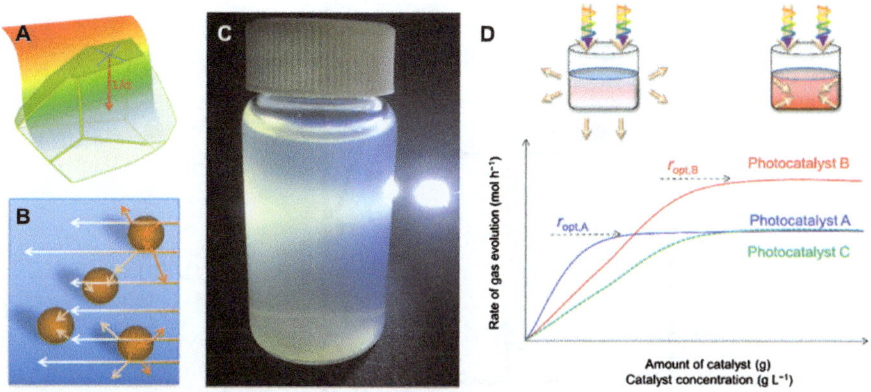

Figure 7.10 (A) Light absorption in particles. α^{-1}: optical penetration depth. (B) Scattering in particle suspension. (C) Tyndall effect in colloidal suspension of $K_4Nb_6O_{17}$ particles. (D) H_2 evolution *versus* photocatalyst concentration.
Reprinted with permission from M. Qureshi and K. Takanabe, Insights on Measuring and Reporting Heterogeneous Photocatalysis: Efficiency Definitions and Setup Examples. *Chem. Mater.* 2016, Copyright 2016 American Chemical Society.

7.9 Recombination

At steady state, the charge carrier generation rate in a light absorbing particle becomes equal to the sum of the rates of carrier recombination and their reaction rate with the redox system. Electron–hole recombination is the most important loss mechanism in excitonic energy conversion systems, including solar cells and in particulate photocatalysts.[83,92,93] It occurs through radiative and non-radiative processes in the bulk phase of the absorber, in the depletion region, and at surface defects (Figure 7.11).[80,94–96] Because recombination regulates the steady state concentrations of usable photogenerated charge carriers, Δn and Δp, it also controls the driving force for water electrolysis (Figure 7.9).

The electron–hole recombination rate, R, is related to the recombination lifetime, τ, of the charge carriers, by eqn (7.6):

$$\Delta R = \frac{\Delta n}{\tau} \tag{7.6}$$

where

$$\Delta n(t) = \Delta n(0) \cdot e^{\left(-\frac{t}{\tau}\right)} \tag{7.7}$$

This rate can be measured using microwave conductivity,[97] or transient absorption spectroscopy,[98] by following the concentrations of excess electrons and holes after turning off the light (eqn (7.7)). The decay time (or lifetime) is defined as the point when $1/e$ of the excess carriers have disappeared.[93]

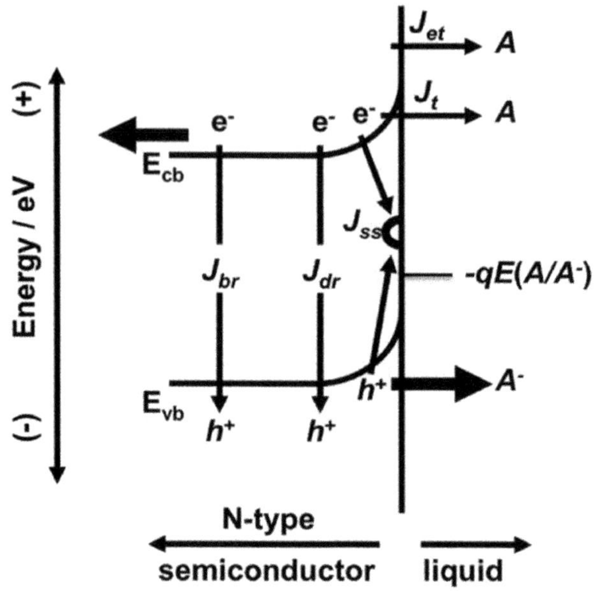

Figure 7.11 (A) Recombination pathways for photoexcited carriers in a semiconductor PEC.[74,80] The arrows signify bulk recombination (J_{br}), depletion-region recombination (J_{dr}), and surface recombination (J_{ss}). Additional loss mechanisms due to undesired charge transfer are also shown. Electron tunneling through and over the barrier produce the current densities (J_t) and (J_{et}). Electron collection by the back contact and hole collection by the redox couple (*e.g.*, oxidation of water to O_2) are desired processes shown by thick black arrows.
Reprinted with permission from N. S. Lewis, Chemical control of charge transfer and recombination at semiconductor photoelectrode surfaces, *Inorg. Chem.* 2005, 44 (20), 6900–6911, Copyright 2005 American Chemical Society.[80]

7.9.1 Auger Recombination

In particulate photocatalysts, non-radiative recombination processes are often the dominating effect. They follow an Auger, Shockley–Read–Hall, or surface/interface mechanism. Auger recombination refers to annihilation of an electron–hole pair by transferring energy to a third free charge carrier and subsequent emission of phonons during thermalization of the third free charge carrier). This process cannot be avoided in semiconductors. The rate (eqn (7.8)) is given by the product of three charge carriers and an Auger recombination constant C ($\sim 10^{-30}\,\mathrm{cm^6\,s^{-1}}$).[93] The Auger recombination rates for electrons and holes are:

$$R_{\mathrm{A(e)}} = C_{A(\mathrm{e})} \cdot n^2 \cdot p \tag{7.8}$$

$$R_{\mathrm{A(h)}} = C_{A(\mathrm{h})} \cdot p^2 \cdot n \tag{7.9}$$

and the Auger lifetime is:

$$\tau_{Aug} = \frac{1}{C \cdot (n_0^2 + p_0^2 + \Delta n^2)} \tag{7.10}$$

Because it involves the interaction between three charge carriers, Auger recombination is most dominant for highly doped semiconductors, when values for n_0 or p_0 are large, or at very high illumination intensity.

7.9.2 Shockley–Read–Hall Recombination

Shockley–Read–Hall (SRH) recombination refers to annihilation of an electron–hole pair *via* a deep defect state in the lattice (Figure 7.12).

The SRH recombination rate depends on the trap state concentration, N_t, the trap state energy, E_t, the trap capture-cross section, $\sigma_{e/h}$, and the thermal

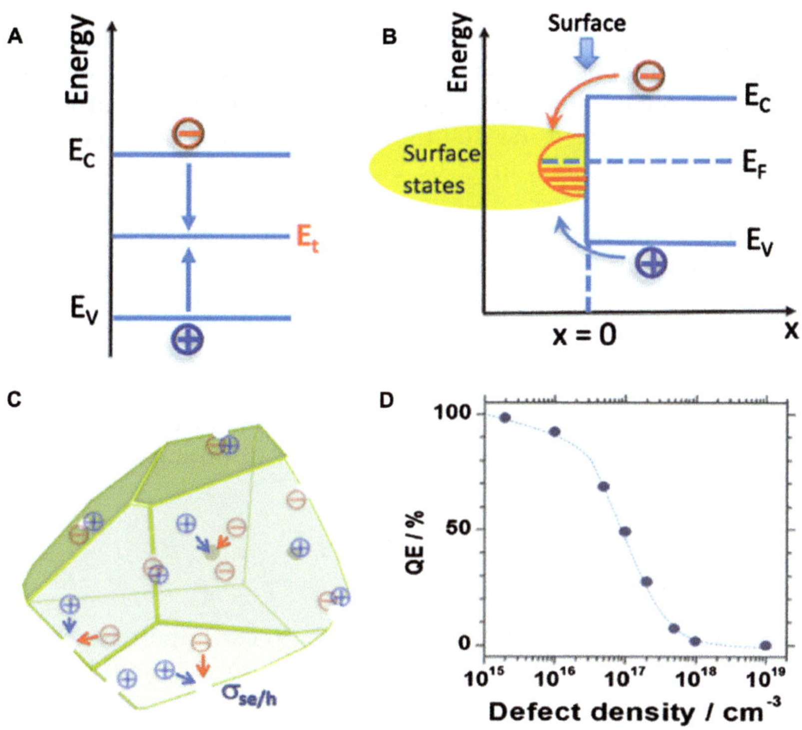

Figure 7.12 Non-radiative recombination mechanisms. (A) SRH recombination at deep trap site. (B) Surface recombination. With permission from Thomas Dittrich, Helmholtz Center for Materials and Energy, Berlin. (C) SRH and surface recombination at particle. (D) The relationship between defect density and quantum yield from numerical modeling.[99] Reproduced from ref. 99 with permission of The Chemical Society of Japan.

velocities of the charge carriers, $\nu_{e/h}$. Trap states near the absorber mid-gap energy cause the fastest recombination because they are equally accessible to electrons and holes. Assuming that the capture rates are at maximum, the minimum SRH lifetime for electrons and holes is:

$$\tau_{e/h} = \frac{1}{\sigma_{e/h} \cdot \nu_{e/h} \cdot N_t} \tag{7.11}$$

The need to suppress SRH is the reason for the high purity requirement of silicon photovoltaic cells where especially Cu and Fe contaminants need to be avoided through specialized handling. In contrast, most particle photocatalysts are prepared from commercially available precursors of 99.9% or lower purity, which may affect their performance. The influence of defects on photocatalytic activity and electron–hole lifetimes is commonly acknowledged in the literature.[100–105] For example, Maeda *et al.* attributed the performance increase of thermally annealed GaN:ZnO (5.9% instead of *ca.* 2.5% at 420–440 nm for the non-annealed form) to a reduction of lattice defects.[59] However, quantitative studies on the relationship between defects and recombination are rare. Hisatomi *et al.*, used numerical modeling to determine the relation between the defect concentration of a particle and its quantum efficiency for water splitting (Figure 7.12D). The analysis shows that defect concentrations must be below 10^{16} cm^{-3} to ensure QE values above 90%.

Maruyama *et al.* used transient absorption spectroscopy to observed the electron lifetimes in $NaTaO_3$ water splitting PCs doped with *Ca*, Sr, Ba, and La.[106] They found that the rate of electron–hole recombination was affected by the nanometer-scale topography of the photocatalyst surface. Wu *et al.* employed surface photovoltage spectroscopy and photoluminescence spectroscopy to observe the effect of thermal defects in g-C_3N_4.[107] Two types of defects were found to be responsible for the diminished ability of the material to photocatalytically evolve hydrogen and to produce a photovoltage under illumination.

7.9.3 Surface Recombination

Surface and interfacial recombination are the most important energy loss mechanisms in particle OWS photocatalysts (Figure 7.12B and C). Surface recombination is promoted by the large density of surface states involved in water reduction and oxidation.[85] Also, the specific surface area of photocatalytic particles is several orders of magnitude above that of thin film absorbers in photovoltaic cells. Generally, the rate of surface recombination, $R_S(n/p)$, depends on the specific surface area, the surface concentration of electrons and holes, n_s/p_s, the surface defect concentration N_{st}, the capture cross sections σ_n and σ_p, and on the thermal velocity ν_{th} of the charge carriers. The latter three parameters can be grouped together as the surface recombination velocities s_n and s_p for electrons and holes, respectively, and expressed as:

$$s_{n/p} = \sigma_{se/sh} \cdot \nu_{th} \cdot N_{st} \tag{7.12}$$

Here, s_n and s_p represents the velocities at which carriers that impinge on the surface recombine as a result of surface-state trapping events. For a flat absorber slab of thickness d, the surface recombination lifetime for electrons and hole can be defined as:

$$\tau_{Surf(e/h)} = \frac{d}{2 \cdot s_{n/p}} \tag{7.13}$$

This shows that the surface recombination lifetime is inversely proportional to the dimension of the absorber. For PCs and PECs, reduction of surface defects is key to the optimization of OWS activity.[80] It can be achieved through several strategies. The simplest one is to reduce the specific surface area of the absorber by using large particles with well-defined facets. For $SrTiO_3$ particles, a clear correlation between overall water splitting performance and surface faceting has been established.[108–110] Also, Mo-doped $BiVO_4$ particles in the water splitting sheet[16] have well defined facets and microscale dimensions, which contributes to the 1.1% STH of that system.

Surface treatments also can reduce the surface defect concentration.[112] From photovoltaics it is well known that treatment of silicon with HF reduces surface recombination and improves the performance of photovoltaic devices.[96] This 'passivation' works by removing detrimental Si–O mid gap states from a SiO_2 surface layer. Growing a Si_3N_4 overlayer on silicon achieves the same effect. For photoelectrodes, passivation layers provide both protection against photocorrosion and efficiency enhancements.[113] For Fe_2O_3 and $BiVO_4$ photoanodes it has been shown that application of Al_2O_3[114] or SnO_2,[115] over- and underlayers[116] also improves the performance, which was attributed to reduction of surface recombination.[117] For ALD-deposited hematite, Hamann's group showed that annealing at 800 °C can remove surface states that are involved in recombination and Fermi level pinning.[118] This treatment boosted the open circuit voltage of the hematite/H_2O system by 0.4 V. Wang's group showed in 2015 that increased photoanodic performance could be achieved in hematite, by exposing the preformed electrode to acidic solutions.[119] The improvement was attributed to a reduction of surface defects induced by the dissolution (of Fe_2O_3) and deposition (of FeOOH) under the acidic conditions. In 2017, Smith and coworkers reported that prolonged exposure of a $BiVO_4$ photoanode to light and alkaline conditions could improve the photoelectrochemical water oxidation activity of the material. This 'photocharging effect' was attributed to reduced recombination resulting from modification of the $BiVO_4$ surface.[120] Interfacial state passivation at the $BiVO_4$–CoO_x interface was achieved by insertion of a 6 nm NiO layer (Figure 7.13).[111] This enhanced the photocurrent

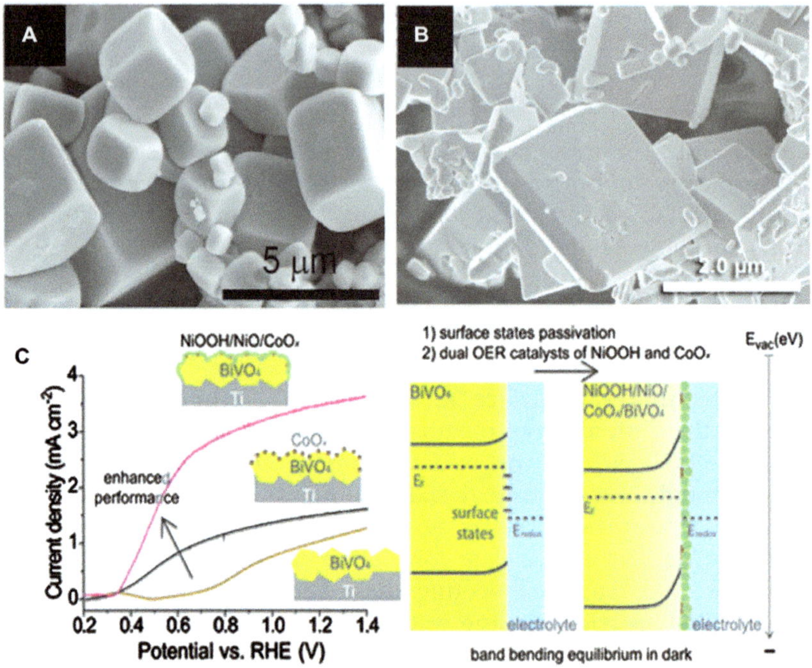

Figure 7.13 (A) SEM image of $SrTiO_3$ particles generated by flux synthesis. Reproduced from ref. 108 with permission from The Royal Society of Chemistry. (B) SEM of Mo:$BiVO_4$ particles in tandem sheet. Reprinted by permission from Macmillan Publishers Ltd: Nature Materials (Q. Wang, T. Hisatomi, Q. Jia, H. Tokudome, M. Zhong, C. Wang, Z. Pan, T. Takata, M. Nakabayashi, N. Shibata, Y. Li, I. D. Sharp, A. Kudo, T. Yamada and K. Domen, Scalable water splitting on particulate photocatalyst sheets with a solar-to-hydrogen energy conversion efficiency exceeding 1%, *Nat. Mater.* 2016, 15, 611–615).[16] Copyright 2016. (C) The effect of a NiO passivation layer on photocurrent and interfacial recombination. Reprinted with permission from M. Zhong, T. Hisatomi, Y. Kuang, J. Zhao, M. Liu, A. Iwase, Q. Jia, H. Nishiyama, T. Minegishi, M. Nakabayashi, N. Shibata, R. Niishiro, C. Katayama, H. Shibano, M. Katayama, A Kudo, T. Yamada and K. Domen, Surface Modification of CoOx Loaded BiVO4 Photoanodes with Ultrathin p-Type NiO Layers for Improved Solar Water Oxidation, *J. Am. Chem. Soc.* 2015, 137 (15), 5053–5060, Copyright 2015 American Chemical Society.[111]

by a factor of three. The NiO layer was also thought to produce an *n–p*-junction for improved extraction of photoholes. Similarly, coating hematite with a thin $Ni_xFe_{2-x}O_3$ layer led to greatly enhanced IPCEs in the 400–550 nm wavelength range, which was attributed to enhanced hole extraction and water oxidation electrocatalysis.[121] In another case, the coating of CdS/Au nanorods with polyethylenimine (PEI) was observed to increase the apparent quantum yield for photocatalytic proton reduction. This was attributed to surface passivation.[122] A similar enhancement of activity was observed upon addition of a CdS surface layer to CdS nanorods.[123]

For particulate PCs the use of passivation layers is not well established because there are limited methods available that allow a conformal coating over the entire particle surface.

7.9.4 Radiative Recombination

As defects are eliminated through surface treatment and materials purification, radiative recombination can become a limiting factor for the electron–hole lifetime and for the efficiency of excitonic solar energy conversion.[92] Radiative recombination occurs by direct reaction of an electron and a hole and leads to luminescence (Figure 7.14). This is the inverse of electron–hole generation by absorption of a photon. According to eqn (7.14), the radiative recombination rate is proportional to the free electron

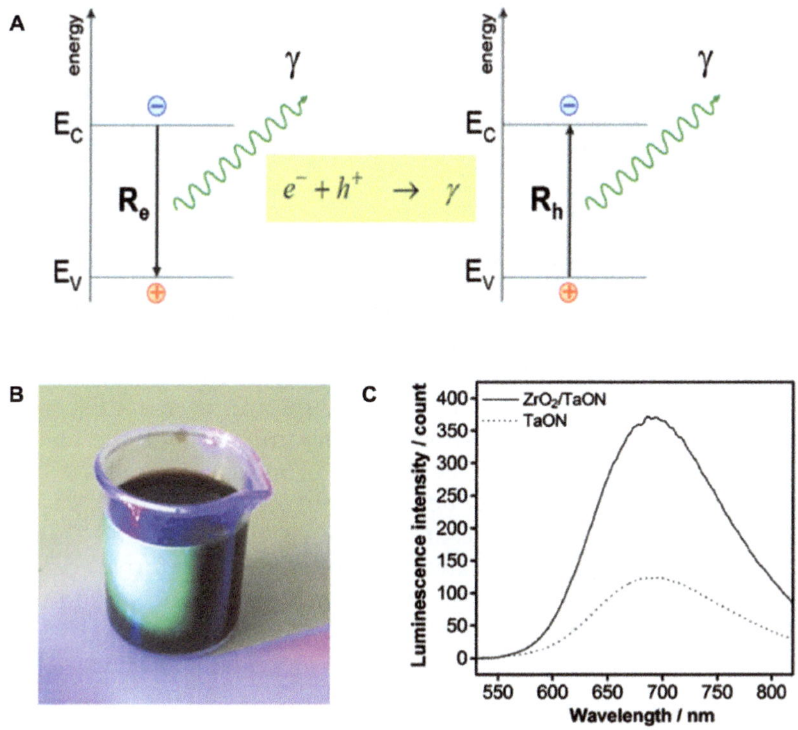

Figure 7.14 Radiative recombination leading to emission of photons. (A) Energy schemes. From Dittrich.[93] (B) Photoluminescence of a carbon dot sol in water. Courtesy by Mauricio Alves De Melo Junior (UC Davis). (C) PL increase of TaON, due to addition of ZrO_2.
Reprinted with permission from K. Maeda, M. Higashi, D. L. Lu, R. Abe and K. Domen, Efficient Nonsacrificial Water Splitting Through Two-Step Photoexcitation by Visible Light Using a Modified Oxynitride as a Hydrogen Evolution Photocatalyst, *J. Am. Chem. Soc.* 2010, 132 (16), 5858–5868, Copyright 2010 American Chemical Society.[34]

concentration, n, and hole concentration, p, along with a material dependent recombination constant, B. For example, $B = 3\times10^{-15}$ cm^3 s^{-1} for silicon and $B = 2\times10^{-10}$ cm^3 s^{-1} for gallium arsenide. Thus, the radiative recombination rate, R_{Rad}, can be expressed as:

$$R_{Rad} = B \cdot n \cdot p \tag{7.14}$$

The radiative recombination lifetime then is:

$$\tau_{Rad} = \frac{1}{B \cdot (n_0 + p_0)} \tag{7.15}$$

This means that radiative recombination is enhanced in highly doped semiconductors with large n_0 or p_0. For comparison, the Auger recombination lifetime in eqn (7.10) is proportional to n_0^{-2} and p_0^{-2}.

7.9.5 Overall Lifetime

In the presence of the various recombination processes, the overall lifetime of an electron–hole pair is limited by the recombination process with the shortest lifetime (eqn (7.16)).

$$\frac{1}{\tau_{tot}} = \frac{1}{\tau_{Rad}} + \frac{1}{\tau_{Auger}} + \frac{1}{\tau_{SHR}} + \frac{1}{\tau_{surf}} \tag{7.16}$$

Because of defect recombination, many metal oxides have lifetimes on the ps and ns timescale (Fe_2O_3: 3 ps, WO_3: 1–9 ns),[97] orders of magnitude smaller than for classical photovoltaic absorbers, such as silicon, cadmium telluride, or gallium arsenide, in which micro- and even millisecond lifetimes have been measured.[124] This short lifetime explains the absence of room temperature photoluminescence (PL) in most PCs. Improved PL can sometimes be observed upon reduction of surface defects, in as $TaON/ZrO_2$ composites (Figure 7.14C), for example.[34] Also, for Ta_3N_5 and Ge_3N_4 it is observed that higher PL correlates with higher PC performance.[34,125] In general, low values of the overall lifetime τ_{tot} reduce the performance of metal oxides in water splitting applications.

7.10 Charge Transport

In order for water electrolysis to occur, photochemical charge carriers need to move to the particle surfaces. Charge transport in semiconductors can occur by drift or by diffusion. Drift describes the motion of charge carriers under an applied or built-in electric field (space charge layer) and is characterized by the mobility μ_d (in units of cm^2 V^{-1} s^{-1}) of the charge carriers, with the drift velocity ν_d in response to an electric field, E, given by:

$$\mu_d = \nu_d / E \tag{7.17}$$

For regular semiconductors μ_d is between 1 and 1000 $cm^2 V^{-1} s^{-1}$, *e.g.* for silicon (1450 $cm^2 V^{-1} s^{-1}$) or CdTe (60 $cm^2 V^{-1} s^{-1}$).[86] Transition metal oxides, especially those of the 3d elements, often have much lower mobility resulting from poor overlap of the d-orbitals and are considered Mott insulators.[126] In these materials charge transport occurs not by drift but by small polaron hopping. For example, W: $BiVO_4$ and Ti: Fe_2O_3 have room temperature mobilities of 10^{-4} and 10^{-2} $cm^2 V^{-1} s^{-1}$ that are 3–7 orders of magnitude smaller than those found for main group element semiconductors.[127] This affects the photocurrent of these materials and their ability to form rectifying junctions (see below).

In the absence of an electric field, charge carriers move by diffusion. The diffusion coefficient, D, of a charge carrier is related to its mobility by the Einstein–Smoluchowski relation:[128]

$$D = \mu_d \cdot k_B \cdot T \cdot \frac{1}{q} \tag{7.18}$$

Because diffusion is slower than drift, recombination becomes a significant competition. The mean free diffusion length, L, describes the average distance a charge carrier can diffuse before it recombines after the carrier lifetime τ (here $q = 2$, 4, or 6 for one-, two-, or three-dimensional diffusion) (eqn (7.19)):

$$\bar{L^2} = q \cdot D \cdot \tau \tag{7.19}$$

For intrinsic semiconductors, often $L_e > L_h$ because of the larger diffusion constant of electrons compared to holes. For example, Si has $D_e = 49$ $cm^2 s^{-1}$ and $D_h = 13$ $cm^2 s^{-1}$ (calculated from mobilities, $\mu_e = 1900$ $cm^2 V^{-1} s^{-1}$, and $\mu_h = 500$ $cm^2 V^{-1} s^{-1}$ at 298 K) and using the assumption that $\tau_e = \tau_h = 10^{-6}$ s, $L_e = 98$ μm and $L_h = 51$ μm for one-dimensional diffusion ($q = 2$). Upon doping, L and τ values often increase for the majority carriers but decrease for the minority carriers.

Diffusion and small polaron transport increase with increasing temperature. This may explain the positive temperature gradient observed in photocatalytic water splitting with $Rh_{2-y}Cr_yO_3$-loaded Zn:Ga_2O_3[99] and for the Rh:$SrTiO_3$/$BiVO_4$ tandem systems.[16] In the 278–322 K temperature interval, the apparent activation energy of the overall water splitting reaction was 8 kJ mol^{-1} for the former and 18 kJ mol^{-1} for the latter. For Zn:Ga_2O_3, the activation energy was attributed to charge transport, and to the presence of Schottky barriers at the SC-metal surface in the tandem.

Because complete charge carrier collection is only possible when the particle radius $r < L_e$, L_h (Figure 7.15A), absorbers are often made into nanoparticles.[12,130] This 'nanoscaling' strategy is applied in particular to first row transition metal oxides because of their small minority carrier diffusion lengths.[126,131] The impact of the absorber size on photocatalytic activity has been experimentally verified with $KCa_2Nb_3O_{10}$[132,133] and with Fe_2O_3

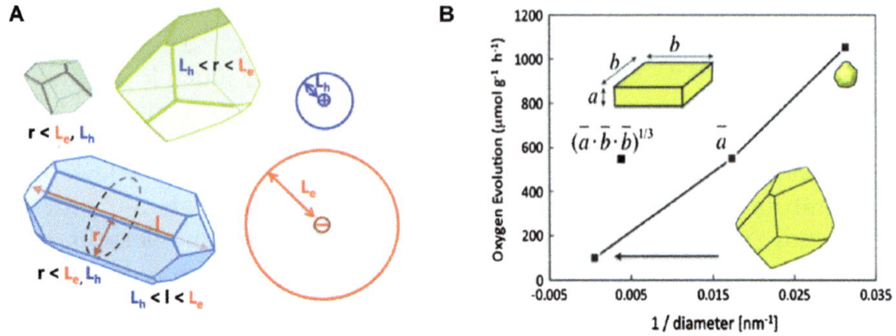

Figure 7.15 (A) Charge transport by diffusion (L_e: electron diffusion length; L_h: hole diffusion length). Particle size and shape determine extraction of majority and minority carriers. (B) Photocatalytic water oxidation rates with WO_3 particles of variable shape and size.
From *Topics in Catalysis*, Size and Morphology of Suspended WO_3 Particles Control Photochemical Charge Carrier Extraction and Photocatalytic Water Oxidation Activity, 59, 2016, 750–756, K. A. Newton and F. E. Osterloh, © Springer Science + Business Media New York 2016, with permission of Springer.[129]

dispersions, which catalyze hydrogen and oxygen evolution respectively.[102] The observed inverse size-reactivity correlations were attributed to improved hole extraction in the smaller nanocrystals. Also, shape-dependent carrier extraction was recently confirmed in the author's lab for WO_3 nanoparticles (Figure 7.15B).[129] It was found that charge transport along the smallest particle dimension controls the rates for photocatalytic water oxidation. In another example,[134] photocurrent measurements on wedge-shaped Si photoelectrodes were employed to demonstrate the effect of the absorber thickness on minority carrier extraction.

Under ideal conditions, the diffusion distance L should exceed the light penetration depth of a material. Based on the absorption coefficient, α, of a material, a *critical diffusion length*, L_{crit}, can be defined as:

$$L_{crit} = \frac{3}{\alpha} \tag{7.20}$$

Complete extraction of all photogenerated carriers and 95% absorption of the incident photons is possible if the critical diffusion length is three times the size of the light penetration depth α^{-1}. Most metal oxides do not fulfill this important criterion, which limits their photocatalytic efficiency. As mentioned above, particles with sizes smaller than α^{-1} generate fewer excitons because the particles absorb fewer photons. In the presence of surface recombination, this reduces the steady state concentration of electron–hole pairs and the free energy that can be drawn from the photocatalyst (see Figure 7.9) and utilized for overall water splitting.

Interparticle charge transport is also limiting the performance of particle films, as in the photocatalytic sheet, for example.[16] Charge transport in particle films usually occurs by thermally activated hopping and by electron tunneling (Figure 7.16).[135–139] The rate depends on the interparticle distance and the electrostatic charging energy of the donor acceptor nanocrystal couple.[140,141]

The problem of slow charge transport over macroscopic distances seriously affects the analytical value of photoelectrochemical measurements with particle films. Here the large Ohmic resistance of the films reduces the photocurrent to microamperes per cm^2 (Figure 7.16B). This makes it difficult to evaluate the photovoltage of the particles and to obtain insight into the factors that control photochemical charge separation and recombination in particle absorbers.

7.11 Charge Separation

In many inorganic semiconductors, the Coulomb binding energy of excitons is below the thermal energy and separation into free charge carriers occurs readily at room temperature. However, further separation of these free carriers is necessary to suppress electron–hole recombination by one of the mechanisms above. In photoelectrochemical cells, carrier separation is typically achieved by solid–liquid junctions or through the application of an electric bias.[144] For suspended particles, a chemical bias is often applied in the form of sacrificial reagents, that remove either photoholes or photoelectrons from the particle. The electrochemical or chemical expended this way reduces the net energy conversion achieved with the photocatalysts.[81,145] For efficient solar energy conversion, PCs must be able to drive the water splitting reaction without any applied chemical or electrochemical bias. This requires the presence of a mechanism for carrier separation.

7.11.1 Junctions

Solid–solid[146] and solid–liquid junctions[73,147–149] are central to the operation of PV, PEC and PC devices.[58,150,151] The use of junctions for charge separation in single absorber and tandem photocatalysts is shown in Figure 7.17.[29] Each PC needs a solid–liquid or solid–solid junction for the extraction of minority carriers and an Ohmic contact for the extraction of majority carriers. The efficiency of the junctions is controlled by several variables, including the built-in potential, φ_{bi}, the carrier concentrations, n/p, and their mobility, μ.

The *built-in potential* φ_{bi} determines the potential drop through the space charge layer and with it the electric field. The larger this field, the larger the driving force for minority carrier transfer out of the semiconductor. Generally, φ_{bi} arises from the electrochemical potential difference between the donor and the acceptor, *i.e.* E_F and E^0 for a solid–liquid interface. However, often the built-in potential is reduced by Fermi level pinning.[118,152,153]

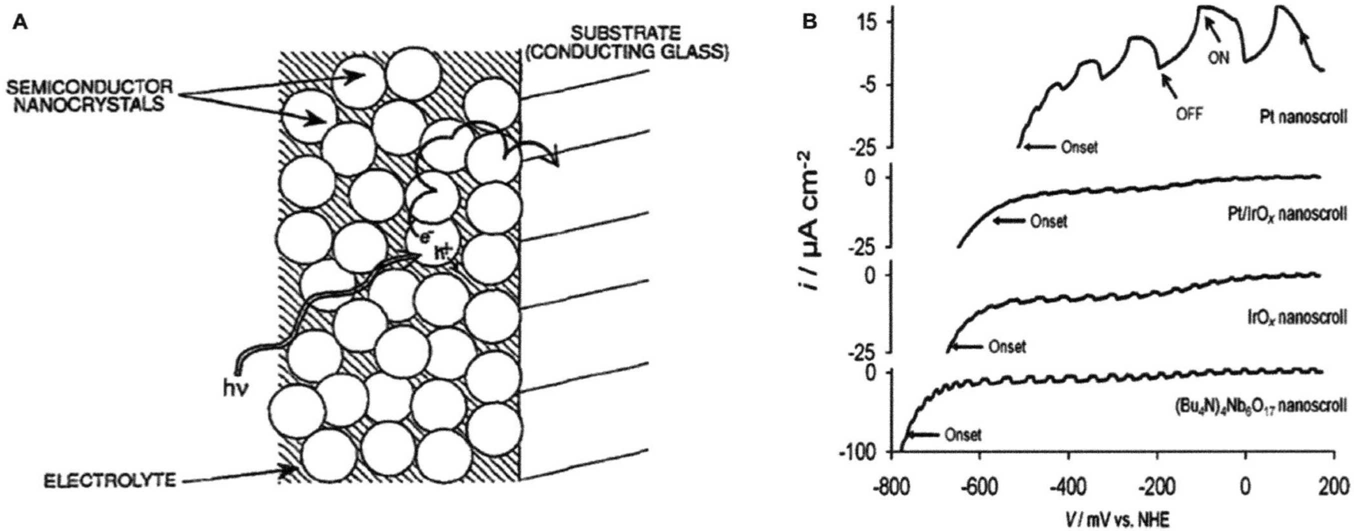

Figure 7.16 (A) Electron transport in particle films. Reproduced from ref. 142 by permission of The Electrochemical Society. (B) Photocurrent scans for niobate nanocrystal film electrodes immersed in aqueous methanol solution.
Reproduced from ref. 143 with permission from John Wiley and Sons, © 2011 Wiley-VCH Verlag GmbH& Co. KGaA, Weinheim.

This occurs when charge carriers are trapped in surface states that reduce the Fermi energy of the semiconductor that is felt at the surface (Figure 7.12B). In photoelectrochemical cells, this problem can be avoided by replacing the solid–liquid junctions with solid–solid junctions. For example, fabrication of an $n+$Si overlayer on the surface of a *p*-Si nanowire array increases the built-in potential of the photocathode and the open circuit voltage for proton reduction.[154] The same strategy was used for n–p^+-Si/n-WO_3 microwire array tandem photoelectrodes.[155] The depletion width W of a junction determines the distance over which the electric field can separate charge carriers in the semiconductor. This parameter can be approximated with eqn (7.21), which is a solution of the Poisson equation.

$$W = \sqrt{\frac{2\varepsilon\varepsilon_0}{eN_D}(\varphi_{bi})} \tag{7.21}$$

Here, N_D is the donor concentration and φ_{bi} the potential drop across the space charge layer. For example, phosphorous doped *n*-silicon with $N_D = 10^{16}$ cm^{-3} and $\varphi_{bi} = 0.3$ V support a depletion layer width of $W = 200$ nm. If the donor concentration is large ($\sim 10^{20}$ cm^{-3}), the space charge layer will be confined to the interface. This situation is encountered with hematite, where the depletion layer width is estimated between 1.5 and 10 nm, depending on the preparation of the material and the built-in potential.[156,157] This means only charge carriers near the surface are separated effectively from the bulk. Eqn (7.21) also explains why charge separation in nanoparticles is more difficult than in microparticles. In the former, the particle radius r is smaller than W, which weakens the local electric field and causes the bands to stay essentially flat (Figure 7.18).[73,137,156,158] Rectifying behavior is then no longer observed and charge separation is controlled by the charge carrier concentration at the interface, by the built-in potential, and by the overlap of the wavefunctions at the donating and receiving phase.[142,159] In other cases, a depletion layer does not form because charge transfer between the semiconductor and the redox couple is too slow. This generally applies to charge transfer between semiconductors and water. However, charge transfer and junction formation can be sped up by the addition of electrocatalysts.

Figure 7.18 illustrates another problem of PCs. In the absence of an inherent asymmetry, the band bending at the solid–liquid surface confines majority carriers to the center, where they cannot be extracted. To avoid this problem, PCs need to be designed to have well defined channels for majority and minority carrier extraction. That means their morphology needs to be intermediate between that of Figures 7.17 and 7.18.

While traditional semiconductors (GaN,[160] InP,[161] silicon,[162] *etc.*) readily form junctions with rectifying properties, most metal oxides do not. That is because the depletion layer width in metal oxides is too small to allow formation of a junction. However, there are exceptions to this rule. Rectifying junctions can be observed in *p*-Cu_2O:Sb contacts,[163] in *p*-$CuAlO_2$,[164] and *p*-NiO thin films,[165] or in *n*-ZnO/GaN[166] and *n*-ZnO/metal contacts.[167] It also

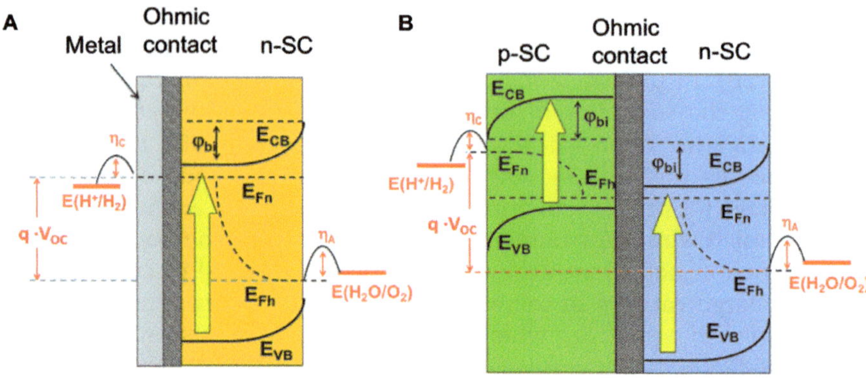

Figure 7.17 Charge and potential distributions for solid–liquid junctions in a (A) single absorber and (B) dual absorber PC. Under illumination the electric field in the space charge region sweeps minority charge carriers towards the liquid.

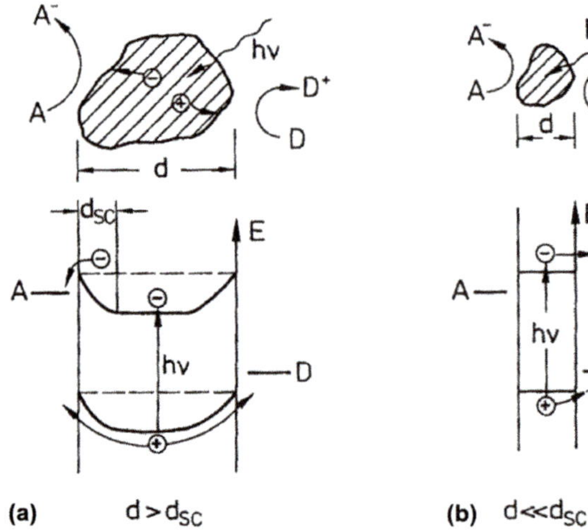

Figure 7.18 Charge and potential distributions for (a) large and (b) small particles. Reproduced from Photoinduced charge transfer processes at semiconductor electrodes, 1994, R. Memming, © Springer-Verlag 1994, with permission of Springer.[73]

helps for the metal oxides to be present as thin films, to allow charge transport and an observable current.

Even though junctions are central to the operation of PCs, relatively little is known about the electric potential and charge distribution in suspended particles. Transient optical spectroscopy is among the few techniques that are able to probe excites states of particles in suspension. It provides information about carrier lifetimes and photochemical charge separation in suspended particles.[168] Photochemical charge transfer between semiconductors, metal oxides, and metals is generally fast (2–50 ps) and strongly dependent on the built-in potential of the junction and the relative sizes of the particles.[169–174]

Electrochemical measurements on particle films are complicated by the large transfer resistance through the particle layer, as described in Figure 7.16. This Ohmic potential drop across the film obscures the potential drop of the particle–liquid junction. Recently, Domen demonstrated that charge transfer problems can be circumvented by evaporating metal (Ti, Nb, Au) electrode layers directly onto particle films. This approach (Figure 7.19) is known as particle transfer method.[175] Particle transfer has been applied to several *n*- and *p*-type absorbers, including $La_5Ti_2CuS_5O_7$,[175–177] Al-doped $SrTiO_3$,[109] $SrNbO_2N$[178] and a $BiVO_4$:Mo/$SrTiO_3$:Rh tandem.[16] It has been shown that photocurrents of 7 mA cm^{-2} are possible in such films (Figure 7.19C).

Boettcher's group used a related approach to probe the potential distribution at the semiconductor–cocatalyst interface. They evaporated gold electrodes on top of TiO_2/cocatalyst films in order to independently measure current and potential at the electrocatalyst and semiconductor surfaces. It was found that ion-permeable electrocatalysts such as $Ni(OH)_2$ or NiOOH gave 'adaptive' junctions, whose built-in potential changed with the oxidation level of the cocatalyst. In contrast, dense, ion-impermeable IrO_x cocatalysts yielded constant-built-in potential 'buried' junctions.[179]

Another way to overcome the Ohmic resistance problem is surface photovoltage spectroscopy, which measures the contact potential change during illumination of a dry particle film (Figure 7.20A).[180,181] Because of the larger time scale of the measurement (>10 s) even slow carrier separation can be observed. Studies on films of $HCa_2Nb_3O_{10}$ nanosheets,[182] CdSe quantum dots,[153] g-C_3N_4 particles,[107] Fe_2O_3 nanorods,[183] WO_3 particles, $SrTiO_3$:Rh nanocrystals,[184] $BiVO_4$ particles,[185] on various substrates (gold, FTO, ITO, *p*-silicon) have provided information about built-in potentials, majority carrier type, and defects. For metal oxide particle films, charge transport occurs by diffusion of majority carriers from the nanocrystals to the conductive support to produce a photovoltage that increases with light intensity, film thickness, and whose limit is determined by the band edge offset of the semiconductor-charge acceptor configuration (Figure 7.20B). Despite large built-in voltages at the interface, no evidence for the presence of space charge layers was seen. In contrast, films of CdSe dots show a particle size dependent behavior.[153] Small dots support diffusive charge transport, while large dots allow for drift within a 150 nm wide space charge layer at the ITO–CdSe interface (Figure 7.20C and D). The built-in potential (300 mV) was found to be 1.0 V less than the theoretical value for this configuration, which was attributed to Fermi level pinning at the ITO–CdSe interface.

Additionally, Kelvin probe microscopy has been used to probe junctions at particles. Ida *et al.* layered individual *p*-type NiO and *n*-type calcium niobate nanosheets on each other and measured their photocatalytic H_2 evolution from aqueous methanol solution under UV illumination.[186] Charge

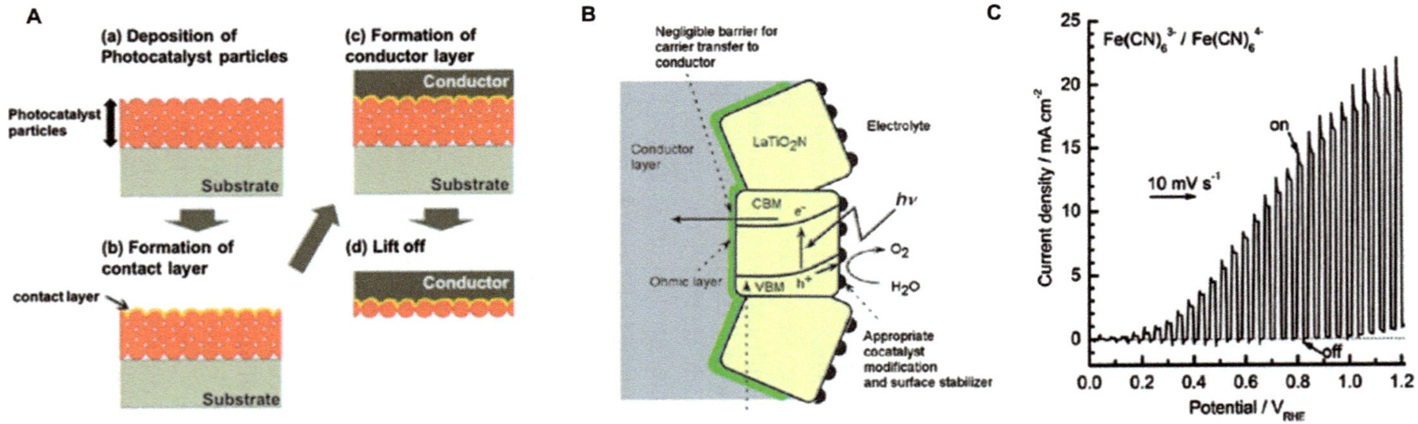

Figure 7.19 (A) The particle transfer method (schematic). (B) Charge separation scheme and (C) photocurrent for $LaTiO_2N$ in the presence of $[Fe(CN)_6]^{3-/4-}$.
Reproduced from ref. 175 with permission from The Royal Society of Chemistry.

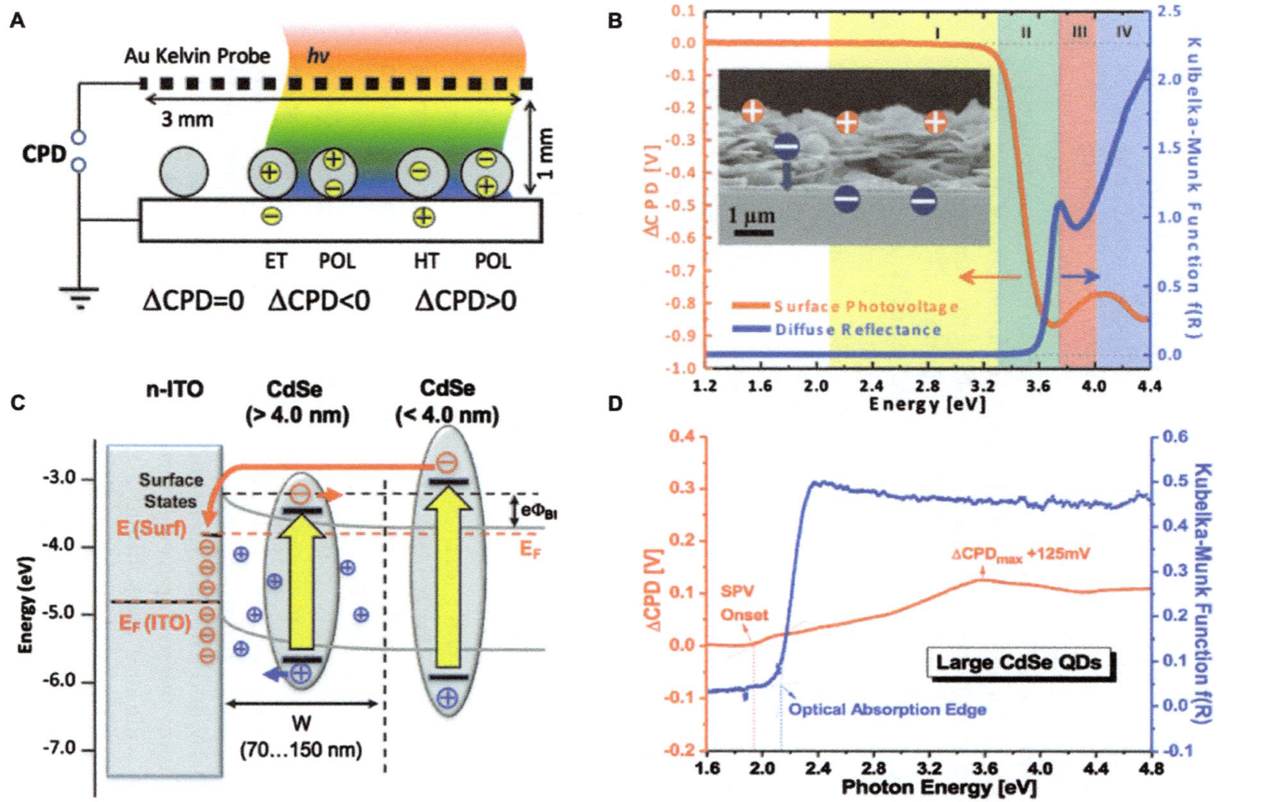

Figure 7.20 Surface photovoltage spectroscopy with particle films. (A) Schematic measurement configuration. Reproduced from ref. 184 with permission from The Royal Society of Chemistry. (B) Spectrum for $HCa_2Nb_3O_{10}$ film on gold substrate. Reprinted with permission from J. Zhao, F. E. Osterloh, Photochemical Charge Separation in Nanocrystal Photocatalyst Films – Insights from Surface Photovoltage Spectroscopy. *J. Phys. Chem. Lett* 2014, 5, 782–786. American Chemical Society Copyright 2016. (C) Charge transfer in CdSe quantum dot films (schematic). (D) SPV spectrum for large CdSe dots showing space charge layer controlled charge transfer. Reprinted with permission from J. Zhao, B. A. Nail, M. A. Holmes and F. E. Osterloh, Use of Surface Photovoltage Spectroscopy to Measure Built-in Voltage, Space Charge Layer Width, and Effective Band Gap in CdSe Quantum Dot Films, *The Journal of Physical Chemistry Letters,* 2016, 3335–3340, Copyright 2016 American Chemical Society.[153]

separation on the nanoscale was confirmed with photochemical labeling, and Kelvin probe microscopy revealed a potential gradient across the photocatalyst.

7.11.2 Electric Dipoles

Surface dipoles are another source of electric fields that can affect carrier movement and separation (Figure 7.21A). Such dipoles arise from polar bonds at the surface of inorganic materials, from water,[187–189] ligands,[112,190] or from specifically absorbed ions.[191,192] In water, the surface potentials of metal oxides, including TiO_2 [193] and Fe_2O_3 [194,195] are primarily controlled by proton/hydroxide adsorption, whereas the surface potentials of II/VI, III/V, and group IV semiconductors are more susceptible to adsorption of soft ions, including sulfur,[196] HS^-,[191,197] HTe^-,[198,199] and Cl^-.[192] The ions often bind to specific crystal surfaces.[198,199] In 1992, Sander *et al.* used UV and X-ray photoelectron spectroscopy to observe a 0.9 eV energy shift from surface adsorbed water on cleaved $CuInSe_2$ surfaces.[187] A similar energy shift was observed for H_2O adsorbed to silicon at 100 K.[188] These observations were attributed to a preferred orientation of the H_2O dipole.

It has also been shown that surface dipoles can control photochemical reactions. For example, on illuminated rutile TiO_2, Pt was found to deposit mainly on the 110 face and PbO_2 mostly on the rutile TiO_2 011 facet.[200] This selectivity was attributed to accumulation of electrons on the 110 facet and of holes on the 011 facet, respectively. For $SrTiO_3$, scanning Kelvin probe microscopy was used together with X-ray photoelectron spectroscopy (XPS) to correlate surface potential with the local Ti^{4+} ion concentration. Titanium ion-rich surfaces were positively charged, while ${SrO_3}^{4-}$ terminated surfaces were negatively charged.[201] The positive surfaces promoted reaction of silver ions with photoelectrons, whereas the negative face promoted deposition of PbO_2 by reaction of Pb^{2+} with photoholes. Similarly, photodeposition of metal salts onto $BiVO_4$ microcrystals allowed facet-selective modification with metals and metal oxides (Figure 7.21B and C).[15]

Ferroelectric materials have a non-symmetrical ion distribution in the crystal structure that can produce a spontaneous electric polarization (Figure 7.21A)[202] and that can aid carrier separation.[203] For example, photovoltaic cells made from the ferroelectric $BiFeO_3$ have high open-circuit voltages of 0.8–0.9 V and external quantum efficiencies up to 10%.[204] Based on piezoelectric force microscopy, each ferroelectric domain in the material was 100–300 nm wide and contributed a potential of 10 mV. The domain orientation could be reversed by the application of an external electric field. For $BaTiO_3$, photochemical labeling experiments confirm that ferroelectric domains can guide the accumulation of photochemical charge carriers and promote photocatalytic reactions.[202,203,205,206] However, it is not yet clear if this effect can be used to enhance photocatalytic water splitting.

Gradient doping (Figure 7.21A) provides an alternative way to generate a potential energy gradient inside of small crystals. This approach was

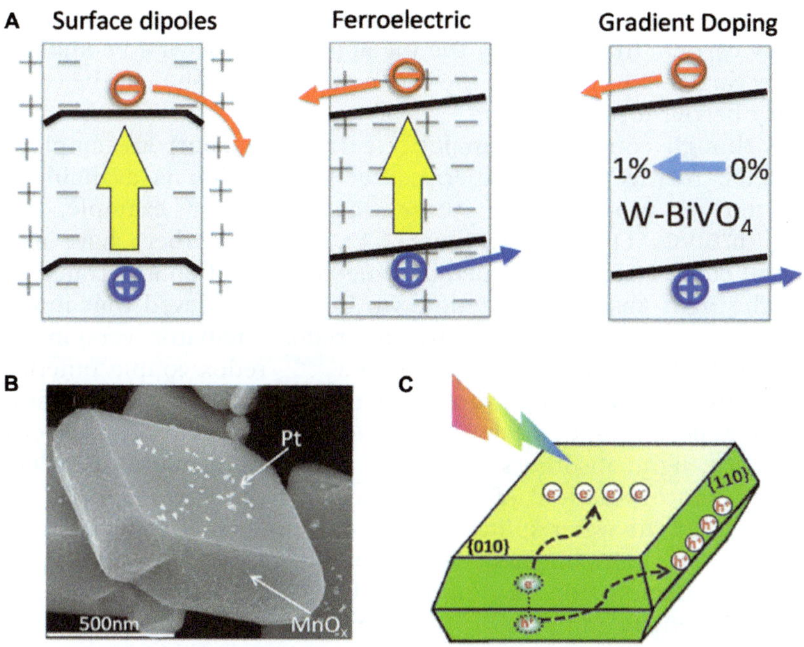

Figure 7.21 (A) Effects of surface dipoles and of ferroelectric dipoles on the carrier movement in particles. (B) $BiVO_4$ crystals after photochemical reaction with Pt and Mn salts. (C) Photochemical charge separation on $BiVO_4$ surface.
Reproduced from ref. 15 with permission from The Royal Society of Chemistry.

demonstrated by van de Krol's group for tungsten-doped $BiVO_4$ films.[207] A tungsten concentration gradient inside the films moves photoelectrons towards the electrode and holes towards the solid–liquid interface. The method may be restricted to defect-tolerant semiconductors, such as $BiVO_4$, and it has not yet been applied to particulate photocatalysts.

7.11.3 Ohmic Contacts

In water splitting photocatalysts, Ohmic contacts are needed for efficient majority carrier transfer between two absorbers in tandem configuration, or at the interface between an *n*-type semiconductor and its proton reduction cocatalyst (Figure 7.17). Ohmic contacts normally form when a metal with a reducing work function (aluminum, gallium) is combined with an *n*-semiconductor, or when a metal with an oxidizing workfunction is combined with a *p*-semiconductor. These conditions minimize the formation of a depletion layer. Because they transmit both majority and minority carriers, Ohmic contacts unfortunately catalyze electron–hole recombination. For this reason, the Ohmic contact area in photovoltaic panels is kept small.

An alternative solution is to introduce a back surface field[93] that prevents minority carriers from entering the metal (where they would combine with majority carriers). By adjusting the doping concentration near the contact, a Schottky barrier forms.

Even though contacts for majority carrier transport are employed in many PCs, relatively little experimental information is available about the electrical properties of theses contacts. For example, in the $SrTiO_3$:Rh/$BiVO_4$ tandem in Figure 7.22, both absorbers have a direct electrical connection.[208] The detailed mechanism of charge transport between the light absorbers is unknown, but it is believed that interfacial $Rh^{3+/4+}$ states are involved.[32] In the redox mediator version of the $SrTiO_3$:Rh/$BiVO_4$ Tandem, a reversible $Fe^{3+/2+}$ redox couple functions as Ohmic contact between the two absorbers,[70] and in $SrTiO_3$:Cr/Ta and WO_3 (the first reported tandem) a I^-/IO_3^- redox couple is used to electrically connect the light absorbers.[209,210] For the tandem made of $BiVO_4$ and Ru/$SrTiO_3$, incorporation of photoreduced graphene oxide as Ohmic contact tripled the water splitting efficiency under visible-light irradiation (Figure 7.22).[211] Analogously, in the $CuGaS_2$/Pt–RGO–TiO_2 tandem, reduced graphene oxide was found to boost the performance, although issues with activity and stability were encountered.[212]

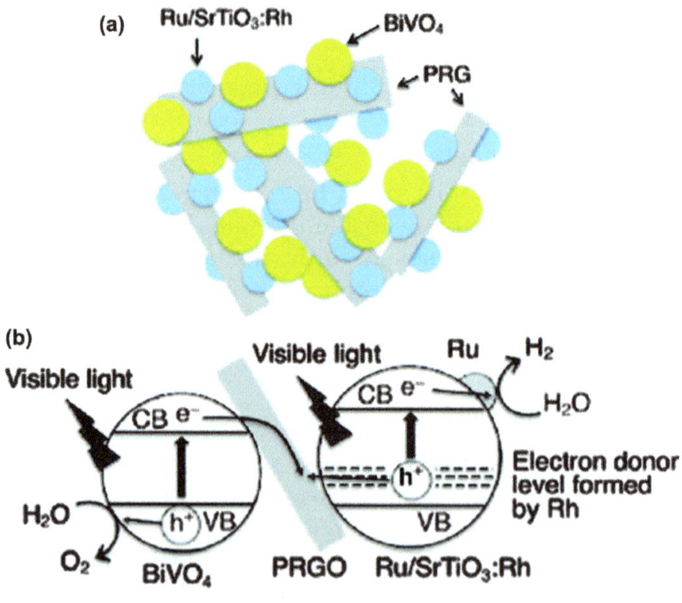

Figure 7.22 Ohmic contacts in photocatalysts. (a) Rh:$SrTiO_3$/$BiVO_4$ tandem. Reprinted with permission from A. Iwase, Y. H. Ng, Y. Ishiguro, A. Kudo and R. Amal, Reduced Graphene Oxide as a Solid-State Electron Mediator in Z-Scheme Photocatalytic Water Splitting under Visible Light, *J. Am. Chem. Soc.* 2011, 133 (29), 11054–11057, Copyright 2011 American Chemical Society.[211]

In the photocatalyst sheet, Wang *et al.* used vapor-deposited or colloidal gold to establish an Ohmic contact between $SrTiO_3$:Rh and $BiVO_4$ absorbers.[16] The water splitting activity of the printed sheet increased with Au colloid concentration up to 40 wt%, after which it declined, possible due to shading effects. Thermal annealing increased the activity which was attributed to improved electric contact between Au and the semiconductors. In a derivative of this tandem, the gold was replaced with a carbon contact.[71] The large workfunction of C (5.2 eV) was found to be important for reducing the Schottky-barrier with the $SrTiO_3$:La,Rh absorber and to promote majority carrier charge transfer from $BiVO_4$.

7.12 Charge Transfer Reactions at the Cocatalyst–Liquid Interface

Solid–liquid charge transfer (CT) occurs between the semiconductor and the liquid and between a cocatalyst and the liquid. The anodic or cathodic current (i) for a generic redox process at the cocatalyst–liquid interface can be described by the Butler Volmer (BV) equation:

$$I = AFk^0 c(\mathrm{ox})^* \exp\left[-\frac{\alpha F(E-E^0)}{RT}\right] - c(\mathrm{red})^* \exp\left[\frac{(1-\alpha)F(E-E^0)}{RT}\right] \quad (7.22)$$

Here k^0 is the standard rate constant, F is the Faraday constant, $c(\mathrm{ox})^*$ and $c(\mathrm{red})^*$ are the concentrations of the oxidized and reduced species at the electrode, α is the transfer factor, E^0 is the standard reduction potential, and E is the Fermi level of the cocatalyst.[213] We recently used a modified BV equation to describe the rate dependence of photocatalytic proton reduction on the quasi Fermi energy of CdSe quantum dots. Here, E_{Fn} is controlled by the quantum confinement effect,[214] and the hydrogen evolution rate is found to correlate with the thermodynamic driving force for electron transfer across the solid–liquid interface. Charge transfer was mediated by cadmium nanoparticles formed at the CdSe quantum dots under illumination.

According to the BV equation, the current is proportional to the interface area, A. This is the reason for the higher performance of nanostructured electrocatalysts.[12,40] Because the mechanism of the water oxidation and reduction reactions is a function of the cocatalyst material and the pH,[215] the exchange rate constant k^0 in eqn (7.22) varies strongly among systems.[216,217]

McCrory *et al.* have studied the performance and stability of electrocatalysts for the oxygen evolution reaction (OER) and the hydrogen evolution reaction (HER).[40] These results are described in Chapter 5 by McCrory *et al.* in this book. Many of today's electrocatalytic materials[40,216,217] have been used in partial water splitting in the presence of sacrificial reagents. For the hydrogen evolution reaction (HER), Pt is by the far the most commonly used cocatalyst in PCs, followed by Pd, Ru and NiO.[11] On the other hand, MoS_2,[218,219] Ni_2P[220] and CoP[41] have never been used in PCs for overall water splitting, but can promote HER in the presence of sacrificial electron donors.[221,222] For the oxygen evolution reaction (OER), RuO_2 and IrO_2 are the most frequently used cocatalysts.[11] Even though Co–Pi, CoO_x, and layered hydroxides[223] have been identified as OER electrocatalysts,[40,224] their application to PCs is restricted to systems with sacrificial acceptors.[225,226] Interestingly, the application of OER catalysts in overall water splitting PCs is rare. For example, all of the reported $SrTiO_3$:Rh tandems[16,32,70,208–210,227–231] employ cocatalyst-free $BiVO_4$ for water oxidation. The C_3N_4/C-dot PC (16%, 420 nm) reported by Liu *et al.* (featured in Figure 7.8) also does not use any metal-containing cocatalysts.[62] However, in this system it is not clear if the C-dots act as cocatalyst or secondary light absorbers. In contrast, many of the WO_3 based tandems, including Pt–$SrTiO_3$: Cr/Ta–Pt–WO_3 (2.7 eV)[232] $MgTa_2O_{6-x}N_y$/TaON–WO_3/Pt tandem,[37] and ZrO_2–TaON/Pt and WO_3/Pt (AQE) 6.3% (420.5 nm)[34] require Pt as a water oxidation cocatalyst.[210] In single absorber systems, like Rh/Cr_2O_3–GaN:ZnO, water oxidation cocatalysts are typically not used.[44] In 2000 Maeda reported an enhancement of the activity when Mn_3O_4 NPs were co-loaded as OER catalyst.[233] However, the AQE of the dual cocatalyst system (1%) was lower than that of GaN:ZnO champion devices (5.8%) employing Rh/Cr_2O_3 alone.[59] Also, the UV active PC NiO/$NaTaO_3$:La achieves its highest activity of 56% AQE at 270 nm without a OER cocatalyst.[234] If the NiO HER catalyst is replaced with a IrO_2 OER catalyst, the activity is markedly lower, but above that of the unmodified $NaTaO_3$:La.[235]

The reason for the absence of OER cocatalysts in PCs for overall water splitting is not entirely clear. It is possible that OER cocatalysts interfere with charge transfer between the water oxidation and reduction subsystems, or across the junction at the particle–liquid interface. Additional studies of this issue are required.

7.13 Charge Transfer Reactions at Semiconductor–Liquid Interfaces

Charge transfer reactions between a semiconductor and a redox couple in the liquid can be understood with Marcus–Gerischer theory.[73,76,80,147,236] In the dark, charge transfer is dominated by majority carriers as shown for the *n*-type semiconductor in Figure 7.23.

The cathodic current density, due to electron transfer from the conduction band containing n_s electrons at the surface, to a redox species of concentration, c_{ox}, is

$$j_c^- = ek_{c,\mathrm{red}}n_s c_{\mathrm{ox}} \tag{7.23}$$

Here, e, is the electron charge, and $k_{c,\mathrm{red}}$ is the rate constant for reduction. The anodic current, which occurs in competition, due to electron transfer from a redox species into the semiconductor conduction band with an effective density of states, N_C, is

$$j_c^+ = ek_{c,\mathrm{ox}}N_C c_{\mathrm{red}} \tag{7.24}$$

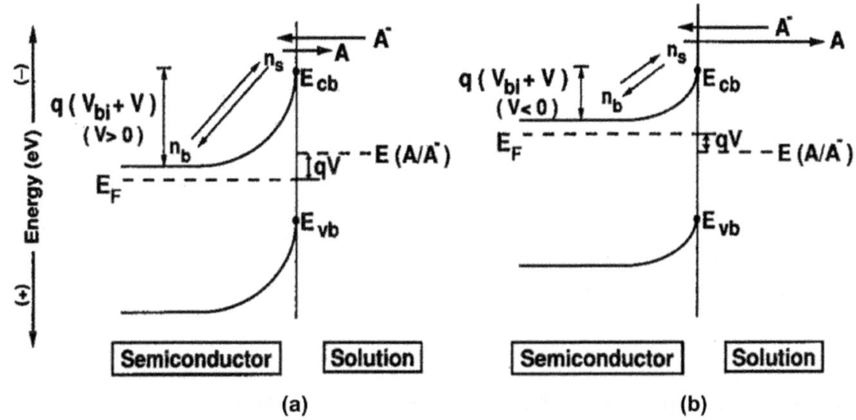

Figure 7.23 Energy scheme for equilibrium charge transfer at interface of *n*-semiconductor in the dark with (a) oxidizing or (b) reducing applied bias. Adapted with permission from N. S. Lewis, Chemical control of charge transfer and recombination at semiconductor photoelectrode surfaces, *Inorg. Chem.* 2005, 44 (20), 6900–6911, Copyright 2005 American Chemical Society.[80] See also ref. 149.

In dynamic equilibrium, both current densities equal each other and there is no net change in the redox system.

$$j_0 = j_c^+ = -\ j_c^+ \tag{7.25}$$

Here, j_0, is defined as the exchange current density. At equilibrium, the Fermi level of the semiconductor equals the redox potential in solution, $E_F = eE^0$, and the surface electron concentration, $n_s = n_{s0}$. When an electric bias V is applied, the surface concentration of electrons, n_s, changes from the equilibrium value, n_{s0}, to

$$n_s = n_{s0} \cdot \exp\left[-\frac{\mathrm{eV}}{\mathrm{kT}}\right] \tag{7.26}$$

and the rates of the forward and backward redox reactions change, as shown in Figure 7.23. Eqn (7.27) (the 'Shockley diode equation') describes the current density, j, through a semiconductor–liquid junction in response to an applied voltage V. Here, $j_0 = en_{s0}k_{c,\mathrm{red}}c_{\mathrm{ox}}$.

$$j = -j_0\left[\exp\left(-\frac{qV}{kT}\right) - 1\right] \tag{7.27}$$

When the semiconductor–liquid junction is illuminated, minority carriers are generated, which transfer to the solution through the valence band (or the conduction band in case of a p-SC). Assuming the majority concentration stays constant, and further, that all minority carriers are swept by the field to the solid–liquid interface, eqn (7.27) transforms into

$$j = j_{\text{ph}} - j_0 \left[\exp\left(-\frac{qV}{kT} \right) - 1 \right] \tag{7.28}$$

where j_{ph} is the photocurrent density from minority carrier (hole) injection into the liquid. According to the Gärtner model (eqn (7.29)), j_{ph} can be estimated on the basis of the absorption coefficient α of the semiconductor, the photon flux, ϕ_{ph}, and the depletion layer width W.[85,237,238]

$$j_{\text{ph}} = \mathrm{e}\phi_{\text{ph}}(1 - e^{-\alpha W}) \tag{7.29}$$

In the ideal limit, α is sufficiently large to absorb all photons in the depletion layer, so they can be separated completely. Then j_{ph} can become equal to the photon flux. One important assumption of the Gärtner model is that all minority carriers are transferred rapidly across the interface. This normally only applies to solid state junctions and to fast redox couples that follow an outer-sphere CT mechanism.[80] Significant deviations from ideal charge transfer behavior in eqn (7.28) and (7.29) are therefore expected for photoelectrochemical water oxidation and reduction. These reactions follow an inner-sphere CT mechanism and are slow. Under these conditions, significant minority carriers build up in the space charge region and at the surface, which changes the band bending, the flatband potential, and which leads to sub-band gap states that promote electron–hole recombination. By dividing the experimental value of j_{ph} by the absorbed photon flux, ϕ_{ph}, the quantum efficiency, QE, of the photon-driven process, can be defined as:

$$\text{QE} = \frac{j_{\text{ph}}}{e\phi_{\text{ph}}} \tag{7.30}$$

The formation of sub-gap states at the solid–liquid interface is the reason for the low quantum efficiency of many photoelectrodes and photocatalysts. It explains why photochemical water reduction and oxidation typically require significant applied bias potentials. This bias serves to increase the band bending and to prevent majority carriers from reaching the surface, where they would recombine with trapped minority carriers.

Because photoelectrochemical measurements on suspended photocatalysts are difficult, most of the information about charge transfer kinetics of water splitting PCs comes from electrochemical studies on photoelectrodes. For example, electrochemical measurements on Fe_2O_3 water oxidation photoanodes confirm that interfacial states involved in water electrolysis also promote surface recombination (see Figure 7.12).[239] For hematite, the anodic photocurrent is found to strictly depend on the fraction of photogenerated holes that reach the semiconductor surface.[240] A similar transport limitation was found for photochemical proton reduction at $CuBi_2O_4$ surfaces.[241] A scanning electrochemical microscopy study by the

Bard group concluded that for $BiVO_4$:Mo/W, >90% of the excited electron – hole pairs underwent bulk-recombination before reaching the electrode surface, another 8% underwent surface recombination, and only 1% of the absorbed photons were used for water oxidation.[242]

7.13.1 Controlling the Back Reaction

Many HER cocatalysts promote the electrochemical oxygen reduction reaction (ORR) in competition to proton reduction, causing the formation of superoxide, and eventually water. This well-known 'back reaction' (the reverse of the water splitting reaction) puts a limit on the O_2/H_2 pressure that can be attained in a single compartment water splitting system. In single compartment reactors it requires continuous removal of the H_2/O_2 products to drive the forward reaction, which is known as pressure bias.[10] Especially platinum catalysts have a very low oxygen reduction overpotential and for this reason must be avoided in water splitting PCs.[34,228] The problem is attenuated in gold,[243] NiO, and RuO_2 cocatalysts, which are also less active for HER.[56,243,244] In 2007, Kudo and coworkers established Fe^{3+} complexes as suppressants for the oxygen reduction reaction on Pt.[228] Especially $[Fe(SO_4)(H_2O)_5]^+$ and $[Fe(OH)(H_2O)_5]^{2+}$ were found to be effective. A similar reactivity was observed for carbonates[103] and for iodide.[245] According to XPS, the latter forms a thin layer of platinum iodide on the Pt cocatalyst. This is what allows many tandem systems to function with Pt HER cocatalysts.

In 2006, Maeda *et al.* discovered that the back reaction can be suppressed with a $Rh_{2-y}Cr_yO_3$ mixed oxide cocatalyst.[61] This cocatalyst was loaded onto GaN:ZnO by chemical co-deposition of $Na_3RhCl_6 \cdot H_2O$ and $Cr(NO_3)_3 \cdot 9H_2O$, followed by thermal annealing in air.[246] According to TEM, $Rh_{2-y}Cr_yO_3$ forms amorphous nanoparticles on the semiconductor surface (Figure 7.24). A core-shell version of this catalyst was prepared by sequential photodeposition of $Na_3RhCl_6 \cdot 2H_2O$ and K_2CrO_4. This version contains a Cr_2O_3 layer (Figure 7.24) around a Rh(0) nanoparticle core. Its selective HER function relies on the Cr_2O_3 layer slowing down O_2 diffusion to the proton reduction site.[247,248]

The Rh/Cr_2O_3 cocatalyst system has been used with a variety of particulate and film absorbers. With InGaN/GaN nanowire heterostructures, it allows OWS with AQE 13% (440–450 nm) and under green-light irradiation (up to 560 nm).[250] On $SrTiO_3$:Ga, Na, Al particles, the cocatalyst supports H_2/O_2 evolution from water with quantum efficiency of 30% at 360 nm.[108,251] The related Ru/Cr_2O_3 core-shell system was used in the $SrTiO_3$:La,Rh/Au/$BiVO_4$:Mo photocatalyst sheet and found to promote H_2/O_2 evolution at higher H_2/O_2 pressures (STH 1.1%), where the back reaction is thermodynamically favored.[16] On $(Zn_{1+x}Ge)(N_2O_x)$, the $Rh_{2-y}Cr_yO_3$ cocatalyst supported overall water splitting with a AQE of 0.20% at 420 nm.[252] In an extension of the core shell approach, Takata *et al.*[253] employed photodeposition of *in-situ* generated Ti, Ta, Nb peroxo complexes for the deposition of amorphous oxyhydroxides onto Rh_2O_3-loaded

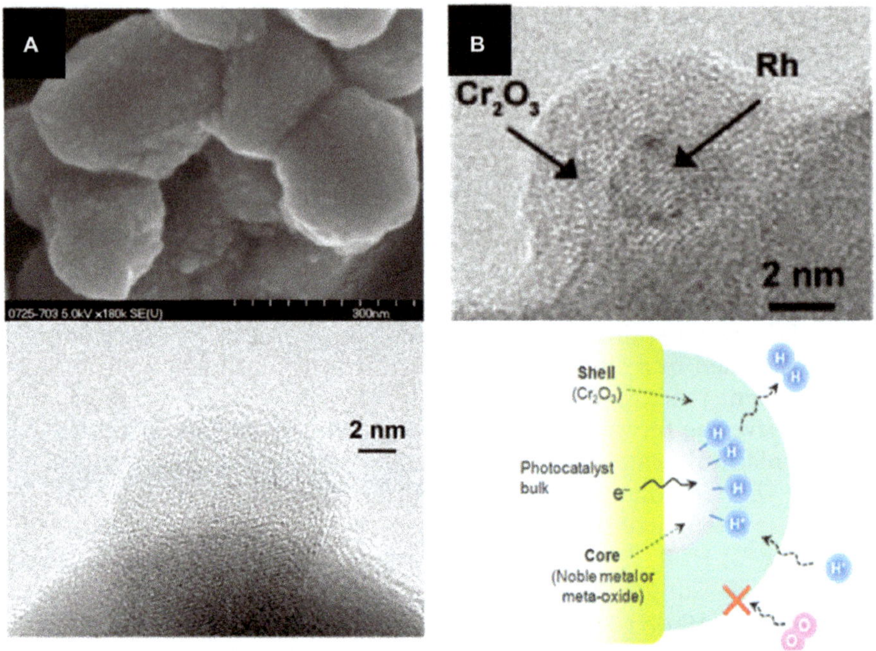

Figure 7.24 Comparison of $Rh_{2-y}Cr_yO_3$ and Rh/Cr_2O_3 cocatalysts. (A) TEM of $Rh_{2-y}Cr_yO_3$ (bottom) deposited onto GaN:ZnO (SEM, top). Reprinted by permission from Macmillan Publishers Ltd: Nature (K. Maeda, K. Teramura, D. L. Lu, T. Takata, N. Saito, Y. Inoue and K. Domen, Photocatalyst releasing hydrogen from water – Enhancing catalytic performance holds promise for hydrogen production by water splitting in sunlight, *Nature* 2006, 440 (7082), 295–295, copyright 2006.[61] (B). TEM of Rh/Cr_2O_3 core shell system (top) and mechanism of oxygen exclusion (bottom). Reprinted with permission from K. Maeda and K. Domen, New non-oxide photocatalysts designed for overall water splitting under visible light, *J. Phys. Chem. C* 2007, 111 (22), 7851–7861, Copyright 2007 American Chemical Society.[249]

$SrTiO_3$:Sc (Figure 7.25). The coatings were found to promote OWS, which was attributed to their ability to selectively block O_2 from accessing the HER reduction sites.

7.13.2 Photocorrosion

The protection of sensitive semiconductors against anodic or cathodic corrosion is central to their long term stability and operation.[112] In a landmark discovery, Hu *et al.* showed that amorphous TiO_2 coatings grown onto *n*-Si (or GaAs and GaP) by atomic layer deposition followed by deposition of a Ni cocatalyst allowed stable OER at >30 mA cm^{-2} and 100% Faradaic yield without significant photocorrosion of the light absorber.[254] This approach also works for silicon nanowire arrays.[42] Similar protective effects were later

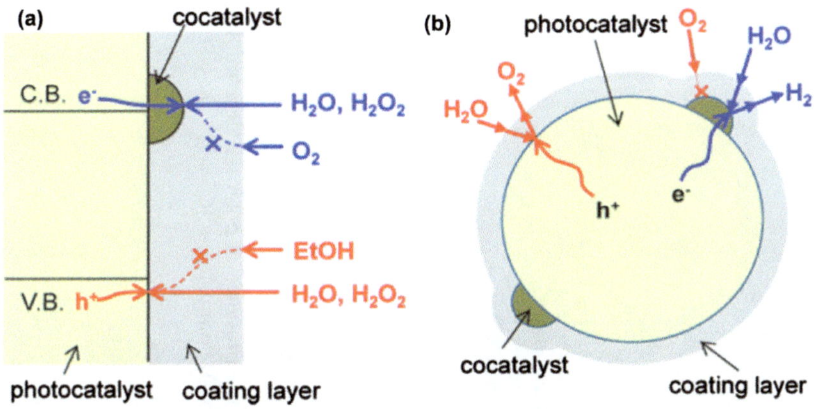

Figure 7.25 Schematics of (a) permeation behavior, and (b) reaction mechanism for overall water splitting on core-shell-structured photocatalyst. Reprinted with permission from T. Takata, C. Pan, M. Nakabayashi, N. Shibata and K. Domen, Fabrication of a Core–Shell-Type Photocatalyst *via* Photodeposition of Group IV and V Transition Metal Oxyhydroxides: An Effective Surface Modification Method for Overall Water Splitting, *J. Am. Chem. Soc.* 2015, 137 (30), 9627–9634, Copyright 2015 American Chemical Society.[253]

established for NiO[43] and $NiCo_2O_4$ layers.[255] These examples illustrate the potential of protective coatings for the stabilization of group V and III/V semiconductor materials for water oxidation. Analogous approaches have not yet been used for the protection of PCs.

7.13.3 Electrolyte Effects and pH

It is well known that the surface potential of suspended particles is affected by specifically adsorbed ions.[87,256] In pure water the most important ions are hydroxide and hydronium ions, and their effect on metal oxides, including TiO_2[193] and Fe_2O_3[194,195] gives rise to the well-known 59 mV pH^{-1} variation of the semiconductor flatband potential (the Fermi energy at the surface) with the solution pH (Figure 7.26A).[189] Adsorbed cations move the band edge to more positive potential and anions move the band edge to more reducing potential. Phosphate, silicate and fluoride ions are often strongly adsorbing[257] and can determine the redox stability of many minerals.[189] On the other hand, adsorption of soft ions, incl. HS^-,[191,192,196,197] and HTe^-,[198,199] occurs on II/VI, III/V, and group IV semiconductors. By changing the semiconductor Fermi level at the surface, these ions can modify open circuit voltage of photoelectrochemical cells.[192] Analogously, potential determining ions also control the built-in potentials of photocatalyst particles.

The effect of potential-determining ions on the charge, surface potential, and proton reduction activity was experimentally verified for cation-modified $KCa_2Nb_3O_{10}$ nanosheets ($C^+ = H^+$, K^+, Sr^{2+}) which are HER

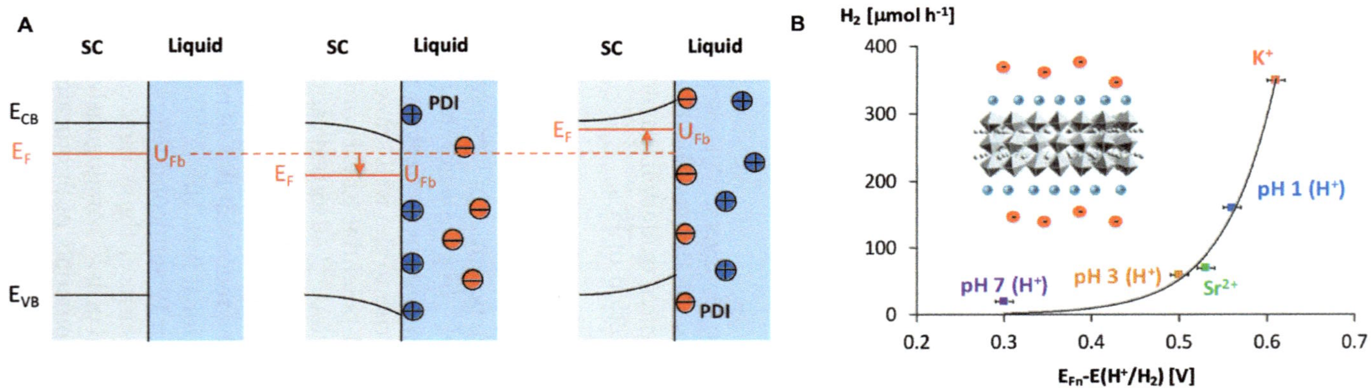

Figure 7.26 (A) Dependence of semiconductor flatband potential, U_{FB}, (Fermi level at surface) on the presence of potential determining ions (PDI) in the solution. (B) Hydrogen evolution rates for cation-modified $KCa_2Nb_3O_{10}$ nanosheets ($C^+ = H^+$, K^+, Sr^{2+}) *versus* the built-in potential with regard to the proton reduction potential.[258]
Reproduced from ref. 258 with permission from The Royal Society of Chemistry.

photocatalysts in the presence of a sacrificial donor.[258] Hydrogen evolution rates from the illuminated nanosheets are plotted in Figure 7.26B against the thermodynamic driving force for proton reduction $F(E_{Fn} - E^0)$, measured by photoelectrochemistry. The data can be fitted with the linear free energy relationship (LFER), confirming that the ions modify the thermodynamic driving force for proton reduction.

7.13.4 Theoretical Modeling

In 2011, US President Barack Obama launched the Materials Genome Initiative in order to promote computational methodologies to predict, screen, and optimize materials for various applications.[259] Since then, research efforts in this area have generated structural, optical, electronic, and stability information on over 70 000 compounds (https://www.materialsproject.org), including many potential photocatalysts. A range of 2D layered nitrides and chalcogenides and several oxides, including Ca_2PbO_4, Cu_2PbO_2, $AgGaO_2$, $AgInO_2$, and $NaBiO_3$ were predicted as possible visible light responsive photoelectrodes.[260,261] Combining the calculations with high throughput experimental techniques[262–266] has identified a range of new vanadates (α- and β-$Cu_2V_2O_7$, γ-$Cu_3V_2O_8$, and $Cu_{11}V_6O_{26}$) with 1.2–2.8 eV optical bandgaps as photoanodes for water oxidation.[267–270] Potentially, these materials can also be used in photocatalysts for OWS.

In addition to materials discovery, theoretical studies also bring about a more quantitative understanding of the performance of photoelectrochemical systems. Calculations on the limiting STH efficiency of single absorber and tandem PEC configurations were described in Section 7.4.

Some of the earliest theoretical studies of the behavior of photoelectrodes date back to work by Alan Bard's group in the 70s.[271] Digital simulations of photochemical water oxidation with hematite and titania electrodes accounted for generation, recombination, and transport of charge carriers. The photocurrent behavior of *n*-germanium were also simulated and compared to experimental results.[272] More recent theory efforts have been applied to individual photocatalyst particles. For example, Hisatomi *et al.*[99] used a numerical modeling approach to estimate quasi-Fermi levels for $LaTiO_2N$ particles under light irradiation. The effect of defects and built-in potentials on the quantum yield for water splitting was also evaluated. It was found that the QE increases by two orders of magnitude as defect densities decrease from 1×10^{18} to 1×10^{16} cm^{-3} (Figure 7.12). The work identified (i) low defect density, (ii) moderate *n*-type semiconductivity, (iii) OEC kinetics, and (iv) asymmetric electric field inside the light absorber as limiting factors for the PCs. Wang *et al.* recently employed a semiconductor device simulator (AFORS-HET)[273] to model charge carrier concentrations, electric fields, and photocurrents inside of a $SrTiO_3$:La,Rh/Carbon/$BiVO_4$:Mo tandem photocatalyst. It was found that conductive materials with large work functions, such as C (5.2 eV) and Au (5.1 eV), promote charge (hole) transfer from $SrTiO_3$:La,Rh to $BiVO_4$:Mo because these elements form Schottky junctions.

On the other hand, $BiVO_4$:Mo did not form appreciable Schottky-type barriers, because the depletion layer width exceeds the dimension of the particle.[71] For $La_5Ti_2CuS_5O_7$ (LTC), similar calculations confirmed the critical role of the metal back contact (Ti or Au) on the formation of Ohmic contact or a rectifying junction, respectively.[177] Depending on the back contact, LTC can function as a photoanode or as a photocathode.[274] Garcia-Esparza *et al.* used a numerical simulation approach to consider the effects of space charge layer, particle size, absorbance, cocatalyst distribution, and other parameters on the STH of a Ta_3N_5 photocatalyst (Figure 7.27).[275] The model shows that random positions of cocatalysts on the absorber do reduce the efficiency. Also, while the solid–liquid junction promotes water oxidation, any Schottky barriers at the cocatalyst–SC interface hinder electron transport and proton reduction. For semiconductor powders of standard purity, charge separation was found to be limited to a narrow space charge layer ($W<10$ nm) due to the high concentration of donors.

Maldonado's group used photoelectrochemical measurements combined with numerical simulations to evaluate the effect of dopant concentrations on the efficiency of semiconductor nanowire photoelectrodes.[276] Carrier collection was found to depend on the value r/W, as defined by the equilibrium barrier height and dopant density. Similar results were obtained for GaP nanowires.[277]

The modeling of particle photoreactor systems has focused on pollutant degradation.[278] In 2013, Mueses *et al.*[279] estimated the effective quantum yield of photocatalytic degradation of water contaminants in heterogeneous pilot-scale solar photoreactors employing slurry suspensions of TiO_2-P25. The kinetic analysis considered Langmuir–Hinshelwood absorption kinetics, quantum yields, and the incident photon flux. The model agreed with experimental results on the photocatalytic degradation of various model pollutants. A refined analysis was published in 2016.[280] For photoelectrochemical cells, a multi-physics numerical model made it possible to compare the efficiency of side-by-side photoelectrodes and back-to-back photoelectrodes and to determine the best dimensions for electrodes, membranes and electrolyte compartments.[281] The model considered charge and species conservation, fluid flow, and electrochemical processes in the photoelectrochemical cell.

7.13.5 Promising Absorber Materials

Promising photoelectrode materials for solar water splitting were recently compiled by Kevin Sivula and Roel van de Krol on the basis of their electronic structure, charge transport, separation, and photostability under reductive or oxidative bias.[282] These and other materials are summarized in Table 7.1. Because of their greater stability, photoanode materials for oxygen evolution consist of either oxides, oxynitrides, or nitrides. The introduction of nitrogen into the lattice reduces the band gap and improves the transport properties, but also increases the potential for anodic photocorrosion.[44]

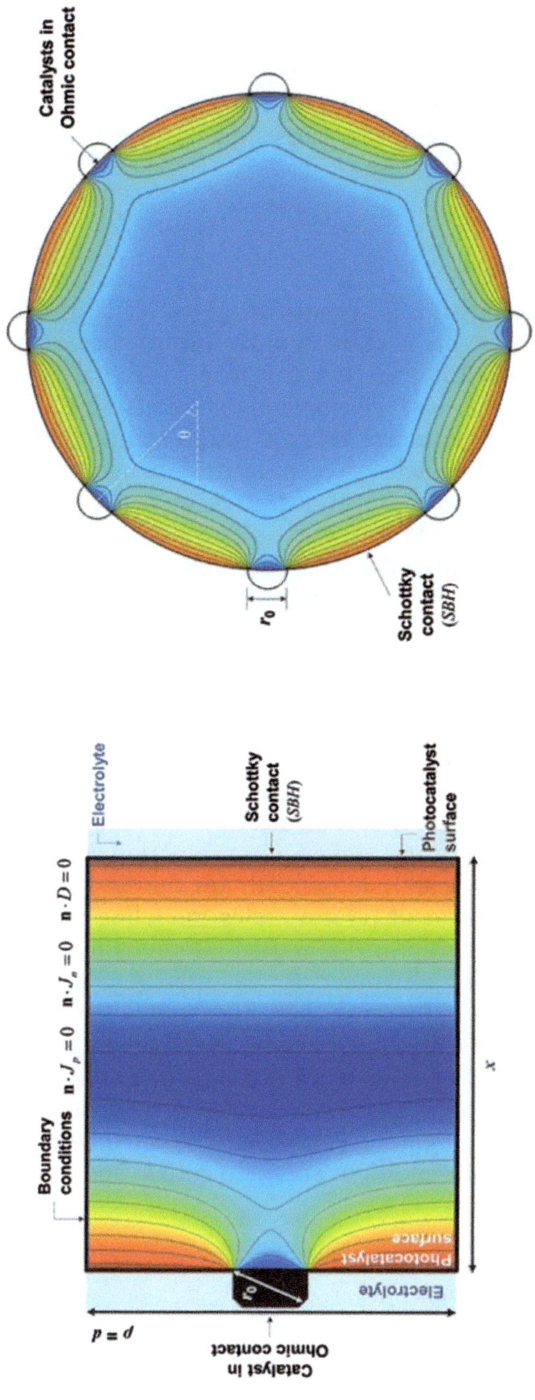

Figure 7.27 Potential distribution across semiconductor particles in contact with electrolyte and cocatalyst. Reproduced from ref. 275 with permission from The Royal Society of Chemistry.

Photocathode materials for hydrogen evolution are dominated by copper oxides and chalcogenides, but selected chalcogenides and phosphides of indium, gallium, and tungsten are also listed. So far, only a subset of the listed materials ($BiVO_4$, $SrTiO_3$, TaO_xN_y, $LaTiO_2N$, $BaTaO_2N$, $CuInS_2$, and $CuGaS_2$) have been tested in PCs for overall water splitting. Future experimentation will be required to evaluate the full potential of these materials as particle photocatalysts.

7.14 Conclusion

In order to meet the demands for clean energy, water splitting photocatalysis needs to be conducted on a scale comparable to that of natural photosynthesis. Particle based approaches can meet this goal, as demonstrated by phytoplankton bacteria which currently generate 50% of all solar fuel. Indeed, some visible light active PCs are known today that promote solar water splitting with efficiencies exceeding that of natural photosynthesis. However, none of them yet perform near their theoretical efficiency limit of 14.4%. That means that for PCs, 'there is plenty of room at the top', to revise Richard Feynman's famous statement. Research on artificial photosynthesis with particles will hopefully decrease this gap in the coming years. The identification of new absorber materials with suitable optical and electronic properties and chemical stability will be a key challenge in this area. Theoretical discovery approaches such as the materials genome initiative combined with high throughput characterization methods will play a central role in these efforts. As new materials come online, stability testing will become more important, because without long term stability, PCs will be useless. More basic research on the photochemical processes at particle interfaces, including charge transfer and defect recombination is also needed. The design of superior photocatalysts depends on the understanding of these factors and the application of photovoltaics modeling software to particle systems is an important step in this regard. Methods for the synthesis and modification of particles also need to be improved, so that passivation layers, Ohmic contacts, junctions, and cocatalysts can be fabricated in controllable ways. Acid-stable cocatalysts based on inexpensive materials need to be developed, as well as cocatalysts that do not catalyze the back reaction. Lastly, laboratory scale demonstrations of visible light driven water splitting reactors that integrate gas separation systems will help bring this technology into the public spotlight and help garner financial support for its further development.

Acknowledgements

This material is based upon work supported by the National Science Foundation under SusChEM 1464938 and by the Department of Energy under DE-SC0015329. Any opinions, findings, and conclusions or recommendations expressed in this material are those of the author and do not necessarily reflect the views of the Funding Agencies.

References

1. A. C. Nielander, M. R. Shaner, K. M. Papadantonakis, S. A. Francis and N. S. Lewis, A taxonomy for solar fuels generators, *Energy Environ. Sci.*, 2015, **8**(1), 16–25.
2. D. Gust, T. A. Moore and A. L. Moore, Solar Fuels via Artificial Photosynthesis, *Acc. Chem. Res.*, 2009, **42**(12), 1890–1898.
3. M. R. Wasielewski, Photoinduced Electron-Transfer in Supramolecular Systems for Artificial Photosynthesis, *Chem. Rev.*, 1992, **92**(3), 435–461.
4. J. Ronge, T. Bosserez, D. Martel, C. Nervi, L. Boarino, F. Taulelle, G. Decher, S. Bordiga and J. A. Martens, Monolithic Cells for Solar Fuels, *Chem. Soc. Rev.*, 2014, **43**, 7963–7981.
5. J. W. Ager, M. R. Shaner, K. A. Walczak, I. D. Sharp and S. Ardo, Experimental demonstrations of spontaneous, solar-driven photoelectrochemical water splitting, *Energy Environ. Sci.*, 2015, **8**(10), 2811–2824.
6. DOE, *Basic Research Needs for Solar Energy Utilization*, Department of Energy, 2005. http://science.energy.gov/bes/news-and-resources/reports/.
7. N. S. Lewis and D. G. Nocera, Powering the Planet: Chemical Challenges in Solar Energy Utilization, *Proc. Natl. Acad. Sci. U. S. A.*, 2006, **103**(43), 15729–15735.
8. B. D. James; G. N. Baum; J. Perez and K. N. Baum, http://www1.eere.energy.gov/hydrogenandfuelcells/pdfs/pec_technoeconomic_analysis.pdf.
9. B. A. Pinaud, J. D. Benck, L. C. Seitz, A. J. Forman, Z. B. Chen, T. G. Deutsch, B. D. James, K. N. Baum, G. N. Baum, S. Ardo, H. L. Wang, E. Miller and T. F. Jaramillo, Technical and Economic Feasibility of Centralized Facilities for Solar Hydrogen Production via Photocatalysis and Photoelectrochemistry, *Energy Environ. Sci.*, 2013, **6**(7), 1983–2002.
10. D. M. Fabian, S. Hu, N. Singh, F. A. Houle, T. Hisatomi, K. Domen, F. E. Osterloh and S. Ardo, Particle Suspension Reactors and Materials for Solar-Driven Water Splitting, *Energy Environ. Sci.*, 2015, **8**, 2825–2850.
11. F. E. Osterloh, Inorganic Materials as Catalysts for Photochemical Splitting of Water, *Chem. Mater.*, 2008, **20**(1), 35–54.
12. F. E. Osterloh, Inorganic Nanostructures for Photoelectrochemical and Photocatalytic Water Splitting, *Chem. Soc. Rev.*, 2013, **42**(6), 2294–2320.
13. K. Maeda, K. Teramura, N. Saito, Y. Inoue, H. Kobayashi and K. Domen, Overall water splitting using (oxy)nitride photocatalysts, *Pure Appl. Chem.*, 2006, **78**(12), 2267–2276.
14. D. Kang, T. W. Kim, S. R. Kubota, A. C. Cardiel, H. G. Cha and K.-S. Choi, Electrochemical Synthesis of Photoelectrodes and Catalysts

for Use in Solar Water Splitting, *Chem. Rev.*, 2015, **115**(23), 12839–12887.

15. R. G. Li, H. X. Han, F. X. Zhang, D. G. Wang and C. Li, Highly efficient photocatalysts constructed by rational assembly of dual-cocatalysts separately on different facets of BiVO4, *Energy Environ. Sci.*, 2014, **7**(4), 1369–1376.
16. Q. Wang, T. Hisatomi, Q. Jia, H. Tokudome, M. Zhong, C. Wang, Z. Pan, T. Takata, M. Nakabayashi, N. Shibata, Y. Li, I. D. Sharp, A. Kudo, T. Yamada and K. Domen, Scalable water splitting on particulate photocatalyst sheets with a solar-to-hydrogen energy conversion efficiency exceeding 1%, *Nat. Mater.*, 2016, **15**, 611–615.
17. Y. Umena, K. Kawakami, J. R. Shen and N. Kamiya, Crystal structure of oxygen-evolving photosystem II at a resolution of 1.9 angstrom, *Nature*, 2011, **473**(7345), U55–U65.
18. R. W. Baker, Future Directions of Membrane Gas Separation Technology, *Ind. Eng. Chem. Res.*, 2002, **41**(6), 1393–1411.
19. M. Carta, R. Malpass-Evans, M. Croad, Y. Rogan, J. C. Jansen, P. Bernardo, F. Bazzarelli and N. B. McKeown, An Efficient Polymer Molecular Sieve for Membrane Gas Separations, *Science*, 2013, **339**(6117), 303–307.
20. F. Dionigi, P. C. K. Vesborg, T. Pedersen, O. Hansen, S. Dahl, A. Xiong, K. Maeda, K. Domen and I. Chorkendorff, Gas phase photocatalytic water splitting with Rh2-yCryO3/GaN:ZnO in [small mu]-reactors, *Energy Environ. Sci.*, 2011, **4**(8), 2937–2942.
21. M. A. Modestino, K. A. Walczak, A. Berger, C. M. Evans, S. Haussener, C. Koval, J. S. Newman, J. W. Ager and R. A. Segalman, Robust production of purified H-2 in a stable, self-regulating, and continuously operating solar fuel generator, *Energy Environ. Sci.*, 2014, **7**(1), 297–301.
22. Y. Sasaki, H. Kato and A. Kudo, [Co(bpy)3]3 + /2 + and [Co(phen)3]3 + /2 + Electron Mediators for Overall Water Splitting under Sunlight Irradiation Using Z-Scheme Photocatalyst System, *J. Am. Chem. Soc.*, 2013, **135**(14), 5441–5449.
23. W. W. Dunn, Y. Aikawa and A. J. Bard, Semiconductor Electrodes 35. Slurry Electrodes Based on Semiconductor Powder Suspensions, *J. Electrochem. Soc.*, 1981, **128**(1), 222–224.
24. K. Domen, S. Naito, M. Soma, T. Onishi and K. Tamaru, Photocatalytic Decomposition of Water-Vapor on an Nio-Srtio3 Catalyst, *J. Chem. Soc., Chem. Commun.*, 1980, (12), 543–544.
25. K. Kalyanasundaram and M. Gratzel, Cyclic Cleavage of Water into H-2 and O-2 by Visible-Light with Coupled Redox Catalysts, *Angew. Chem., Int. Ed. Engl.*, 1979, **18**(9), 701–702.
26. D. Duonghong, E. Borgarello and M. Gratzel, Dynamics of Light-Induced Water Cleavage in Colloidal Systems, *J. Am. Chem. Soc.*, 1981, **103**(16), 4685–4690.
27. A. Mills and S. LeHunte, An overview of semiconductor photocatalysis, *J. Photochem. Photobiol., A*, 1997, **108**(1), 1–35.

28. A. Mills and G. Porter, Photosensitized Dissociation of Water Using Dispersed Suspensions of N-TypeSemiconductors., *J. Chem. Soc., Faraday Trans. 1*, 1982, **78**, 3659–3669.
29. A. J. Nozik, Photochemical Diodes, *Appl. Phys. Lett.*, 1977, **30**(11), 567–569.
30. A. J. Nozik, Photoelectrochemistry – Applications to Solar-Energy Conversion, *Annu. Rev. Phys. Chem.*, 1978, **29**, 189–222.
31. R. Abe, T. Takata, H. Sugihara and K. Domen, Photocatalytic overall water splitting under visible light by TaON and WO3 with an IO3 (-)/ I- shuttle redox mediator, *Chem. Commun.*, 2005, (30), 3829–3831.
32. Y. Sasaki, H. Nemoto, K. Saito and A. Kudo, Solar Water Splitting Using Powdered Photocatalysts Driven by Z-Schematic Interparticle Electron Transfer without an Electron Mediator, *J. Phys. Chem. C*, 2009, **113**(40), 17536–17542.
33. R. Abe, Recent Progress on Photocatalytic and Photoelectrochemical Water Splitting under Visible Light Irradiation, *J. Photochem. Photobiol., C*, 2010, **11**(4), 179–209.
34. K. Maeda, M. Higashi, D. L. Lu, R. Abe and K. Domen, Efficient Nonsacrificial Water Splitting Through Two-Step Photoexcitation by Visible Light Using a Modified Oxynitride as a Hydrogen Evolution Photocatalyst, *J. Am. Chem. Soc.*, 2010, **132**(16), 5858–5868.
35. A. Kudo, Z-Scheme Photocatalyst Systems for Water Splitting under Visible Light Irradiation, *MRS Bull.*, 2011, **36**(1), 32–38.
36. K. Maeda, Z-Scheme Water Splitting Using Two Different Semiconductor Photocatalysts, *ACS Catal.*, 2013, **3**(7), 1486–1503.
37. S. Chen, Y. Qi, T. Hisatomi, Q. Ding, T. Asai, Z. Li, S. S. K. Ma, F. Zhang, K. Domen and C. Li, Efficient Visible-Light-Driven Z-Scheme Overall Water Splitting Using a MgTa2O6-xNy/TaON Heterostructure Photocatalyst for H2 Evolution., *Angew. Chem. Int. Ed.*, 2015, **54**(29), 8498–8501.
38. K. Sayama, R. Yoshida, H. Kusama, K. Okabe, Y. Abe and H. Arakawa, Photocatalytic Decomposition of Water into H-2 and O-2 by a Two-Step Photoexcitation Reaction Using a WO3 Suspension Catalyst and an Fe3 + /Fe2 + Redox System, *Chem. Phys. Lett.*, 1997, **277**(4), 387–391.
39. R. Sathre, J. B. Greenblatt, K. Walczak, I. D. Sharp, J. C. Stevens, J. W. Ager and F. A. Houle, Opportunities to improve the net energy performance of photoelectrochemical water-splitting technology, *Energy Environ. Sci.*, 2016, **9**(3), 803–819.
40. C. C. L. McCrory, S. Jung, I. M. Ferrer, S. M. Chatman, J. C. Peters and T. F. Jaramillo, Benchmarking Hydrogen Evolving Reaction and Oxygen Evolving Reaction Electrocatalysts for Solar Water Splitting Devices, *J. Am. Chem. Soc.*, 2015, **137**(13), 4347–4357.
41. E. J. Popczun, C. G. Read, C. W. Roske, N. S. Lewis and R. E. Schaak, Highly Active Electrocatalysis of the Hydrogen Evolution Reaction by Cobalt Phosphide Nanoparticles, *Angew. Chem.*, 2014, **126**(21), 5531–5534.

42. M. R. Shaner, S. Hu, K. Sun and N. S. Lewis, Stabilization of Si microwire arrays for solar-driven H2O oxidation to O2(g) in 1.0 M KOH(aq) using conformal coatings of amorphous TiO2, *Energy Environ. Sci.*, 2015, **8**(1), 203–207.
43. K. Sun, M. T. McDowell, A. C. Nielander, S. Hu, M. R. Shaner, F. Yang, B. S. Brunschwig and N. S. Lewis, Stable Solar-Driven Water Oxidation to O2(g) by Ni-Oxide-Coated Silicon Photoanodes, *J. Phys. Chem. Lett.*, 2015, **6**(4), 592–598.
44. T. Ohno, L. Bai, T. Hisatomi, K. Maeda and K. Domen, Photocatalytic Water Splitting Using Modified GaN:ZnO Solid Solution under Visible Light: Long-Time Operation and Regeneration of Activity, *J. Am. Chem. Soc.*, 2012, **134**(19), 8254–8259.
45. J. R. Bolton, S. J. Strickler and J. S. Connolly, Limiting and Realizable Efficiencies of Solar Photolysis of Water, *Nature*, 1985, **316**(6028), 495–500.
46. O. K. Varghese and C. A. Grimes, Appropriate strategies for determining the photoconversion efficiency of water photo electrolysis cells: A review with examples using titania nanotube array photoanodes, *Sol. Energy Mater. Sol. Cells*, 2008, **92**(4), 374–384.
47. S. Hu, C. X. Xiang, S. Haussener, A. D. Berger and N. S. Lewis, An Analysis of the Optimal Band Gaps of Light Absorbers in Integrated Tandem Photoelectrochemical Water-Splitting Systems, *Energy Environ. Sci.*, 2013, **6**(10), 2984–2993.
48. M. S. Prévot and K. Sivula, Photoelectrochemical Tandem Cells for Solar Water Splitting, *J. Phys. Chem. C*, 2013, **117**(35), 17879–17893.
49. L. C. Seitz, Z. Chen, A. J. Forman, B. A. Pinaud, J. D. Benck and T. F. Jaramillo, Modeling Practical Performance Limits of Photoelectrochemical Water Splitting Based on the Current State of Materials Research, *ChemSusChem*, 2014, **7**(5), 1372–1385.
50. M. C. Hanna and A. J. Nozik, Solar conversion efficiency of photovoltaic and photoelectrolysis cells with carrier multiplication absorbers, *J. Appl. Phys.*, 2006, **100**(7), 074510.
51. R. E. Rocheleau and E. L. Miller, Photoelectrochemical production of hydrogen: Engineering loss analysis, *Int. J. Hydrogen Energy*, 1997, **22**(8), 771–782.
52. K. T. Fountaine, H. J. Lewerenz and H. A. Atwater, Efficiency limits for photoelectrochemical water-splitting, *Nat. Commun.*, 2016, **7**, 13706.
53. American Society for Testing and Materials (ASTM) Terrestrial Reference Spectra for Photovoltaic Performance Evaluation. Reference Solar Spectral Irradiance: Air Mass 1.5.
54. F. E. Osterloh, Maximum Theoretical Efficiency Limit of Photovoltaic Devices: Effect of Band Structure on Excited State Entropy, *J. Phys. Chem. Lett.*, 2014, 3354–3359.
55. A. J. Bard, Photoelectrochemistry and Heterogeneous Photocatalysis at Semiconductors, *J. Photochem.*, 1979, **10**(1), 59–75.
56. A. Kudo and Y. Miseki, Heterogeneous Photocatalyst Materials for Water Splitting, *Chem. Soc. Rev.*, 2009, **38**(1), 253–278.

57. P. Zhou, J. G. Yu and M. Jaroniec, All-Solid-State Z-Scheme Photocatalytic Systems, *Adv. Mater.*, 2014, **26**(29), 4920–4935.
58. S. J. A. Moniz, S. A. Shevlin, D. J. Martin, Z.-X. Guo and J. Tang, Visible-light driven heterojunction photocatalysts for water splitting – a critical review, *Energy Environ. Sci.*, 2015, **8**, 731–759.
59. K. Maeda, K. Teramura and K. Domen, Effect of post-calcination on photocatalytic activity of (Ga1 – xZnx)(N1 – xOx) solid solution for overall water splitting under visible light, *J. Catal.*, 2008, **254**(2), 198–204.
60. K. Maeda, K. Teramura, D. L. Lu, T. Takata, N. Saito, Y. Inoue and K. Domen, Characterization of Rh-Cr mixed-oxide nanoparticles dispersed on (Ga1-xZnx)(N1-xOx) as a cocatalyst for visible-light-driven overall water splitting, *J. Phys. Chem. B*, 2006, **110**(28), 13753–13758.
61. K. Maeda, K. Teramura, D. L. Lu, T. Takata, N. Saito, Y. Inoue and K. Domen, Photocatalyst releasing hydrogen from water – Enhancing catalytic performance holds promise for hydrogen production by water splitting in sunlight, *Nature*, 2006, **440**(7082), 295.
62. J. Liu, Y. Liu, N. Liu, Y. Han, X. Zhang, H. Huang, Y. Lifshitz, S.-T. Lee, J. Zhong and Z. Kang, Metal-free efficient photocatalyst for stable visible water splitting via a two-electron pathway, *Science*, 2015, **347**(6225), 970–974.
63. Z. G. Zou, J. H. Ye, K. Sayama and H. Arakawa, Direct splitting of water under visible light irradiation with an oxide semiconductor photocatalyst, *Nature*, 2001, **414**(6864), 625–627.
64. Z. G. Zou and H. Arakawa, Direct water splitting into H-2 and O-2 under visible light irradiation with a new series of mixed oxide semiconductor photocatalysts, *J. Photochem. Photobiol., A*, 2003, **158**(2–3), 145–162.
65. L. Liao, Q. Zhang, Z. Su, Z. Zhao, Y. Wang, Y. Li, X. Lu, D. Wei, G. Feng, Q. Yu, X. Cai, J. Zhao, Z. Ren, H. Fang, F. Robles-Hernandez, S. Baldelli and J. Bao, Efficient solar water-splitting using a nanocrystalline CoO photocatalyst, *Nat. Nano*, 2014, **9**(1), 69–73.
66. M. Hara, T. Kondo, M. Komoda, S. Ikeda, K. Shinohara, A. Tanaka, J. N. Kondo and K. Domen, Cu2O as a photocatalyst for overall water splitting under visible light irradiation, *Chem. Commun.*, 1998, (3), 357–358.
67. P. E. de Jongh, D. Vanmaekelbergh and J. J. Kelly, Cu2O: a catalyst for the photochemical decomposition of water?, *Chem. Commun.*, 1999, (12), 1069–1070.
68. A. C. Malingowski, P. W. Stephens, A. Huq, Q. Z. Huang, S. Khalid and P. G. Khalifah, Substitutional Mechanism of Ni into the Wide-Band-Gap Semiconductor InTaO4 and Its Implications for Water Splitting Activity in the Wolframite Structure Type, *Inorg. Chem.*, 2012, **51**(11), 6096–6103.
69. W. J. Jo, H. J. Kang, K.-J. Kong, Y. S. Lee, H. Park, Y. Lee, T. Buonassisi, K. K. Gleason and J. S. Lee, Phase transition-induced band edge engineering of BiVO4 to split pure water under visible light, *Proc. Natl. Acad. Sci.*, 2015, **112**(45), 13774–13778.

70. H. Kato, Y. Sasaki, N. Shirakura and A. Kudo, Synthesis of Highly Active Rhodium-Soped SrTiO3 Powders in Z-Scheme Systems for Visible-Light-Driven Photocatalytic Overall Water Splitting, *J. Mater. Chem. A*, 2013, **1**(39), 12327–12333.
71. Q. Wang, T. Hisatomi, Y. Suzuki, Z. Pan, J. Seo, M. Katayama, T. Minegishi, H. Nishiyama, T. Takata, K. Seki, A. Kudo, T. Yamada and K. Domen, Particulate Photocatalyst Sheets Based on Carbon Conductor Layer for Efficient Z-Scheme Pure-Water Splitting at Ambient Pressure, *J. Am. Chem. Soc.*, 2017, **139**(4), 1675–1683.
72. A. J. Bard and M. A. Fox, Artificial Photosynthesis – Solar Splitting of Water to Hydrogen and Oxygen, *Acc. Chem. Res.*, 1995, **28**(3), 141–145.
73. R. Memming, Photoinduced Charge-Transfer Processes at Semiconductor Electrodes and Particles, in *Electron Transfer I*, 1994, vol. 169, pp. 105–181.
74. M. G. Walter, E. L. Warren, J. R. McKone, S. W. Boettcher, Q. X. Mi, E. A. Santori and N. S. Lewis, Solar Water Splitting Cells, *Chem. Rev.*, 2010, **110**(11), 6446–6473.
75. L. M. Peter, Dynamic Aspects of Semiconductor Photoelectrochemistry, *Chem. Rev.*, 1990, **90**(5), 753–769.
76. M. D. Archer and A. J. Nozik, Nanostructured and photoelectrochemical systems for solar photon conversion, in *Series on Photoconversion of Solar Energy*, Imperial College Press; World Scientific Pub. Co., London, Singapore; Hackensack, NJ, 2008, vol. 3 [Online], p. xx, pp. 760.
77. M. A. Butler and D. S. Ginley, Principles of photoelectrochemical, solar-energy conversion, *J. Mater. Sci.*, 1980, **15**(1), 1–19.
78. A. J. Nozik and R. Memming, Physical chemistry of semiconductor-liquid interfaces, *J. Phys. Chem.*, 1996, **100**(31), 13061–13078.
79. T. Hisatomi, J. Kubota and K. Domen, Recent advances in semiconductors for photocatalytic and photoelectrochemical water splitting, *Chem. Soc. Rev.*, 2014, **43**(22), 7520–7535.
80. N. S. Lewis, Chemical control of charge transfer and recombination at semiconductor photoelectrode surfaces, *Inorg. Chem.*, 2005, **44**(20), 6900–6911.
81. F. E. Osterloh, Photocatalysis versus Photosynthesis: A Sensitivity Analysis of Devices for Solar Energy Conversion and Chemical Transformations, *ACS Energy Lett.*, 2017, 445–453.
82. H. Gerischer, Stability Of Semiconductor Electrodes Against Photodecomposition, *J. Electroanal. Chem.*, 1977, **82**(1–2), 133–143.
83. P. Würfel, *Physics of Solar Cells*, Wiley-VCH, Weinheim, 2005, p. 244.
84. R. Krol, Principles of Photoelectrochemical Cells, in *Photoelectrochemical Hydrogen Production*, 2012, vol. 102, pp. 13–67.
85. L. M. Peter, Photoelectrochemistry: From Basic Principles to Photocatalysis, in *Photocatalysis: Fundamentals and Perspectives*, The Royal Society of Chemistry, 2016, ch. 1, pp. 1–28.
86. O. Madelung, *Semiconductors: Data Handbook*, Springer, Berlin, 3rd edn, 2004, p. 691.

87. Y. Xu and M. A. A. Schoonen, The absolute energy positions of conduction and valence bands of selected semiconducting minerals, *Am. Mineral.*, 2000, **85**(3–4), 543–556.
88. L. I. Berger, Optical Properties of Selected Inorganic and Organic Solids, in *CRC Handbook of Chemistry and Physics*, ed. D. R. Lide, CRC Press/Taylor and Francis, Boca Raton, FL, 88th edn, 2008.
89. M. Qureshi and K. Takanabe, Insights on Measuring and Reporting Heterogeneous Photocatalysis: Efficiency Definitions and Setup Examples, *Chem. Mater.*, 2017, **29**(1), 158–167.
90. A. Polman and H. A. Atwater, Photonic design principles for ultrahigh-efficiency photovoltaics, *Nat. Mater.*, 2012, **11**(3), 174–177.
91. M. D. Kelzenberg, S. W. Boettcher, J. A. Petykiewicz, D. B. Turner-Evans, M. C. Putnam, E. L. Warren, J. M. Spurgeon, R. M. Briggs, N. S. Lewis and H. A. Atwater, Enhanced absorption and carrier collection in Si wire arrays for photovoltaic applications, *Nat. Mater.*, 2010, **9**(3), 239–244.
92. W. Shockley and H. J. Queisser, Detailed Balance Limit Of Efficiency Of P-N Junction Solar Cells, *J. Appl. Phys.*, 1961, **32**(3), 510–519.
93. T. Dittrich, *Materials Concepts for Solar Cells*, Imperial College Press, London, 2015, p. xxxiii, 516 pp.
94. N. S. Lewis, Mechanistic Studies of Light-Induced Charge Separation at Semiconductor Liquid Interfaces, *Acc. Chem. Res.*, 1990, **23**(6), 176–183.
95. N. S. Lewis, Frontiers of research in photoelectrochemical solar energy conversion, *J. Electroanal. Chem.*, 2001, **508**(1–2), 1–10.
96. E. Yablonovitch, D. L. Allara, C. C. Chang, T. Gmitter and T. B. Bright, Unusually Low Surface Recombination Velocity on Silicon and Germanium Surfaces, *Phys. Rev. Lett.*, 1986, **57**(2), 249–252.
97. F. F. Abdi, T. J. Savenije, M. M. May, B. Dam and R. van de Krol, The Origin of Slow Carrier Transport in BiVO4 Thin Film Photoanodes: A Time-Resolved Microwave Conductivity Study, *J. Phys. Chem. Lett.*, 2013, **4**(16), 2752–2757.
98. S. R. Pendlebury, A. J. Cowan, M. Barroso, K. Sivula, J. H. Ye, M. Gratzel, D. R. Klug, J. W. Tang and J. R. Durrant, Correlating long-lived photogenerated hole populations with photocurrent densities in hematite water oxidation photoanodes, *Energy Environ. Sci.*, 2012, **5**(4), 6304–6312.
99. H. Takashi, M. Tsutomu and D. Kazunari, Kinetic Assessment and Numerical Modeling of Photocatalytic Water Splitting toward Efficient Solar Hydrogen Production, *Bull. Chem. Soc. Jpn.*, 2012, **85**(6), 647–655.
100. U. Diebold, The surface science of titanium dioxide, *Surf. Sci. Rep.*, 2003, **48**(5–8), 53–229.
101. C. Y. Cummings, F. Marken, L. M. Peter, A. A. Tahir and K. G. U. Wijayantha, Kinetics and mechanism of light-driven oxygen evolution at thin film alpha-Fe2O3 electrodes, *Chem. Commun.*, 2012, **48**(14), 2027–2029.

102. T. K. Townsend, E. M. Sabio, N. D. Browning and F. E. Osterloh, Photocatalytic Water Oxidation with Suspended alpha-Fe2O3 Particles – Effects of Nanoscaling, *Energy Environ. Sci.*, 2011, **4**, 4270–4275.
103. H. Arakawa, Water Photolysis by TiO2 Particles-Significant Effect of Na2CO3 Addition on Water Splitting, in *Photocatalysis Science and Technology*, ed. M. Kaneko and I. Okura, Springer, New York, 2002, pp. 235–248.
104. K. Saito, K. Koga and A. Kudo, Lithium niobate nanowires for photocatalytic water splitting, *Dalton Trans.*, 2011, **40**(15), 3909–3913.
105. S. C. Yan, L. J. Wan, Z. S. Li and Z. G. Zou, Facile temperature-controlled synthesis of hexagonal Zn(2)GeO(4) nanorods with different aspect ratios toward improved photocatalytic activity for overall water splitting and photoreduction of CO(2), *Chem. Commun.*, 2011, **47**(19), 5632–5634.
106. M. Maruyama, A. Iwase, H. Kato, A. Kudo and H. Onishi, Time-Resolved Infrared Absorption Study of NaTaO3 Photocatalysts Doped with Alkali Earth Metals, *J. Phys. Chem. C*, 2009, **113**(31), 13918–13923.
107. P. Wu, J. Wang, J. Zhao, L. Guo and F. E. Osterloh, Structure defects in g-C3N4 limit visible light driven hydrogen evolution and photovoltage, *J. Mater. Chem. A*, 2014, **2**(47), 20338–20344.
108. Y. Ham, T. Hisatomi, Y. Goto, Y. Moriya, Y. Sakata, A. Yamakata, J. Kubota and K. Domen, Flux-mediated doping of SrTiO3 photocatalysts for efficient overall water splitting, *J. Mater. Chem. A*, 2016, **4**(8), 3027–3033.
109. Y. Ham, T. Minegishi, T. Hisatomi and K. Domen, SrTiO3 Photoanode Prepared by Particle Transfer Method for Oxygen Evolution from Water with High Quantum Efficiencies, *Chem. Commun.*, 2016, 5011–5014.
110. H. Kato, M. Kobayashi, M. Hara and M. Kakihana, Fabrication of SrTiO3 exposing characteristic facets using molten salt flux and improvement of photocatalytic activity for water splitting, *Catal. Sci. Technol.*, 2013, **3**(7), 1733–1738.
111. M. Zhong, T. Hisatomi, Y. Kuang, J. Zhao, M. Liu, A. Iwase, Q. Jia, H. Nishiyama, T. Minegishi, M. Nakabayashi, N. Shibata, R. Niishiro, C. Katayama, H. Shibano, M. Katayama, A. Kudo, T. Yamada and K. Domen, Surface Modification of CoOx Loaded BiVO4 Photoanodes with Ultrathin p-Type NiO Layers for Improved Solar Water Oxidation, *J. Am. Chem. Soc.*, 2015, **137**(15), 5053–5060.
112. N. Guijarro, M. S. Prevot and K. Sivula, Surface modification of semiconductor photoelectrodes, *Phys. Chem. Chem. Phys.*, 2015, **17**(24), 15655–15674.
113. S. Hu, N. S. Lewis, J. W. Ager, J. Yang, J. R. McKone and N. C. Strandwitz, Thin-Film Materials for the Protection of Semiconducting Photoelectrodes in Solar-Fuel Generators, *J. Phys. Chem. C*, 2015, **119**(43), 24201–24228.
114. F. Le Formal, N. Tetreault, M. Cornuz, T. Moehl, M. Gratzel and K. Sivula, Passivating Surface States on Water Splitting Hematite

Photoanodes with Alumina Overlayers, *Chem. Sci.*, 2011, **2**(4), 737–743.

115. R. L. Spray, K. J. McDonald and K.-S. Choi, Enhancing Photoresponse of Nanoparticulate α-Fe2O3 Electrodes by Surface Composition Tuning, *J. Phys. Chem. C*, 2011, **115**(8), 3497–3506.
116. Y. Q. Liang, T. Tsubota, L. P. A. Mooij and R. van de Krol, Highly Improved Quantum Efficiencies for Thin Film BiVO4 Photoanodes, *J. Phys. Chem. C*, 2011, **115**(35), 17594–17598.
117. D. K. Zhong, S. Choi and D. R. Gamelin, Near-Complete Suppression of Surface Recombination in Solar Photoelectrolysis by "Co-Pi" Catalyst-Modified W:BiVO4, *J. Am. Chem. Soc.*, 2011, **133**(45), 18370–18377.
118. O. Zandi and T. W. Hamann, Enhanced Water Splitting Efficiency Through Selective Surface State Removal, *J. Phys. Chem. Lett.*, 2014, **5**(9), 1522–1526.
119. J.-W. Jang, C. Du, Y. Ye, Y. Lin, X. Yao, J. Thorne, E. Liu, G. McMahon, J. Zhu, A. Javey, J. Guo and D. Wang, Enabling unassisted solar water splitting by iron oxide and silicon, *Nat. Commun.*, 2015, **6**, 7447.
120. B. J. Trzesniewski, I. A. Digdaya, T. Nagaki, S. Ravishankar, I. Herraiz-Cardona, D. A. Vermaas, A. Longo, S. Gimenez and W. A. Smith, Near-complete suppression of surface losses and total internal quantum efficiency in BiVO4 photoanodes, *Energy Environ. Sci.*, 2017, **10**(6), 1517–1529.
121. W. Cheng, J. He, Z. Sun, Y. Peng, T. Yao, Q. Liu, Y. Jiang, F. Hu, Z. Xie, B. He and S. Wei, Ni-Doped Overlayer Hematite Nanotube: A Highly Photoactive Architecture for Utilization of Visible Light, *J. Phys. Chem. C*, 2012, **116**(45), 24060–24067.
122. Y. Ben-Shahar, F. Scotognella, N. Waiskopf, I. Kriegel, S. Dal Conte, G. Cerullo and U. Banin, Effect of Surface Coating on the Photocatalytic Function of Hybrid CdS–Au Nanorods, *Small*, 2015, **11**(4), 462–471.
123. P. Tongying, F. Vietmeyer, D. Aleksiuk, G. J. Ferraudi, G. Krylova and M. Kuno, Double heterojunction nanowire photocatalysts for hydrogen generation, *Nanoscale*, 2014, **6**(8), 4117–4124.
124. M. X. Tan, C. N. Kenyon, O. Kruger and N. S. Lewis, Behavior of Si photoelectrodes under high level injection conditions .1. Steady-state current-voltage properties and quasi-Fermi level positions under illumination, *J. Phys. Chem. B*, 1997, **101**(15), 2830–2839.
125. K. Maeda, N. Nishimura and K. Domen, A precursor route to prepare tantalum (V) nitride nanoparticles with enhanced photocatalytic activity for hydrogen evolution under visible light, *Appl. Catal. A: Gen.*, 2009, **370**(1–2), 88–92.
126. M. N. Huda, M. M. Al-Jassim and J. A. Turner, Mott Insulators: An Early Selection Criterion For Materials For Photoelectrochemical H(2) Production, *J. Renewable Sustainable Energy*, 2011, **3**(5), 053101-1–053101-10.
127. A. J. E. Rettie, W. D. Chemelewski, D. Emin and C. B. Mullins, Unravelling Small-Polaron Transport in Metal Oxide Photoelectrodes, *J. Phys. Chem. Lett.*, 2016, **7**(3), 471–479.

128. L. I. Berger, Properties of Semiconductors, in *CRC Handbook of Chemistry and Physics*, ed. D. R. Lide, CRC Press/Taylor and Francis, Boca Raton, FL, 88 edn (Internet Version 2008), 2008.
129. K. A. Newton and F. E. Osterloh, Size and Morphology of Suspended WO3 Particles Control Photochemical Charge Carrier Extraction and Photocatalytic Water Oxidation Activity, *Top. Catal.*, 2016, **59**(8), 750–756.
130. F. E. Osterloh, Nanoscale Effects in Water Splitting Photocatalysis, *Top. Curr. Chem.*, 2015, **371**, 105–142.
131. P. A. Cox, *Transition Metal Oxides: An Introduction to their Electronic Structure and Properties*, Clarendon Press; Oxford University Press, Oxford; New York, 2010.
132. E. M. Sabio, R. L. Chamousis, N. D. Browning and F. E. Osterloh, Correction: Photocatalytic Water Splitting with Suspended Calcium Niobium Oxides: Why Nanoscale is Better than Bulk – A Kinetic Analysis, *J. Phys. Chem. C*, 2012, **116**(35), 19051.
133. E. M. Sabio, R. L. Chamousis, N. D. Browning and F. E. Osterloh, Photocatalytic Water Splitting with Suspended Calcium Niobium Oxides: Why Nanoscale is Better than Bulk – A Kinetic Analysis, *J. Phys. Chem. C*, 2012, **116**(4), 3161–3170.
134. R. A. Pala, A. J. Leenheer, M. Lichterman, H. A. Atwater and N. S. Lewis, Measurement of minority-carrier diffusion lengths using wedge-shaped semiconductor photoelectrodes, *Energ. Environ. Sci.*, 2014, **7**(10), 3424–3430.
135. S. Sodergren, A. Hagfeldt, J. Olsson and S. E. Lindquist, Theoretical-Models for the Action Spectrum and the Current-Voltage Characteristics of Microporous Semiconductor-Films in Photoelectrochemical Cells, *J. Phys. Chem.*, 1994, **98**(21), 5552–5556.
136. A. Hagfeldt, U. Bjorksten and S. E. Lindquist, Photoelectrochemical Studies of Colloidal TiO2-Films – The Charge Separation Process Studied by Means of Action Spectra in the UV Region, *Sol. Energy Mater. Sol. Cells*, 1992, **27**(4), 293–304.
137. A. Hagfeldt and M. Gratzel, Light-Induced Redox Reactions in Nanocrystalline Systems, *Chem. Rev.*, 1995, **95**(1), 49–68.
138. J. Bisquert and V. S. Vikhrenko, Interpretation of the time constants measured by kinetic techniques in nanostructured semiconductor electrodes and dye-sensitized solar cells, *J. Phys. Chem. B*, 2004, **108**(7), 2313–2322.
139. A. Hagfeldt, H. Lindstrom, S. Sodergren and S. E. Lindquist, Photoelectrochemical Studies of Colloidal TiO2 Films – The Effect of Oxygen Studied by Photocurrent Transients, *J. Electroanal. Chem.*, 1995, **381**(1–2), 39–46.
140. B. Abeles, P. Sheng, M. D. Coutts and Y. Arie, Structural and Electrical Properties of Granular Metal-Films, *Adv. Phys.*, 1975, **24**(3), 407–461.
141. R. H. Terrill, T. A. Postlethwaite, C. H. Chen, C. D. Poon, A. Terzis, A. D. Chen, J. E. Hutchison, M. R. Clark, G. Wignall, J. D. Londono,

R. Superfine, M. Falvo, C. S. Johnson, E. T. Samulski and R. W. Murray, Monolayers in three dimensions: NMR, SAXS, thermal, and electron hopping studies of alkanethiol stabilized gold clusters., *J. Am. Chem. Soc.*, 1995, **117**(50), 12537–12548.
142. G. Hodes, I. D. J. Howell and L. M. Peter, Nanocrystalline Photoelectrochemical Cells – A New Concept In Photovoltaic Cells, *J. Electrochem. Soc.*, 1992, **139**(11), 3136–3140.
143. T. K. Townsend, E. M. Sabio, N. D. Browning and F. E. Osterloh, Improved Niobate Nanoscroll Photocatalysts for Partial Water Splitting, *ChemSusChem*, 2011, **4**(2), 185–190.
144. R. Krol, Photoelectrochemical Measurements, in *Photoelectrochemical Hydrogen Production*, ed. R. van de Krol and M. Grätzel, Springer US, 2012, vol. 102, pp. 69–117.
145. Z. B. Chen, T. F. Jaramillo, T. G. Deutsch, A. Kleiman-Shwarsctein, A. J. Forman, N. Gaillard, R. Garland, K. Takanabe, C. Heske, M. Sunkara, E. W. McFarland, K. Domen, E. L. Miller, J. A. Turner and H. N. Dinh, Accelerating materials development for photoelectrochemical hydrogen production: Standards for methods, definitions, and reporting protocols, *J. Mater. Res.*, 2010, **25**(1), 3–16.
146. W. Shockley, The Theory of P-N Junctions in Semiconductors and P-N Junction Transistors, *Bell Syst. Tech. J.*, 1949, **28**(3), 435–489.
147. H. Gerischer, The Impact of Semiconductors on the Concepts of Electrochemistry, *Electrochim. Acta*, 1990, **35**(11–12), 1677–1699.
148. Y. V. Pleskov and Y. Y. Gurevich, *Semiconductor Photoelectrochemistry*, Consultants Bureau, New York, 1986, p. xxv, 422 pp.
149. M. X. Tan; P. E. Laibinis; S. T. Nguyen; J. M. Kesselman; C. E. Stanton and N. S. Lewis, Principles and Applications of Semiconductor Photoelectrochemistry, in *Progress in Inorganic Chemistry*, Wiley, 1994, vol. 41, pp. 21–144.
150. K. Rajeshwar, N. R. de Tacconi and C. R. Chenthamarakshan, Semiconductor-based composite materials: Preparation, properties, and performance, *Chem. Mater.*, 2001, **13**(9), 2765–2782.
151. R. Marschall, Semiconductor Composites: Strategies for Enhancing Charge Carrier Separation to Improve Photocatalytic Activity, *Adv. Funct. Mater.*, 2014, **24**(17), 2421–2440.
152. A. J. Bard, A. B. Bocarsly, F. R. F. Fan, E. G. Walton and M. S. Wrighton, The Concept of Fermi Level Pinning at Semiconductor-Liquid Junctions – Consequences for Energy-Conversion Efficiency and Selection of Useful Solution Redox Couples in Solar Devices, *J. Am. Chem. Soc.*, 1980, **102**(11), 3671–3677.
153. J. Zhao, B. A. Nail, M. A. Holmes and F. E. Osterloh, Use of Surface Photovoltage Spectroscopy to Measure Built-in Voltage, Space Charge Layer Width, and Effective Band Gap in CdSe Quantum Dot Films, *J. Phys. Chem. Lett.*, 2016, 3335–3340.
154. S. W. Boettcher, E. L. Warren, M. C. Putnam, E. A. Santori, D. Turner-Evans, M. D. Kelzenberg, M. G. Walter, J. R. McKone,

B. S. Brunschwig, H. A. Atwater and N. S. Lewis, Photoelectrochemical Hydrogen Evolution Using Si Microwire Arrays, *J. Am. Chem. Soc.*, 2011, **133**(5), 1216–1219.

155. M. R. Shaner, K. T. Fountaine, S. Ardo, R. H. Coridan, H. A. Atwater and N. S. Lewis, Photoelectrochemistry of Core-Shell Tandem Junction n-p(+)-Si/n-WO3 Microwire Array Photoelectrodes, *Energy Environ. Sci.*, 2014, **7**(2), 779–790.
156. I. Cesar, K. Sivula, A. Kay, R. Zboril and M. Graetzel, Influence of Feature Size, Film Thickness, and Silicon Doping on the Performance of Nanostructured Hematite Photoanodes for Solar Water Splitting, *J. Phys. Chem. C*, 2009, **113**(2), 772–782.
157. O. Zandi, A. R. Schon, H. Hajibabaei and T. W. Hamann, Enhanced Charge Separation and Collection in High-Performance Electrodeposited Hematite Films, *Chem. Mater.*, 2016, **28**(3), 765–771.
158. B. Oregan, J. Moser, M. Anderson and M. Gratzel, Vectorial Electron Injection into Transparent Semiconductor Membranes and Electric-Field Effects on the Dynamics of Light-Induced Charge Separation, *J. Phys. Chem.*, 1990, **94**(24), 8720–8726.
159. K. Tvrdy, P. A. Frantsuzov and P. V. Kamat, Photoinduced electron transfer from semiconductor quantum dots to metal oxide nanoparticles, *PNAS*, 2011, **108**(1), 29–34.
160. Y. J. Dong, B. Z. Tian, T. J. Kempa and C. M. Lieber, Coaxial Group III-Nitride Nanowire Photovoltaics, *Nano Lett.*, 2009, **9**(5), 2183–2187.
161. M. H. M. van Weert, A. Helman, W. van den Einden, R. E. Algra, M. A. Verheijen, M. T. Borgstrom, G. Immink, J. J. Kelly, L. P. Kouwenhoven and E. Bakkers, Zinc Incorporation via the Vapor-Liquid-Solid Mechanism into InP Nanowires, *J. Am. Chem. Soc.*, 2009, **131**(13), 4578.
162. T. J. Kempa, B. Z. Tian, D. R. Kim, J. S. Hu, X. L. Zheng and C. M. Lieber, Single and Tandem Axial p-i-n Nanowire Photovoltaic Devices, *Nano Lett.*, 2008, **8**(10), 3456–3460.
163. S. K. Baek, Y. H. Kwon, J. H. Shin, H. S. Lee and H. K. Cho, Low-Temperature Processable High-Performance Electrochemically Deposited p-Type Cuprous Oxides Achieved by Incorporating a Small Amount of Antimony, *Adv. Funct. Mater.*, 2015, **25**(32), 5214–5221.
164. K. Tonooka, H. Bando and Y. Aiura, Photovoltaic effect observed in transparent p–n heterojunctions based on oxide semiconductors, *Thin Solid Films*, 2003, **445**(2), 327–331.
165. E. Thimsen, A. B. F. Martinson, J. W. Elam and M. J. Pellin, Energy Levels, Electronic Properties, and Rectification in Ultrathin p-NiO Films Synthesized by Atomic Layer Deposition, *J. Phys. Chem. C*, 2012, **116**(32), 16830–16840.
166. H. Asıl, E. Gür, K. Çınar and C. Coşkun, Electrochemical growth of n-ZnO onto the p-type GaN substrate: p-n heterojunction characteristics, *Appl. Phys. Lett.*, 2009, **94**(25), 253501.

167. S. Chatman and K. M. Poduska, The Effect of Synthesis Conditions and Humidity on Current-Voltage Relations in Electrodeposited ZnO-Based Schottky Junctions, *ACS Appl. Mater. Interfaces*, 2009, **1**(3), 552–558.
168. K. Wu and T. Lian, Quantum confined colloidal nanorod heterostructures for solar-to-fuel conversion, *Chem. Soc. Rev.*, 2016, **45**(14), 3781–3810.
169. J. E. Evans, K. W. Springer and J. Z. Zhang, Femtosecond Studies of Interparticle Electron-Transfer in a Coupled CdS-TiO2 Colloidal System, *J. Chem. Phys.*, 1994, **101**(7), 6222–6225.
170. C. Nasr, S. Hotchandani, W. Y. Kim, R. H. Schmehl and P. V. Kamat, Photoelectrochemistry of composite semiconductor thin films. Photosensitization of SnO2/CdS coupled nanocrystallites with a ruthenium polypyridyl complex, *J. Phys. Chem. B*, 1997, **101**(38), 7480–7487.
171. P. A. Sant and P. V. Kamat, Interparticle electron transfer between size-quantized CdS and TiO2 semiconductor nanoclusters, *Phys. Chem. Chem. Phys.*, 2002, **4**(2), 198–203.
172. V. Subramanian, E. E. Wolf and P. V. Kamat, Catalysis with TiO2/gold nanocomposites. Effect of metal particle size on the Fermi level equilibration, *J. Am. Chem. Soc.*, 2004, **126**(15), 4943–4950.
173. I. Robel, M. Kuno and P. V. Kamat, Size-dependent electron injection from excited CdSe quantum dots into TiO2 nanoparticles, *J. Am. Chem. Soc.*, 2007, **129**(14), 4136–4137.
174. C. H. Chuang, T. L. Doane, S. S. Lo, G. D. Scholes and C. Burda, Measuring Electron and Hole Transfer in Core/Shell Nanoheterostructures, *ACS Nano*, 2011, **5**(7), 6016–6024.
175. T. Minegishi, N. Nishimura, J. Kubota and K. Domen, Photoelectrochemical properties of LaTiO2N electrodes prepared by particle transfer for sunlight-driven water splitting, *Chem. Sci.*, 2013, **4**(3), 1120–1124.
176. J. Liu, T. Hisatomi, G. Ma, A. Iwanaga, T. Minegishi, Y. Moriya, M. Katayama, J. Kubota and K. Domen, Improving the Photoelectrochemical Activity of La5Ti2CuS5O7 for Hydrogen Evolution by Particle Transfer and Doping, *Energy Environ. Sci.*, 2014, **7**(7), 2239–2242.
177. G. Ma, Y. Suzuki, R. B. Singh, A. Iwanaga, Y. Moriya, T. Minegishi, J. Liu, T. Hisatomi, H. Nishiyama, M. Katayama, K. Seki, A. Furube, T. Yamada and K. Domen, Photoanodic and photocathodic behaviour of La5Ti2CuS5O7 electrodes in the water splitting reaction, *Chem. Sci.*, 2015, **6**(8), 4513–4518.
178. H. Urabe, T. Hisatomi, T. Minegishi, J. Kubota and K. Domen, Photoelectrochemical properties of SrNbO2N photoanodes for water oxidation fabricated by the particle transfer method, *Faraday Discuss.*, 2014, **176**, 213–223.
179. F. D. Lin and S. W. Boettcher, Adaptive semiconductor/electrocatalyst junctions in water-splitting photoanodes, *Nat. Mater.*, 2014, **13**(1), 81–86.
180. L. Kronik and Y. Shapira, Surface Photovoltage Phenomena: Theory, Experiment, and Applications, *Surf. Sci. Rep.*, 1999, **37**(1–5), 1–206.

181. L. Kronik and Y. Shapira, Surface Photovoltage Spectroscopy of Semiconductor Structures: At the Crossroads of Physics, Chemistry and Electrical Engineering, *Surf. Interface Anal.*, 2001, **31**(10), 954–965.
182. J. Zhao and F. E. Osterloh, Photochemical Charge Separation in Nanocrystal Photocatalyst Films – Insights from Surface Photovoltage Spectroscopy, *J. Phys. Chem. Lett.*, 2014, **5**, 782–786.
183. T. L. Shelton, N. Harvey, J. Wang and F. E. Osterloh, Photochemistry of hematite photoanodes under zero applied bias, *Appl. Catal. A: Gen.*, 2016, **521**, 168–173.
184. J. Wang, J. Zhao and F. E. Osterloh, Photochemical charge transfer observed in nanoscale hydrogen evolving photocatalysts using surface photovoltage spectroscopy, *Energy Environ. Sci.*, 2015, **8**, 2970–2976.
185. Y. Yang, J. Wang, J. Zhao, B. A. Nail, X. Yuan, Y. Guo and F. E. Osterloh Photochemical Charge Separation at Particle Interfaces: The n-BiVO4–p-Silicon System. *ACS Appl. Mater. Interfaces*, 2015, **10** (7), 5959–5964.
186. S. Ida, A. Takashiba, S. Koga, H. Hagiwara and T. Ishihara, Potential Gradient and Photocatalytic Activity of an Ultrathin p–n Junction Surface Prepared with Two-Dimensional Semiconducting Nanocrystals, *J. Am. Chem. Soc.*, 2014, **136**(5), 1872–1878.
187. M. Sander, W. Jaegermann and H. J. Lewerenz, Site-specific surface interaction of adsorbed water and halogens on copper indium selenide (CuInSe2) surfaces, *J. Phys. Chem. C*, 1992, **96**(2), 782–790.
188. D. Schmeisser, A comparative study of O2, H2 and H2O adsorption on Si(100), *Surf. Sci.*, 1984, **137**(1), 197–210.
189. G. E. Brown, V. E. Henrich, W. H. Casey, D. L. Clark, C. Eggleston, A. Felmy, D. W. Goodman, M. Gratzel, G. Maciel, M. I. McCarthy, K. H. Nealson, D. A. Sverjensky, M. F. Toney and J. M. Zachara, Metal oxide surfaces and their interactions with aqueous solutions and microbial organisms, *Chem. Rev.*, 1999, **99**(1), 77–174.
190. S. Y. Yang, D. Prendergast and J. B. Neaton, Tuning Semiconductor Band Edge Energies for Solar Photocatalysis via Surface Ligand Passivation, *Nano Lett.*, 2012, **12**(1), 383–388.
191. D. S. Ginley and M. A. Butler, Flatband Potential of Cadmium-Sulfide (Cds) Photoanodes and Its Dependence on Surface Ion Effects, *J. Electrochem. Soc.*, 1978, **125**(12), 1968–1974.
192. P. Singh, R. Singh, R. Gale, K. Rajeshwar and J. Dubow, Surface-Charge and Specific Ion Adsorption Effects in Photoelectrochemical Devices, *J. Appl. Phys.*, 1980, **51**(12), 6286–6291.
193. L. Kavan, M. Gratzel, S. E. Gilbert, C. Klemenz and H. J. Scheel, Electrochemical and photoelectrochemical investigation of single-crystal anatase, *J. Am. Chem. Soc.*, 1996, **118**(28), 6716–6723.
194. S. U. M. Khan and J. Akikusa, Photoelectrochemical splitting of water at nanocrystalline n-Fe2O3 thin-film electrodes, *J. Phys. Chem. B*, 1999, **103**(34), 7184–7189.

195. R. J. Atkinson, A. M. Posner and J. P. Quirk, Adsorption of Potential-Determining Ions at Ferric Oxide-Aqueous Electrolyte Interface, *J. Phys. Chem.*, 1967, **71**(3), 550–558.
196. D. Meissner, R. Memming and B. Kastening, Photoelectrochemistry of Cadmium-Sulfide .1. Reanalysis of Photocorrosion and Flat-Band Potential, *J. Phys. Chem.*, 1988, **92**(12), 3476–3483.
197. K. W. Frese and D. G. Canfield, Adsorption of Hydroxide and Sulfide Ions on Single-Crystal n-CdSe Electrodes, *J. Electrochem. Soc.*, 1984, **131**(11), 2614–2618.
198. D. Lincot and J. Vedel, Adsorption of Telluride Ions on Cadmium Telluride - Consequences for Photoelectrochemical Cells, *J. Phys. Chem.*, 1988, **92**(14), 4103–4110.
199. H. Minoura, T. Watanabe, T. Oki and M. Tsuiki, Effects of Dissolved Cd2 + and S2- Ions on Flatband Potential of CdS Electrode in Aqueous-Solution, *Jpn. J. Appl. Phys.*, 1977, **16**(5), 865–866.
200. T. Ohno, K. Sarukawa and M. Matsumura, Crystal faces of rutile and anatase TiO2 particles and their roles in photocatalytic reactions, *New J. Chem.*, 2002, **26**(9), 1167–1170.
201. Y. Zhu, P. A. Salvador and G. S. Rohrer, Controlling the Relative Areas of Photocathodic and Photoanodic Terraces on the SrTiO3(111) Surface, *Chem. Mater.*, 2016, **28**(14), 5155–5162.
202. L. Li, P. A. Salvador and G. S. Rohrer, Photocatalysts with internal electric fields, *Nanoscale*, 2014, **6**(1), 24–42.
203. S. Y. Yang, J. Seidel, S. J. Byrnes, P. Shafer, C. H. Yang, M. D. Rossell, P. Yu, Y. H. Chu, J. F. Scott, J. W. Ager, L. W. Martin and R. Ramesh, Above-bandgap voltages from ferroelectric photovoltaic devices, *Nat. Nanotechnol.*, 2010, **5**(2), 143–147.
204. S. Y. Yang, L. W. Martin, S. J. Byrnes, T. E. Conry, S. R. Basu, D. Paran, L. Reichertz, J. Ihlefeld, C. Adamo, A. Melville, Y. H. Chu, C. H. Yang, J. L. Musfeldt, D. G. Schlom, J. W. Ager and R. Ramesh, Photovoltaic effects in BiFeO3, *Appl. Phys. Lett.*, 2009, **95**(6), 062909.
205. J. L. Giocondi and G. S. Rohrer, Spatially selective photochemical reduction of silver on the surface of ferroelectric barium titanate, *Chem. Mater.*, 2001, **13**(2), 241–242.
206. N. V. Burbure, P. A. Salvador and G. S. Rohrer, Photochemical Reactivity of Titania Films on BaTiO3 Substrates: Origin of Spatial Selectivity, *Chem. Mater.*, 2010, **22**(21), 5823–5830.
207. F. F. Abdi, L. H. Han, A. H. M. Smets, M. Zeman, B. Dam and R. van de Krol, Efficient Solar Water Splitting by Enhanced Charge Separation in a Bismuth Vanadate-Silicon Tandem Photoelectrode, *Nat. Commun.*, 2013, **4**, 2195.
208. Q. Jia, A. Iwase and A. Kudo, BiVO4-Ru/SrTiO3:Rh Composite of Z-Scheme Photocatalyst for Solar Water Splitting, *Chem. Sci.*, 2014, **5**, 1513–1519.
209. K. Sayama, K. Mukasa, R. Abe, Y. Abe and H. Arakawa, A new photocatalytic water splitting system under visible light irradiation

mimicking a Z-scheme mechanism in photosynthesis, *J. Photochem. Photobiol., A*, 2002, **148**(1–3), 71–77.

210. K. Sayama, K. Mukasa, R. Abe, Y. Abe and H. Arakawa, Stoichiometric water splitting into H-2 and O-2 using a mixture of two different photocatalysts and an IO3-/I- shuttle redox mediator under visible light irradiation, *Chem. Commun.*, 2001, (23), 2416–2417.
211. A. Iwase, Y. H. Ng, Y. Ishiguro, A. Kudo and R. Amal, Reduced Graphene Oxide as a Solid-State Electron Mediator in Z-Scheme Photocatalytic Water Splitting under Visible Light, *J. Am. Chem. Soc.*, 2011, **133**(29), 11054–11057.
212. K. Iwashina, A. Iwase, Y. H. Ng, R. Amal and A. Kudo, Z-Schematic Water Splitting into H-2 and O-2 Using Metal Sulfide as a Hydrogen-Evolving Photocatalyst and Reduced Graphene Oxide as a Solid-State Electron Mediator, *J. Am. Chem. Soc.*, 2015, **137**(2), 604–607.
213. A. J. Bard and L. R. Faulkner, *Electrochemical Methods: Fundamentals and Applications*, John Wiley, New York, 2nd edn, 2001, p. 571.
214. J. Zhao, M. A. Holmes and F. E. Osterloh, Quantum Confinement Controls Photocatalysis – A Free Energy Analysis for Photocatalytic Proton Reduction at CdSe Nanocrystals, *ACS Nano*, 2013, **7**(5), 4316–4325.
215. H. Dau, C. Limberg, T. Reier, M. Risch, S. Roggan and P. Strasser, The Mechanism of Water Oxidation: From Electrolysis via Homogeneous to Biological Catalysis, *ChemCatChem*, 2010, **2**(7), 724–761.
216. S. Trasatti, Work Function, Electronegativity, and Electrochemical Behavior of Metals 3. Electrolytic Hydrogen Evolution in Acid Solutions, *J. Electroanal. Chem.*, 1972, **39**(1), 163–184.
217. S. Trasatti, Electrocatalysis by Oxides – Attempt at a Unifying Approach, *J. Electroanal. Chem.*, 1980, **111**(1), 125–131.
218. T. F. Jaramillo, K. P. Jorgensen, J. Bonde, J. H. Nielsen, S. Horch and I. Chorkendorff, Identification of active edge sites for electrochemical H-2 evolution from MoS2 nanocatalysts, *Science*, 2007, **317**(5834), 100–102.
219. B. Hinnemann, P. G. Moses, J. Bonde, K. P. Jorgensen, J. H. Nielsen, S. Horch, I. Chorkendorff and J. K. Norskov, Biornimetic hydrogen evolution: MoS2 nanoparticies as catalyst for hydrogen evolution, *J. Am. Chem. Soc.*, 2005, **127**(15), 5308–5309.
220. E. J. Popczun, J. R. McKone, C. G. Read, A. J. Biacchi, A. M. Wiltrout, N. S. Lewis and R. E. Schaak, Nanostructured Nickel Phosphide as an Electrocatalyst for the Hydrogen Evolution Reaction, *J. Am. Chem. Soc.*, 2013, **135**(25), 9267–9270.
221. S. Cao, Y. Chen, C.-C. Hou, X.-J. Lv and W.-F. Fu, Cobalt phosphide as a highly active non-precious metal cocatalyst for photocatalytic hydrogen production under visible light irradiation, *J. Mater. Chem. A*, 2015, **3**(11), 6096–6101.
222. F. A. Frame and F. E. Osterloh, CdSe-MoS2: A Quantum Size-Confined Photocatalyst for Hydrogen Evolution from Water under Visible Light, *J. Phys. Chem. C*, 2010, **114**, 10628–10633.

223. J. A. Seabold and K. S. Choi, Efficient and Stable Photo-Oxidation of Water by a Bismuth Vanadate Photoanode Coupled with an Iron Oxyhydroxide Oxygen Evolution Catalyst, *J. Am. Chem. Soc.*, 2012, **134**(4), 2186–2192.
224. X. Deng and H. Tüysüz, Cobalt-Oxide-Based Materials as Water Oxidation Catalyst: Recent Progress and Challenges, *ACS Catal.*, 2014, **4**(10), 3701–3714.
225. J. Wang and F. E. Osterloh, Limiting Factors for Photochemical Charge Separation in BiVO4/Co3O4, a Highly Active Photocatalyst for Water Oxidation in Sunlight, *J. Mater. Chem. A*, 2014, **2**, 9405–9411.
226. T. Hisatomi, C. Katayama, Y. Moriya, T. Minegishi, M. Katayama, H. Nishiyama, T. Yamada and K. Domen, Photocatalytic oxygen evolution using BaNbO2N modified with cobalt oxide under photoexcitation up to 740 nm, *Energy Environ. Sci.*, 2013, **6**(12), 3595–3599.
227. H. Kato, M. Hori, R. Konta, Y. Shimodaira and A. Kudo, Construction of Z-scheme type heterogeneous photocatalysis systems for water splitting into H-2 and O-2 under visible light irradiation, *Chem. Lett.*, 2004, **33**(10), 1348–1349.
228. H. Kato, Y. Sasaki, A. Wase and A. Kudo, Role of iron ion electron mediator on photocatalytic overall water splitting under visible light irradiation using Z-scheme systems, *Bull. Chem. Soc. Jpn.*, 2007, **80**(12), 2457–2464.
229. Y. Sasaki, A. Iwase, H. Kato and A. Kudo, The effect of co-catalyst for Z-scheme photocatalysis systems with an Fe3+/Fe2+ electron mediator on overall water splitting under visible light irradiation, *J. Catal.*, 2008, **259**(1), 133–137.
230. S. Okunaka, H. Tokudome and R. Abe, Z-scheme Water Splitting into H-2 and O-2 under Visible Light over Photocatalyst Panels Consisting of Rh-doped SrTiO3 and BiVO4 Fine Particles, *Chem. Lett.*, 2016, **45**(1), 57–59.
231. T. M. Suzuki, A. Iwase, H. Tanaka, S. Sato, A. Kudo and T. Morikawa, Z-scheme water splitting under visible light irradiation over powdered metal-complex/semiconductor hybrid photocatalysts mediated by reduced graphene oxide, *J. Mater. Chem. A*, 2015, **3**(25), 13283–13290.
232. R. Abe, K. Sayama and H. Sugihara, Development of new photocatalytic water splitting into H-2 and O-2 using two different semiconductor photocatalysts and a shuttle redox mediator IO3 (-)/I-, *J. Phys. Chem. B*, 2005, **109**(33), 16052–16061.
233. K. Maeda, A. K. Xiong, T. Yoshinaga, T. Ikeda, N. Sakamoto, T. Hisatomi, M. Takashima, D. L. Lu, M. Kanehara, T. Setoyama, T. Teranishi and K. Domen, Photocatalytic Overall Water Splitting Promoted by Two Different Cocatalysts for Hydrogen and Oxygen Evolution under Visible Light, *Angew. Chem., Int. Ed. Engl.*, 2010, **49**(24), 4096–4099.
234. H. Kato, K. Asakura and A. Kudo, Highly Efficient Water Splitting into H-2 and O-2 over Lanthanum-Doped NaTaO3 Photocatalysts with

High Crystallinity and Surface Nanostructure, *J. Am. Chem. Soc.*, 2003, **125**(10), 3082–3089.
235. A. Iwase, H. Kato and A. Kudo, A novel photodeposition method in the presence of nitrate ions for loading of an iridium oxide cocatalyst for water splitting, *Chem. Lett.*, 2005, **34**(7), 946–947.
236. R. A. Marcus, Chemical + Electrochemical Electron-Transfer Theory, *Annu. Rev. Phys. Chem.*, 1964, **15**, 155–196.
237. W. W. Gärtner, Depletion-Layer Photoeffects in Semiconductors, *Phys. Rev.*, 1959, **116**(1), 84–87.
238. S. Soedergren, A. Hagfeldt, J. Olsson and S.-E. Lindquist, Theoretical Models for the Action Spectrum and the Current-Voltage Characteristics of Microporous Semiconductor Films in Photoelectrochemical Cells, *J. Phys. Chem. C*, 1994, **98**(21), 5552–5556.
239. B. Klahr, S. Gimenez, F. Fabregat-Santiago, T. Hamann and J. Bisquert, Water Oxidation at Hematite Photoelectrodes: The Role of Surface States, *J. Am. Chem. Soc.*, 2012, **134**(9), 4294–4302.
240. B. M. Klahr and T. W. Hamann, Current and Voltage Limiting Processes in Thin Film Hematite Electrodes, *J. Phys. Chem. C*, 2011, **115**(16), 8393–8399.
241. S. P. Berglund, F. F. Abdi, P. Bogdanoff, A. Chemseddine, D. Friedrich and R. van de Krol, Comprehensive Evaluation of $CuBi_2O_4$ as a Photocathode Material for Photoelectrochemical Water Splitting, *Chem. Mater.*, 2016, **28**(12), 4231–4242.
242. H. S. Park, K. C. Leonard and A. J. Bard, Surface Interrogation Scanning Electrochemical Microscopy (SI-SECM) of Photoelectrochemistry at a W/Mo-BiVO4 Semiconductor Electrode: Quantification of Hydroxyl Radicals during Water Oxidation, *J. Phys. Chem. C*, 2013, **117**(23), 12093–12102.
243. A. Iwase, H. Kato and A. Kudo, Nanosized Au particles as an efficient cocatalyst for photocatalytic overall water splitting, *Catal. Lett.*, 2006, **108**(1–2), 6–9.
244. K. Yamaguti and S. Sato, Photolysis of Water over Metallized Powdered Titanium-Dioxide, *J. Chem. Soc. Farad. Trans. I*, 1985, **81**, 1237–1246.
245. R. Abe, K. Sayama and H. Arakawa, Significant effect of iodide addition on water splitting into H2 and O2 over Pt-loaded TiO2 photocatalyst: suppression of backward reaction, *Chem. Phys. Lett.*, 2003, **371**(3–4), 360–364.
246. K. Maeda, K. Teramura, H. Masuda, T. Takata, N. Saito, Y. Inoue and K. Domen, Efficient Overall Water Splitting under Visible-Light Irradiation on (Ga1-xZnx)(N1-xOx) Dispersed with Rh − Cr Mixed-Oxide Nanoparticles: Effect of Reaction Conditions on Photocatalytic Activity, *J. Phys. Chem. B*, 2006, **110**(26), 13107–13112.
247. K. Maeda, K. Teramura, D. L. Lu, N. Saito, Y. Inoue and K. Domen, Noble-metal/Cr2O3 core/shell nanoparticles as a cocatalyst for photocatalytic overall water splitting, *Angew. Chem., Int. Ed. Engl.*, 2006, **45**(46), 7806–7809.

248. K. Maeda, K. Teramura, D. L. Lu, N. Saito, Y. Inoue and K. Domen, Roles of Rh/Cr2O3 (core/shell) nanoparticles photodeposited on visible-light-responsive (Ga1-xZnx)(N1-xOx) solid solutions in photocatalytic overall water splitting, *J. Phys. Chem. C*, 2007, **111**(20), 7554–7560.
249. K. Maeda and K. Domen, New non-oxide photocatalysts designed for overall water splitting under visible light, *J. Phys. Chem. C*, 2007, **111**(22), 7851–7861.
250. M. G. Kibria, H. P. T. Nguyen, K. Cui, S. Zhao, D. Liu, H. Guo, M. L. Trudeau, S. Paradis, A.-R. Hakima and Z. Mi, One-Step Overall Water Splitting under Visible Light Using Multiband InGaN/GaN Nanowire Heterostructures, *ACS Nano*, 2013, **7**(9), 7886–7893.
251. T. Takata and K. Domen, Defect Engineering of Photocatalysts by Doping of Aliovalent Metal Cations for Efficient Water Splitting, *J. Phys. Chem. C*, 2009, **113**(45), 19386–19388.
252. Y. G. Lee, K. Teramura, M. Hara and K. Domen, Modification of (Zn1 + xGe)(N2Ox) solid solution as a visible light driven photocatalyst for overall water splitting, *Chem. Mater.*, 2007, **19**(8), 2120–2127.
253. T. Takata, C. Pan, M. Nakabayashi, N. Shibata and K. Domen, Fabrication of a Core–Shell-Type Photocatalyst via Photodeposition of Group IV and V Transition Metal Oxyhydroxides: An Effective Surface Modification Method for Overall Water Splitting, *J. Am. Chem. Soc.*, 2015, **137**(30), 9627–9634.
254. S. Hu, M. R. Shaner, J. A. Beardslee, M. Lichterman, B. S. Brunschwig and N. S. Lewis, Amorphous TiO2 coatings stabilize Si, GaAs, and GaP photoanodes for efficient water oxidation, *Science*, 2014, **344**(6187), 1005–1009.
255. L. Chen, J. Yang, S. Klaus, L. J. Lee, R. Woods-Robinson, J. Ma, Y. Lum, J. K. Cooper, F. M. Toma, L.-W. Wang, I. D. Sharp, A. T. Bell and J. W. Ager, p-Type Transparent Conducting Oxide/n-Type Semiconductor Heterojunctions for Efficient and Stable Solar Water Oxidation, *J. Am. Chem. Soc.*, 2015, **137**(30), 9595–9603.
256. M. A. Butler and D. S. Ginley, Prediction of Flatband Potentials at Semiconductor-Electrolyte Interfaces from Atomic Electronegativities, *J. Electrochem. Soc.*, 1978, **125**(2), 228–232.
257. F. J. Hingston, R. J. Atkinson, A. M. Posner and J. P. Quirk, Specific Adsorption of Anions, *Nature*, 1967, **215**(5109), 1459–1461.
258. R. L. Chamousis and F. E. Osterloh, Use of Potential Determining Ions to Control Energetics and Photochemical Charge Transfer of a Nanoscale Water Splitting Photocatalyst, *Energy Environ. Sci.*, 2014, **7**(2), 736–743.
259. A. Jain, S. P. Ong, G. Hautier, W. Chen, W. D. Richards, S. Dacek, S. Cholia, D. Gunter, D. Skinner, G. Ceder and K. A. Persson, Commentary: The Materials Project: A materials genome approach to accelerating materials innovation, *APL Mater.*, 2013, **1**(1), 011002.
260. A. K. Singh, K. Mathew, H. L. Zhuang and R. G. Hennig, Computational Screening of 2D Materials for Photocatalysis, *J. Phys. Chem. Lett.*, 2015, **6**(6), 1087–1098.

261. I. E. Castelli, F. Hüser, M. Pandey, H. Li, K. S. Thygesen, B. Seger, A. Jain, K. A. Persson, G. Ceder and K. W. Jacobsen, New Light-Harvesting Materials Using Accurate and Efficient Bandgap Calculations, *Adv. Energy Mater.*, 2015, **5**(2), 1400915.
262. M. Woodhouse and B. A. Parkinson, Combinatorial discovery and optimization of a complex oxide with water photoelectrolysis activity, *Chem. Mater.*, 2008, **20**(7), 2495–2502.
263. M. Woodhouse, G. S. Herman and B. A. Parkinson, Combinatorial approach to identification of catalysts for the photoelectrolysis of water, *Chem. Mater.*, 2005, **17**(17), 4318–4324.
264. M. Woodhouse and B. A. Parkinson, Combinatorial approaches for the identification and optimization of oxide semiconductors for efficient solar photoelectrolysis, *Chem. Soc. Rev.*, 2009, **38**(1), 197–210.
265. T. F. Jaramillo, S. H. Baeck, A. Kleiman-Shwarsctein, K. S. Choi, G. D. Stucky and E. W. McFarland, Automated electrochemical synthesis and photoelectrochemical characterization of Zn1-xCoxO thin films for solar hydrogen production, *J. Comb. Chem.*, 2005, **7**(2), 264–271.
266. J. E. Katz, T. R. Gingrich, E. A. Santori and N. S. Lewis, Combinatorial synthesis and high-throughput photopotential and photocurrent screening of mixed-metal oxides for photoelectrochemical water splitting, *Energy Environ. Sci.*, 2009, **2**(1), 103–112.
267. Q. Yan, J. Yu, S. K. Suram, L. Zhou, A. Shinde, P. F. Newhouse, W. Chen, G. Li, K. A. Persson, J. M. Gregoire and J. B. Neaton, Solar fuels photoanode materials discovery by integrating high-throughput theory and experiment, *Proc. Natl. Acad. Sci.*, 2017, **114**(12), 3040–3043.
268. L. Zhou, Q. Yan, J. Yu, R. J. R. Jones, N. Becerra-Stasiewicz, S. K. Suram, A. Shinde, D. Guevarra, J. B. Neaton, K. A. Persson and J. M. Gregoire, Stability and self-passivation of copper vanadate photoanodes under chemical, electrochemical, and photoelectrochemical operation, *Phys. Chem. Chem. Phys.*, 2016, **18**(14), 9349–9352.
269. L. Zhou, Q. Yan, A. Shinde, D. Guevarra, P. F. Newhouse, N. Becerra-Stasiewicz, S. M. Chatman, J. A. Haber, J. B. Neaton and J. M. Gregoire, High Throughput Discovery of Solar Fuels Photoanodes in the CuO–V2O5 System, *Adv. Energy Mater.*, 2015, **5**(22), 1500968.
270. Q. Yan, G. Li, P. F. Newhouse, J. Yu, K. A. Persson, J. M. Gregoire and J. B. Neaton, Mn2V2O7: An Earth Abundant Light Absorber for Solar Water Splitting, *Adv. Energy Mater.*, 2015, **5**(8), 1401840.
271. D. Laser and A. J. Bard, Semiconductor electrodes .9. Digital-simulation of relaxation of photogenerated free carriers and photocurrents, *J. Electrochem. Soc.*, 1976, **123**(12), 1837–1842.
272. D. Laser and A. J. Bard, Semiconductor electrodes .8. Digital-simulation of open-circuit photopotentials, *J. Electrochem. Soc.*, 1976, **123**(12), 1833–1837.
273. R. Varache, C. Leendertz, M. E. Gueunier-Farret, J. Haschke, D. Muñoz and L. Korte, Investigation of selective junctions using a newly

developed tunnel current model for solar cell applications, *Sol. Energy Mater. Sol. Cells*, 2015, **141**, 14–23.

274. M. Kodera, J. Wang, B. A. Nail, J. Liu, H. Urabe, T. Hisatomi, M. Katayama, T. Minegishi, F. E. Osterloh and K. Domen, Investigation of charge separation in particulate oxysulfide and oxynitride photoelectrodes by surface photovoltage spectroscopy, *Chem. Phys. Lett.*, 2017, **683**, 140–144.
275. A. T. Garcia-Esparza and K. Takanabe, A simplified theoretical guideline for overall water splitting using photocatalyst particles, *J. Mater. Chem. A*, 2016, **4**(8), 2894–2908.
276. K. Hagedorn, C. Forgacs, S. Collins and S. Maldonado, Design Considerations for Nanowire Heterojunctions in Solar Energy Conversion/Storage Applications, *J. Phys. Chem. C*, 2010, **114**(27), 12010–12017.
277. J. M. Foley, M. J. Price, J. I. Feldblyum and S. Maldonado, Analysis of the operation of thin nanowire photoelectrodes for solar energy conversion, *Energy Environ. Sci.*, 2012, **5**(1), 5203–5220.
278. H. I. De Lasa; B. Serrano and M. Salaices, *Photocatalytic Reaction Engineering*, Springer, New York, 2005, p. xii, 187 pp.
279. M. A. Mueses, F. Machuca-Martinez and G. Li Puma, Effective quantum yield and reaction rate model for evaluation of photocatalytic degradation of water contaminants in heterogeneous pilot-scale solar photoreactors, *Chem. Eng. J.*, 2013, **215–216**, 937–947.
280. F. Machuca-Martinez; M. A. Mueses; J. Colina-Marquez and G. L. Puma, Photocatalytic Reactor Modeling, in *Photocatalysis: Fundamentals and Perspectives*, The Royal Society of Chemistry, 2016, ch. 16, pp. 388–424.
281. S. Haussener, C. Xiang, J. M. Spurgeon, S. Ardo, N. S. Lewis and A. Z. Weber, Modeling, simulation, and design criteria for photoelectrochemical water-splitting systems, *Energy Environ. Sci.*, 2012, **5**(12), 9922–9935.
282. K. Sivula and R. van de Krol, Semiconducting materials for photoelectrochemical energy conversion, *Nat. Rev. Mater.*, 2016, **1**, 15010.
283. J. K. Cooper, S. Gul, F. M. Toma, L. Chen, Y.-S. Liu, J. Guo, J. W. Ager, J. Yano and I. D. Sharp, Indirect Bandgap and Optical Properties of Monoclinic Bismuth Vanadate, *J. Phys. Chem. C*, 2015, **119**(6), 2969–2974.
284. Y. Li, L. Zhang, A. Torres-Pardo, J. M. González-Calbet, Y. Ma, P. Oleynikov, O. Terasaki, S. Asahina, M. Shima, D. Cha, L. Zhao, K. Takanabe, J. Kubota and K. Domen, Cobalt phosphate-modified barium-doped tantalum nitride nanorod photoanode with 1.5% solar energy conversion efficiency, *Nat. Commun.*, 2013, **4**, 2566.
285. S. Khan, M. J. M. Zapata, M. B. Pereira, R. V. Goncalves, L. Strizik, J. Dupont, M. J. L. Santos and S. R. Teixeira, Structural, optical and photoelectrochemical characterizations of monoclinic Ta3N5 thin films, *Phys. Chem. Chem. Phys.*, 2015, **17**(37), 23952–23962.
286. S. Chen, S. Shen, G. Liu, Y. Qi, F. Zhang and C. Li, Interface Engineering of a CoOx/Ta3N5 Photocatalyst for Unprecedented Water

Oxidation Performance under Visible-Light-Irradiation, *Angew. Chem. Int. Ed.*, 2015, **54**(10), 3047–3051.

287. S. Zollner, A. A. Demkov, R. Liu, P. L. Fejes, R. B. Gregory, P. Alluri, J. A. Curless, Z. Yu, J. Ramdani, R. Droopad, T. E. Tiwald, J. N. Hilfiker and J. A. Woollam, Optical properties of bulk and thin-film SrTiO3 on Si and Pt, *J. Vac. Sci. Technol., B*, 2000, **18**(4), 2242–2254.
288. P. L. Washington, H. C. Ong, J. Y. Dai and R. P. H. Chang, Determination of the optical constants of zinc oxide thin films by spectroscopic ellipsometry, *Appl. Phys. Lett.*, 1998, **72**(25), 3261–3263.
289. M. S. Prévot, N. Guijarro and K. Sivula, Enhancing the Performance of a Robust Sol–Gel-Processed p-Type Delafossite CuFeO2 Photocathode for Solar Water Reduction, *ChemSusChem*, 2015, **8**(8), 1359–1367.
290. K. Sekizawa, T. Nonaka, T. Arai and T. Morikawa, Structural Improvement of CaFe2O4 by Metal Doping toward Enhanced Cathodic Photocurrent, *ACS Appl. Mater. Interfaces*, 2014, **6**(14), 10969–10973.
291. Gunawan, W. Septina, S. Ikeda, T. Harada, T. Minegishi, K. Domen and M. Matsumura, Platinum and indium sulfide-modified CuInS2 as efficient photocathodes for photoelectrochemical water splitting, *Chem. Commun.*, 2014, **50**(64), 8941–8943.
292. G. Sharma, Z. Zhao, P. Sarker, B. A. Nail, J. Wang, M. Huda and F. Osterloh, Electronic structure, photovoltage, and photocatalytic hydrogen evolution with p-CuBi2O4 nanocrystals, *J. Mater. Chem. A*, 2016, **4**, 2936–2942.
293. T. Kato, Y. Hakari, S. Ikeda, Q. Jia, A. Iwase and A. Kudo, Utilization of Metal Sulfide Material of (CuGa)1−xZn2xS2 Solid Solution with Visible Light Response in Photocatalytic and Photoelectrochemical Solar Water Splitting Systems, *J. Phys. Chem. Lett.*, 2015, **6**(6), 1042–1047.
294. A. Iwase, S. Yoshino, T. Takayama, Y. H. Ng, R. Amal and A. Kudo, Water Splitting and CO2 Reduction under Visible Light Irradiation Using Z-Scheme Systems Consisting of Metal Sulfides, CoOx-Loaded BiVO4, and a Reduced Graphene Oxide Electron Mediator, *J. Am. Chem. Soc.*, 2016, **138**(32), 10260–10264.
295. Q. Huang, Z. Ye and X. Xiao, Recent progress in photocathodes for hydrogen evolution, *J. Mater. Chem. A*, 2015, **3**(31), 15824–15837.
296. P. K. Nayak and D. Cahen, Updated Assessment of Possibilities and Limits for Solar Cells, *Adv. Mater.*, 2014, **26**(10), 1622–1628.

CHAPTER 8

Degradation of Semiconductor Electrodes in Photoelectrochemical Devices: Principles and Case Studies

JOHANNA EICHHORN,[a] GUIJI LIU[a,b] AND FRANCESCA M TOMA*[a,b]

[a] Chemical Sciences Division, Lawrence Berkeley National Laboratory, Berkeley, CA 94720, USA; [b] Joint Center for Artificial Photosynthesis, Lawrence Berkeley National Laboratory, Berkeley, CA 94720, USA
*Email: fmtoma@lbl.gov

8.1 Introduction

In the simplest version, a photoelectrochemical (PEC) cell consists of a semiconducting light absorber and an electrolyte solution. The functions of the light absorber include photon absorption, as well as charge generation, separation, and transport towards the liquid/solid interface (*i.e.* electrolyte/semiconductor). The liquid electrolyte is a concentrated solution that provides the reactant (*e.g.* water), and connects electrically the anodic and cathodic reaction sites by enabling ionic transport. Typically, liquid electrolytes are strongly acidic or basic in order to improve the ionic conductivity and to decrease polarization losses. However, these conditions create an aggressive environment, which can have an adverse effect on the stability of the photoactive material.[1,2]

Energy and Environment Series No. 22
Integrated Solar Fuel Generators
Edited by Ian D. Sharp, Harry A. Atwater and Hans-Joachim Lewerenz

Published by the Royal Society of Chemistry, www.rsc.org

The instability of semiconductors under operating conditions is one of the major challenges of photoelectrochemical devices, yet one of the least understood.[2,3] A typical photoelectrochemical reaction, such as water electrolysis, involves partial reactions with multiple charge transfers, which may generate reactive intermediates. These reactive intermediates, together with accumulation and trapping of minority carriers at the surface of the semiconductor, may weaken bonds at the semiconductor surface, thus inducing instability. Indeed, while these phenomena are necessary steps for a (photo)-electrochemical reaction to efficiently take place, weakening bonds can induce atoms to leave the surface as solvated ions and change the intrinsic nature of the semiconductor surface. Unfortunately, the side reactions that lead to corrosion of semiconductors are the same that proceed in parallel with those necessary for efficient water electrolysis, and a competition between the two processes is always present. In order to promote photoelectrochemical water splitting, it is important that the corresponding reduction and oxidation potentials are energetically favorable with respect to unwanted side reactions.[4,5]

In this chapter, we will review general thermodynamic and kinetic aspects of photoelectrode stability at the liquid/solid interface. We will then provide specific examples, and discuss titanium dioxide (TiO_2), cuprous oxide (Cu_2O), and bismuth vanadate ($BiVO_4$) semiconductors to explain corrosion mechanisms occurring during photoelectrochemical water splitting.

8.2 Thermodynamic and Kinetic Requirements for Material Stability

PEC device performance and stability are strongly influenced by the efficiency of the charge transfer across solid/solid (*e.g.* semiconductor/electrical contact, semiconductor/catalyst), as well as the liquid/solid (*e.g.* light absorber/electrolyte), interfaces. Losses at these interfaces can favor photocarrier recombination, which reduces efficiency, and unwanted reactions such as (photo)corrosion, which reduce stability. While thermodynamic considerations provide a starting point for estimating material stability and operational durability, careful examination of kinetic factors is required for a comprehensive description of these important issues. In the following section, fundamental conditions for (photo)corrosion are discussed by establishing energy relations, as well as highlighting thermodynamic considerations and kinetic aspects. As a model system, we consider a non-elemental semiconductor (MX, with M the electropositive and X the electronegative components), whose band gap is assumed to be wide enough to prevent any considerable rate of thermal generation of minority carriers.

8.2.1 Thermodynamic Aspects

Decomposition of a semiconductor in contact with an electrolyte solvent (sol) can occur *via* electrochemical reduction or oxidation reactions, depending on the type of electronic reactants. Specifically, electrons induce reductive decomposition and holes cause oxidative decomposition. During reductive decomposition (eqn (8.1)), a stoichiometric quantity of electrons (ze^-) reduces the more electropositive component to a lower oxidation state, M, while the more electronegative component is solvated in the electrolyte as X^{z-}_{sol} (dissolution).[4]

$$MX + ze^- + \mathrm{sol} \rightarrow M + X^{z-}_{\mathrm{sol}} \tag{8.1}$$

During anodic decomposition (eqn (8.2)), a stoichiometric quantity of holes (zh^+) oxidizes the more electropositive component, and the more electronegative component is left in elementary form X.

$$MX + zh^+ + \mathrm{sol} \rightarrow M^{z+}_{\mathrm{sol}} + X \tag{8.2}$$

For both reactions an electrochemical redox potential can be defined ${}_{\mathrm{n}}\epsilon_{\mathrm{dec}}$ for cathodic decomposition by electrons and ${}_{\mathrm{p}}\epsilon_{\mathrm{dec}}$ for anodic decomposition by holes. The energy levels of the semiconductor are referenced to the vacuum level. The energy levels of the anodic (${}_{\mathrm{n}}E_{\mathrm{dec}}$) and cathodic (${}_{\mathrm{p}}E_{\mathrm{dec}}$) decomposition reactions with respect to the vacuum are calculated according to

$${}_{\mathrm{n}}E_{\mathrm{dec}} = +e \cdot {}_{\mathrm{n}}\epsilon_{\mathrm{dec}} + E_{\mathrm{ref}} \tag{8.3}$$

$${}_{\mathrm{p}}E_{\mathrm{dec}} = -e \cdot {}_{\mathrm{p}}\epsilon_{\mathrm{dec}} + E_{\mathrm{ref}} \tag{8.4}$$

where e is the unit electric charge and E_{ref} is the Fermi level of the reference electrode with respect to the vacuum level.[6] The work function for the standard hydrogen electrode (E_{ref}) has a value of −4.4 eV.[7–9] Therefore, the characteristic energy levels of the semiconductor and its decomposition potentials can be plotted on a common energy scale.[6]

8.2.1.1 Decomposition by Majority Carriers under Dark Conditions

In the absence of illumination and at room temperature, electrolytic decomposition reactions are mostly caused by majority carriers.[4] The concentration of majority carriers present at the surface that is necessary to favor the decomposition process can be estimated by the relation between the decomposition level (${}_{\mathrm{n}}E_{\mathrm{dec}}$, ${}_{\mathrm{p}}E_{\mathrm{dec}}$) and the conduction ($E_{\mathrm{C}}$) or valence ($E_{\mathrm{V}}$) energy levels at the semiconductor/electrolyte interface. Specifically, for ${}_{\mathrm{n}}E_{\mathrm{dec}} < E_{\mathrm{C}}$ or ${}_{\mathrm{p}}E_{\mathrm{dec}} > E_{\mathrm{V}}$, even relatively low surface concentrations of electrons or holes allow the decomposition reaction to take place (Figure 8.1(A)). Cu_2O, for example, fulfills the criteria for decomposition and can already be reduced at a very low electron concentration at the surface. However, with ${}_{\mathrm{n}}E_{\mathrm{dec}}$ (${}_{\mathrm{p}}E_{\mathrm{dec}}$) within the conduction (valence) band of the semiconductor, decomposition occurs only when the majority carriers are accumulated to such an extent that the semiconductor is degenerate at the surface, which

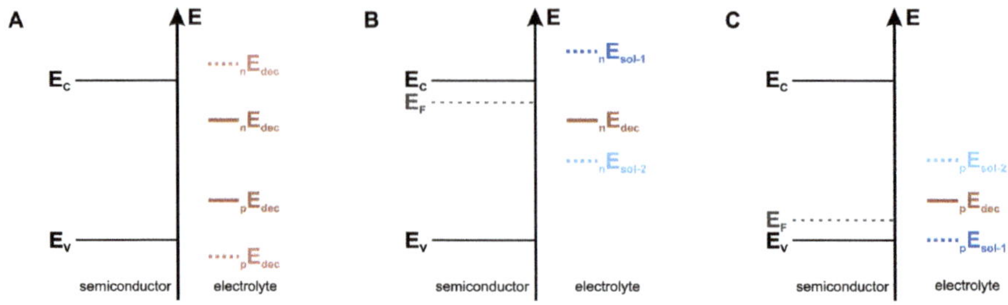

Figure 8.1 Schematic energy diagram of band edges $E_{C,V}$ and decomposition levels $_{n,p}E_{dec}$ under flat band conditions. (A) General thermodynamic criteria for stability (light red) and instability (dark red) against decomposition. (B) Anodic decomposition of n-type semiconductors occurs for $_nE_{dec} < E_F$ and $_nE_{dec} < {}_nE_{sol\text{-}1}$. The semiconductor can be stabilized for $_nE_{dec} > {}_nE_{sol\text{-}2}$. (C) Cathodic decomposition of p-type semiconductors occurs for $_pE_{dec} > E_F$ and $_pE_{dec} > {}_pE_{sol\text{-}1}$. The semiconductor can be stabilized for $_pE_{dec} < {}_pE_{sol\text{-}2}$.

requires, for example, a large external voltage.[6] For example, in TiO_2 the decomposition level $_nE_{dec}$ is above E_C. Thus, the decomposition by majority carriers is only possible for high electron concentrations at the surface.[4]

In the presence of an electrolyte, the decomposition of the semiconductor competes with the redox reaction of the solvent or other species present in solution. Both reactions compete with each other and have to be considered for the evaluation of the stability/instability of the semiconductor.[4] In this context, the positions of the decomposition potentials of the semiconductor and of the other possible redox active species (*e.g.* the solvent itself) relative to the semiconductor band potentials are decisive. For simplicity, it is instructive to consider the case of flat band potentials with no equilibration of the Fermi levels of the semiconductor and of the redox electrolyte. An n-type semiconductor will be thermodynamically unstable, if under flat band conditions the Fermi level is above the decomposition potential $_nE_{dec}$ (Figure 8.1(B)).[4,10] The semiconductor will be thermodynamically stabilized by the decomposition of the solvent $_nE_{sol}$, if the decomposition potential of the solvent $_n\mathrm{E}_{sol\text{-}2}$ is located below $_nE_{dec}$. For $_nE_{sol\text{-}1}$ above $_nE_{dec}$, the semiconductor will be unstable. Similar considerations can be applied to the decomposition of p-type semiconductors by holes (Figure 8.1(C)). For a p-type semiconductor the decomposition will be thermodynamically possible, if E_F is located below $_pE_{dec}$. The decomposition can be prevented if the solvent decomposition potential $_pE_{sol\text{-}2}$ is above the decomposition potential of the semiconductor $_pE_{dec}$.

8.2.1.2 *Photo-induced Decomposition by Minority Carriers under Illumination*

Under dark conditions, when a semiconductor is put in contact with an electrolyte, a band banding is induced by the equilibration of the Fermi level of the semiconductor and that one of the electrolyte. Whereas under dark conditions decomposition is driven mainly by majority carriers, electron/hole pairs are generated under illumination, altering the equilibrium concentration of carriers. Due to the increased number of holes (electrons) in an n-type (p-type) semiconductor under illumination, photo-induced decomposition can be caused *via* minority carriers. In an n-type (p-type) material, photogenerated holes (electrons) drift towards the liquid/solid interface, while the electrons (holes) move to the bulk of the semiconductor. Typically, the distributions of electrons and holes in the illuminated semiconductor are approximated by separate quasi-Fermi levels ${}_{p}E_{F}{}^{*}$ for holes and ${}_{n}E_{F}{}^{*}$ for electrons (Figure 8.2). With increasing illumination intensity, ${}_{p}E_{F}{}^{*}$ and ${}_{n}E_{F}{}^{*}$ will approach the band edges, but can seldom exceed the range of the band gap of the semiconductor arising from the very steep increase of the recombination rate at high illumination intensities.[6] The deviation from equilibrium is locally different and mainly depends on the light intensity, the absorption coefficient of the semiconductor, and the transport properties. At the surface, the consumption of minority carriers due to redox reactions reduces the respective quasi-Fermi level deviation relative to the equilibrium level.[6]

In contrast to the case for minority carriers, the quasi-Fermi level of the majority carriers remains close to its equilibrium level in the bulk even under illumination.[4] Since only the quasi-Fermi level of the minority carriers deviates largely from this equilibrium value, we will discuss the conditions of photodecomposition caused by the reaction with minority carriers. Similar to the decomposition by majority carriers, the position of the decomposition level with respect to the band edge energy levels can be used as indication of stability against photodecomposition.

For an n-type semiconductor, the quasi-Fermi levels are schematically shown in Figure 8.2. For ${}_{n}E_{dec} > {}_{n}E_{F}{}^{*}$, the semiconductor is stable against cathodic photodecomposition induced by electrons in the absence of an

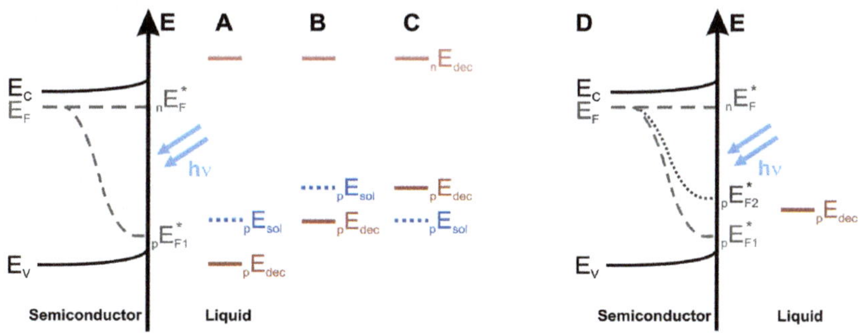

Figure 8.2 Schematic band diagram of the n-type semiconductor–electrolyte interface under illumination. Susceptibility of the semiconductor to photodecomposition: (A) stable, (B) protected by solvent, and (C) unstable. (D) Influence of low (dotted line) and high (dashed line) illumination intensities on the thermodynamic stability.

external voltage or critically high illumination intensities. For the anodic photodecomposition induced by holes, several different scenarios can be depicted. For ${}_{\mathrm{p}}E_{\mathrm{dec}} < {}_{\mathrm{p}}E_{\mathrm{F}}{}^{*}$, the minority carriers can react only with the electrolyte resulting in a stable semiconductor (Figure 8.2(A)). For ${}_{\mathrm{p}}E_{\mathrm{dec}} > {}_{\mathrm{p}}E_{\mathrm{F}}{}^{*}$, anodic photodecomposition by holes is possible. However, for ${}_{\mathrm{p}}E_{\mathrm{sol}} > {}_{\mathrm{p}}E_{\mathrm{dec}}$, the oxidation reaction of the electrolyte is thermodynamically preferred, and the semiconductor will be thermodynamically stable against photo-induced oxidation (Figure 8.2(B)). By contrast, for ${}_{\mathrm{p}}E_{\mathrm{sol}} < {}_{\mathrm{p}}E_{\mathrm{dec}}$, the photo-induced decomposition of the semiconductor is favorable (Figure 8.2(C)). As mentioned above, when the illumination intensity increases, the quasi-Fermi levels shift energetically further away from the bulk Fermi level, thus their alignments change with respect to the decomposition levels.[11] Figure 8.2(D) illustrates how the thermodynamic stability depends on the illumination intensity, where ${}_{\mathrm{p}}E_{\mathrm{F}1}{}^{*}$ is the quasi-Fermi energy level at higher illumination intensity and ${}_{\mathrm{p}}E_{\mathrm{F}2}{}^{*}$ is the quasi-Fermi energy level at lower illumination intensity.

For photoelectrochemical applications polycrystalline materials are often used. These materials exhibit structural defects such as grain boundaries and dislocations that can play a decisive role in the decomposition process.[12] While the ideal material may be stable according to thermodynamic considerations, defective regions can be characterized by a different decomposition level and they can act as initiation sites for the degradation process of the semiconductor. These simple considerations already indicate that for good decomposition suppression the decomposition energy levels of an ideal lattice, as well as of defectives sites in and on the crystal, have to be considered. For a better prediction of stability, accurately described surface defects have to be implemented in the stability models and the corresponding decomposition energy levels need to be estimated.[12]

Wang *et al.* applied these thermodynamic considerations to semiconducting metal oxides for solar water splitting by comparing valence/conduction band edges and oxidation/reduction levels relative to the water redox potentials.[13] For solar water splitting, ${}_{\mathrm{n}}E_{\mathrm{sol}}$ corresponds to the energy level for water reduction $\phi(H^+/H_2)$, whereas ${}_{\mathrm{p}}E_{\mathrm{sol}}$ represents the energy level for water oxidation $\phi(O_2/H_2O)$. The materials' stability against photocorrosion is predicted by calculating thermodynamic oxidation and reduction potentials in the presence of an electrolyte solution. This approach combines *ab initio* calculations of compound formation energy and band alignment with experimental results from electrochemistry, representing a good approximation of how to predict materials stability under operating conditions.[13]

In addition to the above considerations, pH conditions and the applied electrochemical potentials strongly influence the stability of semiconducting materials. Depending on these conditions, different reactions are favored and soluble species or self-passivation layers can be formed. In this context, Pourbaix diagrams are typically used to depict the equilibrium between solution and solid stable phases as a function of the electrochemical potential and of the pH value of the solvent.[2,14,15] Through thermodynamic calculations, it is possible to predict possible (electro)chemical reactions and the predominantly formed species. While these diagrams are powerful for providing information about the thermodynamically stable species at a given potential and pH, kinetics considerations involved in the corrosion process are more complex and require careful studies. Generally, Pourbaix diagrams represent a good approximation for pure elements. However, multi-element semiconductors (*e.g.* metal oxides) are difficult to accurately describe through potential-pH diagrams.

8.2.2 Kinetic Aspects

The above thermodynamic considerations provide general guidelines to assess the stability of semiconductor/electrolyte systems. However, to predict stability with more confidence, it is necessary to consider reaction kinetics, which requires detailed knowledge of all individual reaction steps for both the corrosion reaction and the targeted photoelectrochemical reaction, *e.g.* water splitting.[4,5,11,13] Thermodynamically stable materials are resilient to decomposition independent of kinetic processes. However, thermodynamically unstable materials can be rendered kinetically stable, for example by forming a protective oxide layer, thus preventing further oxidation.[13] Thermodynamic stability considerations are based on the change of Gibbs free energy of the overall reaction, which is an average over all the different intermediate steps. However, (electro)chemical reactions typically proceed in several steps, each having a different rate constant and a different free energy consumption. Therefore, the overall reaction rate is limited by the step with the largest free energy consumption and the highest free energy barrier.

In the following section, we discuss how the splitting of the overall decomposition potential in a two-step reaction modifies the susceptibility towards photodecomposition. In most multi-electron transfer reactions, the first reaction step acts as the rate-determining step.[4,11] Figure 8.3 shows the decomposition energy levels of holes ${}_{p}E_{dec\text{-}1}$ and ${}_{p}E_{dec\text{-}2}$ for two subsequent reaction steps that contribute to the total potential of a decomposition reaction ${}_{p}E_{dec}{}^{0}$. In this example, the potential of the first reaction step is located below the decomposition potential of the solvent ${}_{p}E_{sol}$. Accordingly, the decomposition reaction of the solvent is favorable, which is in contrast to the thermodynamic stability prediction based on the average decomposition potential ${}_{p}E_{dec}{}^{0}$ of the total reaction. Since the equilibrium reaction rates depend exponentially on the free energy difference, the first step of the semiconductor decomposition will proceed at a much smaller rate than the decomposition of the solvent, which limits the overall corrosion process.[4,11]

8.3 Degradation Mechanisms of Semiconductor Materials

Under the strong reductive and oxidative conditions involved in photoelectrochemical reactions, semiconducting materials with interesting

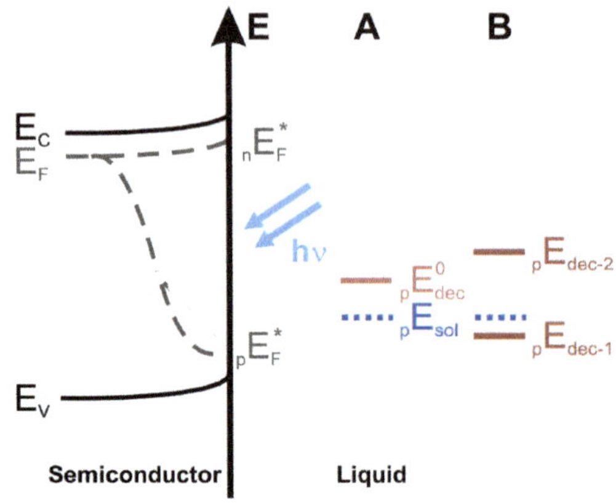

Figure 8.3 Schematic energy diagram for an illuminated semiconductor in contact with a solvent. The total decomposition reaction $_pE_{dec}{}^0$ is split into two separate partial reactions $_pE_{dec\text{-}1}$ and $_pE_{dec\text{-}2}$. Their corresponding decomposition levels are located below and above the decomposition potential of the solvent $_pE_{sol}$.

absorption properties and favorable band alignment are often subject to degradation. Degradation processes can lead to undesirable changes of the material properties such as chemical composition, absorption behavior, surface properties, charge carrier generation and transport.[1] Thus, long-term operation may affect performance, lifetime, and reliability of the entire photoelectrochemical device. In the following section, we discuss corrosion, intercalation, hydroxylation, and chemical destabilization as the most common examples of degradation phenomena at the liquid/solid interface.

8.3.1 Corrosion

Compared to the corrosion of elemental metals, the corrosion of semiconductors is even more complex. Many semiconductors contain more than one element, and corrosion occurs *via* multiple reaction mechanisms, each governed by different electrochemical equilibria. Furthermore, the number of electrons and holes participating in the photoelectrochemical processes can vary depending on the electronic properties of the semiconductor, the composition of the solution, the illumination intensity, and the biasing conditions.[2] Overall, three different types of corrosion at the liquid/solid interface can be distinguished: chemical, electrochemical, and photoelectrochemical corrosion.

Chemical corrosion leads to degradation or dissolution of the semiconducting electrode caused by the chemical attack of the electrolyte. Different from electrochemical and photoelectrochemical corrosion, chemical corrosion mechanism does not require a net charge transfer across the liquid/solid interface, thereby occurring also under dark conditions. Electrochemical corrosion is based on reaction with majority charge carriers, whereas photoelectrochemical corrosion involves photogenerated minority charge carriers. The thermodynamics aspects of these types of corrosions are described in the previous sections.

8.3.2 Intercalation and Hydroxylation

The performance of PEC devices is also affected by proton intercalation and hydroxylation. In the presence of free electrons at or near the surface, protons from the electrolyte can intercalate into the semiconductor lattice.[16] For non-porous TiO_2 films, for example, proton intercalation induces shallow trap levels for electrons with a long residence time, which strongly affects the activation energy for electron diffusion and slows down the transport of photogenerated charges.[17] At the liquid/solid interface, surface hydroxylation considers the continuous adsorption and desorption of protons and/or hydroxide species at the semiconductor surface.[1,16,18] The net total charge adsorbed at the surface depends on the pH of the solution and the Brønsted acidity of the surface.[16] Surface hydroxylation can affect the charge distribution at the liquid/solid interface, and the potential distribution.

8.3.3 Chemical Destabilization

Chemical destabilization is a kinetic effect which prevents the formation of a self-passivating layer on the surface of the photoelectrode and consequently results in the dissolution of bulk material. Chemical destabilization is attributed to the accumulation of photoexcited charge carriers in the lattice, which leads to changes in the semiconductor crystallinity, grain size, or composition.[1] For the specific case of bismuth vanadate photoelectrode, the degradation process is discussed in more detail in Section 8.4.3.

8.4 Investigation of Material Instability

State-of-the-art operational lifetimes for prototype PEC devices are around 2500 sun-hours, which are much shorter than commercially relevant operational times (~30 000 sun-hours).[1,19] Furthermore, operational stability is often much lower than this state-of-the-art limit and is frequently not reported. To overcome present stability limitations and to develop strategies for improving durability, reliability, and efficiency, a comprehensive understanding of degradation mechanisms as well as their impact on the device performance is required. The most commonly accepted figure of merit to quantify device stability and device lifetime is the variation of the photocurrent as a function of time. Photoelectrochemical measurements such as chronoamperometry or cyclic voltammetry are used to test the device performance under dark conditions as well as under illumination. Additionally, structural, morphological, and compositional changes that contribute to the corrosion process have to be considered. In order to provide a complete picture of the corrosion process, a combination of different, complementary characterization techniques is required to diagnose the underlying degradation mechanisms. Changes in the crystalline phase and unit cell dimensions are determined by X-ray diffraction (XRD). The morphological evolution of the semiconductor can be observed using microscopy techniques such as scanning electron microcopy (SEM) or atomic force microscopy (AFM). Rutherford backscattering spectrometry (RBS) and X-ray photoelectron spectroscopy (XPS) provide access to changes in the chemical composition in the bulk and at the surface, respectively. In addition, these measurements can be complemented by inductively coupled plasma mass spectroscopy (ICP-MS), which measures the dissolved species concentration in the electrolyte.

In the following sections, we describe how the use of different characterization techniques, such as those mentioned above, can help to identify the degradation mechanisms of commonly used p- and n-type metal oxides such as Cu_2O, TiO_2, and $BiVO_4$.

8.4.1 Cuprous Oxide

Cuprous oxide (Cu_2O) is a non-toxic, earth abundant, p-type semiconductor with suitable band alignment for water and CO_2 reduction, a direct band gap in the visible range of about 2 eV,[20,21] and a theoretical maximum photocurrent density of $-14.7\ mA\,cm^{-2}$.[22] These characteristics make Cu_2O a suitable photocathode material in photoelectrochemical devices. However, Cu_2O is very unstable under operating conditions due to the availability of different oxidation states in copper. The instability is related to the fact that the Cu^+ cations exhibit an intermediate valence state between Cu^0 and Cu^{2+}. Thus, the Cu^+ cation can be further oxidized to a higher valence state (Cu^+ to Cu^{2+}). For example, in comparison to Cu_2O, $CuWO_4$ with Cu^{2+} exhibits a much lower oxidative decomposition level, since both cations are in the highest valence state. Accordingly, cations that are not in the highest valence state have the tendency to be susceptible to oxidation and to be unstable.[13]

The corresponding energy levels for anodic (to CuO) and cathodic (to Cu) decomposition of Cu_2O forming Cu^{2+} and Cu^0 valence states are located above/below $\phi(O_2/H_2O)/\phi(H^+/H_2)$, in the middle of the band gap (Figure 8.4(A)).[6,13] Wang *et al.* even reported ${}_{n}E_{dec} < {}_{p}E_{dec}$ which is in contrast to most other materials. Typically, ${}_{n}E_{dec}$ is located above ${}_{p}E_{dec}$, since the oxidative decomposition takes electrons from the valence band and the reductive decomposition adds electrons to the conduction band. In the case of Cu_2O, the reverse order of ${}_{n}E_{dec}$ and ${}_{p}E_{dec}$ enables the spontaneous generation of electron/hole pairs without illumination causing the material to decompose even under dark conditions.[13]

The Pourbaix diagram for copper (Figure 8.4(B)) reveals that copper tends to oxidize, and dissolves at neutral and positive applied bias depending on the pH of the electrolyte. Cu_2O is mainly generated within a narrow potential window and in a pH range of 8–12. In addition, at positive biases at

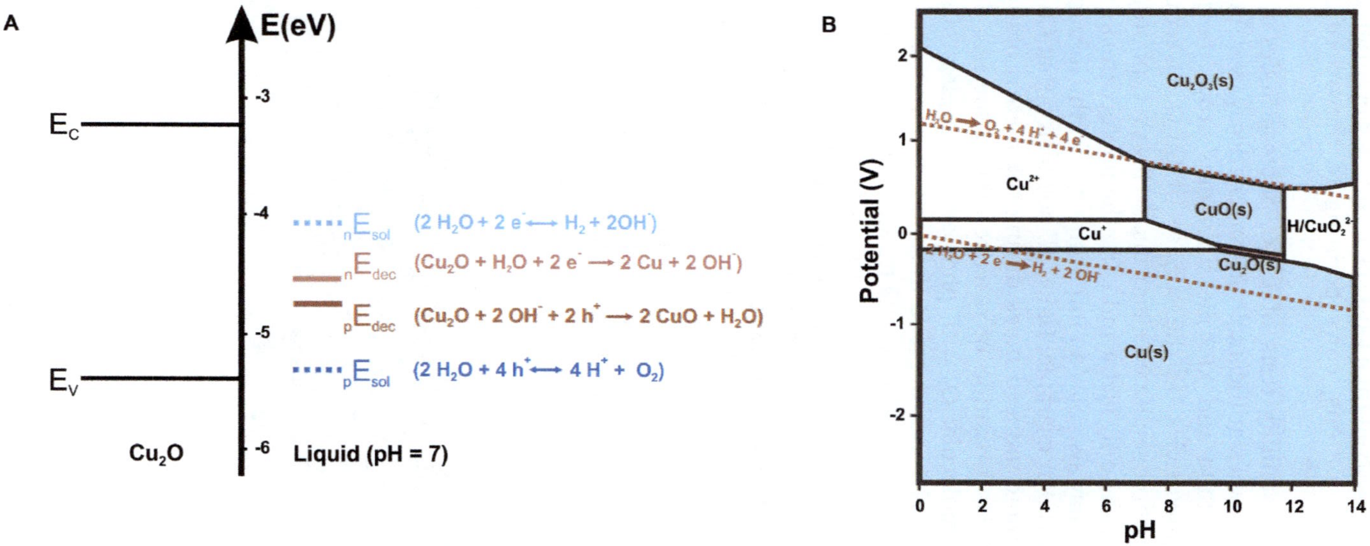

Figure 8.4 (A) Schematic energy diagram of the Cu_2O/electrolyte interface. Redox potentials of the relevant reactions for hydrogen and oxygen production, as well as decomposition of Cu_2O in neutral aqueous solution (pH = 7). (B) Pourbaix diagram of copper generated using 10^{-8} mol kg^{-1} concentration or aqueous species. The diagram is calculated using the hybrid calculated-experimental Pourbaix formalism of the Materials Project.[40,50] The formation of stable solid compounds are highlighted in blue, Cu_2O in red, and stable ions in aqueous solution in white.

near-neutral to basic pH, copper is present as CuO. Indeed, a reversible anodic current can be observed for a Cu_2O electrode at potentials above 0 V *vs.* RHE under dark conditions, corresponding to the oxidation of the Cu_2O.[23]

Chronoamperometry stability tests of Cu_2O photocathodes under illumination commonly show a significant photocurrent decay over time.[23] After PEC characterizations, SEM and XPS measurements prove the formation of elemental Cu nanoparticles on the surface of Cu_2O by reductive decomposition through photogenerated electrons.[22] However, photogenerated electrons might not be the only factor inducing the photocorrosion of Cu_2O. Photogenerated holes may also play an important role in the initiation of the photocorrosion process. For example, Dai *et al.* postulated that Cu_2O can be oxidized to CuO by photo-induced holes during the methyl orange photodegradation reaction *via* formation of reactive oxygen species, due to the reaction of adsorbed molecular O_2 with photogenerated electrons.[24] While the reactive oxygen species can degrade the methyl orange dye, the accumulated photogenerated holes can oxidate Cu_2O to CuO. Chen *et al.* further demonstrated that Cu^{2+} ions were present after 1 h irradiation, which resulted from the oxidation of Cu_2O by photogenerated holes.[25] In addition, morphological changes are observed during the transformation from Cu_2O to CuO. Specifically, the transition between the presence of a dense structure to a network of leaf-like crystals plays a critical role in the degradation of its photocurrent (Figure 8.5).[26]

The role of photogenerated holes in the photocorrosion process can be investigated in more detail by comparing photoelectrochemical performance with front *versus* back illumination. Specifically, under back illumination, chronoamperometry tests show an improved stability of Cu_2O

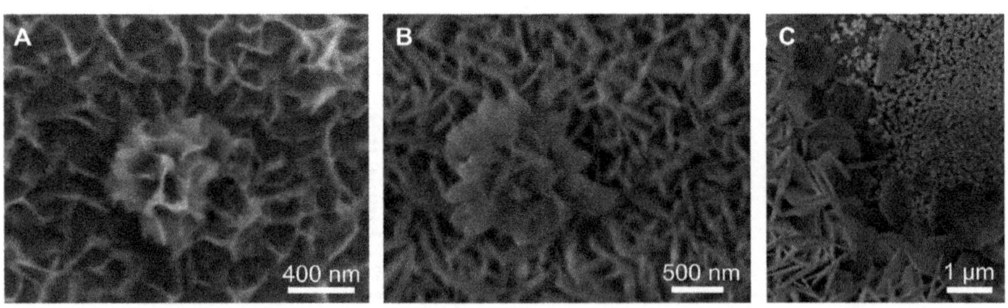

Figure 8.5 SEM images of morphological changes of Cu_2O after cyclic electrochemical polarization. (A) The initial stage of the morphological transformation involves the formation of a loose network of thin leaf structures on top of the grains after five cycles. (B) The leaf structure is the final form observed after 20 cycles. In the center of (A) is a structure that is likely a precursor of the flower structure seen in (B). (C) Leaf-like structures were also obtained after potentiostatic UV-Vis illuminated conditions. Reprinted with permission from L. Wu, L. Tsui, N. Swami and G. Zangari, *J. Phys. Chem. C*, 2010, **114**, 11551, Copyright 2010 American Chemical Society.[26]

and a higher photocurrent signal after 10 min of operation than under front illumination. Notably, in this configuration the travel distance of electrons to the Cu_2O/electrolyte interface is much longer than that one of holes to the back electrode, thereby indicating a more adverse effect of holes on the Cu_2O stability compared to electrons.[27] To date, the deactivation mechanism of Cu_2O as a photocathode under illumination is still unclear. It will be highly important to investigate in more detail the degradation of Cu_2O under operating conditions with advanced measurement techniques and methods.

8.4.2 Titanium Dioxide

Titanium dioxide (TiO_2) is an extensively investigated n-type semiconductor as a photoanode for water oxidation (Figure 8.7(A)).[6,13] Depending on the phase, the band gap changes slightly from 3.0 eV for rutile and 3.2 for anatase TiO_2,[28–30] with relative maximum photocurrent density ranging between 1.8 and 1.1 mA cm^{-2}.

The cathodic decomposition level ${}_nE_{dec}$ of TiO_2 is located above the conduction band edge E_C, which prevents electrons from reducing the semiconductor (Figure 8.6(A)). The anodic decomposition level ${}_pE_{dec}$ is located above the valence band edge E_V, and it is clearly lower than the energy level for water oxidation $\phi(O_2/H_2O)$.[6,10,13] For high illumination intensity, ${}_pE_F{}^*$ can be lower than the energy level for decomposition ${}_pE_{dec}$ of TiO_2. Under these conditions, as depicted in Figure 8.2(B), both oxidation of the solvent and of the semiconductor are possible. However, the much faster reaction kinetics of the oxygen evolution reaction can prevent TiO_2 from anodic decomposition.

The Pourbaix diagram for titanium (Figure 8.6(B)) predicts that TiO_2 is thermodynamically stable within a large potential range and for any pH value. Under these reaction conditions, the dissolution kinetics are slow, however, they will be more pronounced at more reducing potentials and in more acidic pH ranges.[2]

Although TiO_2 has the potential to be a very stable material under operating conditions, and several reports show its promising resilience as a protective layer,[31–33] Li *et al.* found that TiO_2 photoanodes, including rutile nanowires, anatase nanotubes, and P25 nanoparticle films, suffer from substantial photocurrent decay in neutral, as well as basic, electrolyte solution.[34] Cyclic voltammetry measurements of rutile TiO_2 before and after 12 h stability testing result in a decreased photocurrent density of about 0.1 mA cm^{-2} with a constant onset potential. Chronoamperometry tests reveal significant photocurrent decay of 20%, 14%, and 50% for rutile, anatase nanotubes, and P25 nanoparticle TiO_2, respectively, over 12 h. Thus, these experiments show that TiO_2 is not as stable as expected. This finding is in agreement with thermodynamic considerations (Figure 8.6(A)), as the formation of higher oxidation states of Ti species may be favored over the water oxidation reaction.

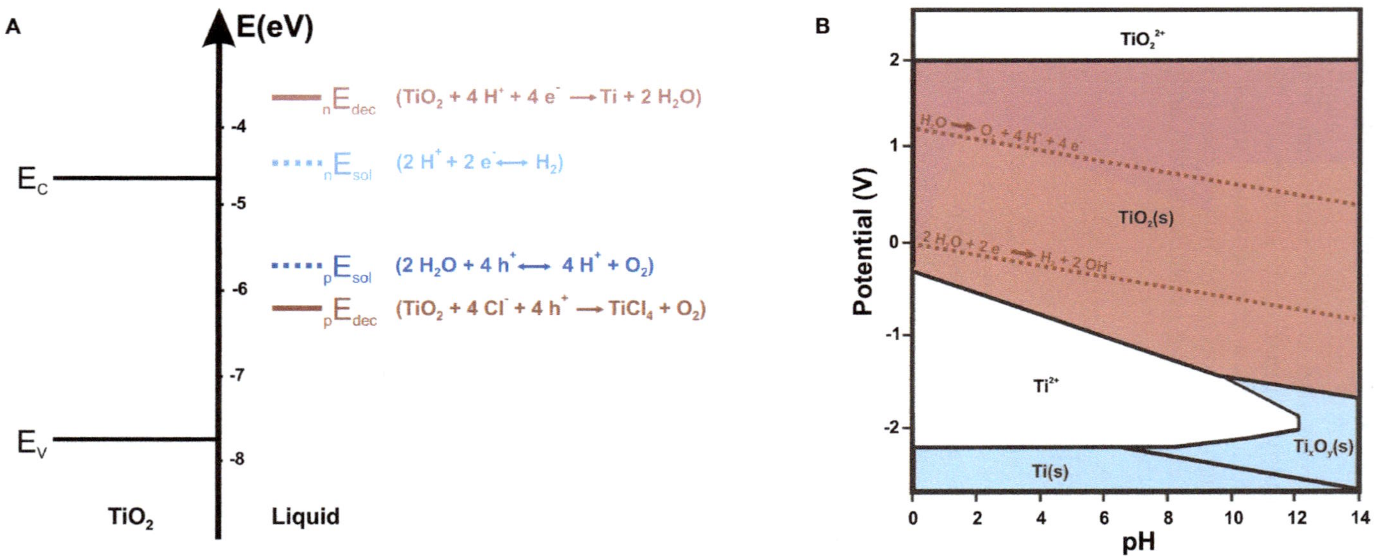

Figure 8.6 (A) Schematic energy diagram showing the correlation between band edges of TiO_2 and energy level for the decomposition $_{n,p}E_{dec}$ and redox reactions $_{n,p}E_{sol}$. (B) Pourbaix diagram of titanium generated using 10^{-8} mol kg^{-1} concentration for aqueous species. The diagram is calculated using the hybrid calculated-experimental Pourbaix formalism of the Materials Project.[40,50] The formation of stable solid compounds are highlighted in blue, TiO_2 in red, and stable ions in aqueous solution in white.

Figure 8.7 SEM images of a TiO_2 photoanode before and after 12 h stability test. Reprinted with permission from Y. Yang, Y. Ling, G. Wang, T. Liu, F. Wang, T. Zhai, Y. Tong and Y. Li, *Nano Lett.*, 2015, **15**, 7051, Copyright 2015 American Chemical Society.[34]

XRD measurements before and after stability testing do not show any changes in the X-ray diffraction spectra, excluding modification in the chemical composition or crystal phase. However, SEM images reveal a reduced thickness of 0.4 μm, surface roughening, and reconstruction on the TiO_2 surface as the oxygen evolution reaction progresses (Figure 8.7), thereby supporting the degradation hypothesis of TiO_2.

Control experiments applying a constant current density of 0.8 mA cm^{-2} under dark conditions clearly show no morphological changes, thus allowing assignment of the instability of TiO_2 to photo-induced corrosion initiated by surface trapped holes. The surface reconstruction can be attributed to the production of active O^- anions by photogenerated holes and the adsorbed O^{2-} anions that occupy gradually the vacancies of TiO_2 film surface/subsurface. In acidic solutions, because of the difficulty of oxidizing hydronium and the limited hydroxide ion concentration, active O^- anions cannot be produced efficiently, thus the electrodes are quite photostable.[35] The newly formed surface layer significantly slows down the surface kinetics for water oxidation and results in the observed decay of the photocurrent.

8.4.3 Bismuth Vanadate

Bismuth vanadate ($BiVO_4$) is a promising photoanode material for photoelectrochemical water splitting due to its favorable band alignment to the water oxidation potential, and a band gap of about 2.5 eV.[36,37] It can generate a large photovoltage (>1 V), and has a theoretical maximum photocurrent density of about 7 mA cm^{-2}.[36,38,39] Calculations of band edges and decomposition levels by Wang *et al.* predict thermodynamic stability of $BiVO_4$ against anodic decomposition.[13] Indeed, the anodic decomposition level $_pE_{dec}$ is reported to be located above E_V but slightly lower than $\phi(O_2/H_2O)$. Thus, the water oxidation should be favored and $BiVO_4$ may be relatively stable against oxidative decomposition.

However, evidence from our group proves unexpected changes in the photoelectrochemical performance of $BiVO_4$ thin films deposited by means of different techniques on fluorine-doped tin oxide (FTO) coated glass substrates. The *J–E* curves under near-neutral (Figure 8.8(A)) and alkaline (B) pHs show an anodic shift in the photocurrent onset potential of about 0.1 V, and a reduction in both the current density and the fill factor.[38] In addition, chronoamperometry measurements reported in Figure 8.8(C) reveal a decrease of the photocurrent density and of the catalytic activity for water oxidation over time. These experimental results already suggest photocorrosion of the $BiVO_4$ film.

In addition, we demonstrate that these functional changes are accompanied by morphological changes of the examined electrodes. $BiVO_4$ thin films are studied *ex-situ* by SEM comparing pristine and degraded samples using similar test conditions as discussed before. Figure 8.9(A) shows the topography of a pristine sample with grain sizes between 50 and 200 nm. After stability testing at pH 6.8 (Figure 8.9(B)), the grains become rounded and voids form at intergrain boundaries. After stability testing at pH 12.3 (Figure 8.9(C)), almost the entire $BiVO_4$ layer is dissolved and large regions of the underlying FTO back contact are exposed.

Changes in the chemical composition at the grain surface as well as in the bulk material are also detected and determined by means of XPS and RBS, respectively. For pristine $BiVO_4$ samples, RBS measurements reveal near-stoichiometric composition in the bulk. However, on the surface, a non-stoichiometric composition is detected by XPS. The surface exhibits an excess of Bi (23.5 at.%) relative to V (16.3 at.%), which might affect the stability against photocorrosion. After chronoamperometric stability testing in alkaline environments, the surface is enriched with Bi (25.3/26.7 at.% at pH 6.8/12.3) and is slightly depleted of V (15.4/14.0 at.% at pH 6.8/12.3). Notably, similar surface and bulk compositions are found after testing $BiVO_4$ photoanodes under dark conditions at an applied bias of 1.23 V *vs.* RHE. These experimental results suggest extensive chemical etching and indicate chemical instability of $BiVO_4$ thin films even under dark conditions.

Complementary to XPS and RBS measurements, the chemical composition of the electrolyte solution is analyzed after stability testing of $BiVO_4$ by means of ICP-MS, which confirms the loss of about stoichiometric quantities of Bi and V into solution. Thus, the determined degradation rates for both Bi and V under illumination and applied bias of 1.23 *vs.* RHE are similar. However, compared to the degradation rates under dark conditions, the degradation under illumination is faster by approximately an order of magnitude. These results indicate that dissolution of the $BiVO_4$ in electrolyte solution is accelerated under illumination and with increasing pH at applied bias.

As already discussed before, materials stability can be predicted by electrochemical potential-pH diagrams (Figure 8.10(A)). Although it is challenging to predict Pourbaix diagrams for complex oxides, the Pourbaix

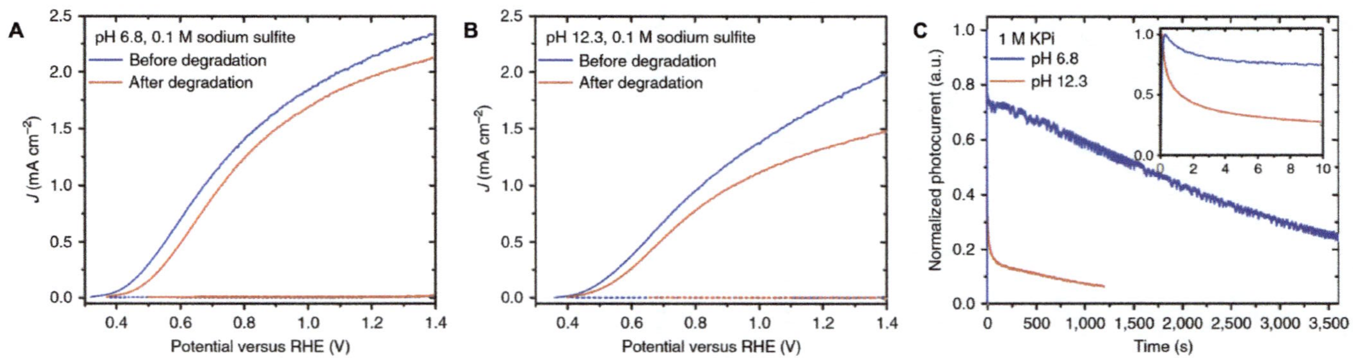

Figure 8.8 Cyclic voltammetry measurements of $BiVO_4$ thin films in solution containing 0.1 M sulfite (A) at pH 6.8 and (B) at pH 12.3 under illumination (solid line) and under dark conditions (dotted line) before (blue) and after (red) stability testing. (C) Chronoamperometry measurements of pristine $BiVO_4$ thin films in pure KPi solution (no sulfite present) at pH 6.8 (blue) and pH 12.3 (red) at a fixed applied bias of 1.23 V *vs.* RHE.
Reprinted by permission from Macmillan Publishers Ltd: *Nature Communications* (ref. 38) Copyright 2016.

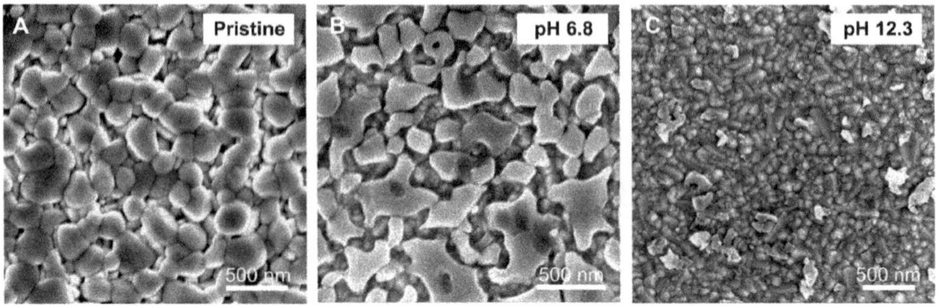

Figure 8.9 SEM plan view images of (A) pristine, (B) degraded at pH 6.8, and (C) degraded at pH 12.3 $BiVO_4$ thin films.

diagram of $BiVO_4$ was calculated with the Materials Project approach.[40] While, according to these calculations, $BiVO_4$ is found to be stable under very limited conditions, the formation of bismuth oxide stable phases are predicted under a wide range of pHs and positive biases. Interestingly, XPS reveals Bi enrichment at the surface following stability testing, which could support the formation of a self-passivating bismuth oxide layer at the surface. However, ICP-MS measurements detect the presence of Bi ions in the electrolyte solution, thereby leading to the conclusion that dissolution of bulk material occurs and the possible formation of bismuth oxide would not prevent corrosion. The formation of a self-passivating Bi–O phase at the surface would require the dissolution of vanadium (*i.e.* 50% of the cations from $BiVO_4$), thus rendering the Bi–O layer highly defective and unstable. Furthermore, the structural rearrangement processes, which are necessary to form bismuth oxide, might be kinetically hindered under room temperature conditions. These experimental considerations allow us to refine our calculations and understand what prevents the possible formation of bismuth oxides. The Pourbaix diagram without bismuth oxide stable phases (Figure 8.10(B)), which were obtained from calculation, support the presence of bismuth soluble species that could more likely form under operating conditions. Interestingly, these considerations are strongly related to and dependent on the surface termination of this material. While the formation of a self-passivating Bi–O phase remains kinetically inaccessible under operating conditions, the initial surface termination as a result of the synthetic process utilized to make the pristine $BiVO_4$ can lead to different degradation rate.

The enhanced corrosion rate of $BiVO_4$ under illumination deserves a separate discussion. In this material, photogenerated holes can accumulate at the valence band maximum, which consists of Bi 6s hybridized with O 2p orbitals.[41] Within monoclinic $BiVO_4$, previous theoretical studies by Kweon *et al.* already reported positive charge carriers localization on BiO_6 polyhedra. The hole charge density is distributed with 22% on Bi cation and 8% on each O anions and balance of charge located outside of the

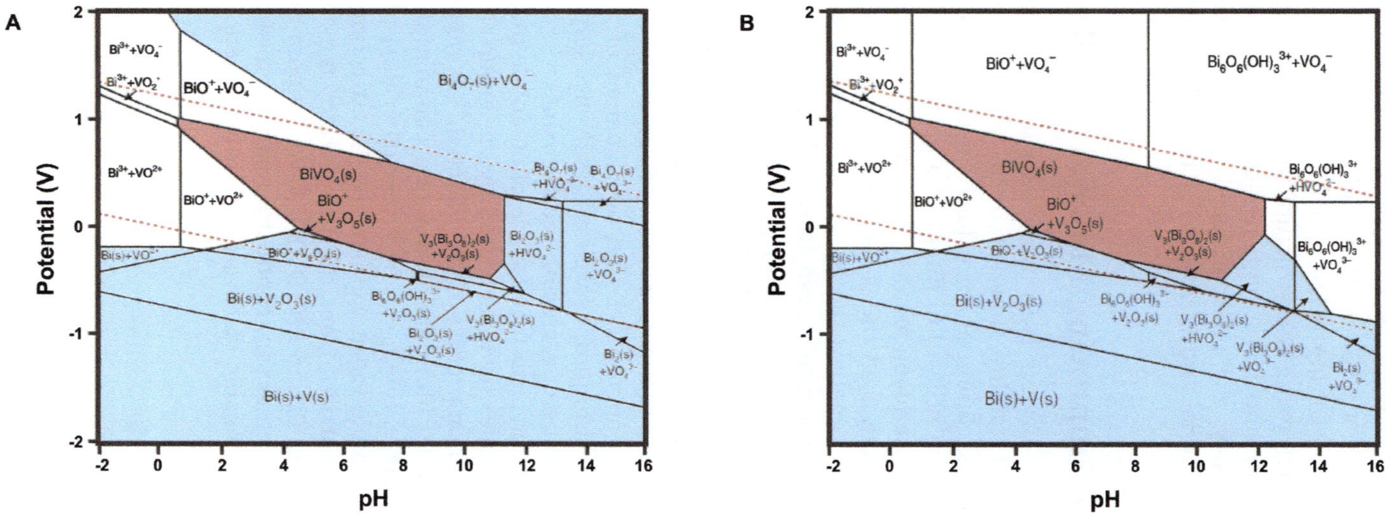

Figure 8.10 (A) Calculated pourbaix diagram of 50–50% Bi–V system in aqueous solution using the Materials Project structure predictor. For the calculation a Bi and V ion concentration of 10^{-5} mol kg^{-1} is assumed. (B) Pourbaix diagram from (A) with Bi_2O_3 and Bi_4O_7 phases removed. The upper/lower red dotted line represents the potential for the oxygen evolution reaction (OER) and for the hydrogen evolution reaction (HER), respectively. The blue regions highlight stable solid compounds, the red region can be assigned to $BiVO_4$, and the white region mark stable ions in aqueous solution.
Reprinted by permission from Macmillan Publishers Ltd: *Nature Communications* (ref. 38) Copyright 2016.

polyhedron.[42] By considering the impact of the modified charge density, we hypothesize a scenario in which Bi^{3+} oxidizes to Bi^{5+}, a potentially stable oxidation state for Bi. To assess the stability of the formed oxide framework, we predict the resulting oxide structure as $BiVO_5$ containing Bi^{5+} and V^{5+} valence states using the structure predictor algorithm.[43,44] However, $BiVO_5$ is unstable and has a high tendency to decompose into $BiVO_4$ and $\frac{1}{2}$ O_2. Accordingly, the photo-induced oxidation event on Bi further destabilizes the material and increases the rate of corrosion under operating conditions.

8.5 Strategies for Improving Material Stability

The main challenge of PEC devices is to provide good performance while ensuring long term material stability. However, metal oxide semiconductors suffer from degradation, dissolution or corrosion. Currently, trial-and-error approaches have mainly been utilized to integrate a protective coating and stabilize otherwise unstable semiconductors. For example, semiconductors are often coupled with a catalyst on the surface that can lower the kinetic barriers of the reaction, and enables the actual charge transfer to the reaction site in the electrolyte. The catalyst can act directly as a protection layer mitigating the direct semiconductor/electrolyte interface.

However, to improve the desired target for stability and enable technology impact of PEC devices, insights into degradation mechanisms and improved understanding of the reaction processes will aid the discovery of tailored protective coatings and novel semiconductors with improved stability.[2,45] Multimodal characterization approaches and development of *in situ* and *operando* characterization techniques, as well as of advance level of theory can lead to deep understanding of corrosion mechanisms.

For $BiVO_4$, which is a prototypical metal oxide photoanode material, several characterization techniques have been used by our group to provide a detailed study of the dissolution process of this material.[38] As we proved that corrosion is accelerated under illumination due to accumulation of photogenerated holes at the surface, $BiVO_4$ can be protected against photocorrosion by preventing the accumulation and enhancing the extraction efficiency of photogenerated charge carriers from the near-surface region *via* integration of catalysts such as CoPi and FeO(OH).[46–49] We propose that similar studies can be conducted on other semiconductor materials to shed light on the corrosion processes and enable discovery of highly efficient and stable PEC devices.

Acknowledgements

This book chapter preparation and composition is based on work supported by the Laboratory Directed Research and Development Program of Lawrence Berkeley National Laboratory under US Department of Energy contract number DE-AC02-05CH11231 (JE and FMT), and by the Joint Center for

Artificial Photosynthesis, a DOE Energy Innovation Hub, supported through the Office of Science of the US Department of Energy under Award Number DE-SC0004993 (GL). The cited original work of FMT *et al.* is supported by and cited original works by Joint Center for Artificial Photosynthesis, a DOE Energy Innovation Hub, supported through the Office of Science of the US Department of Energy under Award Number DE-SC0004993.

References

1. F. Nandjou and S. Haussener, *J. Phys. Appl. Phys.*, 2017, **50**, 124002.
2. S. Hu, N. S. Lewis, J. W. Ager, J. Yang, J. R. McKone and N. C. Strandwitz, *J. Phys. Chem. C*, 2015, **119**, 24201.
3. A. Shinde, D. Guevarra, G. Liu, I. D. Sharp, F. M. Toma, J. M. Gregoire and J. A. Haber, *ACS Appl. Mater. Interfaces*, 2016, **8**, 23696.
4. H. Gerischer, *J. Vac. Sci. Technol.*, 1978, **15**, 1422.
5. L. A. Harris and R. Wilson, *Annu. Rev. Mater. Sci.*, 1978, **8**, 99.
6. H. Gerischer, *J. Electroanal. Chem. Interfacial Electrochem.*, 1977, **82**, 133.
7. A. J. Bard, R. Parsons and J. Jordan, *Standard Potentials in Aqueous Solution*, CRC Press, 1985, vol. 6.
8. S. Trasatti, *Pure Appl. Chem.*, 1986, **58**, 955.
9. H. Reiss and A. Heller, *J. Phys. Chem.*, 1985, **89**, 4207.
10. A. J. Bard, *J. Electrochem. Soc.*, 1977, **124**, 1706.
11. H. J. Lewerenz, *J. Electrochem. Soc.*, 2014, **161**, H3117.
12. K. W. Frese, M. J. Madou and S. R. Morrison, *J. Phys. Chem.*, 1980, **84**, 3172.
13. S. Chen and L.-W. Wang, *Chem. Mater.*, 2012, **24**, 3659.
14. P. Acevedo-Pena, J. Vazquez-Arenas, R. Cabrera-Sierra, L. Lartundo-Rojas and I. Gonzalez, *J. Electrochem. Soc.*, 2013, **160**, C277.
15. M. Pourbaix, *Atlas of Electrochemical Equilibria in Aqueous Solutions*, National Association of Corrosion Engineers, 1974.
16. R. Van de Krol and M. Grätzel, *Photoelectrochemical Hydrogen Production*, Springer, 2012, vol. 90.
17. A. F. Halverson, K. Zhu, P. T. Erslev, J. Y. Kim, N. R. Neale and A. J. Frank, *Nano Lett.*, 2012, **12**, 2112.
18. H. Tamura, K. Mita, A. Tanaka and M. Ito, *J. Colloid Interface Sci.*, 2001, **243**, 202.
19. X. Zhou, R. Liu, K. Sun, K. M. Papadantonakis, B. S. Brunschwig and N. S. Lewis, *Energy Environ. Sci.*, 2016, **9**, 892.
20. A. Musa, T. Akomolafe and M. Carter, *Sol. Energy Mater. Sol. Cells*, 1998, **51**, 305.
21. R. N. Briskman, *Sol. Energy Mater. Sol. Cells*, 1992, **27**, 361.
22. A. Paracchino, V. Laporte, K. Sivula, M. Grätzel and E. Thimsen, *Nat. Mater.*, 2011, **10**, 456.

23. P. E. de Jongh, D. Vanmaekelbergh and J. J. Kelly, *J. Electrochem. Soc.*, 2000, **147**, 486.
24. Z. Zheng, B. Huang, Z. Wang, M. Guo, X. Qin, X. Zhang, P. Wang and Y. Dai, *J. Phys. Chem. C*, 2009, **113**, 14448.
25. Y. Li, X. Yun, H. Chen, W. Zhang and Y. Li, *Phys. Chem. Chem. Phys.*, 2016, **18**, 7023.
26. L. Wu, L. Tsui, N. Swami and G. Zangari, *J. Phys. Chem. C*, 2010, **114**, 11551.
27. X. Chang, T. Wang, P. Zhang, Y. Wei, J. Zhao and J. Gong, *Angew. Chem., Int. Ed.*, 2016, **55**, 8840.
28. D. O. Scanlon, C. W. Dunnill, J. Buckeridge, S. A. Shevlin, A. J. Logsdail, S. M. Woodley, C. R. A. Catlow, M. J. Powell, R. G. Palgrave, I. P. Parkin, G. W. Watson, T. W. Keal, P. Sherwood, A. Walsh and A. A. Sokol, *Nat. Mater.*, 2013, **12**, 798.
29. K. Rajeshwar, N. R. de Tacconi and C. R. Chenthamarakshan, *Chem. Mater.*, 2001, **13**, 2765.
30. T. Luttrell, S. Halpegamage, J. Tao, A. Kramer, E. Sutter and M. Batzill, *Sci. Rep.*, 2014, **4**, 4043.
31. A. Didden, P. Hillebrand, B. Dam and R. van de Krol, *Int. J. Photoenergy*, 2015, 1.
32. J. Choi, J. T. Song, H. S. Jang, M.-J. Choi, D. M. Sim, S. Yim, H. Lim, Y. S. Jung and J. Oh, *Electron. Mater. Lett.*, 2017, **13**, 57.
33. M. T. McDowell, M. F. Lichterman, J. M. Spurgeon, S. Hu, I. D. Sharp, B. S. Brunschwig and N. S. Lewis, *J. Phys. Chem. C*, 2014, **118**, 19618.
34. Y. Yang, Y. Ling, G. Wang, T. Liu, F. Wang, T. Zhai, Y. Tong and Y. Li, Nan*o* Lett., 2015, **15**, 7051.
35. J. L. Cao, Z. C. Wu and J. Q. Zhang, *J. Electroanal. Chem.*, 2006, **595**, 71.
36. J. K. Cooper, S. Gul, F. M. Toma, L. Chen, Y.-S. Liu, J. Guo, J. W. Ager, J. Yano and I. D. Sharp, *J. Phys. Chem. C*, 2015, **119**, 2969.
37. Y. Park, K. J. McDonald and K.-S. Choi, *Chem. Soc. Rev.*, 2013, **42**, 2321.
38. F. M. Toma, J. K. Cooper, V. Kunzelmann, M. T. McDowell, J. Yu, D. M. Larson, N. J. Borys, C. Abelyan, J. W. Beeman, K. M. Yu, J. Yang, L. Chen, M. R. Shaner, J. Spurgeon, F. A. Houle, K. A. Persson and I. D. Sharp, *Nat. Commun.*, 2016, **7**, 12012.
39. Y. Liang, T. Tsubota, L. P. A. Mooij and R. van de Krol, *J. Phys. Chem. C*, 2011, **115**, 17594.
40. A. Jain, S. P. Ong, G. Hautier, W. Chen, W. D. Richards, S. Dacek, S. Cholia, D. Gunter, D. Skinner, G. Ceder and K. A. Persson, *APL Mater.*, 2013, **1**, 011002.
41. J. K. Cooper, S. Gul, F. M. Toma, L. Chen, P.-A. Glans, J. Guo, J. W. Ager, J. Yano and I. D. Sharp, *Chem. Mater.*, 2014, **26**, 5365.
42. K. E. Kweon and G. S. Hwang, *Phys. Rev. B*, 2013, **87**, 205202.
43. G. Hautier, C. Fischer, V. Ehrlacher, A. Jain and G. Ceder, *Inorg. Chem.*, 2011, **50**, 656.

44. A. Jain, G. Hautier, C. J. Moore, S. Ping Ong, C. C. Fischer, T. Mueller, K. A. Persson and G. Ceder, *Comput. Mater. Sci.*, 2011, **50**, 2295.
45. J. Yang, J. K. Cooper, F. M. Toma, K. A. Walczak, M. Favaro, J. W. Beeman, L. H. Hess, C. Wang, C. Zhu, S. Gul, J. Yano, C. Kisielowski, A. Schwartzberg and I. D. Sharp, *Nat. Mater.*, 2016, **16**, 335.
46. F. F. Abdi, L. Han, A. H. M. Smets, M. Zeman, B. Dam and R. van de Krol, *Nat. Commun.*, 2013, **4**, 2195.
47. J. A. Seabold and K.-S. Choi, *J. Am. Chem. Soc.*, 2012, **134**, 2186.
48. L. Chen, F. M. Toma, J. K. Cooper, A. Lyon, Y. Lin, I. D. Sharp and J. W. Ager, *ChemSusChem*, 2015, **8**, 1066.
49. T. W. Kim and K.-S. Choi, *Science*, 2014, **343**, 990.
50. K. A. Persson, B. Waldwick, P. Lazic and G. Ceder, *Phys. Rev. B*, 2012, **85**, 235438.

New Materials and Components

CHAPTER 9

High Throughput Experimentation for the Discovery of Water Splitting Materials

JOHN M. GREGOIRE,*[a,†] DAVID A. BOYD,[a,b] DAN GUEVARRA,[a] JOEL A. HABER,[a] RYAN JONES,[a] KEVIN KAN,[a] MARTIN MARCIN,[a] PAUL F. NEWHOUSE,[a] ANIKETA SHINDE,[a] EDWIN SOEDARMADJI,[a] SANTOSH K. SURAM[a] AND LAN ZHOU[a]

[a] Joint Center for Artificial Photosynthesis, California Institute of Technology, 1200 E. California Blvd., Pasadena, CA 91125, USA; [b] Division of Physics, Mathematics & Astronomy, California Institute of Technology, Pasadena, CA 91125, USA
*Email: gregoire@caltech.edu

9.1 Mission-driven Materials Discovery: Introduction and Strategies

9.1.1 High Throughput Screening for Specific Device Components and Operating Conditions

As highlighted in several chapters of this book, and perhaps most compellingly in the device simulation work, development of a high efficiency solar

[†]Co-authors are listed in alphabetical order.

Energy and Environment Series No. 22
Integrated Solar Fuel Generators
Edited by Ian D. Sharp, Harry A. Atwater and Hans-Joachim Lewerenz

Published by the Royal Society of Chemistry, www.rsc.org

water splitting device is most directly achieved by employing a tandem light absorber and an electrolyte with pH near either 0 or 14. By virtue of being integrated into a multi-project research program with device-oriented objectives, the high throughput experimentation (HTE) project launched with this device framework as a guiding principle, and consequently any materials exploration is related to either the anode (dioxygen evolving) or cathode (dihydrogen evolving) side of the device. The distinction between PV-electrolysis and photo-electrochemistry is another device-level concept that was critical in defining the functional materials of interest. The most compelling materials challenges specific to solar water splitting involve photoelectrochemical processes in which a semiconductor–liquid junction accelerates charge separation, enabling extraction of appreciable photocurrent from a semiconductor that is not of sufficient quality to be competitive for photovoltaic-based generation of electricity. As a solid state materials research group, the four device sub-components provided by this concept for the device architecture were electrocatalysts and photoelectrocatalysts for both the oxygen evolution reaction (OER) and hydrogen evolution reaction (HER). While discovery of an ideal photoelectrocatalyst would circumvent the need for an electrocatalyst for the corresponding reaction, we established a consensus that optimizing device efficiency would likely involve coupling an electrocatalyst to enhance the activity of a photoelectrocatalyst, particularly since known high-efficiency light absorbers were not efficient electrocatalysts and *vice versa*, with Fe-based (hydr)oxides being a notable exception to this rule.

Additional strategic decisions for defining the scope of materials discovery efforts were made based on prospects for making transformative discoveries. Given the availability of several HER electrocatalysts with overpotentials near 0.1 V at 10 mA cm^{-2}, the discovery of superior HER electrocatalysts would have limited impact on device efficiency and would only be transformative upon demonstration of stable operation for a 10^3-h time scale, which is largely incompatible with high throughput research. Since the analogous overpotentials for state-of-the-art OER electrocatalysts were 3–4 times larger and catalyst deactivation was commonly observed on a much shorter time scale, this reaction was chosen as the initial focus for electrocatalyst discovery experiments. A similar situation existed for photoelectrocatalysts with high performance photocathodes based on p-Si setting a high bar for HER photoelectrocatalysts, but even more compelling was the general lack of high-performance photoanodes with a sub-2.8 eV band gap. Focusing on the anode side of the device for photoelectrocatalyst research offered the added benefit of leveraging capabilities—both OER electrocatalysts and photoelectrocatalysts require materials with extreme resistance to oxidation, making metal oxides the most promising class of materials. By considering operational conditions and device components with critically-missing materials, the five-year plan for HTE research focused on the discovery of photoanode materials for operation in either pH 0 or pH 14 electrolytes.

9.1.2 General Strategies for Constructing Experimental Screening Pipelines

A high-level concept for an HTE pipeline is shown in Figure 9.1. The central section of the schematic represents an experimental flow that begins with synthesizing materials libraries and proceeds with screening those materials for the desired functionally and characterizing their chemical and physical properties. The inclusion of "data informatics and distribution" as a component of the experimental pipeline is intended to highlight the critical role that data plays, as indeed the utility of an experimental pipeline is ultimately determined by the utility of its data. The illustrations surrounding this central pipeline in Figure 9.1 denote additional critical aspects, including the necessity for data collection to be performed for each experiment and the principle that the most effective design of materials libraries includes guidance from both previous experiments and theory. The final component of the pipeline is perhaps the most critical and addresses a hard-learned reality in materials discovery—reporting hits from high throughput screening is not sufficient for making substantial impact in a given research field. Due to practical barriers for directed researchers to adopt a line of experimentation with recently-discovered materials, hits from an HTE pipeline must be benchmarked against state-of-the-art materials when possible. In addition, the directed research to further understand the hit materials should commence using materials synthesized using the HTE synthesis tools. Since the HTE pipeline synthesis and screening tools typically differ from traditional techniques, performing traditional experiments using an HTE-synthesized material is the most direct way to validate a discovery and garner the attention of the broader research community. To a large degree, the success of an HTE pipeline hinges upon these validation experiments, creating a critical design principle that the screening experiments emulate

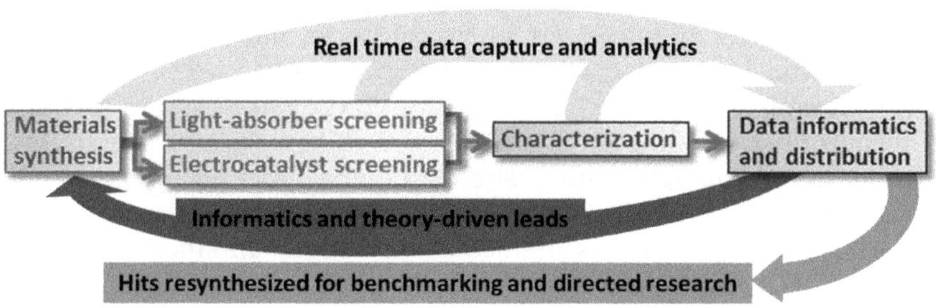

Figure 9.1 High-level concepts of an HTE pipeline where a series of synthesis, screening, characterization, and data techniques are made to operate using standard library formats and with commensurate experimental throughput. The additional critical components for operating the pipeline are also shown: real-time data capture, using previous data and guidance from theory, and synthesizing materials for validating discoveries *via* traditional methods.

the conditions of traditional experiments and device operation to the greatest extent possible.

When considering pipeline throughput (typically measured in number of materials per day), adhering to the pipeline design principles requires that at least one screening experiment emulates the conditions and data quality of traditional experiments. The engineering task then becomes to design an instrument that performs such experiments at the highest possible throughput. If a higher throughput is desired, the options are to create multiple copies of the screening instrument and operate them in parallel, or create less-stringent screening experiments that down-select the materials. In fact, several down-selecting screening experiments can be linked into a "tiered screening pipeline." Since the screening experiment(s) that emulate traditional methods must generally be serial measurements (one material measured at a time), parallel screening measurements can play an important role in initial material screening. While we highlight some parallel screening methods below, we generally found the serial screening techniques to be sufficiently-high-throughput with a sufficiently-high material discovery rate to mitigate the operation of parallel screening techniques. Another critical strategy for developing the pipeline is that the screening instrument that best emulates traditional techniques should be developed first to gain traction by reproducing known results. Also, the availability of a robust technique is necessary to perform feasibility experiments and quality control for more exploratory, higher-throughput screening instruments. The organization of the sections below follow this logical sequence to some extent. These HTE pipeline concepts are broadly-applicable and provide the key guiding principles for the specific pipelines described below.

9.2 Cross-cutting Capabilities: Materials Synthesis and Data Management

Since screening for (photo)electrochemical performance was identified as the most critical aspect of HTE technology for solar water splitting, the screening component of the pipeline required the most intense instrument and technique development. As illustrated in Figure 9.1, the materials synthesis, materials characterization, and data management components of the pipeline can be developed to be more broadly-applicable. We provide some specific examples of materials characterization below and focus on synthesis and data management in the present section. The initial challenge was to identify materials synthesis techniques that provide sufficient material quality with appreciable throughput. While a number of techniques may meet these requirements, two complementary techniques were chosen for the HTE pipeline. Inkjet printing of composition libraries had proven to be quite powerful for exploring solar fuels materials,[1] and the ability of the technique to deposit arbitrary compositions on a discrete grid of sample locations offered the critical capabilities to vary the dimension (number of elements) and granularity (composition resolution) of composition libraries. Due to concerns with the ability to deposit semiconductor-quality thin films using inkjet printing approaches, the complementary technique of combinatorial co-sputtering was also incorporated into the pipeline, as described in further detail below. In either case, the schema for creating an array of materials to be screened or otherwise characterized is referred to as a "plate map."

9.2.1 Inkjet Printing of Functional Metal Oxides

Due to the variety of material morphologies that can result from inkjet printing (IJP) materials, an initial focus was to develop deposition strategies in which a versatile ink chemistry could be used to deposit mixed-metal precursors that, upon calcination, yielded "optical quality" materials. Specifically, each discrete metal oxide composition sample must be: well-mixed (with respect to the different elements or cations incorporated into the composition sample) by avoiding segregation of the precursors; approximately uniform in size and shape (*e.g.* 1 mm $\pm$ 0.2 mm diameter or edge length); and coat the substrate as uniformly as possible. That is, the samples should appear as thin solid films with minimal irregularities on the micron scale and above, requiring optimization of several printing steps including ink formulation, drop reliability, substrate preparation, and ink drying. To attain a high degree of flexibility in these printing steps, a drop-on-demand printer (Microfab Technologies JetLab4) was chosen due its compatibility with a wide range of ink viscosity, pH value, and salt concentration. To leverage the high solubility of many elemental nitrate and chloride salts in aqueous solvent, we established an ink base of 10–20 vol% dietheylene glycol in water. Generally, this yields solutions with the appropriate rheological properties for reliable drop formation, miscibility, and sufficient organic character (*e.g.* "grease") to hinder salt re-crystallization at early, pre-reaction temperatures of the calcination. To achieve IJP "optical quality" samples, the component inks must be sufficiently intermixed yet confined to the appropriate sample area and/or shape before being calcined. In our experience, this has been difficult to effect using primarily aqueous salt solutions on fluorine-doped tin oxide (FTO) glass surfaces, since surface tension effects can drive the inks to wet non-uniformly, resulting in pooled or segregated regions with widely varying shapes and metal loadings. Photocurrent measurements on such samples are difficult to interpret and obfuscate any composition trends. To attain a homogeneous as-printed state, we developed a scalable, inexpensive preparation of the FTO surface *via* spin coating an agarose (sugar) gel, which behaves as a hydrophilic sponge. Printed drops of precursor solution are absorbed by the gel, whose loading is designed so that it is highly hydrated upon completion of jetting into the sample area. Each sample then appears as a raised region of hydrated gel. Critically, the gel allows the printed precursors to intermix, which proceeds until the sample area starts appreciably increasing through lateral expansion of the hydrated area. Slow air-drying of the samples at ~35 °C effectively stops the lateral diffusion of the sample borders. With fully optimized parameters, the dried sample consists of a layer of mixed salts that uniformly coat the sample area due to mitigation of solution convection during drying, which leads to so-called coffee rings due to precipitation of salts at the perimeter of the solution. The small amount of residual agarose is driven off during sample calcination, during which the metal salts are converted to metal oxides. These films perform well in the optical and photoelectrochemical screening experiments described below.

For rapid evaluation of metal oxide catalysts, the uniform coating of the substrate is less critical, enabling our adoption of a modified commercial, professional quality ink-jet printer for preparation of the library of sample spots. The Epson 4880 StylusPro has a high print resolution (up to 2880×1440 dpi), high print rate, and 8-channel parallel ink delivery capabilities. This printer can deposit 10 000 of our usual 1 mm^2 samples in under 15 min. The disadvantage of using a commercial printer designed for high-quality document printing is that the precursor inks we develop for materials library printing must be compatible with the materials used to construct the printer and printhead, and must also have similar wetting, viscosity, and rheological properties as the color document inks the printer is designed to dispense. This limits the selection of solvent, the pH range to near neutral, and the concentration of metal precursors that can be used. Fortunately, we were able to build upon prior precursor ink-development work for use in this printer.[2]

We use several different plate map designs to investigate different composition-dependent properties. For illustrative purposes, we describe the layout and preparation of the discrete composition libraries currently used to investigate OER catalyst properties as a function of composition. Our standard methodology is to prepare a separate ink for each metal precursor, assign each ink to be one of the colors in the CMYK (cyan, magenta, yellow, and black) color scheme, and print a standard composition library as a color document to mix the inks (metals) in the correct ratios to obtain each sample composition. The standard library document for printing the library is generated once, by taking all combinations of $C+M+Y+K=100\%$ at 5% increments of color saturation. This produces a set of 1771 unique colors which are equivalent to 1771 discrete compositions, including all of the pseudo-unary, -binary, -ternary, and -quaternary compositions at 5 at% intervals for any set of four metals (1 metal (4 samples), 2 metals (114 samples), 3 metals (684 samples), and 4 metals (969 samples)). These discrete color mixtures (compositions) are shown in Figure 9.2a mapped onto a tetrahedron where the vertices are pure C, M, Y, and K. In the actual document used to print the composition library, shown in Figure 9.2b, these same color mixtures (compositions) are randomly placed on a grid of 1×1 mm^2 samples on a 2 mm pitch to prevent convolution of the effect of composition on activity with spatial effects arising from sample processing or temporal effects arising from instabilities in sequential electrochemical measurements.

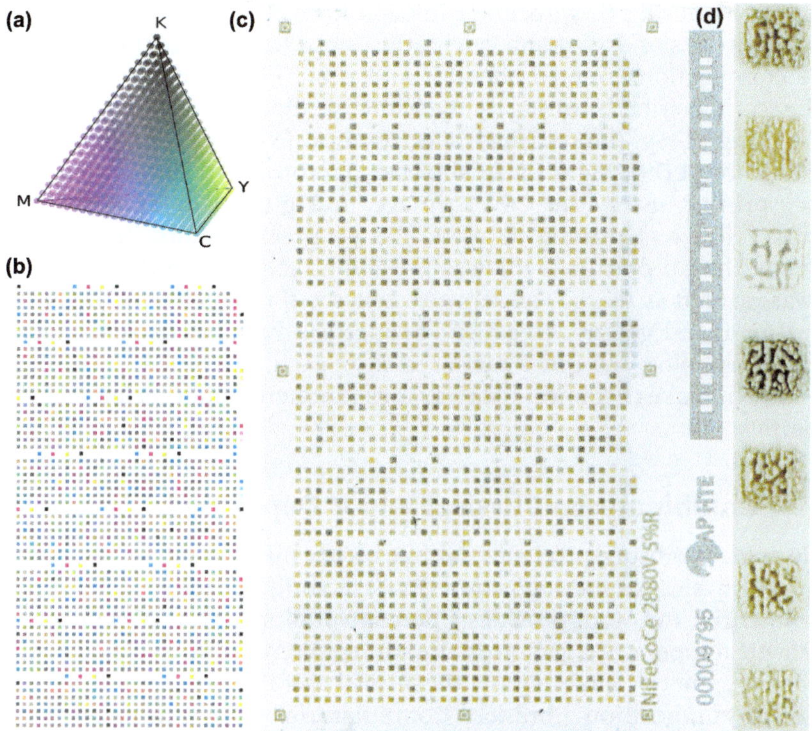

Figure 9.2 Methodology used to produce discrete composition libraries using a high speed commercial printer. (a) The discrete compositions at 5 at% steps shown as mixtures of four colors (cyan, magenta, yellow, and black, *i.e.* CMYK) shown on a quaternary composition tetrahedron. (b) The same colored spots (compositions) randomly distributed as 1×1 mm^2 spots on a 2 mm pitch. (c) The mixed-metal oxide material library generated by printing the document in (b) using equal molar, single metal inks, with the metals assigned to colors as Ni = C, Fe = M, Co = Y, and Ce = K. (d) A zoomed-in view of seven 1×1 mm^2 printed spots of various metal ratios illustrating the non-uniform film thickness produced using the high speed printing approach.

Each ink is prepared to contain the same concentration of metal precursor, type and concentration of structure directing agents and other ink modifiers, such that the set of precursor inks have similar viscosity and rheological properties to each other and the droplets of each ink generated by the printer are consistent in volume. We load each precursor ink into separate ink channels of the printer, and then generate composition mixtures of metal oxide precursor solutions by mixing them on the substrate in the appropriate ratios by mixing C, M, Y, and K inks in the correct ratios. The printer has been modified to enable printing onto rigid substrates. Our substrate of choice for electrochemical measurements under OER conditions is 10 cm by 15 cm plates of Tec-7 F:SnO_2 (FTO)-coated soda-lime glass. After printing the precursor ink solutions, the libraries are processed through a series of drying and calcination steps to generate the mixed-metal oxide compositions free of organics and other contaminants. The fidelity of the composition printing is typically verified at several composition spots using X-ray fluorescence spectroscopy (XRF) or energy dispersive spectroscopy (EDS). Figure 9.2c shows a photo of a resultant composition library printed using the CMYK image in Figure 9.2b in which the Ni precursor ink was printed as cyan, the Fe precursor ink was printed as magenta, the Co precursor ink was printed as yellow, and the Ce precursor ink was printed as black. The samples generated using this method are well suited for investigation of catalytic properties, but the discontinuous and rough morphology of the material within each sample spot (as seen in Figure 9.2d) make them non-ideal for measurement of photoelectrochemical properties.

9.2.2 Combinatorial Physical Vapor Deposition

While inkjet printing produces samples with throughput beyond the capability of physical vapor deposition (PVD) techniques, the custom-built HTE combinatorial PVD tool provides fine control of synthesis parameters such as growth temperature, gas pressure, atmosphere, film thickness, and deposition geometry, enabling synthesis of carefully designed, high-quality thin film composition libraries. Combinatorial PVD techniques typically produce thin film libraries whose composition varies as a function of position across the substrate using either mask deposition or co-sputtering techniques.[3] The mask deposition uses robotically controlled physical masks to control deposition gradients for each deposition source, providing custom designed composition profiles across the library. The mask deposition produces a multilayer film resulting from a series of depositions from different sources, which significantly lowers synthesis throughput and often requires heat treatment to ensure sufficient diffusion for intimate mixing. Sputter co-deposition uses deposition gradients of the different sources to create a continuous composition spread in a single run. It ensures intimate mixing of the library constituents in the as-deposited state, shortens the deposition time, and minimizes system maintenance per composition library.

The HTE PVD tool is designed for combinatorial co-sputtering (Kurt J Lesker, CMS-24), as shown in Figure 9.3, from up to six sources at a time. Six 2-inch KJLC Torus magnetron sputter sources are arranged in 6-fold symmetry at the bottom of chamber and 165 mm away from the center axis of the chamber. Each source has independently computer-controlled shutter and *in-situ* motorized gun tilt assembly, allowing 0 to 24° confocal tilt. The sputtering sources operate under either DC or RF power depending on the type of the material being deposited. The substrate holder accommodates a single 100 mm substrate and the holder assembly is equipped with a boron nitride (BN) coated graphite serpentine filament to provide uniform

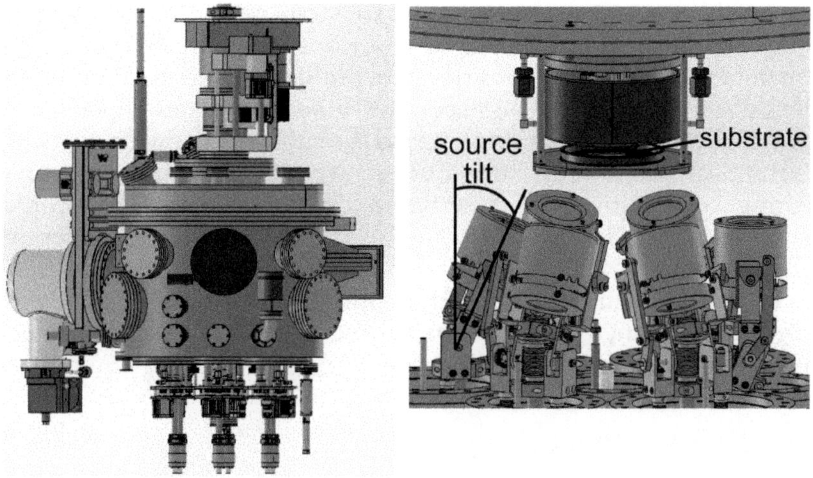

Figure 9.3 Schematic of the HTE co-sputtering PVD system including an external view of the entire chamber and an in-vacuum view of six magnetron sputtering sources with independent motorized tilt and motorized substrate platen height.
This figure is based on work reported in ref. 5.

radiative heating of the substrate, with a temperature range of 25–800 °C. The whole substrate platen assembly is installed on a motorized rotation and translation stage, which allows the source-to-substrate distance to vary. A helium-cooled cryopump (CTI Cryo-Torr 8F) is used to maintain base pressure in the 10^{-8} Torr range. Four separate process gas lines are attached to the chamber through a common gas injection valve, and each process gas flow rate is controlled by pre-calibrated mass flow controllers (MFC) (MKS, 1179A Mass-Flo®) with pressure measured using a capacitance manometer (MKS, 626B BARATRON®). In addition to the dedicated MFC per gas supply, two process gases have additional manually operated variable leak valves (Granville-Philips, series 203) parallel with the respective MFCs, allowing partial pressure tuning at the 1 μTorr level. This capability allows controlled reactive sputtering synthesis of broad classes of materials such as metal oxides, nitrides, carbides, borides, and oxynitrides. Specifically, synthesis of oxynitrides (*e.g.* β-TaON) by deposition rate heuristics based on oxygen and nitrogen partial pressures has been demonstrated.[4]

During co-sputtering, the metal or cation composition spread of the library is typically controlled by tuning source deposition powers and deposition geometry, in particular the distance and angle between the substrate and each deposition source. Since all of these parameters are tunable in the HTE deposition system, we incorporated a 3D deposition profiling capability that involves replacing the substrate platen assembly with a flange containing a deposition rate platen. This assembly contains three 6 MHz Au-coated quartz crystals, which are coplanar with the substrate in the substrate platen and are positioned at the equivalent substrate locations of 0, 28, and 45 mm from the substrate center. At fixed gun power, rotating this deposition rate platen and translating its vertical position in the chamber provides 3D deposition rate profiles that are used in the design of deposition experiments because, unlike some masked deposition techniques, the composition variation across the substrate using co-deposition is not arbitrarily tunable. In Suram *et al.*[5] we present specific examples of designing composition libraries based on these deposition profiles to address specific research needs. As an alternative to this experimental approach to deposition profiling, recent progress has also been made by Bunn *et al.*[6] where in a Nelder–Mead optimization procedure is adapted to determine sputtering conditions for desired deposition rate profiles using a continuum-based sputter model.

9.2.3 Thermal Processing

Given the diverse materials systems of interest for solar fuels materials discovery and the suite of deposition strategies outlined above, high throughput thermal processing with broad flexibility has been an essential capability for HTE research. A basic but critical instrument for thermal processing is the box oven, which is used to calcine IJP composition libraries in air. Greater environmental control is enabled *via* the custom-built library annealing system (LAS) that conducts static sub-atmospheric (total pressure) anneals in O_2 (oxidation), N_2 (nitridation), Ar/H_2 (forming gas, reduction) or Ar (inert) with pressure control down to 30 Torr. The system consists of an array of four tube furnaces, each with a 6 foot long, 6 inch diameter quartz tube and a stable heating zone in excess of 20 inches, which can accommodate up to 48 4×6 inch library plates mounted in custom quartz racks. The LAS enables processing temperatures from ambient up to 1100 °C and includes an active air-cooling system that speeds cool-down and allows for multiple process runs per day to maximize sample throughput. The custom LAS control software incorporates intuitive access to manual controls (gases, valves, vacuum) as well as a selection of automatic sub-routines and processing recipes with detailed attention to operational safety.

To access a different time-temperature processing regime, we also developed rapid thermal processing (RTP) for composition libraries. The HTE RTP can process a glass plate or silicon substrate with IJP samples or PVD materials and can flow gases (oxygen, nitrogen, argon, hydrogen) at atmospheric pressure while ramping to temperatures between 400 and 1000 °C with soak times up to hundreds of seconds. Our capability of rapid processing allows HTE to obtain a range of highly specific phase transformations through precise rapid thermal annealing profiles. RTP is used to very quickly expose IJP samples to higher temperatures than the underlying glass plate, which enables crystallization events to proceed without significant degradation of the underlying FTO conducting layer or softening of the glass plate. The RTP has also been utilized in crystallizing oxynitride films from reactive sputtered deposition. For example, amorphous Ta–O–N thin films deposited in a mixed O_2/N_2 reactive sputtering have been crystallized in β-TaON through post-deposition RTP in an N_2 atmosphere. The brief temperature excursion is sufficient to enable local solid state diffusion without equilibration with the atmosphere so that the anion stoichiometry is fixed by the PVD deposition conditions.[4] This suite of thermal processing tools and techniques provides access to a wide range of materials for screening experiments.

9.2.4 Data Management

Establishment of a custom experimental pipeline requires development of specific data management, processing, and visualization tools. In particular, exploration of high order composition spaces poses unique challenges for statistical analysis and data visualization. We recently summarized our work in statistical analysis and data informatics,[7] and here we summarize the data management tools required for our high throughput experimental pipeline. As noted in Figure 9.1, data management must be tightly integrated with the pipeline experiments, and here we discuss low-overhead data management for instruments that are in the prototyping stage to those operating at consistent throughput.

Due to evolving science objectives, there are frequent modifications to the synthesis, screening, and characterization instruments. In order to manage these changes, sufficient flexibility has to be built into an experimental high-throughput data processing workflow and data management system. The various experiments span a large range of data volume and data formats, and the mediation of experiment down-selection *via* a central data management system requires data processing in quasi-real-time with automated routines. To ensure smooth, cost-effective, day-to-day operation, the system design has to take into account constraints imposed by storage, computing, and human resources. From the start, both raw and processed data are designed and formatted to be easily accessible by all members of the research team with varying degrees of data handling and programming skills. This design allows non-critical and *ad-hoc* data processing to be distributed based on the researchers' particular needs, interests, and comfort in handling the data. However, special attention must be given to ensure these processes are integrated back into the high-throughput pipeline once they become mature to handle a larger data volume. Figure 9.4 illustrates a data management strategy that enables the requisite flexibility. Data management for each pipeline experiment is designed with a common structure but implemented so that it can run independently, with development of linkages among the sub-databases mirroring the experimental linkages in the pipeline. While the data formats and technical description of the data management system will be described elsewhere, we note the importance of adopting a data management strategy as the very first step of building a high-throughput experimental pipeline.

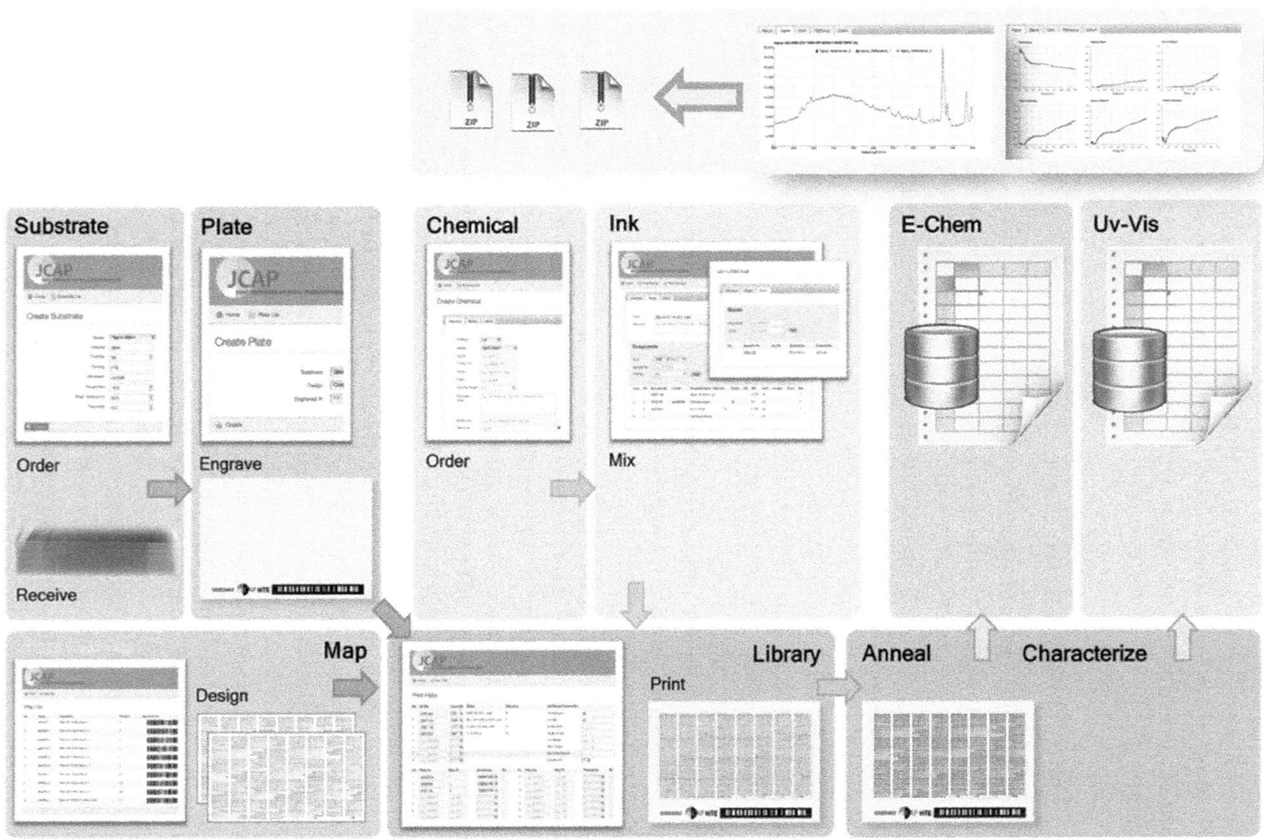

Figure 9.4 This illustration of the data management strategy features database sub-components that are developed alongside the custom engineering of a new high throughput instrument or technique, and as the experiments are assembled into an experimental pipeline, the cross-referencing and high-level database indexing proceeds.

9.3 Experimental Pipeline for Discovering OER Electrocatalysts

9.3.1 The Scanning Droplet Cell and Its Deployment for Electrocatalyst Discovery

The scanning droplet cell (SDC) is a critical component of the experiment screening pipeline for solar fuels materials. The instrument consists of a small volume, three-electrode cell which delivers electrolyte solution to the working electrode in the form of an isolated droplet, thereby increasing serial measurement throughput over traditional gasket-sealed designs.[8] Continuous siphoning at fixed collection points around the droplet achieves isolation of a single composition sample (typically 1 mm^2) as the active working electrode. Custom control software automates measurements through control of the potentiostat (Gamry G 300), translation stages (Dover LM-600), and siphon pumps (Masterflex C/L) in addition to auxiliary TTL outputs (Galil DMC-4143) for illumination shutters (Ocean Optics INLINE-TTL) or LED drivers (Thorlabs LEDD1B) in the case of photoelectrochemical experiments. The control software collates per-sample potentiostat data with camera images and/or transmission spectra. While many of the results and citations included in this chapter involve measurements from this instrument, the design and operation continues to evolve for improved droplet stability, chemical compatibility, and ease of maintenance.

The cell typically operates with continuous electrolyte flow at approximately 0.01 mL s^{-1} *via* gravity-feed from a gas-purged solution. Continuous flow mitigates mass transport issues such as gas bubble nucleation over the working electrode for gas-evolving reactions. The scanning head is made from poly(methyl methacrylate) (PMMA) for compatibility with aqueous electrolytes from pH 0 (sulfuric acid) to pH 14 (sodium hydroxide) with PEEK connection ports. The top of the scanning head contains three vertical ports: one coaxial with the bottom droplet port for fiber optic insertion, and two side ports for electrolyte input and differential pumping. Differential pumping diverts some electrolyte flow across the counter electrode (typically Pt wire) to keep counter electrode reaction products from the working electrode effluent. The majority of electrolyte exits vertically through the bottom droplet port onto the working electrode sample. Siphons made from PEEK-coated silica tubing collect the solution at three contact points on the working electrode surface. Figure 9.5 depicts an overall schematic of the SDC head unit and solution transfer devices. Under optimal wetting conditions, the collection point effectively pins the droplet diameter. The reference electrode (typically Ag/AgCl) connection is made through a long-taper glass pipette capillary, with a 50 micron tip penetrating the droplet.

Droplet stability under continuous flow is paramount to data quality. Peristaltic pumps (Masterflex C/L) drive the siphons and differential flow. This style of pump has many advantages: compact size, continuous operation, and zero wetted parts. However, the peristaltic action creates periodic

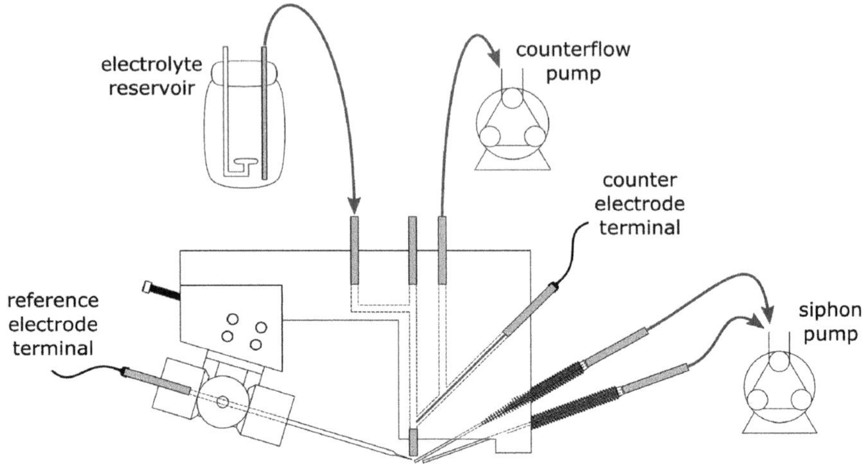

Figure 9.5 Schematic of scanning droplet cell head unit and solution transfer devices.
This figure is based on work reported in ref. 8.

pressure waves in the pump tubing which will affect droplet stability if not dampened properly. Air is an effective dampener for effluent collection, so full submersion of the siphon tip cross-section must be avoided. Assuming a tip cross-section normal to the tube axis, siphon tubes impinging at 25 to 30 degrees with respect to the working electrode surface work well with 0.1 M sodium hydroxide electrolyte. Steeper angles risk tip submersion, while shallower angles wick solution between the tube and working electrode surface, increasing the droplet area and potential for sample cross-talk. In particular, the mean contact angle of 2 μL of 0.1 M sodium hydroxide on FTO glass measured 16 degrees. Siphon tip positions and impingement angles may be adjusted for different substrate-electrolyte wetting conditions. Pressure waves caused by differential pumping on the counter electrode channel are dampened by high-impedance tubing between the scanning head and peristaltic pump. Specific implementations of this instrument in screening of solar fuels materials are described below, starting with the discovery of OER electrocatalysts in pH 14 electrolyte.

Using the modified commercial Epson printer, a number of pseudo-quaternary metal oxide (mixtures of four metals plus oxygen) composition-space libraries have been prepared and screened for electrocatalytic OER activity in 1.0 M NaOH(aq) (pH 14) electrolyte. As illustrated in Figure 9.6, inks suitable for use in this printer have been developed and utilized for 28 metals from across the periodic table, including alkaline earths, early transition metals, late transition metals, lanthanides, and main group metals. SDC screening of the pseudo-quaternary metal oxide libraries have only been reported for two systems to date, (Ni–Fe–Co–Ce)O_x[9] and (Ni–La–Co–Ce)O_x,[10] with publications in progress for the dozens of additional pseudo-quaternary metal oxide compositions. One powerful capability of this printing-based

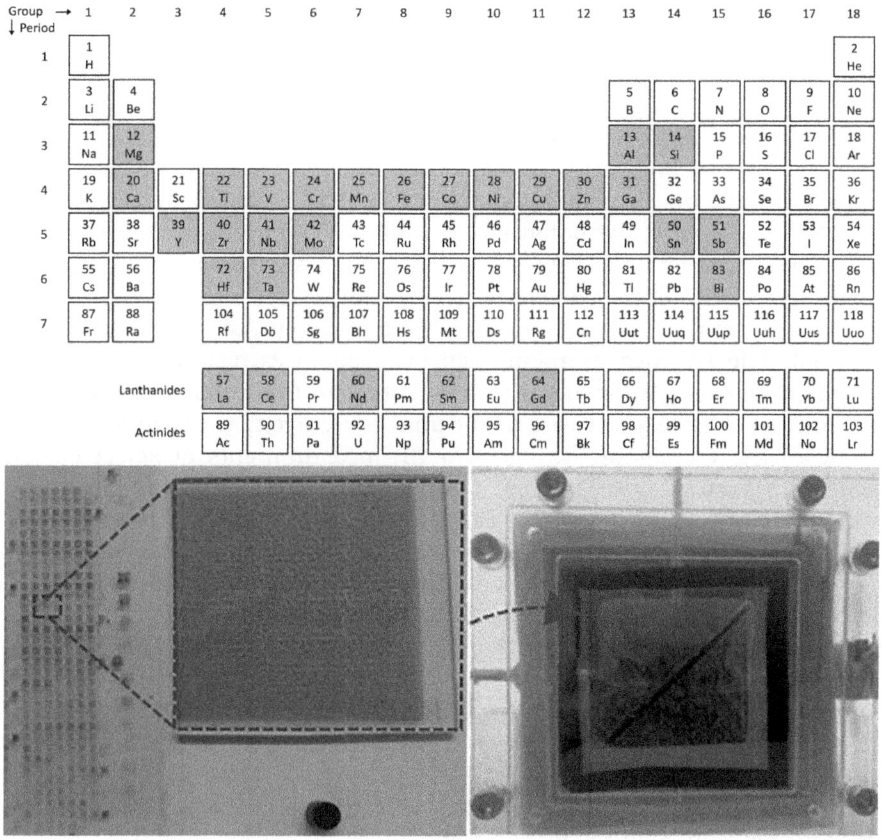

Figure 9.6 (a) Periodic table showing the metals (shaded) for which Epson printer-compatible inks have been prepared and utilized. (b) Different scales and substrates on which catalyst compositions are prepared by printing to enable standard test conditions for performance validation and prototype evaluation (1×1 mm^2 library sample to 50×50 mm^2 electrode for prototype integration).
This figure is based on work reported in ref. 9.

synthesis approach is the ease with which promising catalyst compositions can be reproduced on different substrates and over larger areas. As shown in Figure 9.6, a promising OER catalyst composition discovered by screening 1 mm^2 sample spots in a pseudo-quaternary metal oxide composition space library is readily deposited on glassy carbon rotating disk electrodes to verify performance under standardized electrochemical test conditions, or over larger areas for long duration operation in test-bed reactors in which gas collection is feasible for evaluation of faradaic yields. Notably, all lead compositions discovered using our 1 mm^2 library samples have comparable performance under standard electrochemical test conditions and in prototype devices; thereby validating our approach to electrocatalyst discovery.

There is no universal "best" electrocatalyst for OER. It is, therefore, valuable to carefully consider the final operating condition or conditions for which discovery of a new catalyst composition is desired—this will define the most important figures of merit (FOM) for catalyst performance. The applicable conditions we screened for are directly related to those required for an anion-exchange membrane-containing photoelectrochemical water splitting device: highly alkaline electrolyte (pH 14), current density close to 10 mA cm^{-2}, lowest overpotential possible. We therefore collect electrochemical data directly related to these conditions, such as using 1.0 M NaOH as the electrolyte, collect cyclic voltammograms (CVs), chronopotentiograms (CPs) at 10 mA cm^{-2}, and chronoamperometry (CAs) at 300 mV overpotential. The CV measurements can be used to extract the approximate overpotential over a range of current density, and to measure overpotentials in the absence of capacitive current and any current transients that contribute to the CV signal, we perform CP measurements at select current densities, typically 3, 10, and 20 mA cm^{-2}. In order to visualize how the catalyst performance varies as a function of composition, it is valuable to reduce the performance at each composition to a single scalar value (typically the overpotential at 10 mA cm^{-2}) and then plot that value on a color scale as a function of composition. Typically, performance is a smooth function of composition and optimal compositions are readily identified. However, it is instructive to plot multiple FOMs from each library, as the compositional dependence of the performance (*e.g.* overpotential) generally varies significantly with the specific value selected to evaluate the FOM (*e.g.* current density).

We wish to emphasize that we are able to collect high-quality electrochemical data on each of the thousands of composition samples in the library, not simply screening for a single direct, or indirect, measure of activity. Because of this capability, we are able to relate changes in performance with underlying electrochemical properties, and relate those changes directly to composition changes. For example, the first OER catalyst discovery we reported was in the (Ni–Fe–Co–Ce)O_x composition space, with a newly discovered Ce-rich composition ($Ni_{0.2}Fe_{0.07}Co_{0.30}Ce_{0.43}O_x$) showing comparable overpotential at 10 mA cm^{-2} as the previously known high-performance catalyst composition of $Ni_{0.5}Fe_{0.3}Co_{0.17}Ce_{0.03}O_x$, but significantly lower overpotential at 3 mA cm^{-2}. Examination of the FOM plots of overpotential at 10 and 3 mA cm^{-2} indicated that the regions of high-performance compositions centered on the above compositions were divided by a composition region with relatively poor catalytic performance. This was illustrated by extracting the pseudo-ternary composition space from the pseudo-quaternary composition space, as shown by the dashed triangle inscribed in the tetrahedron inset in Figure 9.7. A custom plate map was generated to produce the compositions lying close to this plane at 3 at% composition steps on a single substrate. Detailed electrochemical characterization (including CVs and CPs at multiple current densities) of this composition space was performed and reported in a subsequent paper.[11]

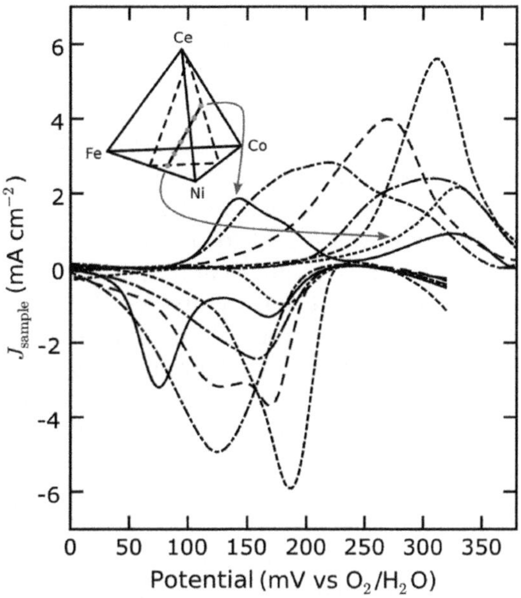

Figure 9.7 Systematic variation in catalyst oxidation/reduction waves extracted from experimental CV curves collected at the compositions indicated in the inset, which are a series of compositions connecting the Ni–Fe–Co and Ni–Co–Ce faces of the Ni–Fe–Co–Ce oxide composition tetrahedron. This figure is based on work reported in ref. 10.

Figure 9.7 shows the oxidation/reduction waves of the catalyst, mathematically extracted from the experimental CV curves collected at six compositions along the pseudo-binary composition line bisecting the two optimal catalyst compositions (shown in the inset tetrahedron). There is a clear compositional dependence upon the potential for the catalyst oxidation and subsequent reduction which is ascribed to the $Ni^{II}/Ni^{III/IV}$ oxidation to form the active OER catalyst sites.

9.3.2 Parallel Screening *via* Bubble Imaging

The most general strategy for designing a parallel screening technique is to identify a proxy for the desired performance that can be imaged using a camera. Since the catalytic reaction at hand involves transformation of aqueous species into gas molecules with limited solubility in the aqueous electrolyte, optical detection of evolved bubbles of O_2 serves as an excellent proxy for electrocatalysis of the OER. This concept was noted for HER electrocatalyst libraries[12] and substantial engineering was required to develop the technique for screening large arrays of materials. The key aspects of our bubble screening platform[13] were: (1) use of a mesh counter electrode parallel to the working electrode to provide a uniform electric field while maintaining sufficient transparency for imaging through the counter electrode; (2) recognition that with sufficient lateral confinement the bubbles from the HER counter reaction can be imaged along with those from the OER; and (3) suspension of a bubble nucleation agent in the electrolyte to yield a uniform bubble size and mitigate super saturation. With these innovations, the bubble screening method was able to identify active catalyst regions but with some significant limitations. Due to Ohmic losses and rampant bubble evolution resulting from libraries with many active electrocatalysts, the technique was found to be most well suited for detecting rare active catalysts in a library of poor catalysts. As described above, many OER electrocatalyst screening experiments involved searching in compositions near that of known electrocatalysts, and that important scientific insights came from mapping advanced parameters such as the effective Tafel slope. As the research focus trended away from exploring adventurous composition spaces, so did the utilization of the bubble screening method.

9.3.3 Screening Libraries with Unstable Catalysts

As shown above, electrochemical experiments on the 10–100 s time scales are sufficient for identifying OER electrocatalysts using composition libraries in which all constituent elements form stable species in the testing electrolyte. When the library contains elements that are unstable under testing conditions, extended electrochemical exposure is necessary to identify and differentiate stable from unstable OER electrocatalysts, particularly when anodic corrosion may be mistaken for anodic electrocatalysis. In materials discovery for water splitting, this situation was most prominent for OER catalyst screening in pH 0 electrolyte. Scanning droplet cell measurements are capable of rapidly assessing the electrocatalytic performance of compositions within a material library, but the serial nature of the instrument limits the experimental throughput for temporally demanding investigations like operational stability. Parallel methods were therefore employed to expose an entire library to electrocatalytic conditions for prolonged periods using our custom-designed parallel electrochemical treatment system (PETS) that minimizes non-uniformities in an electrochemical environment over the library area.

Establishing electrochemical uniformity over large areas required simulation-guided engineering of system design and operation, which proved to be particularly important for mitigating spatial artifacts resulting from resistive losses in the planar working electrode. Both chemical dissolution and electrochemical corrosion of catalyst samples result in metal species leaching into the electrolyte, which alters the chemical potential for continued dissolution, motivating the design of rapid electrolyte flushing to avoid large special variations in chemical potentials. Electrolyte flow in both the catholyte and anolyte can be performed with once-through or recirculated electrolyte from a large reservoir whose volume is chosen so that the molarity of metal species stays below a chosen threshold even if the entire library dissolves. Non-uniformities due to electrode geometry can also affect PETS. Ideally, a large area electrochemical conditioning instrument establishes uniform electric field lines perpendicular to the working electrode surface. This is accomplished by polarizing planar working and counter electrodes in a parallel geometry, thus establishing uniform 2-electrode contact to each material simultaneously. Under this approximation, the addition of a reference electrode provides measurement of the solution potential at a single point to enable pseudo-3-electrode measurements.

A typical procedure for screening an OER electrocatalyst library in strong acid electrolyte entails: (1) initial optical image acquisition of the material library; (2) exposing the library to an operational electrochemical environment in a parallel fashion for an extended period of time (PETS); (3) post-PETS secondary optical image acquisition; (4) comparing the initial and secondary optical images to observe composition samples with significant modification from electrochemical exposure; and (5) performing serial SDC measurements on the electrocatalysts to ascertain activity of the stable or passivated composition samples. The results of this procedure can be further evaluated against the same serial measurements made prior to extended parallel exposure to characterize the loss in activity over the time scale of the PETS experiment.

9.3.4 Materials Characterization for Electrocatalysts

The high-throughput characterization of electrocatalyst libraries comprises a technology gap in the high throughput pipeline. Detailed characterization of several OER catalysts noted above demonstrated that XRD is insufficient for structural characterization due to the nano-scale grain sizes of active catalysts. This motivates nanocharacterization of the catalysts with state-of-the-art transmission electron microscopy (TEM) and synchrotron radiation techniques that are not amenable to high-throughput experiments and are best performed *via* collaborations with world experts.[14] Combinatorial X-ray photoelectron spectroscopy (XPS) has shown promise in providing insights into active catalyst species and characterization of passivation layers of metal oxides both in acid[15] and alkaline[16] electrolytes. While catalyst characterization techniques will continue to be developed, we note that materials discovery for electrocatalysis is an important example of a technology area where screening for performance under device-relevant conditions is relatively easy compared to measuring the relevant material properties that give rise to the performance.

9.4 Experimental Pipeline for Discovering Photoanodes

9.4.1 High Throughput Spectroscopy for Band Gap Screening

Characterization of optical properties of materials is essential for development of efficient materials for several technologies such as transparent conductive oxides, phosphors, pigments, photovoltaics, and photo(electro)catalysts. Specifically, for photovoltaic and solar fuels technologies, light absorbers are a critical functional component, and the band gap energy is a primary FOM for identifying candidate materials. While several studies[17–19] have mapped optical properties of thin-film composition spread libraries, small solid angles of detection due to collection of optical response spectra using fiber optically coupled CCD array-based spectrometers renders these techniques inefficient for rough films or particulate samples, in addition to limiting the throughput of the measurement. By using integrating spheres coupled with a brilliant illumination source placed normal to the sample plate that allows collection of optical response spectra over a large solid angle, and synchronizing the data acquisition with continuous translation of the spheres, we enabled rapid data acquisition on thin-films with varied morphologies at a throughput better than one sample per second.[20] For transparent substrates, we developed a transmission-reflection (TR) instrument that simultaneously collects transmission and reflection spectra using two integrating spheres that are placed above and below the plane of the library plate. For opaque substrates, we developed a diffuse-reflection (DR) instrument that collects diffuse-reflection spectra using an integrating sphere incorporated with an absorbing plug such that the spectra is comprised of light scattered beyond approximately 6.5° of the specular reflection and less than 44° from the substrate normal.

Using an instrument specific definition of the quantity β, which is defined as αL (L is the optical path length) for the TR instrument and α/s (s is the spectral scattering factor, which we assume to be independent of incident photon energy) for the DR instrument, the optical absorption spectra are calculated up to a factor using eqn (9.1) for spectra from the TR instrument and eqn (9.2) for spectra from the DR instrument; where T, R, and DR are fractional transmitted, reflected, and diffusely reflected spectra, respectively.

$$\beta = \alpha L = -\ln(T/(1-R)) \tag{9.1}$$

$$\beta = \alpha/s = (1-DR^2)/(2DR) \tag{9.2}$$

The band gap energy (E_g) is typically estimated from Tauc ($(\beta h\nu)^{1/n}$ *vs.* $h\nu$) plots[21,22] by extrapolating a manually identified linear region (Tauc line segment) that follows the Tauc relationship (eqn (9.3)) to a baseline.

$$\alpha h\nu \propto (h\nu - E_g)^n \tag{9.3}$$

Since manual inspection of the Tauc plots is prohibitive for automated analysis, linear regression in user defined energy ranges and tangents at the maximum in first derivatives of Tauc plots have been proposed as methodologies to identify the Tauc line segment.[23–25] However, these approaches require a user input and/or are sensitive to noise and local optical features. To overcome these issues, we fit the Tauc property (TP) spectra (eqn (9.4)) using a piecewise linear fitting algorithm wherein the discontinuities are constrained to be apart by more than 0.05 eV (to avoid fitting of noisy features), followed by identification of a baseline. The Tauc line segments use constraints on the lengths and slopes of the linear segments that mimic the judgement of an expert scientist.[26] The band gap energy is then estimated as the abscissa at the intersection of the Tauc line segment with the baseline segment.

$$\mathrm{TP} = (\beta h\nu)^{1/n}/\max((\beta h\nu)^{1/n}), \tag{9.4}$$

where the Tauc exponent $n = 2$ and 1/2 for indirect-allowed (*IA*) and direct-allowed (*DA*) band gap transitions, respectively.

As an example of automated band gap estimation from *TP* spectra using this algorithm, we estimate direct-allowed (E_g^{DA}) and indirect-allowed (E_g^{IA}) band gaps of α-$Cu_2V_2O_7$, shown in Figure 9.8. As a rule of thumb, we typically report E_g^{IA} values only when they are more than 0.3 eV below E_g^{DA}, to avoid reporting E_g^{IA} when the *TP* is dominated by the stronger absorption from direct-allowed transitions.

To demonstrate the applicability of this algorithm for automated estimation of band gap energy for various materials, we compare the direct-allowed band gaps estimated by expert scientists and by automated algorithm for 60 optical spectra obtained from composition samples that included metal oxides and metal sulfides with a total of 11 cation elements represented and up to four cations present in each sample. The dataset consisted of 50 TP^{DA} spectra for which the algorithm identified a single E_g^{DA} and 10 spectra for which a E_g^{DA} value was not identified. For each spectrum, majority consensus of three expert scientists was used to establish the ground truth for the presence/absence of a direct band gap. The performance of the automated algorithm in mimicking the expert judgement

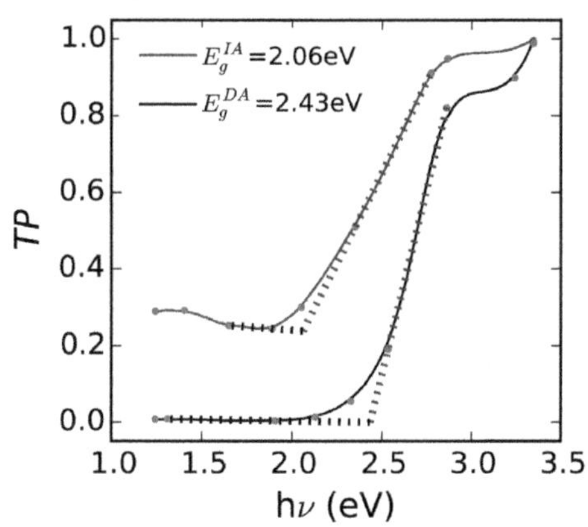

Figure 9.8 Direct-allowed (DA) and indirect-allowed (IA) *TP* spectra for α-$Cu_2V_2O_7$. For both types of Tauc plots, discontinuities in the piecewise linear fit are shown as points with dashed lines showing the Tauc and base line segments, whose intersection is the band gap energy shown in the legend.
This figure is based on work reported in ref. 26.

was well demonstrated by the excellent true-positive (98%) and true-negative (82%) rates for spectra with presence and absence of a ground-truth E_g^{DA}, respectively. For the true-positive samples, the discrepancies were typically within 0.1 eV of the values derived from manual analysis. Larger discrepancies are typically due to a limitation in the source data, for example the extension of the baseline beyond the limit of the spectrometer, creating a more subjective model. This provides an important guide for experiment design, indicating that the lower and higher energy limit of the spectra should extend at least 0.5 eV beyond the band gap energy range of interest.

With these tools in place, the band gap energy and a variety of additional optical parameters can be readily mapped over composition libraries. It is important to note that the band gap energy is the property of a single phase and the vast majority of composition samples in our libraries are mixed-phase samples. Therefore, it is important to consider the composition maps of band gap energy as the representative values for the optically-dominant phase, and composition regions exhibiting a band gap energy of interest must be characterized to understand the underlying phase behavior, as we recently demonstrated in the Bi–V–Fe oxide space.[27]

9.4.2 Colorimetry as a Parallel Screen

To accelerate optical screening of large libraries, optical imaging can be employed to characterize the color of each composition sample with the premise that compositions that appear transparent (black) in visible-spectrum images from consumer imaging devices must have a band gap too large (small) for solar PEC applications. Consumer plate-bed scanners provide rapid imaging *via* translation of a line sensor, which we used to develop a workflow to perform color analysis of composition samples in very high throughput. In particular, imaging of 1 mm^2 samples at a throughput of 22.5 samples per second with spatial resolution of 21 μm was achieved.[28] With respect to band gap screening, we demonstrated that with appropriate image processing and signal averaging, colorimetric screening can successfully identify distinct phase regions and composition samples with band gap energy in the optical range. With our subsequent advances in the full-spectrum optical characterization described above, this parallel screening offered diminishing returns for throughput since the spectra are acquired on-the-fly, such that down-selecting samples from a randomized plate map lowers the data load but not the measurement throughput. An aspect of the colorimetry imaging that remains important is its characterization of the distribution of optical properties within a composition sample. That is, the detection of a distribution of hue values within a composition sample is a strong indication of coarse phase segregation and signifies that the mixed-phase sample is not suitable for characterization of single-phase properties such as band gap energy.

9.4.3 Photoelectrochemistry with the Scanning Droplet Cell

SDC instrumentation is configured for photoelectrochemical experiments by adding an optical fiber (Ocean Optics) to the central port of the cell discussed above. One end of the fiber is connected to an illumination source, while the other was trimmed using a scoring tool and polished before securing it in the cell using PEEK tubing and epoxy. The bare fiber terminates 1 mm above the substrate and is centered within the flowing drop, as seen in Figure 9.9a. Illumination sources include light emitting diodes from Thor Labs and Doric Lenses at wavelengths ranging from 385 nm to 850 nm. A xenon arc lamp (Newport) with filters to optimize the output is used as a white light source. The illumination is toggled during CA and CV measurements in order to calculate values such as photocurrent density, quantum efficiency, maximum power point, photostability, and open circuit potential. Typical measurements begin with no illumination in order to stabilize dark current and then illumination is toggled with repeated 0.5 s light-on, 0.5 s light-off cycles, as seen in Figure 9.9b for a 1800 s CA measurement. The illuminated time should take into account the current transients that occur when the illumination is first turned on, such that the current stabilizes within each light-on event. The instrument is configured for front-side illumination or back-side illumination (for transparent substrates, where the optical fiber is secured underneath the library plate). For screening of photoanode materials, OER photocurrent is a primary FOM, with complementary experiments including modification of the electrolyte by including a hole-acceptor (*e.g.* sodium sulfite) or redox couple (*e.g.* ferricyanide/ferrocyanide).[29,30] Additional techniques employing non-aqueous electrolytes and metallocene junctions have been developed but have not been extensively used in material screening.[31]

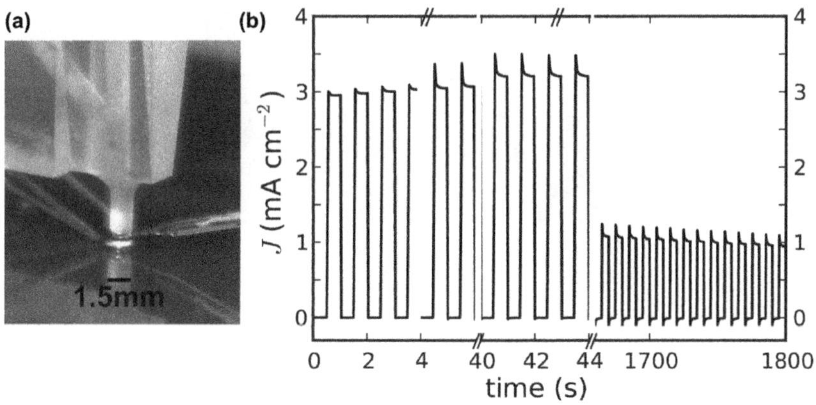

Figure 9.9 (a) Scanning droplet cell configured with an optical fiber to provide toggled-illumination. (b) An example of an 1800 s toggled-illumination CA measurement with illumination cycled off for 0.5 s and on for 0.5 s. This figure is based on work reported in ref. 29.

9.4.4 Material Characterization of Photoanodes: Linking to Theory

The photoelectrocatalytic performance of a material depends on several factors such as electronic character, band gap, and photoelectrochemical stability, and since these factors are properties of a given phase and composition, characterization of the crystal structure is necessary to establish structure-property relationships and link experiments to theory. Computational materials theory approaches rely on crystal structure information as the building block to predict descriptors such as valence band maximum, conduction band minimum, band gap, and hole/electron effective masses, and discovery of these descriptor-property relationships is essential to develop a feedback loop for design of efficient photoelectrodes, emphasizing the need for experimental crystal structure characterization. XRD is the most ubiquitously used characterization technique for crystal structure analysis, and is particularly applicable for dense thin films with thickness >100 nm, as is typically the case for sputter deposited films.

In this context, we discuss structure-property relationships in a $Cu_{1-x}V_xO_z$ composition spread library that was synthesized using reactive co-sputtering. Figure 9.10 shows 4s OER short-circuit photocurrent density overlaid on colored stacked plots of composition-dependent phase concentrations, wherein the phase concentrations were estimated by analyzing the corresponding XRD patterns. These maps reveal that Fingerite, McBirneyite, and Blossite are active phases for OER photocatalysis. Further electronic structure calculations of these phases revealed that the sub-2 eV band gap energies of these phases can be attributed to the low CBM energy due to their

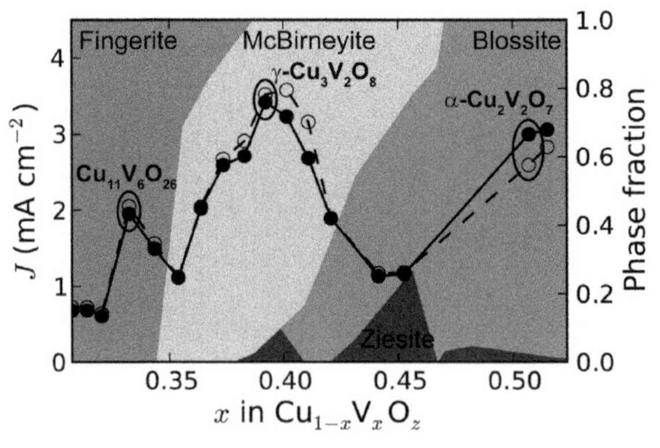

Figure 9.10 Short-circuit photocurrent density measurements in pH 13 electrolyte after 4 s of operation under 385 nm chopped illumination, which is overlaid on a colored stacked plot of phase concentration (right vertical axis) for a $Cu_{1-x}V_xO_z$ library.
This figure is based on work reported in ref. 27 and 29.

predominantly Cu 3d character and possibly high VBM energy due to O 2p and Cu 3d hybridization in the valence band, enabling discovery and mechanistic understanding of low band gap oxide photoanodes in the $CuO–V_2O_5$ system.[27] The discovery of these copper vanadate photoanodes was enabled by an integrated theory-experiment pipeline whose complete description is being prepared for publication.

While synthesis of photoelectrodes onto conducting substrates such as FTO/glass and $Pt/Ti/SiO_2/Si$ enables measurement of PEC properties of the material in a geometry relevant for solar fuels device applications, the presence of a substrate layer creates challenges for very thin films (order of 10 nm or lower) and poorly-crystallized samples for which the XRD signal is dominated by contributions from the substrate. Specifically, for IJP samples deposited on FTO glass substrates, the relatively weak XRD signal from the composition samples compared to that of the substrate results in a limited ability to characterize phase behavior with traditional XRD techniques. In such cases, an alternate crystal structure characterization technique such as Raman spectroscopy, which has a much shallower sampling depth, is preferred.

As an example of how we utilized Raman spectroscopy, the phase behavior of an IJP $Cu_{1-x}V_xO_z$ library was investigated using a Renishaw inVia Reflex Micro Raman spectrometer. The instrument is equipped with a 532 nm laser beam spot expanded into a line of approximately 26×2 μm using a cylindrical lens that is scanned over the desired area using a Renishaw Streamline™ mapping system. Contribution of the substrate to the Raman signal is minimized by dynamically maintaining focus (using surface map capability of Wire 4.1) on the surface of the sample along with limiting the depth of focus to approximately 2 μm using 50× objective (numerical aperture 0.75). Figure 9.11 summarizes the phase map constructed from 100 000 Raman spectra, revealing the distribution of CuO, $Cu_5V_2O_{10}$, β-$Cu_2V_2O_7$, α-$Cu_2V_2O_7$, and α-CuV_2O_6, V_2O_5 phases in the composition library. Combining these phase distribution maps with optical and photoelectrochemical data has enabled establishment of structure-property relationships in this library of photoanodes, demonstrating that excellent photoactivity can be obtained with IJP materials.[30]

In addition to the lab-based techniques discussed above, rapid structure characterization can be enabled by using synchrotron-based measurement techniques coupled with automated data analysis algorithms, which has been pioneered through collaborations noted below in the Acknowledgements. The synchrotron-based high-throughput XRD technique is tailored for structure measurements on combinatorial composition libraries.[32] Automated phase mapping is enabled with a novel factor decomposition algorithm AgileFD[33] that incorporates physical properties of phase diagrams such as Gibbs' phase rule and peak shifting due to alloying, which has been demonstrated for light absorber discovery in the V–Mn–Nb oxide system.[34]

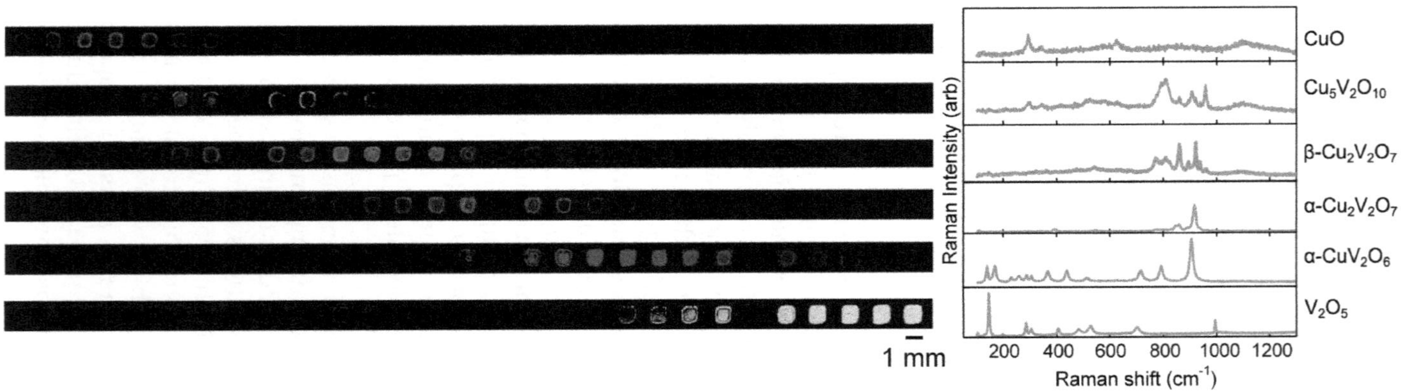

Figure 9.11 Phase distribution mapping for CuO, $Cu_5V_2O_{10}$, β-$Cu_2V_2O_7$, α-$Cu_2V_2O_7$, α-CuV_2O_6, and V_2O_5 phases on 26 samples of ink-jet printed $Cu_{1-x}V_xO_z$. The basis components for the phases are shown along with phase labels on the right, and the spatial distribution of each phase is shown as a series of intensity plots of the same library region where the series of compositions span $x = 0$ to 1 from left to right.
This figure is based on work reported in ref. 30.

9.5 Combining Materials and Techniques for Discovery of Integrated Materials

In order to predict and measure the performance of catalyst/semiconductor combinations for photoelectrodes, we devised instrumentation and designed experiments and analysis by leveraging high-throughput experimentation techniques. The basic strategy includes: (1) characterizing an electrocatalyst library; (2) depositing the same electrocatalyst library on a uniform, thin film light absorber; and (3) characterizing the performance of the material combination as a photoanode. Through this strategy, we discovered that the results from step 1 cannot accurately predict the results from step 3; thus, the catalyst/semiconductor combination must be tested in order to measure device performance of the integrated material.

The instrumentation for the discovery of integrated materials includes the SDC described above with front-side illumination as described in the photoelectrochemistry experiments and a transmission spectroscopy subsystem consisting of an integrating sphere (Spectral Products AT-IS-1) coupled to a spectrometer (Spectral Products SM303). Samples are illuminated by a xenon arc lamp (Newport 66921, 450W), including a water filter to remove infra-red radiation (Oriel Instruments, 6123NS), AM 1.5 filter (Newport, 81094) and neutral density filter (Rolyn Optics, #66.0220 50%T). In order to combine front-side illumination with spectra collection underneath the substrate, the SDC and integrating sphere are fixed to the lab benchtop while the substrate holder uses a translation stage to move the substrate.

The initial demonstration of this approach to combinatorial materials integration included a (Ni–La–Co–Ce)O_x catalyst library with $BiVO_4$ as the light absorber, and the resulting photoanodes were screened for OER photoactivity in pH 13 electrolyte (0.1 M sodium hydroxide and 0.5 M sodium sulfate). The catalyst library with 10 atom% composition steps was deposited on a 10×10 cm^2 FTO-coated glass plate. The 286 compositions in this quaternary composition space were deposited at three different catalyst loadings, referred to as L1 (0.75 nmoles), L2 (1.9 nmoles), and L3 (3.8 nmoles), totaling 858 unique, 1×1 mm^2 square samples. Within each loading section, the 286 compositions were randomly distributed in order to prevent the convolution of spatial or temporal artifacts.

The library of metal oxides was first deposited onto an FTO/glass substrate (no light absorber) to characterize electrocatalytic activity with *in situ* optical transmission measurements. Using the SDC, CA measurements were conducted on each catalyst at an overpotential (η_{OER}) of 0.35 V (0.35 V *vs.* O_2/H_2O or 1.58 V *vs.* RHE) for 25 s. Catalytic current density (J_{cat}) was calculated to be the average current density over the final 1 s of the measurement. The catalysts were illuminated with the white light source during this measurement in order to calculate optical transmission. Transmission spectra were averaged over a 2 s interval at the end of the CA measurement to provide the fractional transmission spectrum (T_{cat}) of each catalyst after equilibration.

Reference transmission spectra were collected from bare regions of FTO-coated glass on the plate and combined with the dark signal from the spectrometer to determine the fractional transmission ($T(\lambda)$) for each spectral measurement. This combination of measurements gives the corresponding optical transmission efficiency of the catalyst, seen in Figure 9.12a for L1 loading, in particular the fractional transmission of photons that are expected to be absorbed by the light absorber on which the catalysts will be integrated:

$$\alpha_{T,\text{cat}} = \int_{\lambda_1}^{\lambda_2} T_{\text{cat}}(\lambda)\boldsymbol{A}_{\text{BVO}}(\lambda)\boldsymbol{N}_{\text{Lamp}}(\lambda)\text{d}\lambda \Big/ \int_{\lambda_1}^{\lambda_2} \boldsymbol{A}_{\text{BVO}}(\lambda)\boldsymbol{N}_{\text{Lamp}}(\lambda)\text{d}\lambda. \qquad (9.5)$$

The wavelength integration range of 390 nm to 600 nm was chosen as the relevant portion of the measurement spectrum with respect to the $BiVO_4$ band gap. $\boldsymbol{A}_{\text{BVO}}$ is the fractional absorption spectrum of the $BiVO_4$ thin film and $\boldsymbol{N}_{\text{Lamp}}$ is the photon flux of the xenon arc lamp. The two quantities J_{cat}

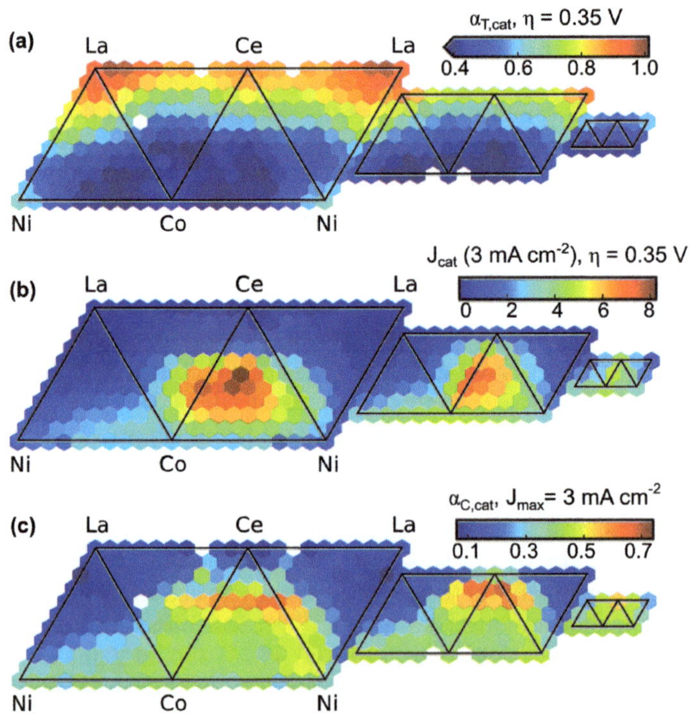

Figure 9.12 Compositions maps of the L1 loading of the catalyst library (Ni–La–Co–Ce)O_x on FTO: (a) $\alpha_{T,\text{cat}}$, the fractional transmission of photons that are expected to be absorbed by $BiVO_4$; (b) J_{cat}, catalytic current density at $\eta = 0.35$ V; and (c) $\alpha_{C,\text{cat}}$, the predicted optical-electrocatalytic efficiency upon integration with $BiVO_4$.
This figure is based on work reported in ref. 35.

and $\alpha_{T,\text{cat}}$ can be combined in order to predict how the integrated photoanode is expected to behave based on intrinsic properties of the catalysts, which we quantify by the combined efficiency

$$\alpha_{C,\text{cat}} = \alpha_{T,\text{cat}}(\text{Min}[J_{\text{cat}}, J_{\max}]/J_{\max}). \tag{9.6}$$

Based on this calculation, catalyst compositions with high $\alpha_{C,\text{cat}}$ are predicted to have high performance when integrated with $BiVO_4$. Composition maps of all three quantities (J_{cat}, $\alpha_{T,\text{cat}}$, and $\alpha_{C,\text{cat}}$) are shown in Figure 9.12 for L1 loading. In order to test if this calculation, or values of J_{cat} and $\alpha_{T,\text{cat}}$, correctly predicts high photoanode performance, a duplicate of the catalyst library was printed on spin-coated, uniform $BiVO_4$ prepared on FTO-glass (step (2)).

The photoanode library was prepared by depositing the same library of metal oxides on the $BiVO_4$ light absorber plate, and the photoelectrochemical properties of the 858 samples were measured utilizing the same electrolyte and SDC as the catalyst samples. Toggled-illumination CV measurements were conducted on each photoanode sample with potential sweep from 1.23 to 0.585 V and back to 1.23 V *vs.* RHE (0 to -0.645 V and back to 0 V *vs.* O_2/H_2O) at a scan rate of 0.02 V s^{-1}, while illumination was toggled at 0.5 Hz (1.34 s on, 0.67 s off). Photocurrent density ($J_{\text{photo}}(V)$) was calculated from the anodic sweep by subtracting the dark current from the illuminated current and using a cubic polynomial fit for each, as shown in Figure 9.13a and b. The $J_{\text{photo}}(V)$ model enables rapid extraction of additional FOMs such as J_{O_2/H_2O} (the photo-generated current density at the Nernstian potential for water oxidation), E_{OC} approximate open-circuit photovoltage, and $P_{\max}$ (maximum electrochemical power generation). The highest $P_{\max}$ for the La-containing library is centered on the $La_{0.2}Co_{0.2}Ce_{0.6}O_x$ region for all three loadings, with the highest for L1 (0.93 mW cm^{-2}), shown in Figure 9.13c.[35] For comparison, bare $BiVO_4$ from a reference sample on this plate (a region of $BiVO_4$ with no catalyst printed) has $P_{\max}$ of 0.065 mW cm^{-2}.

In order to compare the predicted performance ($\alpha_{C,\text{cat}}$) based on the optical-catalytic properties of the catalysts with the integrated photoanode performance ($P_{\max}$), we use a relative log-ratio metric (Γ):

$$\Gamma' = \log_{10}(P_{\max}/\alpha_{C,\text{cat}}),\ \Gamma = \Gamma' - \text{median}(\Gamma'). \tag{9.7}$$

Values of Γ near 0 relate to catalysts whose predicted performance based on combined optical-catalytic efficiency is comparable to actual performance on $BiVO_4$. The composition map of Γ in Figure 9.13d (L1 loading) reveals distinct composition regions (yellow) in which the photoanode far outperforms that of the prediction. The nature of the interface was not included in the prediction, and correspondingly this data analysis and visualization identifies compositions for which the interface to $BiVO_4$ gives rise to enhanced photoelectrochemical performance. This foundational demonstration of high throughput interface discovery can be effectively used to identify photoanodes for implementation into devices and interfaces whose

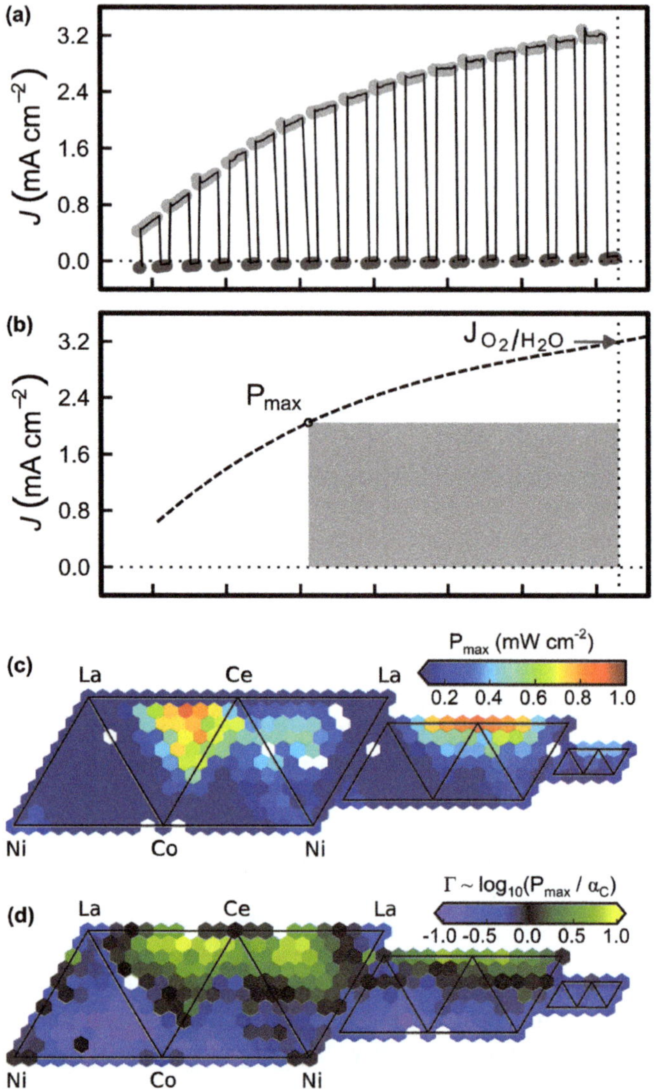

Figure 9.13 Toggled-illumination anodic sweep of a CV (a) used to generate the photocurrent signal (b) *via* polynomial fitting, which enables identification of the maximum electrochemical power generation (P_{max}) and the photo-generated current density at the Nernstian potential for water oxidation (J_{O_2/H_2O}). Composition maps of the L1 loading of the photoanode library (Ni–La–Co–Ce)O_x on $BiVO_4$: (c) P_{max} and (d) Γ, which shows the relative enhancement of Ce-rich coatings compared to the predicted performance shown in Figure 9.12c.
This figure is based on work reported in ref. 29.

detailed characterization may reveal fundamental properties of charge separation in complex light absorber/catalyst/liquid junctions. Further details, validation and scale-up experiments, the performance of additional catalyst loadings, and results from a (Ni–Fe–Co–Ce)O_x library on $BiVO_4$ can be found in our published work.[35–38]

9.6 Lessons Learned and Future Prospects

The high throughput program described in this chapter was closely coupled to directed research efforts and a prototyping effort with device designs and operating conditions that guided, in a crucial manner, the design and validation of new screening instruments. The development and deployment of the high throughput pipeline could not have been done without this tight coupling to these complementary research efforts. The most prominent lesson concerning the design of instruments is that a well-engineered serial screening technique can provide comparable data quality at orders-of-magnitude faster experiment rate than traditional methods, and this is the most prudent approach for high throughput experiments in a basic research effort. Commencing data management with the first measurement from a new instrument is critical for enabling efficient data management in the completed high throughput pipeline, and strategic integration of theory and materials characterization into the experimental pipeline can greatly accelerate materials discovery and understanding.

Acknowledgements

While this chapter was authored by the present high throughput experimentation group, much of the work was made possible by contributions from previous group members. We thank Jian Jin for his initial leadership of the group and establishing many of the initial concepts and instrument designs; Slobodan Mitrovic for his commencement of the inkjet printing, optical spectroscopy, and materials characterization efforts; Earl Cornell for his commencement of data management and experiment automation; Lung-Sheng (Sean) Lin for his commencement of instrument engineering efforts; Xiaonao Liu for her contributions to the development of high speed printing; Chengxiang (CX) Xiang for his contributions to the (photo)electrochemical screening techniques; Meyer (Misha) Pesenson for his contributions to materials informatics and statistical analysis; Natalie Becerra-Stasiewicz for her contributions to materials characterization; and Nathan Lewis and Eric McFarland whose advice was critical in establishing the high throughput pipeline.

We also thank the many collaborators, in particular our theory collaborators Jeffrey Neaton, Kristin Persson, Guo Li, Jie Yu, and Qimin Yan; our materials characterization collaborators Junko Yano, Walter Drisdell, Marco Favaro, Sean Fackler, Ethan Crumlin, Apurva Mehta, Douglas van Campen, Fang Ren, and Christian Kisielowski; our materials integration collaborators

Ian Sharp, Francesca Toma, and Guiji Liu; and our computer science and phase mapping collaborators Carla Gomes, Ronan LeBras, Stefano Ermon, Yexiang Xue, Junwen Bai, Brendan Rappazzo, Richard Bernstein, Johan Bjorck, and R. Bruce van Dover.

This material is based upon work performed by the Joint Center for Artificial Photosynthesis, a DOE Energy Innovation Hub, supported through the Office of Science of the US Department of Energy under Award Number DE-SC0004993. In addition to the financial support, we thank the DOE for their implementation of the hub approach to advancing energy technologies, which uniquely enabled the high throughput research described in this chapter.

References

1. M. Woodhouse and B. A. Parkinson, *Chem. Mater.*, 2008, **20**, 2495–2502.
2. X. Liu, Y. Shen, R. Yang, S. Zou, X. Ji, L. Shi, Y. Zhang, D. Liu, L. Xiao, X. Zheng, S. Li, J. Fan and G. D. Stucky, *Nano Lett.*, 2012, **12**, 5733–5739.
3. R. B. van Dover and L. Schneemeyer, *Macromol. Rapid Commun.*, 2004, **1**, 150–157.
4. L. Zhou, S. K. Suram, N. Becerra-Stasiewicz, S. Mitrovic and K. Kan, *J. Mater. Res.*, 2015, **30**, 2928–2933.
5. S. K. Suram, L. Zhou, N. Becerra-Stasiewicz, K. Kan, R. J. R. Jones, B. M. Kendrick and J. M. Gregoire, *Rev. Sci. Instrum.*, 2015, **86**, 033904.
6. J. K. Bunn, R. Z. Voepel, Z. Wang, E. P. Gatzke, J. A. Lauterbach and J. R. Hattrick-Simpers, *Ind. Eng. Chem. Res.*, 2016, **55**, 1236–1242.
7. S. K. Suram, M. Z. Pesenson and J. M. Gregoire, in *Information Science for Materials Discovery and Design*, ed. T. Lookman, J. F. Alexander and K. Rajan, Springer International Publishing, Cham, 2016, pp. 271–300.
8. J. M. Gregoire, C. Xiang, X. Liu, M. Marcin and J. Jin, *Rev. Sci. Instrum.*, 2013, **84**, 024102.
9. J. A. Haber, Y. Cai, S. Jung, C. Xiang, S. Mitrovic, J. Jin, A. T. Bell and J. M. Gregoire, *Energy Environ. Sci.*, 2014, **7**, 682–688.
10. J. A. Haber, D. Guevarra, S. Jung, J. Jin and J. M. Gregoire, *ChemElectroChem*, 2014, **1**, 1613–1617.
11. J. A. Haber, C. Xiang, D. Guevarra, S. Jung, J. Jin and J. M. Gregoire, *ChemElectroChem*, 2014, **1**, 524–528.
12. A. Ivanovskaya, N. Singh, R.-F. Liu, H. Kreutzer, J. Baltrusaitis, T. Van Nguyen, H. Metiu and E. McFarland, *Langmuir*, 2013, **29**, 480–492.
13. C. Xiang, S. K. Suram, J. A. Haber, D. W. Guevarra, J. Jin and J. M. Gregoire, *ACS Comb. Sci.*, 2014, **16**, 47–52.
14. J. A. Haber, E. Anzenburg, J. Yano, C. Kisielowski and J. M. Gregoire, *Adv. Energy Mater.*, 2015, **5**, 1402307.
15. A. S. Shinde, R. J. Jones, D. Guevarra, S. Mitrovic, N. Becerra, J. A. Haber, J. Jin and J. M. Gregoire, *Electrocatalysis*, 2015, **6**, 229–236.

16. L. Zhou, Q. Yan, J. Yu, R. J. R. Jones, N. Becerra-Stasiewicz, S. K. Suram, A. Shinde, D. Guevarra, J. B. Neaton, K. A. Persson and J. M. Gregoire, *Phys. Chem. Chem. Phys.*, 2016, **18**, 9349–9352.
17. J. D. Perkins, C. W. Teplin, M. F. A. M. van Hest, J. L. Alleman, X. Li, M. S. Dabney, B. M. Keyes, L. M. Gedvilas, D. S. Ginley, Y. Lin and Y. Lu, *Appl. Surf. Sci.*, 2004, **223**, 124–132.
18. M. P. Taylor, D. W. Readey, C. W. Teplin, M. F. A. M. v. Hest, J. L. Alleman, M. S. Dabney, L. M. Gedvilas, B. M. Keyes, B. To, J. D. Perkins and D. S. Ginley, *Meas. Sci. Technol.*, 2005, **16**, 90–94.
19. J. Gregoire, S. Kirby, M. Turk and R. Van Dover, *Thin Solid Films*, 2009, **517**, 1607–1609.
20. S. Mitrovic, E. W. Cornell, M. R. Marcin, R. J. Jones, P. F. Newhouse, S. K. Suram, J. Jin and J. M. Gregoire, *Rev. Sci. Instrum.*, 2015, **86**, 013904.
21. J. Tauc, R. Grigorovici and A. Vancu, *Phys. Status Solidi B*, 1966, **15**, 627–637.
22. J. Tauc, A. Menth and D. L. Wood, *Phys. Rev. Lett.*, 1970, **25**, 749–752.
23. A. Escobedo Morales, E. Sanchez Mora and U. Pal, *Rev. Mex. Fis.*, 2007, **53**, 18–22.
24. A. Y. Anderson, Y. Bouhadana, H.-N. Barad, B. Kupfer, E. Rosh-Hodesh, H. Aviv, Y. R. Tischler, S. Rühle and A. Zaban, *ACS Comb. Sci.*, 2014, **16**, 53–65.
25. N. Ghobadi, *Int. Nano Lett.*, 2013, **3**, 2.
26. S. K. Suram, P. F. Newhouse and J. M. Gregoire, *ACS Comb. Sci.*, 2016, **18**, 673–681.
27. S. K. Suram, P. F. Newhouse, L. Zhou, D. G. Van Campen, A. Mehta and J. M. Gregoire, *ACS Comb. Sci.*, 2016, **18**, 682–688.
28. S. Mitrovic, E. Soedarmadji, P. F. Newhouse, S. K. Suram, J. A. Haber, J. Jin and J. M. Gregoire, *ACS Comb. Sci.*, 2015, **17**, 176–181.
29. L. Zhou, Q. Yan, A. Shinde, D. Guevarra, P. F. Newhouse, N. Becerra-Stasiewicz, S. M. Chatman, J. A. Haber, J. B. Neaton and J. M. Gregoire, *Adv. Energy Mater.*, 2015, **5**, 1500968.
30. P. F. Newhouse, D. A. Boyd, A. Shinde, D. Guevarra, L. Zhou, E. Soedarmadji, G. Li, J. B. Neaton and J. M. Gregoire, *J. Mater. Chem. A*, 2016, **4**, 7483–7494.
31. C. Xiang, J. Haber, M. Marcin, S. Mitrovic, J. Jin and J. M. Gregoire, *ACS Comb. Sci.*, 2014, **16**, 120–127.
32. J. M. Gregoire, D. G. Van Campen, C. E. Miller, R. J. Jones, S. K. Suram and A. Mehta, *J. Synchrotron Radiat.*, 2014, **21**, 1262–1268.
33. Y. Xue, J. Bai, R. Le Bras, B. Rappazzo, R. Bernstein, J. Bjorck, L. Longpre, S. K. Suram, R. B. van Dover, J. Gregoire and C. P. Gomes, in *Proceedings of the Twenty-Ninth AAAI Conference on Innovative Applications*, Association for the Advancement of Artificial Intelligence, 2017, vol. IAAI-17.
34. S. K. Suram, Y. Xue, J. Bai, R. Le Bras, B. Rappazzo, R. Bernstein, J. Bjorck, L. Zhou, R. B. van Dover, C. P. Gomes and J. M. Gregoire, *ACS Comb. Sci.*, 2017, **19**, 37–46.

35. D. Guevarra, A. Shinde, S. K. Suram, I. D. Sharp, F. M. Toma, J. A. Haber and J. M. Gregoire, *Energy Environ. Sci.*, 2016, **9**, 565–580.
36. A. Shinde, D. Guevarra, G. Liu, I. D. Sharp, F. M. Toma, J. M. Gregoire and J. A. Haber, *ACS Appl. Mater. Interfaces*, 2016, **8**, 23696–23705.
37. A. Shinde, G. Li, L. Zhou, D. Guevarra, S. K. Suram, F. M. Toma, Q. Yan, J. A. Haber, J. B. Neaton and J. M. Gregoire, *J. Mater. Chem. A*, 2016, **4**, 14356–14363.
38. A. Shinde, D. Guevarra, J. A. Haber, J. Jin and J. M. Gregoire, *J. Mater. Res.*, 2014, **30**, 442–450.

CHAPTER 10

Membranes for Solar Fuels Devices

DANIEL J. MILLER* AND FRANCES A. HOULE

Lawrence Berkeley National Laboratory, Berkeley, CA, USA
*Email: danieljmiller@lbl.gov

Semipermeable membranes are widely used in gas separation, water purification, electrodialysis, fuel cells, and biomedical applications. Membranes are typically comprised of a thin polymeric material that permits selective transport of one or more chemical species. Membranes for fuel cells and other electrochemical applications have been extensively studied,[1–9] but the requirements for solar fuels devices demand the development of membranes with different properties than those of membranes designed for other applications.[10,11] Traditionally, work on solar fuels devices has focused on individual device components, such as optimization of electrodes for the oxygen evolution and hydrogen evolution reactions. However, the membrane performance is inextricably linked to that of the other components. With JCAP's focus on device and systems level considerations, membrane performance and design emerged as central design criteria that drive many other considerations, such as relevant electrolytes and component geometries. In the most basic solar fuels device architecture, the membrane is immersed in aqueous electrolyte between two planar electrodes, in common with many electrochemical device designs. To ensure that current can flow in the device and maintain overall electroneutrality, the membrane must permit the transport of electrolyte ions. However, to maximize device efficiency, the membrane must also limit the transport of oxidation and reduction products from one electrode to the other. The materials comprising

Energy and Environment Series No. 22
Integrated Solar Fuel Generators
Edited by Ian D. Sharp, Harry A. Atwater and Hans-Joachim Lewerenz

Published by the Royal Society of Chemistry, www.rsc.org

typical membranes impose certain limitations on suitable electrolyte environments (*i.e.*, pH), and therefore conditions for benchmarking and HTE discovery. These considerations are discussed in the chapters by McCrory (Chapter 5) and Gregoire (Chapter 9). The chapters by Weber (Chapter 13) and Xiang (Chapter 11) describe computational modeling and evaluation of prototype electrochemical water splitting devices incorporating membrane separators, further defining their performance requirements. Membrane materials requirements were outlined by Berger *et al.*,[10] and will be discussed in more detail below. The membrane, therefore, has focused the research efforts in many facets of JCAP.

This chapter will outline fundamental topics in membrane structure, characterization, and transport within the context of the key challenges for solar fuels device membranes. The chapter contents will first be framed with a discussion of overall transport challenges facing membranes in solar fuels devices, and then proceed to cover common membrane materials (including commercial membranes), fundamental aspects of ion and neutral molecule transport in membranes, and experimental membrane characterization. Finally, the development of membranes for solar fuels devices within JCAP to date will be discussed.

10.1 Transport Challenges in Membranes for Solar Fuels Devices

Devices for photoelectrochemical production of chemicals and fuels present a number of challenges for membrane performance and design. Solar fuels devices contain a complex mixture of reactants, products, and non-reacting species. A simplified schematic diagram of an electrochemical water-splitting device operating under alkaline conditions is shown in Figure 10.1. A photoabsorber is not shown, but either of the electrodes could be driven with captured solar energy. The device consists of, from left to right, a (photo)cathode where water is reduced to hydrogen gas, aqueous electrolyte, a membrane, aqueous electrolyte, and a (photo)anode where hydroxide ions are oxidized to oxygen gas. Electrons from the hydroxide oxidation are carried from the anode to the cathode, where they facilitate water reduction. In this simplified case, no supporting electrolyte is employed, so hydroxide ions are the primary charge carriers. If the device is operating under acidic conditions, protons would be the primary charge carriers, generated by water oxidation at the anode and consumed by proton reduction to form H_2 at the cathode. This dependence of charge carrier on electrolyte pH is one illustrative example of the connection between membrane performance and device characteristics. At high pH, the membrane must effectively transport negatively-charged hydroxide ions, while at low pH, the membrane must effectively transport positively-charged protons. Such requirements can have implications for the membrane chemistry and stability, as will be discussed later.

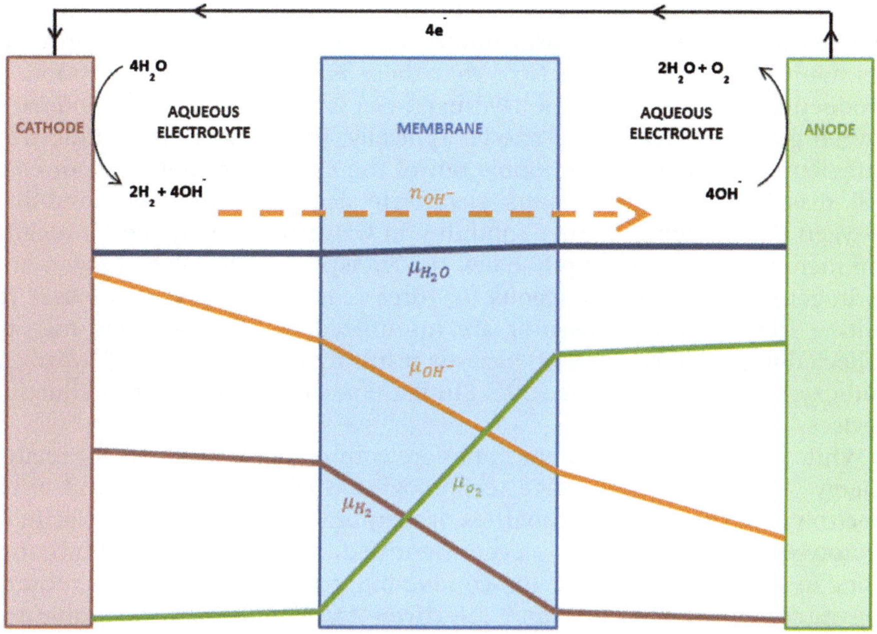

Figure 10.1 Schematic of a water-splitting device operating under alkaline conditions. Qualitative chemical potential gradients are shown for water, hydroxide ion, oxygen gas, and hydrogen gas in the aqueous electrolyte and hydrated membrane. For clarity, a simple electrolyzer is shown; in the case of a photoelectrochemical cell, either the cathode or the anode would be photo-driven in this configuration.

In Figure 10.1, there is no active mixing in the aqueous electrolyte compartments besides that due to motion of any gas bubbles that are formed, and all transport occurs *via* concentration-driven diffusion and/or electric potential-driven migration. Mass transfer boundary layers, which frequently develop near electrode and membrane surfaces, have been neglected for the purposes of this overall view, but their effects have been discussed in detail elsewhere.[12] The membrane dividing the cathode and anode compartments is assumed to be fully hydrated (that is, the membrane has sorbed water from the external solution and swollen to an equilibrium state). All components of the cell are at a uniform temperature.

Qualitative (not-to-scale) gradients in electrochemical potential for water, hydroxide ions, hydrogen gas, and oxygen gas are shown in Figure 10.1. These gradients motivate the transport of chemical species in the device, and may be affected by concentration, electric potential, and other factors.[13] Although water is consumed at a greater rate at the cathode than it is produced at the anode, the amount of water in each of the aqueous electrolyte compartments is assumed to be very large relative to that of the other species. Therefore, only a very small gradient in water electrochemical potential exists between the anode and cathode compartments, and transport of water across the membrane is negligible. Because hydroxide ions are the primary charge carriers, their transport from the cathode to the anode is required to maintain charge neutrality. Note that there is no overall consumption or production of hydroxide ions. Hydrogen gas is evolved at the cathode and oxygen gas is evolved at the anode. Typically, these gases are produced in sufficient quantity so as to bubble out of the electrolyte, but some of each will dissolve into the aqueous electrolyte according to their solubility. (Oxygen has a slightly higher solubility in water than hydrogen.[14]) Ideally, the membrane would entirely block the transport of dissolved oxygen and hydrogen, which is advantageous for three reasons: (1) potential losses resulting from product crossover are minimized, (2) the products may be collected in pure form, and (3) hazards resulting from mixtures of hydrogen and oxygen gas are mitigated.[10,15] Similar concerns exist for CO_2 reduction devices.

While high-performing membranes are commercially available for related energy generation and storage technologies, such as fuel cells and water electrolyzers, the optimal properties of membrane separators for solar fuels prototypes have only recently been identified, as part of early membranes work in JCAP.[10] In general, ion-conducting polymers used as membrane separators must efficiently and selectively transport electrolyte ions and rigorously limit the permeation of neutral species. Minimizing membrane permeability to oxidation and reduction products reduces their loss at the opposing electrode during device operation. Because typical currents are in the range of tens of mA cm^{-2},[10,16] not several A cm^{-2} as required by fuel cells and electrolyzers,[17–19] ionic conductivity need only be sufficient to avoid substantial pH changes in the anode and cathode compartments during operation. Therefore, the principal design criterion for solar fuels device membranes is limiting the transport of water splitting and/or CO_2 reduction products.[10] Their conductivity does not need to be as high as is required for electrolyzers. In the following sections, key considerations for solar fuels membrane design, function and characterization are described.

10.2 Membrane Materials and Structure

Generally, membranes used for solar fuels devices (and fuel cells, among other electrochemical applications) are comprised of dense, non-porous organic polymer films. Porous membranes, which are used more commonly in water purification, are not capable of discriminating among small molecules such as the products of water oxidation or CO_2 reduction.[20] Polymers are large macromolecules synthesized by various techniques where small monomer molecules are repeatedly added to a growing chain. Many polymers are mechanically robust, chemically resilient, and relatively inexpensive compared to inorganic materials, and depending upon the monomers employed, exhibit widely ranging properties that can be tuned to suit a particular application. The molecular weight of polymer chains can reach millions of grams per mole. High mechanical integrity of the membrane film, neither too soft or too brittle, is needed for solar fuels systems to ensure successful membrane performance at the thickness required to minimize resistance. Integrity generally improves with increasing molecular weight due to the many entanglements that occur with long chains. Mechanical integrity may also be improved by the addition of crosslinkers that physically (by strong secondary interactions) or covalently couple two or more chains together.[21]

Another important characteristic of solar fuels membranes is impermeability to electrochemically produced water splitting and CO_2 reduction products. Crystallinity may have a substantial effect on the transport of small molecules through polymeric films because diffusion does not readily occur through crystallites.[21] Polymer chains may be able to crystallize if their monomers are able to self-assemble into tightly packed, ordered structures. Crystalline domains typically exist within an amorphous matrix as only a small fraction of the overall material; chain ends and entanglements prevent perfect ordering of all chains. Crystallites may serve as physical crosslinks, improving overall mechanical integrity. However, highly crystalline materials may be brittle. Sample crystallinity may be readily observed by a number of techniques. Optical birefringence uses polarized light to reveal a high-contrast, specular pattern in crystalline regions.[21] Differential scanning calorimetry (DSC) measures the amount of heat required to raise the temperature of the sample at a specified rate as a function of temperature. At the melting temperature of the crystalline region, the DSC thermogram exhibits a sharp transition that may be quantitatively related to the degree of crystallinity.[22,23] The degree of crystallization in a polymer film may also be readily ascertained by small-angle or wide-angle X-ray scattering experiments, where crystalline samples diffract impinging X-rays at specific angles (whereas amorphous samples uniformly scatter X-rays).[21,24]

Polymer properties can be tuned by using more than one monomer during the polymer synthesis, enabling incorporation of multiple chemical functionalities in a copolymer. Depending upon the order of monomer addition, copolymers may exist as random, alternating, gradient, graft, or block copolymers.[21] Block copolymers can be conceptualized as two or more coupled homopolymers. One potential advantage of block copolymers is the ability to combine the favorable properties of two or more constituent homopolymers in a single covalently coupled chain. Block copolymers comprised of two or more chemically dissimilar homopolymers, such as one polar polymer and one non-polar polymer, may phase separate. Phase separation can create many different morphologies, such as lamellar, spherical, cylindrical, and others, and can be used to control the mechanical and transport properties of membranes.[25] Several techniques have been used to study block copolymer structure. Transmission electron microscopy has traditionally been used to visualize microstructures with the aid of heavy metal staining. More recently, small angle X-ray scattering (SAXS) and small-angle neutron scattering (SANS) have proven effective in elucidating microdomain size and spacing.[25]

The dense, non-porous thin films required for discrimination at the molecular scale can be formed by solution casting, where the solvent is allowed to evaporate from a polymer solution spread on a flat surface, or by melt pressing, where a dry polymer powder is heated and pressed into a homogeneous film. Provided that these films are mechanically robust enough to be handled, they may be used without further mechanical reinforcement (in this case, the films are typically isotropic). If the membrane is very thin or particularly delicate, a backing support layer that provides little or no mass transfer resistance (relative to the mass transfer resistance provided by the selective layer) may be employed.[26] Layered structures may also provide desirable separation characteristics. For example, bipolar membranes consist of alternating layers of positively and negatively charged polymers, and may be used with dissimilar electrolytes contacting each face.[27]

10.3 Commercial Membranes

A variety of ion exchange membranes suitable for electrochemical applications are available in the market.[5] However, the suitability of these materials for a solar fuels application varies. Ion exchange membranes contain charged chemical moieties that are covalently bound to the polymer backbone (fixed charge groups). If these fixed charge groups are positively charged, the membrane is said to be an anion exchange membrane, since anions are electrostatically attracted to the polymer matrix. Conversely, membranes containing negative fixed charge groups are cation exchange membranes. Ion exchange membranes are also commonly referred to as "polyelectrolyte membranes," since these materials contain a relatively high concentration of ions and are, therefore, reminiscent of electrolyte solutions, especially when hydrated with water. (The replacement of liquid electrolytes with polyelectrolytes in some applications, such as batteries, is an area of intense investigation.[28,29]) Synthetic polyelectrolyte membranes have been available commercially since the 1940s, largely targeted at various dialysis and electrolysis applications.[5] Examples of fixed charge groups commonly found in anion exchange membranes include $-NH_3^+$, $-NRH_2^+$, $-NR_2H^+$, $-NR_3^+$, $-PR_3^+$, and $-SR_2^+$, while common cation exchange groups include $-SO_3^-$, $-COO^-$, $-PO_3^{2-}$, and $-PO_3H^-$. Commercial ion exchange membranes may be broadly divided into homogeneous and heterogeneous types. Homogeneous membranes have fixed charge groups evenly distributed throughout polymer matrix. These membranes are commonly prepared by polymerizing functional monomers that already have a fixed charge group (or the precursor to a fixed charge group), or by post-polymerization modification of the polymer with the desired fixed charge group. Heterogeneous membranes have an uneven distribution of fixed charge groups throughout the matrix, which could be the result of spontaneous phase separation of two immiscible polymer phases, or a direct consequence of manufacturing procedures.[30] For example, many heterogeneous membranes contain a material that provides mechanical support to the ion exchange polymer. Several membranes, widely used ion exchange, are fabricated by the "paste method," wherein a non-woven PVC cloth is coated with a thick paste containing styrene monomer functionalized with a neutral precursor to the desired ion exchange moiety, divinylbenzene (a radical polymerization initiator), and finely powdered PVC. The coated fabric is then heated to induce polymerization, and the ion exchange groups are treated to produce a mechanically robust membrane with the desired ion exchange groups.[31] The properties of various commercially available ion exchange membranes are comprehensively reviewed elsewhere.[4,5,7,30–34]

The most widely-studied commercial ion exchange membrane is probably Nafion®, which has found industrial use in the Chlor-alkali electrolysis process.[7,9] Nafion® is a perfluorinated sulfonic acid (PFSA) copolymer introduced in the 1970s by DuPont. Several materials similar to Nafion® have since come to market, including Aciplex® (Asahi Kasei), Flemion® (Asahi Glass), Aquivion® (Solvay Specialty Polymers), and several products from 3M. Nafion® and related materials have charged, sulfonated side chains tethered to a neutral, semicrystalline poly(tetrafluoroethylene) (PTFE) backbone. Figure 10.2 shows the chemical and morphological structure of PFSA polymers used for energy applications.[7]

The charged side chains in PFSA polymers are randomly distributed along the backbone. The primary differences among these materials are the chemical nature and length of the side chain. In all of these materials, the charged side chains tend to phase separate from the uncharged backbone due to the relative hydrophilicity of the side chains and the relative hydrophobicity of the PTFE backbone. The clusters of charged side chains become highly hydrated and swollen when the membrane is in an aqueous or humidified environment; the vast majority of ion and neutral molecule transport occurs through this hydrated phase. In comparison, transport through the hydrophobic, semicrystalline phase is very poor. The hydrated, swollen clusters of side chains are commonly referred to as "hydrated channels," but it is important to remember that these channels are *not* open pores, which would insinuate that transport occurs by convective flow. Instead, transport of ions and solutes occurs by the solution-diffusion mechanism, whereby diffusion is responsible for the movement of ions and solutes through the hydrated phase.[35] The semicrystalline PTFE phase is mechanically robust, thermally stable, and largely chemically inert. The combination of the hydrophobic PTFE phase and the hydrophilic charged phase therefore provides a desirable combination of material stability with efficient transport of solutes and ions.[7]

In addition to the simple cation and anion exchange membranes already described, bipolar membranes, consisting of a cation exchange membrane and an anion exchange membrane laminated together, have found use. Bipolar membranes have been evaluated in fuel cells,[36] electrochemical cells,[37,38] and photoelectrochemical cells.[27,39,40] Because bipolar membranes present different polarities to each chamber of an electrochemical cell, they permit the use of different electrolytes and different pH ranges in adjacent

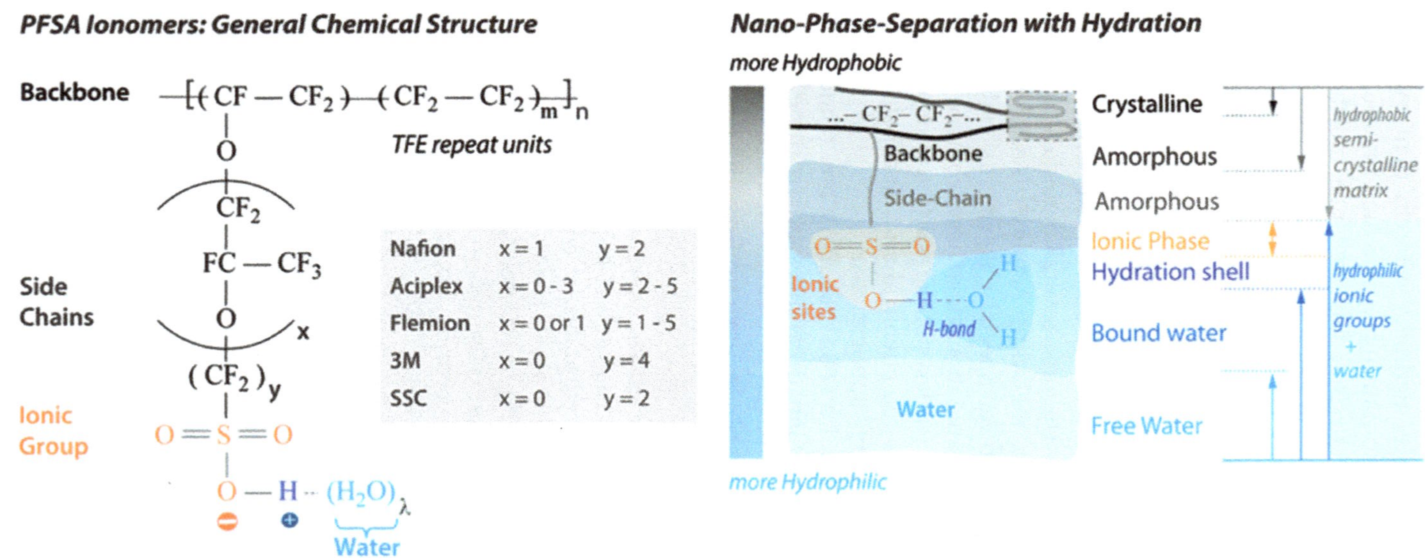

Figure 10.2 Chemical structure of some common PFSA polymers (left). Phase separation of hydrophobic backbone and hydrophilic side chains in PFSA polymers (right).
Reprinted with permission from A. Kusoglu and A. Z. Weber, *Chem. Rev.*, 2017, **117** (3), 987, http://pubs.acs.org/doi/abs/10.1021/acs.chemrev.6b00159. Copyright 2017 American Chemical Society.[7]

compartments of photoelectrochemical devices, so materials used in device construction may be optimized.[27,39] Under forward bias, where the polarity of each side of the membrane opposes the polarity of the electrode facing it, ion migration causes charge to become compensated near the internal interface, allowing co-ions of the anion exchange and cation exchange materials to traverse the membrane, increasing polarization of the device. When the cell is operated in reverse bias mode, where the membrane is oriented such that the polarity of each side of the membrane matches the polarity of the electrode facing it, most ions are driven out of the membrane and water dissociation into protons and hydroxide ions is promoted at the junction of the anion exchange and cation exchange materials.[39] In operation, a large voltage drop is sustained across the two electrode chambers, which comes at a cost of higher cell operating potential and the need to maintain stable bias direction at all times. Nonetheless, when the cell design is optimized for bipolar membrane use through selection of materials and inclusion of stirring to maintain electrolyte homogeneity, water splitting efficiency can reach the same level found for anion- or cation-conducting membranes.[27] A comprehensive discussion of system characteristics that influence performance has been presented in these studies, providing design criteria for discovery of optimized bipolar membrane materials for solar fuels applications. Figure 10.3 illustrates a bipolar membrane operating in a

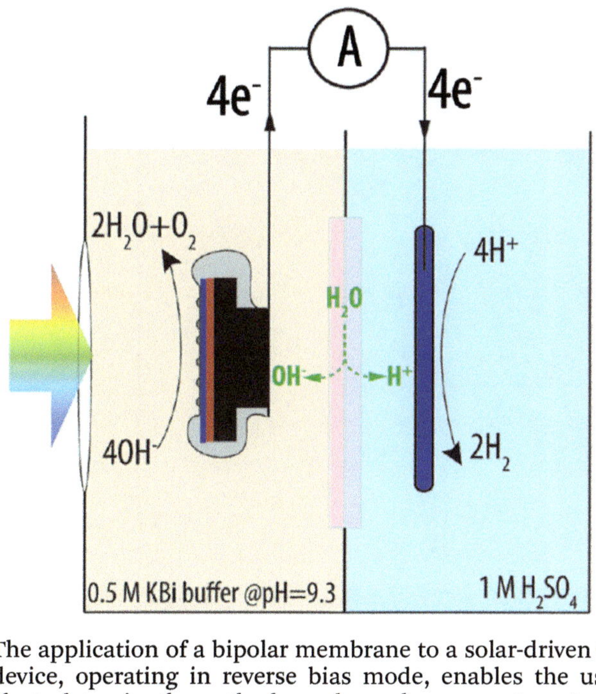

Figure 10.3 The application of a bipolar membrane to a solar-driven water splitting device, operating in reverse bias mode, enables the use of differing electrolytes in the cathode and anode compartments by promoting water dissociation at the interface of cation and anion exchange layers. Reproduced from ref. 27 with permission from John Wiley and Sons, © 2016 WILEY-VCH Verlag GmbH & Co. KGaA, Weinheim.[27]

solar-driven water splitting device with differing electrolytes in the cathode and anode compartments.[27]

10.4 Transport of Solutes in Membranes

Transport of water, ions, and neutral molecules (*e.g.*, protons, hydroxide ions, electrolyte ions, and the products of water splitting and CO_2 reduction) through dense, non-porous polymers is described by the solution–diffusion model.[41] In this model, the solute (water, ion, or neutral molecule) dissolves into the polymer matrix at one face of the membrane, diffuses through the membrane thickness, and elutes at the other face. The driving force for the movement of solute is a gradient in solute electrochemical potential over the membrane; gradients in pressure, temperature, concentration, and electrical potential can all contribute to the overall electrochemical potential gradient.[41] Figure 10.4 shows examples of gradients in electrochemical potential (μ_i) and concentration (c_i) that induce a flux (n_i) of solute i.

The overall permeability (P_i) of a membrane to a species i is given by the product of a thermodynamic term and a kinetic mobility term: the sorption

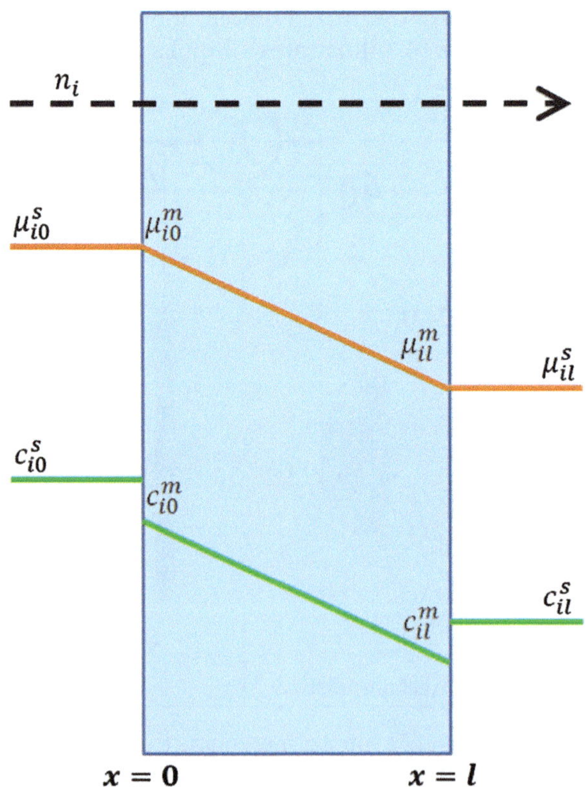

Figure 10.4 Electrochemical potential (μ_i) and concentration (c_i) profiles that drive transport of species i through a dense membrane.

coefficient (K_i) of solute i in the polymer matrix, which describes the amount of solute sorbed into the membrane, and the diffusion coefficient (D_i) of solute i, which describes the movement of the solute through the polymer matrix.[41]

$$P_i = K_i D_i \tag{10.1}$$

Eqn (10.1) suggests that high diffusivities, high solubilities, or both can contribute to high permeabilities, and that a high diffusivity could be counteracted by a low solubility (or *vice versa*).[42] The permeability is a material property and is normalized by the membrane thickness and driving force for solute transport. The selectivity of a membrane for component A over component B, $\alpha_{A/B}$, is a simple ratio of permeabilities:[43]

$$\alpha_{A/B} = \frac{P_A}{P_B} = \left(\frac{K_A}{K_B}\right)\left(\frac{D_A}{D_B}\right) \tag{10.2}$$

The ratios K_A/K_B and D_A/D_B are the solubility selectivity and diffusivity selectivity, respectively, representing the relative contributions of differing solubilities and diffusivities of components A and B to the overall permeability selectivity.[43]

The effective permeability of a membrane to a solute i is given by:

$$P_i = \frac{n_i l}{c_{i0}^{s} - c_{il}^{s}} \tag{10.3}$$

where n_i is the molar flux of solute i through the membrane, l is the thickness of the membrane, c_{i0}^{s} is the concentration of solute i in the solution in contact with the upstream face of the membrane, and c_{il}^{s} is the concentration of solute i in the solution in contact with the downstream face of the membrane.[44] In this simple example, the only contribution to the electrochemical potential gradient that drives solute transport is a difference in solute concentration over the membrane (*i.e.*, there is no applied electrical potential, there are no pressure or temperature differences over the membrane, *etc.*). The flux n_i at steady state and in the absence of convective flow is given by Fick's Law:

$$n_i = -D_i \frac{\mathrm{d}c_i^{m}}{\mathrm{d}x} = D_i \frac{c_{i0}^{m} - c_{il}^{m}}{l} \tag{10.4}$$

where c_i^{m} is the concentration of solute i at distance x through the membrane, c_{i0}^{m} is the concentration of solute i in the membrane at the high concentration face ($x = 0$), c_{il}^{m} is the concentration of solute i in the membrane at the low concentration face ($x = l$), and D_i is the concentration-averaged diffusion coefficient.[44] It is generally easiest to measure membrane permeability when the concentration of solute i in the solution and in the membrane at the upstream face are much greater than its concentration in solution and in the membrane at the downstream face (*i.e.*, $c_{i0}^{s} \gg c_{il}^{s}$ and $c_{i0}^{m} \gg c_{il}^{m}$). In this case, eqn (10.3) and (10.4) may be combined and simplified, yielding:[44]

$$P_i = D_i \frac{c_{i0}^{m}}{c_{i0}^{s}} \tag{10.5}$$

10.5 Solute Sorption

Comparison of eqn (10.1) and (10.5) shows that the sorption coefficient, K_i, of solute i in the membrane is:[44,45]

$$K_i = \frac{c_{i0}^{m}}{c_{i0}^{s}} \tag{10.6}$$

The solubility, therefore, describes the partitioning of solute molecules into the polymer matrix from the external solution, and is influenced by solute properties, secondary interactions between the solute and the polymer chains, as well as polymer morphology.[46] Note that, at thermodynamic equilibrium, the electrochemical potential of solute i in the external solution, μ_{i0}^{s}, is identical to the electrochemical potential of solute i in the membrane at the upstream face, μ_{i0}^{m} (*cf.*, Figure 10.2). A similar equality exists at the downstream face ($\mu_{il}^{m} = \mu_{il}^{s}$). However, depending upon the partitioning of the solute into the membrane from the external solution, a discontinuity may exist in solute concentration between the external solution and the membrane (*i.e.*, c_{i0}^{s} and c_{il}^{s} may not equal c_{i0}^{m} and c_{il}^{m}, respectively). The nature and magnitude of this discontinuity is expressed in the solubility coefficient. In the example shown in Figure 10.2, partitioning from the external solution into the membrane is unfavorable, so $K_i < 1$.

Many polymers used as membranes in energy applications have charged moieties along their backbone, which facilitate conduction of ions through them.[28,29] Uptake of charged solutes by ion exchange polymers is either enhanced or suppressed depending upon the relative charges of the solute(s) and the polymer.[29,43] For example, a negatively charged (cation exchange) polymer (*e.g.*, Nafion®) will generally sorb more cations than anions. To maintain charge neutrality, the sorbed cations will displace cations associated with the negatively charged moieties on the polymer chain, and the degree to which this displacement occurs is based upon the relative association affinities of the cations and the quantity of each of the cations available.[43] Anions will tend to be excluded from the polymer matrix by the negatively charged moieties on the polymer chain, and will balance cations in solution that were displaced from the polymer. This phenomenon is called Donnan exclusion and is affected by, among other factors, pH, external ion concentration, polymer ion exchange capacity (the concentration of charged moieties in the polymer matrix), and the presence of an electrolyte.[20,29,43] Figure 10.5 illustrates the uptake of cations and anions in an uncharged polymer and in a polymer that has fixed anionic charge groups from a salt solution.[43]

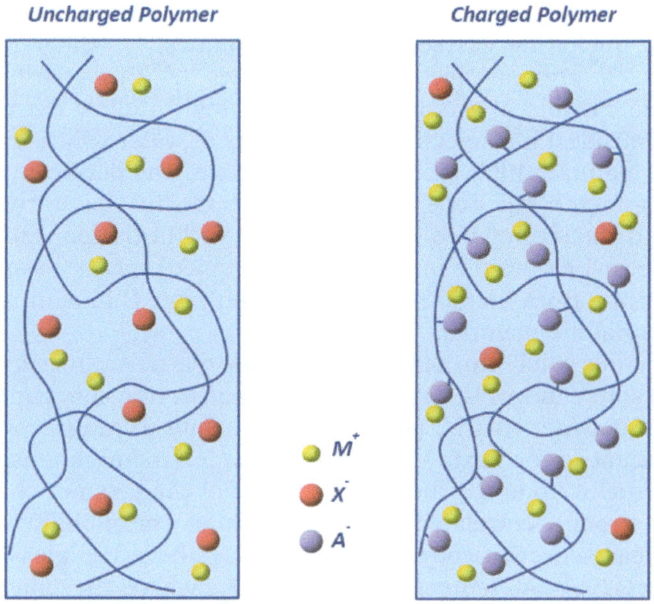

Figure 10.5 Comparison of ion uptake by charged and uncharged polymers from a salt (M^+X^-) solution. The uncharged polymer sorbs equal numbers of cations and anions. The charged polymer, which has fixed cation exchange (negatively charged) groups (A^-), sorbs more cations to balance the fixed negative charges and fewer anions. Shielding of ion uptake by fixed charge groups in the polymer is Donnan exclusion.[43]

These principles are important for water splitting and CO_2 reduction devices; however, the requirements are more complicated in the latter. In addition to the charge-carrying electrolyte ions, many CO_2 reduction products are either charged (*e.g.*, formate, acetate) or are polar (*e.g.*, CO, methanol, ethanol). If the predominant CO_2 reduction products are ionic, their permeation through the membrane could be reduced by using a membrane of similar charge to the product(s). For example, if a CO_2 reduction device produces formic acid, the use of a cation exchange membrane, which will tend to exclude the formate ion by electrostatic repulsion and reduce product crossover, is desirable. Uncharged but polar products present a more difficult challenge, because charged membranes tend to substantially sorb water and will also likely sorb these polar products, which will not be electrostatically excluded.

10.6 Solute Diffusion

One disadvantage of nearly all polymeric membranes is a tradeoff between permeability and selectivity. That is, as the amount of material that can pass through the membrane increases, the capability of the membrane to distinguish among solutes (and, therefore, reject one or more of the solutes) decreases. This empirical tradeoff has been observed for many types of membranes, including gas separation membranes,[46,47] water purification membranes,[20,43] and polyelectrolyte membranes.[43,48] In a solar fuels device, a membrane capable of passing electrolyte ions (to allow current to flow between electrodes) but rejecting the products of water splitting or CO_2 reduction (to maintain high efficiency) is desirable. The permeability and selectivity of a membrane depend strongly on both the size of the diffusing solute(s) and the size of free volume cavities within the polymer matrix.[49] Free volume cavities are the interstitial (unoccupied) spaces among polymer chains resulting from inefficient packing of polymer chains in an amorphous, non-crystalline material.[26]

Movement of a solute through the membrane matrix occurs *via* diffusional jumps. The solute moves through the membrane as transient gaps open and close in the polymer matrix, permitting the solute to jump from one free volume element to the next.[26] Diffusion through a membrane is, therefore, highly dependent upon the molecular motion of the polymer chains which is, in turn, strongly correlated with temperature. The mobility of polymer chains increases dramatically at temperatures above the glass transition temperature (T_g).[42] At temperatures below T_g, where the polymer is described as "glassy," the mobility of polymer chains is substantially retarded. Above T_g, in the "rubbery" regime, chain mobility is much higher. Consequently, the diffusivity of a solute through a polymer at a temperature above T_g is generally much higher than its diffusivity at a temperature below T_g.[42] If the polymer is semicrystalline, it will exhibit a thermal transition corresponding to the melting of crystallites (T_m) that is higher than T_g. Both T_g and T_m are readily determined from DSC experiments.[21]

The molecular chain motion that results in the opening and closing of transient gaps in the polymer matrix is thermally activated.[49] The dependence of the diffusivity (and, by extension, permeability) on temperature is typically described by an Arrhenius relationship:

$$D_i = D_0 \exp\left(-E_{D_i}/RT\right) \tag{10.7}$$

where D_0 is a constant, E_{D_i} is the activation energy of diffusion of component i, R is the ideal gas constant, and T is the bulk temperature.[50–52] In glassy polymers, E_{D_i} is typically independent of temperature, but in rubbery polymers, E_{D_i} can vary with temperature.[42] For many solutes, an increase in E_{D_i} is observed at the glass transition temperature, which might be expected given the appreciable increase in chain mobility as temperature exceeds T_g.[53] It is likely that the performance of a membrane will vary since a solar fuels device is subjected to daily and seasonal temperature variations.[54]

The diffusion coefficient D_i of a solute i in a polymer is directly related to the fractional free volume $\langle v_f \rangle$ of the polymer by:

$$D_i = a_i \exp\left(\frac{-b_i}{\langle v_f \rangle}\right) \tag{10.8}$$

where a_i and b_i are adjustable parameters; a_i is related to a geometric factor and the velocity of solute i, and b_i is related to the Lennard–Jones diameter of solute i.[43,46,55–57] The total free volume in a polymer is the sum of the voids produced by inefficient packing of polymer chains and the transient gaps that open as a result of thermal motion.[26] From eqn (10.8), as $\langle \nu_f \rangle$ increases, so does the diffusivity of a solute of a particular size. However, due to the exponential relationship between $\langle \nu_f \rangle$ and D_i, the diffusivity of a large solute (relatively large b_i) will increase more rapidly than the diffusivity of a small solute (relatively small b_i) as $\langle \nu_f \rangle$ increases, leading to a decrease in the diffusivity selectivity (eqn (10.2)).[43]

The size similarity of charge-carrying ions and the products of water splitting or CO_2 reduction makes the design of selective membranes for solar fuels devices challenging. Table 10.1 presents the diameters of several species of interest, including charge-carrying ions, water splitting products, and some common CO_2 reduction products. In the example shown in Figure 10.1, the principle charge carrier (OH^-) has a larger diameter (3.00 Å) than hydrogen gas (2.89 Å). For this reason, H_2 will likely have a high diffusivity in membranes where the diffusivity of OH^- is high. CO_2 reduction products are typically larger in diameter than hydrogen and oxygen gases, but a similar situation arises if a supporting electrolyte is introduced. For example, buffered carbonate electrolytes have been proposed for CO_2 reduction devices, and these devices are typically operated at moderate pH to

Table 10.1 Diameter of some important solutes in water splitting and CO_2 reduction devices. For ionic species, hydrated radii are presented; for non-ionic species, kinetic diameters are presented.

Solute		Diameter (Å)
Water[58]	H_2O	1.38
Carbon dioxide[26]	CO_2	3.30
Charge carriers		
Proton[58]	H^+	2.82
Hydroxide[58]	OH^-	3.00
Sodium[58]	Na^+	3.58
Potassium[58]	K^+	3.31
Carbonate[58]	CO_3^{2-}	3.94
Water splitting products		
Hydrogen[26]	H_2	2.89
Oxygen[26]	O_2	3.46
CO_2 reduction products		
Carbon monoxide[26]	CO	3.76
Methane[26]	CH_4	3.80
Ethylene[26]	C_2H_4	3.90
Formic acid[59]	CHOOH	4.6
Acetic acid[59]	CH_3COOH	4.4
Methanol[60]	CH_3OH	3.6
Ethanol[60]	C_2H_5OH	4.5

minimize polarization losses and to maximize the concentration of dissolved CO_2 in the aqueous electrolyte.[11] However, some of the most commonly reported CO_2 reduction products (*e.g.*, CO, methane, ethylene, and methanol) have smaller diameters than carbonate (3.94 Å), so those products are likely to readily diffuse through a membrane in which carbonate has a high diffusivity. Fortunately, it appears that artificial photosynthesis devices can operate efficiently with membranes of lower ionic conductivity than those employed in other electrochemical processes.[10] Therefore, membrane permeability to both charge carriers and products may be reduced somewhat before substantially affecting device operation.

10.7 Water Sorption

Water uptake is essential to the function of membranes used for solar fuels applications because it facilitates the transport of charge-carrying electrolyte ions. Hydrophilic polymers, such as those containing polar moieties, tend to take up more water than hydrophobic polymers. Polymers functionalized with fixed charge groups also generally sorb water due to the energetic favorability of surrounding the charged moiety with polar water molecules.[43] Water uptake, w_u, is typically reported as a fraction and is defined as the mass of water sorbed by the polymer divided by the mass of the dry polymer:[45]

$$w_u = \frac{m_w - m_d}{m_d} \tag{10.9}$$

where m_w is the mass of the fully hydrated polymer and m_d is the mass of the dry polymer. The volume fraction of water in the membrane Ø_w^m is related to the water uptake:[45]

$$\text{Ø}_w^m = \frac{w_u}{w_u + \dfrac{\rho_w}{\rho_p}} \tag{10.10}$$

where ρ_w is the density of water, and ρ_p is the density of the dry polymer. This equation assumes volume additivity, which is reasonable in most polymers of interest under typical laboratory conditions.[43] The concentration of water in the membrane at the upstream face, c_{w0}^m, necessary to calculate the water sorption coefficient, K_w, using eqn (10.5) is difficult to measure directly. However, calculation of K_w is straightforward using Ø_w^m:

$$K_w = \frac{\text{ø}_w^m M_w}{c_w^s V_w} \tag{10.11}$$

where M_w is the molar mass of water, c_w^s is the concentration of water in the external solution, and V_w is the molar volume of water.[44]

Water uptake generally increases with an increasing concentration of fixed charge groups and with increasing acidity/basicity.[29] Figure 10.6 shows the

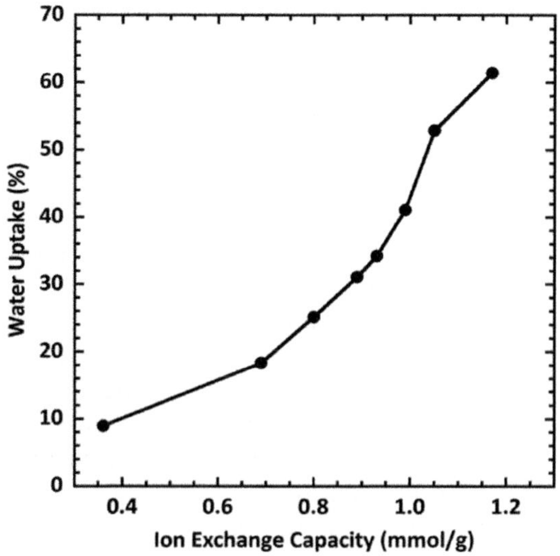

Figure 10.6 Effect of ion exchange capacity on water uptake in a perfluorinated sulfonic acid membrane. The counter-ion neutralizing the sulfonic acid moieties is a proton.
Reproduced from ref. 62 with permission from the Royal Society of Chemistry.[62]

monotonic increase in water uptake of a perfluorinated sulfonic acid membrane (*e.g.*, Nafion®) as the number of sulfonic acid fixed charge groups (here described as the ion exchange capacity, *i.e.*, mmoles of sulfonic acid groups per gram of dry polymer) increases.[61]

Water uptake results in changes to the polymer matrix because water molecules surround the polymer chains. Swelling can result in a loss of mechanical integrity and, in the limit of a critical level of solvation of the polymer chains by water,[63] dissolution of the membrane. Swelling can be restricted by crosslinking the polymer chains, either by covalently tethering chains together or by using block copolymers that phase separate to form physical crosslinks.[43] Sufficient crosslinking can make it possible to form solid, hydrated films from materials that would otherwise dissolve in water (*e.g.*, poly(ethylene glycol)).[64] As a crosslinked polymer sorbs water and begins to swell, the crosslinks constrain swelling; therefore, increasing crosslink density typically decreases water uptake. The external solution concentration can also affect the degree of hydration of a membrane. If the external solution has a high osmotic pressure (*e.g.*, a highly saline solution), a hydrated membrane may "de-swell" as water migrates out of the membrane into the external solution, resulting in a measurable decrease in thickness.[44]

Hydration and swelling of a polymeric membrane can strongly influence the transport of water, ions, or other solutes. As suggested by eqn (10.1), membranes with high water sorption tend to have high water permeability.

Free volume generally increases as a polymer hydrates and swells, so high water sorption also tends to increase solute diffusion coefficients because the large, swollen free volume elements permit less restricted movement through the polymer matrix.[43,57] Increases in diffusivity contribute to increases in overall water permeability in highly swollen polymers. The diffusion coefficient of water, for example, through highly hydrated polymers is often near its self-diffusion coefficient.[43] Membranes with high water sorption also tend to exhibit high solubilities, diffusivities and permeabilities to other solutes.[43] Figure 10.7 illustrates the increasing permeability of Nafion® to water and protons as the membrane water content (moles of water per mole of fixed sulfonic acid groups) increases.[65,66]

The sorption of ions and other hydrophilic solutes, in particular, typically increases with increasing hydration because of favorable hydrophilic interactions between the polymer and a polar solute and/or its hydration shell.[43] Diffusivities increase with increasing free volume and, consequently, overall solute permeability increases. Often, solute selectivity decreases as membranes become highly swollen and they lose the capacity to discriminate among solutes on the basis of differing solubilities or diffusivities.[43]

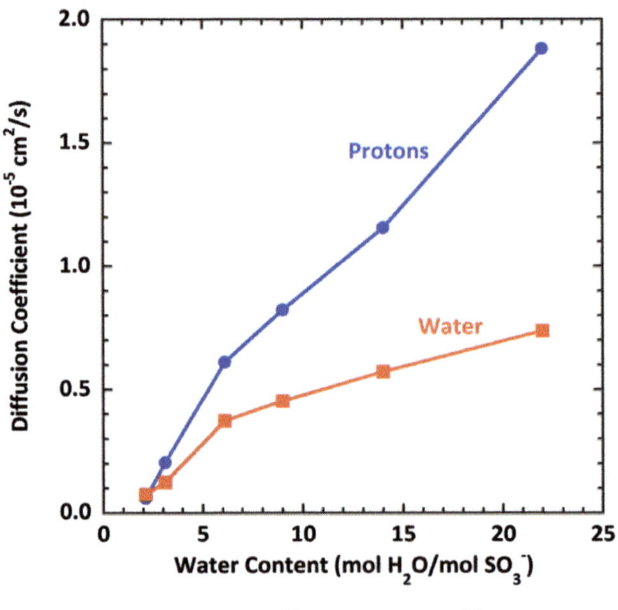

Figure 10.7 The diffusivity of protons[65] (●) and water[66] (■) in Nafion® 117 as a function of membrane hydration. Here, membrane hydration is expressed as the ratio of moles of water in the membrane to moles of fixed sulfonate groups in the membrane.
Reproduced with permission from T. A. Zawodzinski, M. Neeman, L. O. Sillerud and S. J. Gottesfeld, *Phys. Chem.*, 1991, **95** (15), 6040, Copyright 1991 American Chemical Society, and the Journal of the Electrochemical Society, Copyright © 1948, CCC Republication.

The polyelectrolyte membranes employed in solar fuels devices tend to be highly hydrated, as polar water molecules have an affinity for fixed charge groups in the polymer.[43] High levels of hydration, therefore, while facilitating the requisite transport of charge-carrying electrolyte ions, can confound the ability of the membrane to discriminate between charge-carrying ions and the products of water splitting and/or CO_2 reduction.

10.8 Electrical Properties

The capacity of a membrane to permit the transport of ions under an electric field is its conductivity. Membranes used in an electrochemical cell, such as a fuel cell or solar fuels device, must be sufficiently conductive to ions to permit current to flow in the device. Ions must be free to migrate between electrodes to maintain charge neutrality while electrons pass from one electrode to the other. Conductivity may be measured in the plane of the membrane or through the plane of the membrane.[67] Through-plane measurements are the more relevant measurement for device operation, since ions pass through the membrane thickness from one electrolyte reservoir to the other. However, in-plane measurements are generally easier to perform and are, therefore, more frequently reported. Furthermore, in-plane measurements may be made along two different axes of the membrane (longitudinal and transverse). Anisotropy in the conductivity of protons and other ions has been reported to various degrees in commercial ion exchange membranes, as through-plane, in-plane longitudinal, and in-plane transverse measurements often produce different conductivity values.[67]

The ionic conductivity σ of a membrane may be calculated by:

$$\sigma = \frac{l}{\mathscr{R}A} \tag{10.12}$$

where l is the membrane thickness, $\mathscr{R}$ is the resistance of the membrane, and A is the cross-sectional area available for ion transport.[68] Note that the cross-sectional area for ion transport, A, is calculated differently for in-plane and through-plane conductivity measurements. Nyquist plots, obtained *via* AC impedance spectroscopy using small potentials (10 mV), may be used to deduce the membrane resistance.[28] Heterogeneous polyelectrolytes, such as ion-conductive block copolymers, have been the subject of many recent studies.[68–73] In the case of a heterogeneous polyelectrolyte, the ionic conductivity is a function of the volume fraction of the material that comprises the conducting phase, as well as the geometric orientation of the conducting phase (if it adopts an anisotropic morphology). Typically, the relationship between the overall conductivity of the heterogeneous material, σ_h, and the intrinsic conductivity of the neat conducting phase, σ_c, is given by a simple approximation:

$$\sigma_h = f\phi_c\sigma_c \tag{10.13}$$

where f is an adjustable factor accounting for the connectivity and tortuosity of the conducting phase, and ϕ_c is the volume fraction of the conducting phase.[28]

The intrinsic diffusion coefficient, D_{ic} of ion i in the conducting phase may likewise be scaled to give the overall diffusion coefficient D_{ih} of ion i in the heterogeneous material:[28]

$$D_{ih} = fD_{ic} \tag{10.14}$$

Note that the diffusion coefficient does not depend upon the volume fraction of the conducting phase. Both of these approximations assume that no ion transport occurs in the non-conducting phase.[28] In the simplest case, the intrinsic values of the conducting phase within the heterogeneous material equal those of the pure conducting material (*i.e.*, $\phi_c = \phi$ and $D_{ic} = D_i$). Ideal values of f for various block copolymer morphologies, including gyroid, lamellar, cylindrical, and spherical, are tabulated elsewhere.[28] For heterogeneous materials containing more than one conducting block, more complex models are required. Models describing the conductivity of several polymer morphologies as various combinations of series and parallel conductivity additivity have been proposed.[48]

The Nernst–Planck equation reformulates Fick's law of diffusion (eqn (10.3)) to incorporate the effect of an electric field on the non-convective flux (*i.e.*, diffusive transport due to a concentration gradient or drift due to an imposed electric field) of ion i through an electrolyte, such as a solid polyelectrolyte. In the absence of convective flow:

$$n_i = -D_i\left(\frac{dc_i^m}{dx} + \frac{z_i F c_i^m}{RT}\frac{d\psi}{dx}\right) \tag{10.15}$$

where z_i is the valence of ion i, F is Faraday's constant, R is the ideal gas constant, T is the bulk temperature, and ψ is the applied electric potential.[29] The overall diffusion coefficient of a salt (*e.g.*, an electrolyte), $D_{\pm}$, consisting of an anion and cation that must move together to maintain charge neutrality, is given by the mobility-weighted average of cation and anion diffusion coefficients:

$$D_{\pm} = \frac{u_+ D_- + u_- D_+}{u_+ + u_-} \tag{10.16}$$

where D_i is the diffusivity of the cation or anion (i = + or −) and u_i is the mobility of the cation or anion.[28] The mobility of an ion is given by the Nernst–Einstein relationship:[29]

$$u_i = \frac{|z_i| F D_i}{RT} \tag{10.17}$$

Faraday's constant, F, relates the ionic conductivity to the valence, mobility, and concentration of ion i in a mixture of ions:[74]

$$\sigma = F\sum_i |z_i| u_i c_i \tag{10.18}$$

Therefore, the ionic conductivity of a solid polymer is directly related to the diffusivity of ions within it. For a pair of ions (*e.g.*, an electrolyte) moving through an uncharged polymer matrix, combination of eqn (10.17) and (10.18) yields:

$$\sigma = \frac{F^2}{RT}\left(|z_+|c_+^m D_+ + |z_-|c_-^m D_-\right) \tag{10.19}$$

In the case of membranes with fixed charge groups within the polymer matrix, counter-ions may diffuse alone. For example, an anion exchange membrane contains immobile positive charges ($D_+ = 0$), so cations are largely excluded and only anions contribute to membrane conductivity and eqn (10.19) simplifies to:[28]

$$\sigma = \frac{F^2}{RT}|z_-|c_-^{\mathrm{m}} D_- \tag{10.20}$$

As is the case for simple concentration gradient-driven diffusion of ions, conductivity is strongly influenced by membrane hydration and temperature.[67,75] Literature data on the water uptake and proton conductivity of polyelectrolyte fuel cell membranes have been compiled and are shown in Figure 10.8.[48] An empirical upper bound relationship between proton conductivity and water uptake emerges: high conductivities have not been reported for polyelectrolytes with low water uptake. That is, polymeric membranes with a given water uptake typically do not exhibit conductivities that exceed the solid black line. Therefore, while high water uptake does not guarantee high proton conductivity, high water uptake appears to be necessary for high conductivity.[48] Since the diffusion of other solutes may be facilitated by water sorption, the upper bound relationship illustrated in Figure 10.8 emphasizes the challenge of developing membranes capable of high electrolyte ion conductivity that also limit the permeation of other solutes, such as oxidation/reduction products whose permeation reduces device efficiency and product purities.

As described previously, solute diffusion coefficients increase by an Arrhenius relationship as temperature increases (eqn (10.7)). The relationship between ionic conductivity and temperature in glasses is expressed by the Vogel–Tamman–Fulcher (VTF) equation, which reflects the exponential dependence of the diffusivity on temperature described by the Arrhenius equation:

$$\sigma = \alpha \exp\left(\frac{-\beta}{R(T - T_0)}\right) \tag{10.21}$$

where α is an adjustable parameter related to the concentration of ions, and β is an adjustable parameter related to the activation energy of ion motion.[28,76,77] The Vogel temperature, T_0, is related to the glass transition temperature, T_g, and is commonly approximated as $T_0(\mathrm{K}) = T_g(\mathrm{K}) - 50$. However, notable deviations from this rule of thumb have been reported.[28,77,78] The VTF equation captures the interdependence of

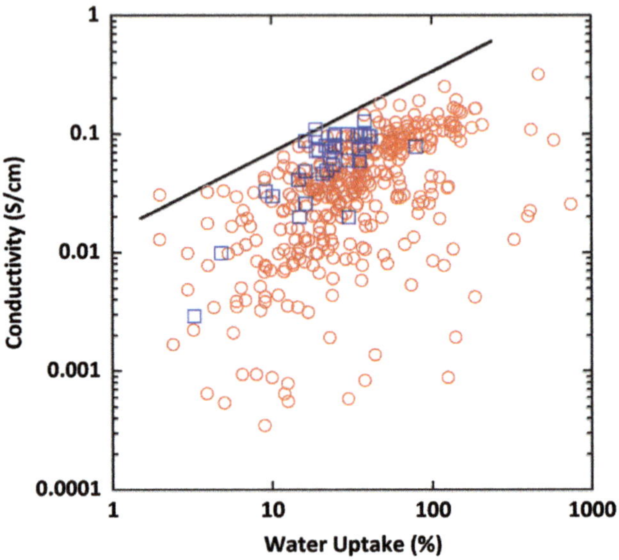

Figure 10.8 Upper bound relationship for conductivity and water uptake in various proton exchange membranes (○), including many perfluorinated sulfonic acid polymers (□). Data were collected from the literature for conductivities in the range of 20–30 °C.
Reprinted from *Journal of Membrane Science*, **302**, L. M. Robeson, H. H. Hwu and J. E. McGrath, Upper bound relationship for proton exchange membranes: Empirical relationship and relevance of phase separated blends, 70–77, Copyright 2007, with permission from Elsevier.[48]

conductivity, ion concentration, and the glass transition temperature. In a polyelectrolyte, T_g is typically a function of the concentration of mobile ions (*i.e.*, not bound to the polymer chains) inside the polymer matrix.[76] Most commonly, T_g decreases with increasing concentration of mobile ions, resulting in a corresponding decrease in T_0. Therefore, as expected, conductivity tends to increase with increasing mobile ion concentration.

Under the influence of an electric field, cations and anions move in opposite directions and do not necessarily carry equal amounts of charge. Therefore, for the purposes of quantifying device performance, it is sometime preferable to consider the amount of charge transported across a membrane rather than the number of solutes transported.[26] The transference number is the fraction of the total current carried by a particular ion (*i.e.*, the contribution to conductivity of a particular ion divided by the total conductivity):[74]

$$t_i = \frac{|z_i| u_i c_i}{\sum_j |z_j| u_j c_j} \tag{10.22}$$

It follows that, for all ions *i*:[74]

$$\sum_i t_i = 1 \tag{10.23}$$

For ion permeation in a charged membrane, the ions that are excluded by electrostatic repulsion (*i.e.*, Donnan exclusion) have very low or negligible transference numbers, while counter-ions to the fixed charge groups have high transference numbers (near unity). Therefore, the non-excluded ions are responsible for carrying nearly all of the charge through a polyelectrolyte membrane.[26] Transference numbers of cations and anions permeating an uncharged membrane may be much closer in magnitude than in the case of a charged membrane. The conductivity, diffusion coefficient, and transference number are, together, the three transport parameters required to fully describe ion motion in electrolytes. Most studies of polyelectrolytes have, however, been limited to measurement of conductivity.[28]

It should be noted that the framework described above was developed for classical, dilute electrolytes, and may not adequately describe the physics associated with ion transport in concentrated polyelectrolytes. For example, ion activity coefficients, which account for deviations from ideal behavior, are frequently not properly accounted for in studies of concentrated systems.[28,79] It is not clear whether the relationships developed for dilute systems may be directly applied to transport in polyelectrolyte membranes. Recent studies of charged membranes have focused on developing predictive models for ion permeability, solubility, and diffusivity in polyelectrolyte membranes by incorporating ion activity coefficients into classical theory to account for non-idealities.[79–81]

10.9 Multicomponent Transport

Solute transport can be dramatically and unpredictably affected by the presence of other species that permeate through the membrane or that interact with the membrane. Here, systems consisting of (1) the dense polymeric membrane and (2) two or more solutes moving through the membrane by the solution-diffusion mechanism are considered. In a water splitting system, O_2 and H_2 are formed on opposite sides of the membrane and will counter-diffuse across it. The permeants produced in CO_2 reduction devices are more complicated. Current CO_2 reduction catalysts, such as copper, typically do not selectively produce a single CO_2 reduction product; instead, a multiplicity of various ionic and neutral product molecules are produced.[82] A complex mixture of solute molecules including water, CO_2, H_2, O_2, supporting electrolyte ions, and CO_2 reduction products may exist within a solar fuels device. Clearly, understanding the implications of multicomponent transport in polymeric membranes is critical in the context of a solar fuels device.

A solute may affect the diffusion of another solute through the membrane, the sorption of another solute into the membrane, or both. Transport of a solute can arise even in the absence of an appreciable driving force when another solute is diffusing through a liquid or polymeric medium. This coupling of fluxes is thought to arise from molecular friction between solutes, where one solute is subject to a strong diffusive driving force and the other solute is not subject to a strong diffusive driving force.[83] Diffusion in multicomponent systems may be modeled using either a generalization of Fick's Law, or a generalization of the Maxwell–Stefan equations. The generalized Maxwell–Stefan equations can be written as follows for one-dimensional transport:

$$d_i = -\sum_{\substack{i=1 \\ j\neq i}}^{n} \frac{x_i x_j}{Đ_{ij}} (v_i - v_j) \qquad (10.24)$$

where d_i is a generalized driving force that results in the diffusion of solute i relative to stationary coordinates (*e.g.*, gradients in concentration, temperature, pressure, *etc.*), x_i is the mole fraction of i in the mixture, $Đ_{ij}$ are multicomponent diffusion coefficients, and v_i is the velocity of solute i relative to stationary coordinates.[20,84–86]

The generalized Maxwell–Stefan approach is typically considered preferable to the generalized Fickian approach for steady-state transport. The diffusion coefficients in the Maxwell–Stefan equations represent the frictional drag between binary pairs of solute molecules.[83] Advantageously, the Maxwell–Stefan diffusion coefficients are symmetrical (meaning that, according to the Onsager reciprocal relations, $Đ_{ij} = Đ_{ji}$, reducing the number of independent diffusivities that must be determined) and are less sensitive to system composition than Fickian diffusion coefficients.[83,85–88] Coupled fluxes, especially in electrolytic systems exhibiting acid/base behavior, have been shown to affect permeation rates of solutes in ternary solution through polyamide membranes.[89] Electrostatic interactions among multiple solutes, especially in mixtures of non-ideal, highly polar solutes (*e.g.*, many electrolytes), can also contribute to unexpected fluxes.[89] Strong coupling of fluxes among solutes can reduce the rejection and selectivity of the membrane, since solutes that might be ordinarily rejected are transported through the membrane.[89] Extensive discussion of the Maxwell–Stefan approach to multicomponent transport, including formulations appropriate for transport through membrane, can be found elsewhere.[20,85–88,90,91]

Solute sorption may also be affected by the presence of other solutes. Fundamental studies of transport of solute mixtures in hydrated membranes are scarce. Early studies suggested that the transport of ion mixtures differs considerably from that of single a single ion pair. In sulfonated, negatively-charged polymers, transport of monovalent ion transport increased in the presence of divalent cations.[29,92] While the mechanisms responsible for this behavior are not well understood, such results could be due to reduced

Donnan exclusion of monovalent ions in the presence of divalent cations. Divalent cations more effectively screen fixed negatively charged sulfonate groups in the polymer than monovalent cations, reducing the exclusion of anions from the polymer matrix. As more anions transport through the polymer, more monovalent cations also transport through the polymer to maintain overall electroneutrality.[92]

While multicomponent transport phenomena have not been extensively studied in the liquid phase, many reports describe competitive sorption phenomena in the gas separation literature. Commonly, the permeability of a gaseous solute is reduced in the presence of other gaseous solutes due to competitive sorption of the solutes in the polymer matrix.[93] Since fewer solutes sorb into the polymer, fewer are able to diffuse through the thickness of the membrane. Therefore, when multiple solutes are present, the permeabilities of one or more of them tend to decrease relative to the permeabilities exhibited by each individual solute.[93] Competitive sorption effects also appear to be exaggerated in glassy polymers with high free volume, where permeabilities of some gases decrease by an order of magnitude or more upon introduction of a more condensable gas.[94] Competitive sorption makes prediction of multicomponent permeabilities from pure component permeability challenging, and highlights the importance of laboratory measurement of multicomponent permeabilities.[95]

10.10 Measurement of Transport Parameters in Membranes

Many techniques for measuring permeability, diffusivity, and solubility of a solute in a membrane are reported in the literature. Here, we briefly survey the most common experimental techniques used to ascertain the transport coefficients, with consideration given to the nature of membranes and solutes typically encountered in electrochemical and solar fuels devices.

The permeability of a membrane may be calculated by measuring the solute flux across the membrane when a gradient in solute (electro)chemical potential is applied across the membrane and then normalizing that flux by the membrane thickness and area available for transport. Solar fuels devices that contain an aqueous electrolyte require solute transport through a hydrated membrane. Diffusion cells are commonly used to measure membrane permeabilities in the laboratory.[57,96] A membrane of known thickness is sandwiched between two diffusion half cells that have an orifice of known dimension. The half cells may be jacketed to permit the use of a recirculating fluid for temperature control. One of the half cells (the donor chamber) is filled with a solution containing the solute of interest at known concentration. The other half cell (the receiver chamber) is filled with a solution of a much lower known concentration (*e.g.*, ultrapure water). The increase in solute concentration in the receiver chamber is measured as the solute

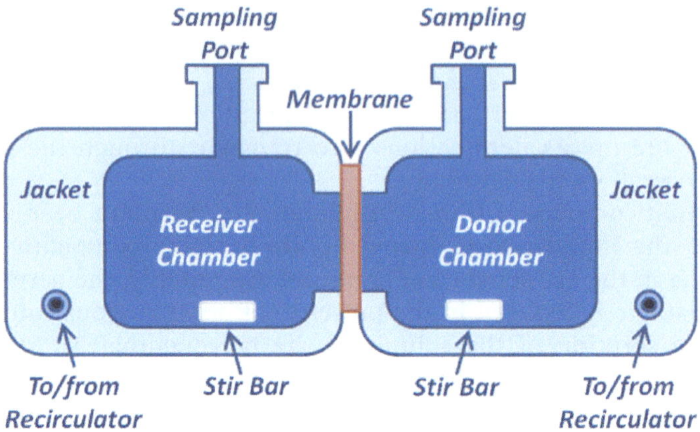

Figure 10.9 Common diffusion cell for measurement of solute transport through a thin membrane. The membrane is clamped between two half cells, the donor chamber containing a solution with the solute of interest, and the receiver chamber initially containing pure water. The half cells may be jacketed to control temperature by circulation of a heating or cooling fluid. The concentration in the receiver chamber is periodically sampled to ascertain time-resolved concentration data.

diffuses through the membrane down the chemical potential gradient between the donor and receiver chambers.[57,96] A typical commercially available diffusion cell is shown in Figure 10.9.

Time-resolved concentration data derived from permeation experiments can then be fit to models to extract membrane permeability. The most well-known model describing solute permeation through hydrated, swollen polymer films was developed by Yasuda *et al.*[57,96] Using a typical diffusion cell setup as described above, the permeability of a polymer film to a solute i can be calculated by:

$$\ln\left[1 - 2\frac{c_{il}^{s}(t)}{c_{i0}^{s}(0)}\right] = -2\frac{AV}{l}P_i t \tag{10.25}$$

where $c_{il}^{s}(t)$ is the concentration of solute i in the receiver chamber at time t, $c_{i0}^{s}(0)$ is the initial concentration of solute i in the donor chamber, A is the cross-sectional area available for solute transport through the polymer film, V is the solution volume in the donor and receiver chambers, l is the thickness of the polymer film, and P_i is the permeability of the polymer to solute i.[44,57,96] Typically, the permeability is extracted by fitting this model to time-resolved concentration data using the permeability as an adjustable parameter. It should be noted that the permeabilities calculated by this technique are not true membrane permeabilities, but include the effects of any boundary layers that develop along the membrane surfaces.[96] For this reason, it is generally recommended that the solutions in each half cell be stirred vigorously.

The method used to measure receiver chamber concentration in a permeation experiment depends upon the nature of the solute(s) of interest. For a charged solute in aqueous solution, conductivity is often used to track the receiver chamber concentration.[44] Conductivity measurements cannot discriminate among multiple charged solutes, so conductivity is most effective when monitoring permeation of a single solute without buffer. Electrochemical techniques have been developed to measure the permeation of dissolved gaseous species such as oxygen.[97] Gas chromatography, ion exchange chromatography, flame atomic adsorption spectroscopy, and high performance liquid chromatography may also be employed.[98] These techniques are usually able to discriminate among multiple solutes, but they all require aliquot sampling from the receiver chamber, and permeation models such as that proposed by Yasuda *et al.*[57,96] may assume a constant receiver chamber volume throughout the experiment. Therefore, aliquots must be very small relative to the overall diffusion cell volume, or the volume change must be taken into account in the permeability calculation. Recent studies have explored the use of *in situ* ATR FTIR to monitor the evolution of solute concentration in the receiver chamber.[99] An *in situ* ATR FTIR probe inserted into the receiver chamber is used to periodically measure the infrared absorbance of the receiver chamber solution, which can be converted to a solute concentration using calibration curves. A significant advantage of this technique is that multicomponent transport of infrared-absorbing species may be characterized in a straightforward way, since spectral deconvolution may be used to calculate the contribution of multiple solutes to the overall solution infrared absorbance.[99]

The calculation of membrane permeability requires that solute transport through the membrane is in a pseudo-steady state regime, so the driving force for permeation must remain approximately constant throughout the measurement. Therefore, the receiver chamber concentration must remain much lower (*i.e.*, often one or two orders of magnitude) than the donor chamber concentration so that changes in the donor chamber concentration are very small relative to the overall concentration gradient.[44]

Many products of water oxidation and CO_2 reduction are gaseous. Aside from the electrochemical technique mentioned above, which is used for membranes in contact with liquid electrolyte,[97] the permeability of polymeric membranes to gases is often measured using a constant volume, variable pressure technique.[100,101] The analyte gas is applied at a fixed pressure on the upstream face of a membrane of known area and thickness, and the pressure increase in a fixed volume downstream of the membrane is monitored with respect to time. The challenge with this measurement technique is that it is best suited to dry gas permeation. Membranes used for solar fuels devices are typically hydrated, and transport of solute gases through dry membranes may not accurately reflect transport through a hydrated membrane. Use of a humidified feed gas stream is challenging because water tends to permeate polymeric membranes much more quickly than many other gases.[102] Water may then condense inside the fixed volume or on pressure transducers, effectively reducing the volume and conflating the measurement of gas permeation. Recently, however, a variant of the constant volume/variable pressure apparatus described above that takes advantage of the rapid transport of water through many polymer films has been reported, which may enable measurements on humidified films.[102]

Overall permeability measurements are useful to characterize membrane performance, but they do not provide insight into the relative contributions of solubility and diffusivity. It is rare that independent measurements of permeability, solubility, and diffusivity would all be made; typically, two of the parameters are measured and the third is calculated from eqn (10.1). In both the liquid-phase and gas-phase permeation experiments described above, a time lag is observed at the beginning of the experiment when a solute is diffusing through the polymer film but it has not yet reached the other side of the membrane.[50] The time required to establish a pseudo-steady state solute concentration gradient through the polymer film is linked to the diffusivity. For rubbery polymers, the diffusion coefficient can be estimated by the following relation:

$$D_i = \frac{l^2}{6\theta} \tag{10.26}$$

where θ is the time lag and l is the membrane thickness.[50,103] For glassy polymers, the relationship between diffusivity and time lag appears more complex. In the gas separation literature, corrections of the above relationship have been proposed based on multiple mechanisms of solute sorption into the polymer.[103,104] Figure 10.10 illustrates the time lag apparent at the start of an experiment measuring diffusive transport.

While the time lag can usually be used to estimate the diffusion coefficient from most routine permeation measurements, a more accurate determination of the diffusivity is generally achieved by calculation from independently measured permeability and solubility values.[101,105] The solubility of non-gaseous solutes in hydrated films is commonly measured by desorption experiments.[45] A polymer film is fully hydrated in ultrapure water, then immersed in a solution of known solute concentration until equilibrium solute partitioning between the polymer and the solution is achieved. The geometry and mass of the film are carefully measured, and then the film is immersed in a known volume of ultrapure water. Because the volume of ultrapure water is much greater than the volume of the film, the solute completely desorbs into the water. The concentration of solute in the water following the desorption step is measured by gas chromatography, high performance liquid chromatography, ion chromatography, flame atomic adsorption spectroscopy, total organic carbon analysis, or other technique based on the chemical nature of the solute.[45] The sorption coefficient K_i of solute i in the polymer may then be calculated as follows:

$$K_i = \frac{c_i^{\mathrm{d}} V_{\mathrm{d}}}{c_i^{\mathrm{s}} V_{\mathrm{p}}} \tag{10.27}$$

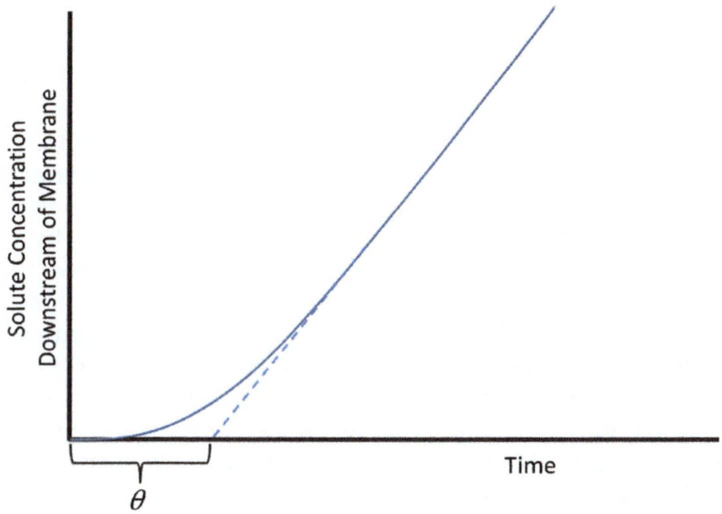

Figure 10.10 Time lag for diffusion. The pseudo-steady state concentration rise is extrapolated to short times to estimate the time lag. The time lag is related to the diffusion coefficient of the solute in the membrane, and arises because the solute must diffuse through the thickness of the membrane before it can be detected on the downstream side.

where c_i^{d} is the concentration of solute i in the water following the desorption step, V_{d} is the volume of the water used for desorption, c_i^s is the concentration of solute i in the initial solution, and V_{p} is the volume of the fully hydrated, swollen polymer.[44]

Solubility is often reported as a unitless partitioning coefficient, although the units are: mol(i)L^{-1} (swollen polymer)/mol(i)L^{-1} (external solution).[43] Concentration may also be measured on a mass basis.

10.11 Phenomena Affecting Transport: Physical Aging and Degradation

A phenomenon closely tied to the non-equilibrium nature of glassy polymers is physical aging. Excess free volume, which is kinetically trapped in the polymer matrix due to reduced chain mobility upon cooling below the glass transition temperature, gradually decreases with time as slow, localized segmental motion permits structural reorganization. As the polymer nears its equilibrium specific volume, physical aging slows. Generally, physical aging results in densification of the polymer and, therefore, a decrease in permeability and an increase in selectivity with time.[46] While any glassy material may be subject to physical aging, polymers with high intrinsic free volume tend to exhibit more severe physical aging effects than polymers with lower intrinsic free volume, with permeabilities sometimes decreasing by over an order of magnitude before physical aging slows.[46]

Degradation of polymer membranes, either by chemical or physical means, can severely impact separation performance and has been observed in many applications.[106–109] Membrane degradation is one of the leading causes of polymer electrolyte membrane fuel cell failure, and has been widely studied.[109–114] Mechanical failure by punctures, cracks, tears, or pinholes severely compromises a membrane's ability to separate solutes, since convective flow through the perforation tends to dominate over diffusive transport.[109] Scission of polymer chains, which can occur by a number of mechanisms and is often exacerbated by high temperature, frequently leads to an increase in permeability and a decrease in rejection.[106,108] In electrochemical systems, the formation of radical species that can attack the polymer chains is considered to be the main contributor to chemical membrane degradation.[109] In addition to the desired oxidation or reduction products, transition metal catalysts can induce formation of peroxides (especially hydrogen peroxide in aqueous electrolyte), which are then capable of attacking the polymer backbone or side chains.[109] Although the formation of hydrogen peroxide in fuel cells is widely acknowledged, recent publications suggest that species other than hydrogen peroxide may also be responsible for membrane degradation.[112–114] Trace transition metal impurities in the electrolyte, which are electrostatically attracted to the fixed charge groups in cation exchange membranes (*e.g.*, Nafion®) and can subsequently catalyze the formation of reactive species, appear to worsen degradation.[109]

In the context of solar fuels devices, degradation due to the ultraviolet portion of the solar spectrum may be a particularly important consideration. Many polymers, especially those containing aromatic moieties, absorb strongly in the UV range, which can lead to dissociation of covalent bonds in the polymer chain.[115] Although UV degradation of polymers is well known, there has been little effort to understand the impact of UV light on the performance of a photo-driven electrochemical solar fuels device. Depending upon device design, the membrane could be subjected to steady UV impingement for an extended period. Despite this knowledge gap regarding the influence of UV light on membrane performance, other durability tests on polyelectrolyte membranes could provide insight into environmental conditions affecting membrane stability. Data from an accelerated durability test on a perfluorinated ionomer membrane in a membrane-electrode assembly are shown in Figure 10.11.[110] These experiments were carried out at 80 °C under dark (non-illuminated) conditions with a humidified H_2 feed gas stream. Membrane degradation was monitored *via* H_2 gas permeation from one side of the membrane to the other. A greater than ten-fold increase in H_2 permeation was observed in 60 days, suggesting changes that could lead to substantial losses in device efficiency.[110]

Water splitting devices are usually operated at either very high or very low pH so that the concentration of charge carriers (H^+ or OH^-) in the electrolyte is high. If a supporting electrolyte is employed, which is neither consumed nor reduced, severe concentration gradients of charge carriers will evolve and eventually the device will no longer be able to support the

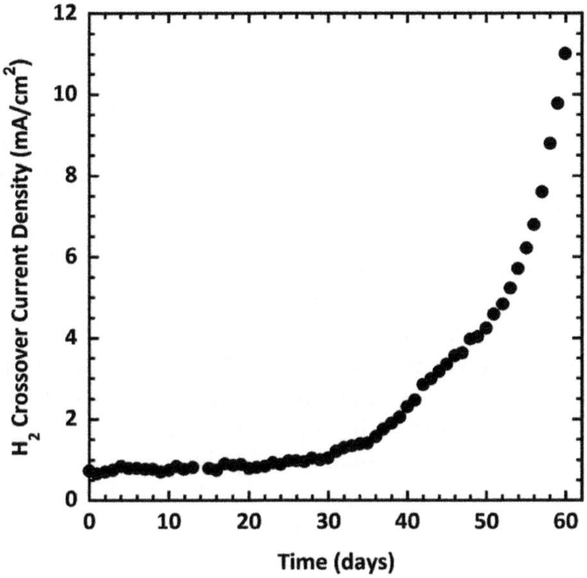

Figure 10.11 Accelerated durability test of a perfluorinated ionomer membrane. Membrane degradation was monitored by H_2 permeation in an electrochemical cell. H_2 gas feeding the cell was humidified at 60 °C and the cell temperature was maintained at 80 °C. H_2 gas permeation was monitored electrochemically by oxidation at a Pt electrode and is reported here as current density.
Reproduced from *Electrochimica Acta*, **51**, M. Inaba, T. Kinumoto, M. Kiriake, R. Umebayashi, A. Tasaka and Z. Ogumi, Gas crossover and membrane degradation in polymer electrolyte fuel cells, 5746–5753, Copyright 2006, with permission from Elsevier.[110]

required current.[15] Operational stability at pH extremes is a significant challenge for membrane development, as few polymers are robust enough to withstand prolonged exposure to highly acidic or alkaline environments. The development of conductive membranes capable of withstanding highly alkaline environments is an area of intense interest.[116–118]

10.12 JCAP Membrane Research

JCAP has studied various aspects of membrane design, development, and implementation, ranging from fundamental synthesis of new polymeric systems to the evaluation of membrane performance in prototype devices. In addition to the effects of polymer composition and structure on membrane conductivity and permeability, membrane properties are also determined by solar fuel device architecture. Three configurations have been considered in JCAP: a microwire structure inspired by earlier work,[119] the face-to-face design shown in Figure 10.3, where the electrodes are physically separated and an aqueous electrolyte is employed to transport charge-carrying ions to and from a membrane,[27,120] and a louvered structure in which the electrodes are mounted back-to-back and the membrane is incorporated as stand-alone film that does not physically separate the electrodes.[121]

The original microwire architecture is a complex structure in which an array of contacted microwires is vertically arranged through a membrane film, with anode and cathode functions on opposite sides. Photogenerated electrons are conducted through the wires, and protons or hydroxide ions are transported through the electrolyte and the supporting membrane.[122] The microwire array structure may offer particular advantages for materials with low carrier mobilities because diffusion distances are relatively short, and for gross defect tolerance such as missing wires, but has stringent requirements for materials integration.[119] JCAP focused on improving microwire properties using a core-shell rather than an end-contacted geometry, where the membrane would be incorporated by filling sub-micron-scale gaps between the wires. Since the properties of polymer films in contact with solid surfaces are strongly influenced by the local geometry and chemical compositions of the materials, a parallel effort focused on evaluating the membrane morphology near interfaces. Nafion®, which is frequently used in those applications, was selected for study because of the substantial body of literature on it.[7,9]

A number of important process–structure–property relations for thin membrane films were identified as a result of work in JCAP. When very thin (160 nm or less), PFSA membranes cast on SiO_2 surfaces are less conductive than bulk membranes due to the confinement of swelling to one dimension, and have strongly thickness-dependent proton conductivity.[123] Below 50 nm, the films are no longer strongly phase-separated, which reduces their ability to form water-filled channels and decreases conductivity by about a factor of 100.[123] Figure 10.12 shows the effect of film thickness and environmental humidity on conductivity. Films less than 50 nm in thickness tended to exhibit lower conductivity than thicker films, especially at lower environmental relative humidities, under which films sorbed less water.[124]

In situ grazing-incidence small-angle X-ray scattering techniques provided additional insights into SiO_2/membrane interfacial interactions. Surface energies affect both ionomer morphology and transport properties.[124] A SiO_2 surface treated with *n*-octyltrichlorosilane is hydrophobic. Thin PFSA films cast on it had sulfonic acid domains that ran parallel to the interface, inhibiting the water uptake required for conductivity and promoting membrane–substrate adhesion. A more hydrophilic, untreated SiO_2 surface, on the other hand, had an isotropic morphology that promoted the uptake of water required for ion transport but had poor adhesion. *Operando* measurements using small-angle X-ray scattering revealed distinct differences in the kinetics of PFSA response to water vapor and liquid water that was interpreted to be due to differences in interfacial interactions.[125] In addition, sub-second morphological changes were observed during liquid water permeation that were suggested to be due to conversion of sulfonic acid domains from a mixture of spherical and cylindrical shapes to exclusively cylindrical ones.

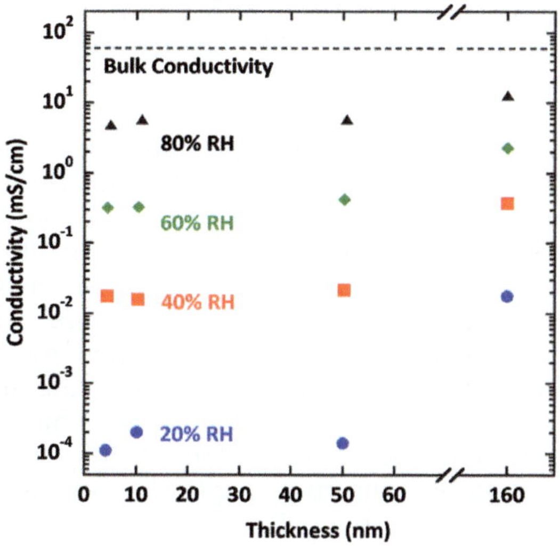

Figure 10.12 Effect of PFSA film thickness and environmental relative humidity (RH) on film conductivity. The dotted line represents the conductivity of bulk PFSA equilibrated with a saturated atmosphere.
Reprinted with permission from M. A. Modestino, A. Kusoglu, A. Hexemer, A. Z. Weber and R. A. Segalman, *Macromolecules*, 2012, **45** (11), 4681, Copyright 2012 American Chemical Society.[124]

It became clear from these studies that the random character of the polymer and the distinct chemical differences between its domains makes precise control and optimization of morphology and transport properties very challenging in even simple flat film environments, let alone the highly chemically heterogeneous microstructured architectures that were considered. In particular, controlling the extent of hydration is challenging. Increasing hydration of sulfonic acid channels increases the membrane's ion conductivity at the expense of its ability to block product permeation. Control over product permeation is particularly important in solar fuels applications, even at the expense of high ion conductivity.[10] Moreover, because it contains negatively charged sulfonate moieties, PFSA most readily permits transport of protons and other cations, while solar fuels devices operate over a range of pH. New acid- and base- stable membrane materials will be required to enable successful optimization of the performance characteristics of solar fuels generators, especially if alkaline electrolytes are to be used.

JCAP has taken a multi-pronged approach to meet these requirements. The majority of the effort has been focused on discovery and development of new membrane materials that have balanced and tunable ion and small molecule transport properties. Block copolymers offer an ideal platform to select and individually optimize the hydrophobic (structural) and hydrophilic (ion-conducting) components and the resulting membrane morphology. Polymer chain compositions and lengths can be varied to tune phase-separated domain sizes ranging from 10 to 100 nm,[25] potentially resulting in controlled water uptake with conductivity in a useful range and reduced product crossover relative to PFSA.

Initial work[72] built on studies of block copolymer platforms blended with an ionic liquid (IL) to create proton-conducting membranes.[76,126–128] Imidazolium bis(trifluoromethylsulfonyl)imide ([Im][TFSI]) was combined with poly(styrene-*b*-2-vinylpyridine) (PS-*b*-P2VP) and poly(styrene-*b*-methylmethacrylate) (PS-*b*-PMMA). The IL permeates the P2VP and PMMA domains, imparting ion-conducting channels to the material. A scaling relationship between IL content and membrane properties was identified over a wide range of compositions (Figure 10.13), and it was shown that the conductivity is insensitive to the details of the resulting morphology if the IL-containing domains are interconnected. When immersed in aqueous electrolyte for extended periods, stability of the blends is a challenge, leading to the alternative strategy of incorporating ILs by copolymerization.[129]

By appropriate selection of the polymerized IL (PIL) it is possible to decrease the small molecule permeability of membranes while still permitting sufficient ionic conductivity through the tethered charge group. A novel block copolymer system containing a protic polymerized ionic liquid block has been synthesized by first preparing poly(styrene-*b*-histamine methacrylamide) (PS-*b*-PHMA), then treating it with trifluoroacetic acid to

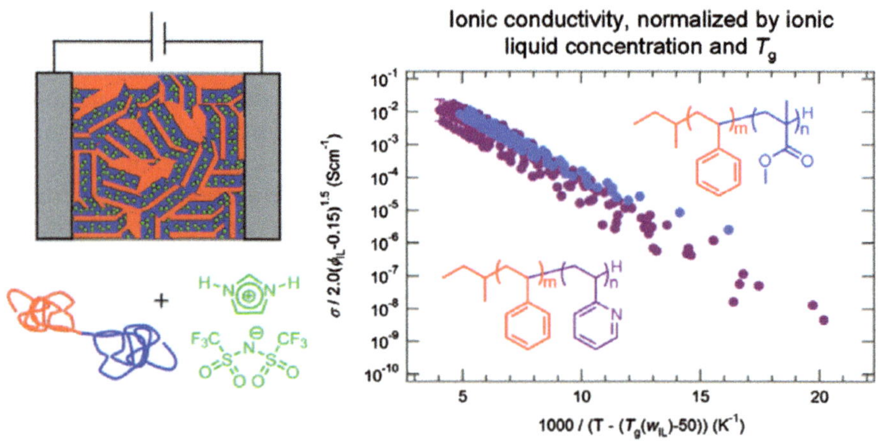

Figure 10.13 The conductivity of imidazolium bis(trifluoromethylsulfonyl)imide ([Im][TFSI]) scales with normalized temperature over a wide compositional range in poly(styrene-*b*-2-vinylpyridine) (PS-*b*-P2VP) and poly(styrene-*b*-methylmethacrylate) (PS-*b*-PMMA) systems. The diagram in the upper left represents poly(methyl methacrylate) conduction pathways (blue) for [Im][TFSI] ions (green) through a non-conducting poly(styrene) matrix in an electrochemical device. Poly(2-vinylpyridene) is shown in purple.
Reprinted with permission from M. L. Hoarfrost and R. A. Segalman, *ACS Macro Lett.* 2012, **1** (8), 937, Copyright 2012 American Chemical Society.[72]

generate the PIL. The resulting membranes had a lamellar morphology and exhibited ionic conductivities greater than 10^{-4} S cm^{-1}, which could be tuned by adjusting the domain size and volume fraction of the conducting block. Protons and trifluoroacetate ions were both present, so the conductivity values corresponded to the overall ionic conductivity of membranes equilibrated in water. A new synthetic route to PIL was also introduced, significantly improving control over the membrane's mechanical properties while retaining its conductivity and low water uptake characteristics.[73]

JCAP has also explored the synthesis and characterization of new anion conducting block copolymers for use in alkaline electrolytes. An initial study used polystyrene as the structural block, with quaternization of the conducting block to create ionizable domains after polymerization.[130] Poly(styrene-*b*-chloromethylstyrene) (PS-*b*-PCMS) was quaternized with either trimethylammonium or *n*-butylimidazolium to compare their properties. It was found that their domain structure, Cl^- conductivity, and water uptake were very similar. Poly(styrene-*b*-(2-acryloxy)ethyltributylphosphonium bromide) diblock copolymers (PS-*b*-STBP) were also prepared to evaluate the performance of tributylphosphonium cations as anion conductors.[131] They were found to have low water uptake and Br^- conductivity in the mS range that increased with increasing domain size. These materials had useful properties in pure water but limited stability in aqueous base. Because of their stable aromatic character and IL properties, cyclopropenium ions have been identified as promising conducting moieties for incorporation in anion-conducting block copolymers. Block copolyelectrolytes using PS and cyclopropenium ions as the hydrophilic block had conductivities in the range of 4 mS cm^{-1}.[132]

JCAP has sought to further improve membrane stability in highly alkaline electrolyte. To this end, poly(styrene-*b*-diallyl dimethyl ammonium chloride) (PS-*b*-PDADMAC) block copolymers were synthesized.[133] First, PDADMAC homopolymers were synthesized and found to be stable in 2M NaOD for 2000 hours with no structural changes detectable by NMR analysis. Based on these promising results, PDADMAC homopolymer precursors were incorporated into block copolymers with polystyrene, employing an anion exchange metathesis technique to overcome the disparate solubilities of PDADMAC and PS. The resultant copolymers were pressed into membrane films, which exhibited hydroxide conductivities of 0.8 mS cm^{-1}.[133]

In addition to discovery of new membrane materials, JCAP has evaluated techniques for modification of membrane properties through process control and has developed a scientific understanding of impact on membrane performance. These techniques include thermal processing to control transport through hydrophobic regions of the membrane (demonstrated for Nafion®),[35] preparation of block copolymer films using partly soluble blends to create artificial free volume and hence controlled transport properties,[134] and development of a new simplified model for control of polymer synthesis kinetics through monomer reactivity ratios, which facilitates design of synthetic strategies.[135]

10.13 Outlook for Membranes in CO_2 Reduction Devices

The relationship between the selective transport of ions and the transport of dissolved neutral product fuels defines performance criteria for a membrane separator in a solar fuels device. Multicomponent, mesoscopically heterogeneous polymer materials have emerged as a versatile means to control both, with deep insights into the structure–property relationships required under photoelectrochemical water splitting conditions.[72,76,123–125,128–130,136–139] These insights may not be applicable to solar CO_2 reduction devices, however, because many polymeric materials plasticize under conditions that produce carbonaceous fuels.[140–142] For example, Nafion® has not been successfully used for efficient direct methanol fuel cells due to a significant crossover current.[143] To incorporate heterogeneity and reduce CO_2 reduction product permeation, JCAP has synthesized multifunctional polymeric materials that contain crystallites impervious to small molecule transport.[144,145] Research on CO_2 separation from gas mixtures has revealed additional routes to stabilization of membrane performance, however the materials are not polyelectrolytes.[146] Moreover, transport of ions and small molecules within polyelectrolytes is an inherently mesoscopic phenomenon, where ions are traversing and interacting within the dynamic structure of the polymer.

A potentially important consideration for membranes used in a CO_2 reduction device is plasticization. Plasticization is an increase in chain mobility in a glassy polymer resulting from sorption of a highly soluble solute. The increased chain mobility permits structural relaxation in non-equilibrium glasses, leading to swelling and a corresponding increase in free volume. This swelling-induced increase in free volume can have a significant impact on solute transport in the polymer.[46] As the concentration of a plasticizing solute inside the membrane increases, so too do the effects of plasticization. Many glassy polymeric membranes are reportedly plasticized by CO_2. The increase in free volume upon plasticization tends to increase permeability and decrease selectivity for all solutes. Increased permeabilities are generally attributed to increased diffusion coefficients resulting from increased free volume, increased chain mobility, or both.[43,46] While reports of carbon dioxide plasticization are most frequently encountered in the gas separation literature (often motivated by natural gas sweetening or carbon dioxide removal from industrial effluent),[46] the effects of plasticization on solar fuels device membranes operating in the presence of CO_2 are not established.

In general, progress in development of suitable membranes for solar fuels applications is hindered by a limited capability to predict membrane behavior in an electrochemical prototype based on the membrane's physical structure and chemical properties. More is known about the requirements than how to meet them. Performance depends on permeability–conductivity tradeoffs due to the expected low operating current densities

(~1 to 10 mA cm^{-2}),[10] a desire for pure product fuels to avoid energetically costly separations, and the requirement to prevent inefficiencies due to catalytic oxidation of product alcohols at the positive electrode. While the problem is challenging for water splitting, additional complexities arise for CO_2 reduction due to the presence of chemical species that interact and compete for sites within the membrane (*e.g.*, bicarbonate formation) and the presence of neutral organics, water and salts in the operating electrolyte. To complicate matters, the membranes interact with their liquid and solid external environments during film preparation, causing deviations from bulk properties.[147–151] They also change over time due to aging, a process that requires quantitative models to predict because it will be influenced not only by the chemical environment, but also the diurnal and seasonal temperature variations expected in device operation. Recent computational studies have shown that temperature swings as large as 100 °C can be expected for water splitting devices deployed in the field.[54] As temperature fluctuates, membranes will densify to some degree because of relaxation of the molecular chains, directly reducing their permeability.[140,152] This effect may be counterbalanced by plasticization, but cannot be neutralized. Intensive studies of membrane properties under these conditions will be very valuable for both fundamental understanding and applications.

List of Symbols

Accepted values or typical units, if applicable, are provided following each definition.

Roman Letters

A	cross-sectional area available for transport [cm^2]
a_i	pre-exponential adjustable factor related to a geometric factor and the velocity of component i [cm^2 s^{-1}]
b_i	adjustable factor related to the Lennard–Jones diameter of component i (unitless)
c_i	concentration of component i [M]
c_i^{d}	concentration of component i in desorption solution [M]
c_i^{m}	concentration of component i in a membrane [M]
c_{i0}^{m}	concentration of component i in the membrane at $x=0$ [M]
c_{il}^{m}	concentration of component i in the membrane at $x=l$ [M]
c_i^{s}	concentration of component i in solution equilibrated with a membrane [M]
c_{i0}^{s}	concentration of component i in solution equilibrated with a membrane at $x=0$ [M]
c_{il}^{s}	concentration of component i in solution equilibrated with a membrane at $x=l$ [M]
$c_{\mathrm{w}}^{\mathrm{s}}$	concentration of water in a solution equilibrated with a membrane [M]
D_0	Arrhenius pre-exponential constant [cm^2 s^{-1}]

D_i	diffusion coefficient of component i [$cm^2\,s^{-1}$]
d_i	generalized force motivating diffusion of component i [m^{-1}]
D_{ic}	diffusion coefficient of component i in a conducting phase in a heterogeneous material [$cm^2\,s^{-1}$]
D_{ih}	overall diffusion coefficient of component i in a heterogeneous material [$cm^2\,s^{-1}$]
$Đ_{ij}$	Maxwell–Stefan diffusion coefficient for the binary pair of components i and j [$cm_2\,s^{-1}$]
E_D	activation energy for diffusion [$J\,mol^{-1}$]
f	adjustable factor accounting for connectivity and tortuosity of a conducting phase in a heterogeneous material (unitless)
F	Faraday's constant (96 482 $C\,mol^{-1}$)
K_i	solubility coefficient of component i [$M\,M^{-1}$] (typically reported as unitless)
l	membrane thickness [μm]
m_d	mass of dry membrane [g]
m_w	mass of fully hydrated membrane [g]
M_w	molar mass of water (18.02 $g\,mol^{-1}$)
n_i	molar flux of component i [$mol\,cm^{-2}\,s^{-1}$]
P_i	permeability coefficient of component i [$cm^2\,s^{-1}$]
P_s	permeability coefficient of salt [$cm^2\,s^{-1}$]
P_w	permeability coefficient of water [$cm^2\,s^{-1}$]
R	ideal gas constant (8.314 $J\,mol^{-1}\,K^{-1}$)
$\mathscr{R}$	resistance of the membrane to ion conduction [S^{-1}]
T	temperature [K]
t	time [s]
T_0	Vogel temperature [K]
T_g	glass transition temperature [K]
t_i	transference number of component i (unitless)
T_m	melting temperature [K]
u_i	mobility of component i [$mol\,m\,N^{-1}\,s^{-1}$]
V	diffusion cell volume [mL]
V_d	volume of desorption solution [mL]
$\langle v_f \rangle$	polymer fractional free volume [$cm^3\,g\,cm^{-3}\,g^{-1}$] (typically reported as unitless)
v_i	velocity of component i [$m\,s^{-1}$]
V_p	volume of fully hydrated, swollen membrane [cm^3]
V_w	molar volume of water [$mL\,mol^{-1}$]
w_u	water uptake of a membrane [$g\,g^{-1}$] (typically reported as unitless)
x	coordinate describing distance through membrane thickness [cm]
x_i	mole fraction of component i [$mol\,mol^{-1}$] (typically reported as unitless)
z_i	valence of component i (unitless)

Greek Letters

α	Arrhenius adjustable parameter related to the concentration of component i [S cm^{-1}]
$\alpha_{A/B}$	membrane selectivity of component A over component B (unitless)
β	Arrhenius adjustable parameter related to the activation energy of diffusion of component i [J mol^{-1}]
θ	time lag for diffusion [s]
μ_i	electrochemical potential of component i [J mol^{-1}]
ρ_p	density of dry polymer [g cm^{-3}]
ρ_w	density of liquid water [g cm^{-3}]
σ	ionic conductivity [S cm^{-1}]
σ_c	ionic conductivity of a conducting phase in a heterogeneous material [S cm^{-1}]
σ_h	overall ionic conductivity of a heterogeneous material [S cm^{-1}]
$\varnothing_c$	volume fraction of a conducting phase in a heterogeneous material [cm cm^{-3}] (typically reported as unitless)
$\varnothing_w^m$	volume fraction of water in a membrane [cm cm^{-3}] (typically reported as unitless)
ψ	applied electric potential [V]

References

1. O. Savadogo, *J. New Mater. Electrochem. Syst.*, 1998, **1**(1), 47.
2. O. Savadogo, *J. Power Sources*, 2004, **127**(1–2), 135.
3. V. Neburchilov, J. Martin, H. Wang and J. Zhang, *J. Power Sources*, 2007, **169**(2), 221.
4. N. P. Berezina, N. A. Kononenko, O. A. Dyomina and N. P. Gnusin, *Adv. Colloid Interface Sci.*, 2008, **139**(1–2), 3.
5. T. Xu, *J. Membr. Sci.*, 2005, **263**(1–2), 1.
6. N. W. DeLuca and Y. A. Elabd, *J. Polym. Sci., Part B: Polym. Phys.*, 2006, **44**(16), 2201.
7. A. Kusoglu and A. Z. Weber, *Chem. Rev.*, 2017, **117**(3), 987.
8. M. A. Hickner, *Mater. Today*, 2010, **13**(5), 34.
9. K. A. Mauritz and R. B. Moore, *Chem. Rev.*, 2004, **104**(10), 4535.
10. A. Berger, R. A. Segalman and J. Newman, *Energy Environ. Sci.*, 2014, **7**(4), 1468.
11. M. R. Singh, E. L. Clark and A. T. Bell, *Phys. Chem. Chem. Phys.*, 2015, **17**(29), 18924.
12. M. A. Modestino, S. M. Hosseini Hashemi, S. Haussener, S. M. H. Hashemi and S. Haussener, *Energy Environ. Sci.*, 2016, **9**(5), 1533.
13. J. Newman, K. E. Thomas-Alyea, *Electrochemical Systems*, 3rd edn, John Wiley & Sons, Inc, Hoboken, New Jersey, 2004.

14. P. Atkins, J. de Paula, *Physical Chemistry*, 7th edn, W.H. Freeman and Company, New York, 2002.
15. M. A. Modestino, K. A. Walczak, A. Berger, C. M. Evans, S. Haussener, C. Koval, J. S. Newman, J. W. Ager and R. A. Segalman, *Energy Environ. Sci.*, 2014, 7(1), 297.
16. C. Xiang, A. Z. Weber, S. Ardo, A. Berger, Y. Chen, R. Coridan, K. T. Fountaine, S. Haussener, S. Hu, R. Liu, N. S. Lewis, M. A. Modestino, M. M. Shaner, M. R. Singh, J. C. Stevens, K. Sun, K. Walczak, *Angew. Chem. Int. Ed.*, 2016, **55**(42), 12974.
17. M. Carmo, D. L. Fritz, J. Mergel and D. Stolten, *Int. J. Hydrogen Energy*, 2013, **38**(12), 4901.
18. A. Kirubakaran, S. Jain and R. K. Nema, *Renewable Sustainable Energy Rev.*, 2009, **13**(9), 2430.
19. S. J. Peighambardoust, S. Rowshanzamir and M. Amjadi, *Int. J. Hydrogen Energy*, 2010, **35**(17), 9349.
20. G. M. Geise, H.-S. Lee, D. J. Miller, B. D. Freeman, J. E. McGrath and D. R. Paul, *J. Polym. Sci., Part B: Polym. Phys.*, 2010, **48**(15), 1685.
21. H. R. Allcock, F. W. Lampe, J. E. Mark, *Contemporary Polymer Chemistry*, 3rd edn, Pearson Education, Inc, Upper Saddle River, NJ, 2003.
22. T. D. Fornes and D. R. Paul, *Polymer*, 2003, **44**(14), 3945.
23. H. Ju, B. D. McCloskey, A. C. Sagle, V. A. Kusuma and B. D. Freeman, *J. Membr. Sci.*, 2009, **330**(1–2), 180.
24. T. D. Gierke, G. E. Munn and F. C. Wilson, *J. Polym. Sci., Polym. Phys. Ed.*, 1981, **19**(11), 1687.
25. F. S. Bates and G. H. Fredrickson, *Annu. Rev. Phys. Chem.*, 1990, **41**, 525.
26. R. W. Baker, *Membrane Technology and Applications*, 3rd edn, John Wiley & Sons, Ltd., West Sussex, United Kingdom, 2012.
27. K. Sun, R. Liu, Y. Chen, E. Verlage, N. S. Lewis and C. Xiang, *Adv. Energy Mater.*, 2016, **6**(13), 1.
28. D. T. Hallinan and N. P. Balsara, *Annu. Rev. Mater. Res.*, 2013, **43**(1), 503.
29. J. Kamcev and B. D. Freeman, *Annu. Rev. Chem. Biomol. Eng.*, 2016, **7**(1), 111.
30. P. Długołecki, K. Nymeijer, S. Metz and M. Wessling, *J. Membr. Sci.*, 2008, **319**(1–2), 214.
31. Y. Mizutani, *J. Membr. Sci.*, 1990, **49**(2), 121.
32. A. Elattar, A. Elmidaoui, N. Pismenskaia, C. Gavach and G. Pourcelly, *J. Membr. Sci.*, 1998, **143**(1–2), 249.
33. J. Veerman, R. M. de Jong, M. Saakes, S. J. Metz and G. J. Harmsen, *J. Membr. Sci.*, 2009, **343**(1–2), 7.
34. G. Merle, M. Wessling and K. Nijmeijer, *J. Membr. Sci.*, 2011, **377**(1–2), 1.
35. C. M. Evans, M. R. Singh, N. A. Lynd and R. A. Segalman, *Macromolecules*, 2015, **48**(10), 3303.
36. M. Ünlü, J. Zhou and P. A. Kohl, *J. Phys. Chem. C*, 2009, **113**(26), 11416.
37. D. A. Vermaas and W. A. Smith, *ACS Energy Lett.*, 2016, **1**(6), 1143.

38. Y. C. Li, D. Zhou, Z. Yan, R. H. Gonçalves, D. A. Salvatore, C. P. Berlinguette and T. E. Mallouk, *ACS Energy Lett.*, 2016, **1**(6), 1149.
39. N. M. Vargas-Barbosa, G. M. Geise, M. A. Hickner and T. E. Mallouk, *ChemSusChem*, 2014, **7**(11), 3017.
40. D. A. Vermaas, M. Sassenburg and W. A. Smith, *J. Mater. Chem. A*, 2015, **3**(38), 19556.
41. J. G. Wijmans and R. W. Baker, *J. Membr. Sci.*, 1995, **107**(1–2), 1.
42. S. Matteucci, Y. Yampolskii, B. D. Freeman, I. Pinnau, in *Materials Science of Membranes for Gas and Vapor Separation*, ed. Y. Yampolskii, I. Pinnau, B. D. Freeman, John Wiley & Sons, West Sussex, England, 2006, pp. 1–48.
43. G. M. Geise, D. R. Paul and B. D. Freeman, *Prog. Polym. Sci.*, 2013, **39**(1), 1.
44. G. M. Geise, B. D. Freeman and D. R. Paul, *J. Membr. Sci*, 2013, **427**, 186.
45. G. M. Geise, L. P. Falcon, B. D. Freeman, D. R. Paul, *J. Memb. Sci.*, 2012, **423–424**, 195.
46. D. F. Sanders, Z. P. Smith, R. Guo, L. M. Robeson, J. E. McGrath, D. R. Paul and B. D. Freeman, *Polymer*, 2013, **54**(18), 4729.
47. L. M. Robeson, *J. Membr. Sci.*, 1991, **62**(2), 165.
48. L. M. Robeson, H. H. Hwu and J. E. McGrath, *J. Membr. Sci.*, 2007, **302**(1–2), 70.
49. M. L. Greenfield and D. N. Theodorou, *Macromolecules*, 1993, **26**(20), 5461.
50. D. R. Paul and A. T. DiBenedetto, *J. Polym. Sci., Part C: Polym. Lett.*, 1965, **10**(1), 17.
51. L. Maldonado, J. C. Perrin, J. Dillet and O. Lottin, *J. Membr. Sci.*, 2012, **389**, 43.
52. H. Mehdizadeh, J. M. Dickson and P. K. Eriksson, *Ind. Eng. Chem. Res.*, 1989, **28**(6), 814.
53. H. Yasuda and T. Hirotsu, *J. Appl. Polym. Sci.*, 1977, **21**(1), 105.
54. J. C. Stevens and A. Z. Weber, *J. Electrochem. Soc.*, 2016, **163**(7), 475484.
55. M. H. Cohen and D. Turnbull, *J. Chem. Phys.*, 1959, **31**(5), 1164.
56. H. Lin and B. D. Freeman, *Macromolecules*, 2006, **39**(10), 3568.
57. H. Yasuda, C. E. Lamaze and L. D. Ikenberry, *Makromol. Chem.*, 1968, **118**(2858), 19.
58. E. R. Nightingale, *J. Phys. Chem.*, 1959, **63**(9), 1381.
59. J. Jae, G. A. Tompsett, A. J. Foster, K. D. Hammond, S. M. Auerbach, R. F. Lobo and G. W. Huber, *J. Catal.*, 2011, **279**(2), 257.
60. H. Wu, Q. Gong, D. H. Olson and J. Li, *Chem. Rev.*, 2012, **112**(2), 836.
61. H. B. Park, B. D. Freeman, Z.-B. Zhang, M. Sankir, J. E. McGrath, *Angew. Chem., Int. Ed.*, 2008, **47**(32), 6019.
62. L. Li, F. Shang, L. Wang, S. Pei and Y. Zhang, *Energy Environ. Sci.*, 2010, **3**(1), 114.
63. L. W. Flanagin, C. L. McAdams, W. D. Hinsberg, I. C. Sanchez and C. G. Willson, *Macromolecules*, 1999, **32**(16), 5337.

64. A. C. Sagle, H. Ju, B. D. Freeman and M. M. Sharma, *Polymer*, 2009, **50**(3), 756.
65. T. A. Zawodzinski, M. Neeman, L. O. Sillerud and S. Gottesfeld, *J. Phys. Chem.*, 1991, **95**(15), 6040.
66. T. A. Zawodzinski, *J. Electrochem. Soc.*, 1993, **140**(4), 1041.
67. K. R. Cooper, *ECS Trans.*, 2011, **41**(1), 1371.
68. M. Singh, O. Odusanya, G. M. Wilmes, H. B. Eitouni, E. D. Gomez, A. J. Patel, V. L. Chen, M. J. Park, P. Fragouli, H. Iatrou, N. Hadjichristidis, D. Cookson and N. P. Balsara, *Macromolecules*, 2007, **40**(13), 4578.
69. A. Panday, S. Mullin, E. D. Gomez, N. Wanakule, V. L. Chen, A. Hexemer, J. Pople and N. P. Balsara, *Macromolecules*, 2009, **42**(13), 4632.
70. M. J. Park and N. P. Balsara, *Macromolecules*, 2010, **43**(1), 292.
71. R. Yuan, A. A. Teran, I. Gurevitch, S. A. Mullin, N. S. Wanakule and N. P. Balsara, *Macromolecules*, 2013, **46**(3), 914.
72. M. L. Hoarfrost and R. A. Segalman, *ACS Macro Lett.*, 2012, **1**(8), 937.
73. G. E. Sanoja, B. C. Popere, B. S. Beckingham, C. M. Evans, N. A. Lynd and R. A. Segalman, *Macromolecules*, 2016, **49**(6), 2216.
74. A. J. Bard, L. R. Faulkner, *Electrochemical Methods: Fundamentals and Applications*, 2nd edn, John Wiley & Sons, Inc, Hoboken, NJ, 2001.
75. C. H. Lee, H. B. Park, Y. M. Lee and R. D. Lee, *Ind. Eng. Chem. Res.*, 2005, **44**(20), 7617.
76. M. L. Hoarfrost and R. A. Segalman, *Macromolecules*, 2011, **44**(13), 5281.
77. J. Sun, G. M. Stone, N. P. Balsara and R. N. Zuckermann, *Macromolecules*, 2012, **45**(12), 5151.
78. M. M. Mok, X. Liu, Z. Bai, Y. Lei and T. P. Lodge, *Macromolecules*, 2011, **44**(4), 1016.
79. J. Kamcev, D. R. Paul and B. D. Freeman, *Macromolecules*, 2015, **48**(21), 8011.
80. J. Kamcev, M. Galizia, F. M. Benedetti, E.-S. Jang, D. R. Paul, B. D. Freeman and G. S. Manning, *Phys. Chem. Chem. Phys.*, 2016, **18**(8), 6021.
81. J. Kamcev, D. R. Paul, G. S. Manning, B. D. Freeman, *ACS Appl. Mater. Interfaces*, 2017, **9**(4), 4044.
82. K. P. Kuhl, E. R. Cave, D. N. Abram and T. F. Jaramillo, *Energy Environ. Sci.*, 2012, **5**(5), 7050.
83. A. Kubaczka, *J. Membr. Sci.*, 2014, **470**, 389.
84. R. B. Bird, W. E. Stewart, E. N. Lightfoot, *Transport Phenomena*, 2nd edn, John Wiley & Sons, Inc, New York, 2002.
85. W. M. Deen, *Analysis of Transport Phenomena*, Oxford University Press, New York, NY, 1998.
86. R. Krishna and J. A. Wesselingh, *Chem. Eng. Sci.*, 1997, **52**(6), 861.
87. J. Newman, *Chem. Eng. Sci.*, 2009, **64**(22), 4796.
88. C. F. Curtiss and R. B. Bird, *Ind. Eng. Chem. Res.*, 1999, **38**(7), 2515.
89. M. Soltanieh and S. Sahebdelfar, *J. Membr. Sci.*, 2001, **183**(1), 15.

90. D. R. Paul, *J. Membr. Sci.*, 2004, **241**(2), 371.
91. C. P. Ribeiro, B. D. Freeman and D. R. Paul, *Polymer*, 2011, **52**(18), 3970.
92. S. G. Kimura, *Ind. Eng. Chem. Prod. Res. Dev.*, 1971, **10**(3), 335.
93. R. T. Chern, W. J. Koros, B. Yui, H. B. Hopfenberg and V. T. Stannett, *J. Polym. Sci., Polym. Phys. Ed.*, 1984, **22**(6), 1061.
94. R. D. Raharjo, B. D. Freeman and E. S. Sanders, *Polymer*, 2007, **48**(20), 6097.
95. W. J. Koros, R. T. Chern, V. Stannett and H. B. Hopfenberg, *J. Polym. Sci., Polym. Phys. Ed.*, 1981, **19**(10), 1513.
96. H. Yasuda, L. D. Ikenberry and C. E. Lamaze, *Makromol. Chem.*, 1969, **125**(3062), 108.
97. A. T. Haug and R. E. White, *J. Electrochem. Soc.*, 2000, **147**(3), 980.
98. K. Broka and P. Ekdunge, *J. Appl. Electrochem.*, 1997, **27**(2), 117.
99. B. M. Carter, B. M. Dobyns, B. S. Beckingham and D. J. Miller, *Polymer*, 2017, **123**, 144.
100. D. R. Paul and D. R. Kemp, *J. Polym. Sci., Polym. Symp.*, 1973, **41**(1), 79.
101. W. J. Koros, D. R. Paul and A. A. Rocha, *J. Polym. Sci., Polym. Phys. Ed.*, 1976, **14**(4), 687.
102. P. J. Roman, F. Detlev, K. Thomas and P. Klaus-Viktor, *J. Membr. Sci.*, 2012, **389**, 343.
103. D. R. Paul and W. J. Koros, *J. Polym. Sci., Polym. Phys. Ed.*, 1976, **14**(4), 675.
104. D. R. Paul, *J. Polym. Sci., Part A*, 1969, **7**, 1811.
105. W. J. Koros and D. R. Paul, *J. Polym. Sci., Polym. Phys. Ed.*, 1976, **14**(10), 1903.
106. T. Knoell, *Ultrapure Water,* April, 2006, 24.
107. A. Simon, L. D. Nghiem, P. Le-Clech, S. J. Khan and J. E. Drewes, *J. Membr. Sci.*, 2009, **340**(1–2), 16.
108. Y. Huang, J. Ly, D. Nguyen and R. W. Baker, *Ind. Eng. Chem. Res.*, 2010, **49**(23), 12067.
109. A. Collier, H. Wang, X. Zi Yuan, J. Zhang and D. P. Wilkinson, *Int. J. Hydrogen Energy*, 2006, **31**(13), 1838.
110. M. Inaba, T. Kinumoto, M. Kiriake, R. Umebayashi, A. Tasaka and Z. Ogumi, *Electrochim. Acta*, 2006, **51**(26), 5746.
111. T. Kinumoto, M. Inaba, Y. Nakayama, K. Ogata, R. Umebayashi, A. Tasaka, Y. Iriyama, T. Abe and Z. Ogumi, *J. Power Sources*, 2006, **158**(2), 1222.
112. V. O. Mittal, H. R. Kunz and J. M. Fenton, *J. Electrochem. Soc.*, 2007, **154**(7), B652.
113. V. O. Mittal, H. R. Kunz and J. M. Fenton, *J. Electrochem. Soc.*, 2006, **153**(9), A1755.
114. V. O. Mittal, H. Russell Kunz and J. M. Fenton, *Electrochem. Solid-State Lett.*, 2006, **9**(6), A299.
115. F. Rodriguez, C. Cohen, C. Ober, L. A. Archer, *Principles of Polymer Systems*, 5th edn, Taylor & Francis, New York, NY, 2003.

116. N. Li, Y. Leng, M. A. Hickner and C. Y. Wang, *J. Am. Chem. Soc.*, 2013, **135**(27), 10124.
117. E. N. Komkova, D. F. Stamatialis, H. Strathmann and M. Wessling, *J. Membr. Sci.*, 2004, **244**(1–2), 25.
118. B. Bauer, H. Strathmann and F. Effenberger, *Desalination*, 1990, **79**(2–3), 125.
119. M. R. Shaner, K. T. Fountaine, S. Ardo, R. H. Coridan, H. A. Atwater and N. S. Lewis, *Energy Environ. Sci.*, 2014, **7**(2), 779.
120. K. A. Walczak, G. Segev, D. M. Larson, J. W. Beeman, F. A. Houle and I. D. Sharp, *Adv. Energy Mater.*, 2017, **7**(13), 1602791.
121. K. Walczak, Y. Chen, C. Karp, J. W. Beeman, M. Shaner, J. Spurgeon, I. D. Sharp, X. Amashukeli, W. West, J. Jin, N. S. Lewis and C. Xiang, *ChemSusChem*, 2015, **8**(3), 544.
122. J. M. Spurgeon, M. G. Walter, J. Zhou, P. A. Kohl and N. S. Lewis, *Energy Environ. Sci.*, 2011, **4**(5), 1772.
123. M. A. Modestino, D. K. Paul, S. Dishari, S. A. Petrina, F. I. Allen, M. A. Hickner, K. Karan, R. A. Segalman and A. Z. Weber, *Macromolecules*, 2013, **46**(3), 867.
124. M. A. Modestino, A. Kusoglu, A. Hexemer, A. Z. Weber and R. A. Segalman, *Macromolecules*, 2012, **45**(11), 4681.
125. A. Kusoglu, M. A. Modestino, A. Hexemer, R. A. Segalman and A. Z. Weber, *ACS Macro Lett.*, 2012, **1**(1), 33.
126. Y. Gu and T. P. Lodge, *Macromolecules*, 2011, **44**(7), 1732.
127. M. L. Hoarfrost, M. Tyagi, R. A. Segalman and J. A. Reimer, *J. Phys. Chem. B*, 2012, **116**(28), 8201.
128. M. L. Hoarfrost, M. S. Tyagi, R. A. Segalman and J. A. Reimer, *Macromolecules*, 2012, **45**(7), 3112.
129. Y. Schneider, M. A. Modestino, B. L. McCulloch, M. L. Hoarfrost, R. W. Hess and R. A. Segalman, *Macromolecules*, 2013, **46**(4), 1543.
130. G. Sudre, S. Inceoglu, P. Cotanda and N. P. Balsara, *Macromolecules*, 2013, **46**(4), 1519.
131. P. Cotanda, G. Sudre, M. A. Modestino, X. C. Chen and N. P. Balsara, *Macromolecules*, 2014, **47**(21), 7540.
132. Y. Jiang, J. L. Freyer, P. Cotanda, S. D. Brucks, K. L. Killops, J. S. Bandar, C. Torsitano, N. P. Balsara, T. H. Lambert and L. M. Campos, *Nat. Commun.*, 2015, **6**, 5950.
133. P. Cotanda, N. Petzetakis, X. Jiang, G. Stone and N. P. Balsara, *J. Polym. Sci., Part A: Polym. Chem.*, 2017, **55**(13), 2243.
134. N. Petzetakis, C. M. Doherty, A. W. Thornton, X. C. Chen, P. Cotanda, A. J. Hill and N. P. Balsara, *Nat. Commun.*, 2015, **6**, 7529.
135. B. S. Beckingham, G. E. Sanoja and N. A. Lynd, *Macromolecules*, 2015, **48**(19), 6922.
136. M. A. Modestino, E. R. Chan, A. Hexemer, J. J. Urban and R. A. Segalman, *Macromolecules*, 2011, **44**(18), 7364.
137. M. A. Modestino, C. A Diaz-Botia, S. Haussener, R. Gomez-Sjoberg, J. W. Ager and R. A. Segalman, *Phys. Chem. Chem. Phys.*, 2013, **15**(19), 7050.

138. J. M. Virgili, M. L. Hoarfrost and R. A. Segalman, *Macromolecules*, 2010, **43**(12), 5417.
139. S. L. McFarlane, B. A. Day, K. McEleney, M. S. Freund and N. S. Lewis, *Energy Environ. Sci.*, 2011, **4**(5), 1700.
140. Y. Huang and D. R. Paul, *Polymer*, 2004, **45**(25), 8377.
141. N. R. Horn and D. R. Paul, *Polymer*, 2011, **52**(7), 1619.
142. M. S. Suleman, K. K. Lau and Y. F. Yeong, *Chem. Eng. Technol.*, 2016, **39**(9), 1604.
143. A. Heinzel and V. M. Barragán, *J. Power Sources*, 1999, **84**(1), 70.
144. C. Gagnard, Y. Germain, P. Keraudren and B. Barrière, *J. Appl. Polym. Sci.*, 2003, **90**(10), 2727.
145. P. Kofinas, R. E. Cohen and A. F. Halasa, *Polymer*, 1994, **35**(6), 1229.
146. S. Wang, X. Li, H. Wu, Z. Tian, Q. Xin, G. He, D. Peng, S. Chen, Y. Yin, Z. Jiang and M. D. Guiver, *Energy Environ. Sci.*, 2016, **9**(6), 1863.
147. G. Coulon, T. P. Russell, V. R. Deline and P. F. Green, *Macromolecules*, 1989, **22**(6), 2581.
148. A. Knoll, A. Horvat, K. S. Lyakhova, G. Krausch, G. J. A. Sevink, A. V. Zvelindovsky and R. Magerle, *Phys. Rev. Lett.*, 2002, **89**(3), 355011.
149. G. Krausch, *Mater. Sci. Eng., R*, 1995, **14**(1–2), 1.
150. A. Menelle, T. P. Russell, S. H. Anastasiadis, S. K. Satija and C. F. Majkrzak, *Phys. Rev. Lett.*, 1992, **68**(1), 67.
151. B. D. Olsen, X. Li, J. Wang and R. A. Segalman, *Soft Matter*, 2009, **5**(1), 182.
152. T. M. Murphy, D. S. Langhe, M. Ponting, E. Baer, B. D. Freeman and D. R. Paul, *Polymer*, 2011, **52**(26), 6117.

Devices and Modelling

CHAPTER 11

Prototyping Development of Integrated Solar-driven Water-splitting Cells

CHENGXIANG XIANG,*[a] KARL WALCZAK,[b] JOEL HABER,[a] RYAN JONES,[a] JEFFREY W. BEEMAN,[b] DAN GUEVARRA,[a] CHRIS KARP,[a] RUI LIU,[a] MATTHEW SHANER,[a] KE SUN,[a] WILLIAM WEST[c] AND LAN ZHOU[a]

[a] Joint Center for Artificial Photosynthesis, California Institute of Technology, Pasadena, CA 91125, USA; [b] Joint Center for Artificial Photosynthesis, Lawrence Berkeley National Laboratory, Berkeley, CA 94720, USA; [c] Jet Propulsion Laboratory, California Institute of Technology, Pasadena, CA 91109, USA

*Email: cxx@caltech.edu

11.1 Introduction

As the level of deployment and utilization of renewable energy sources continues to rise, large-scale, long-term energy storage technologies that can time-shift the usage of energy daily or seasonally start to play a significant role in the overall development and deployment of renewable energies. Solar energy is considered one of the most abundant and environmentally-friendly energy sources. However, the relatively low power density, typically 200 W m^{-2} to 250 W m^{-2}, as well as the intermittent nature of the source, present significant challenges for producing cost-effective electricity, heat, or fuels from sunlight.[1] As discussed throughout this book, one promising route is to store the solar

Energy and Environment Series No. 22
Integrated Solar Fuel Generators
Edited by Ian D. Sharp, Harry A. Atwater and Hans-Joachim Lewerenz

Published by the Royal Society of Chemistry, www.rsc.org

energy in the form of chemical bonds, *i.e.*, artificial photosynthesis.[2–4] In this scheme, the incident photons are converted into energetic electrons and holes, which transport to the catalytic sites to perform the desired fuel-forming reactions efficiently and selectively. Two key components: a power-generating component and a fuel forming component are usually required and need to be optimized for operation. The sunlight-driven power-generating component is often a semiconductor material, in which the photo-generated carriers are separated and transported selectively into two terminals. Various strategies to create asymmetry in semiconductor materials systems have been used, including semiconductor liquid junctions as well as solid-state heterojunctions and homojunctions. The basic principle of the power-generating component is similar to the one in the photovoltaic devices. For that reason, there are a variety of well established photovoltaic materials, *e.g.*, Si, CdTe, CIGS, III–V compound semiconductors, *etc.*, that are technologically relevant for solar-driven water-splitting applications. The fuel-forming component often consists of two electrochemical reactions and a transport medium/mechanism between the two reactions. For a solar-driven water-splitting device, the anode and cathode reactions are the oxygen evolution reaction (OER) and hydrogen evolution reaction (HER), respectively. A robust mechanism for separating the product gasses and for providing necessary ionic transport between the cathode and anode chambers is also required. The power-generating component and the fuel-forming component can be, in principle, connected electrically in series in a modular approach, where a PV panel is wired directly to an electrolyzer unit.[5–12] However, the low capacity factor of PV due to the intermittent nature of the solar resource, and high capability factor and high cost of current electrolyzers, present a significant challenge to produce cost-competitive hydrogen.[13] In contrast, producing fuels directly from sunlight using cost-effective and earth-abundant materials, with scalable processes, offers a unique opportunity and design space for long-term, grid-scale energy storage relative to series connected PV + electrolyzer configurations.[3,14–16]

Rapid prototyping is crucial to the development of an integrated solar-driven water-splitting device, where multiple components and subcomponents need to come together and perform synergistically under realistic operational conditions. The overall solar-to-fuel conversion efficiency, device stability and materials performances for a real-world application can only be evaluated and characterized in a prototype device where all the necessary components are integrated and tested under realistic operational conditions. A range of laboratory-scale (typically <1 cm^2 photoactive area) solar water-splitting devices have been demonstrated.[15,17–19] However, PEC devices with practical life-time and photo-active areas have yet to be demonstrated. This book chapter summarizes key materials, components and device designs that are important to advance this technology, as well as attempts to integrate materials and components at different length scales. One unique feature of this chapter, as well as this book as a whole, is that it does not only report successful device demonstrations but also puts an emphasis on the failed attempts and lessons learned during the original phase of JCAP that emphasized integrated solar-driven water-splitting development. For example, various epoxy materials were tested by JCAP researchers for stability in acid or base electrolytes and their findings are tabulated in this chapter; experimental handling and integration of ion-exchange membranes into an integrated device is discussed in detail; failed attempts on light absorber materials and lessons learned on electrolyte conditions and constraints at the early stage of JCAP are reviewed and discussed as well.

The book chapter is organized as follows: first, the selection of materials and components that constitute efficient, stable, scalable and safe solar fuel devices will be discussed in detail. Then, full device design, characterization and benchmarking will be reviewed and discussed. At the end, a system engineering approach will be introduced and specific examples of hierarchical requirements relating to the solar fuel devices will be given.

11.2 Materials and Components

The critical components for a solar-driven water-splitting prototype include light absorbers, electrocatalyts, membrane separators, electrolyte, chassis, and auxiliary components. Signficant advances have been made in the development of light absorbers, electrocatalysts, and membrane separators, as discussed in previous chapters. In this chapter we will discuss the selection and design consideration of light absorber materials, electrolytes, and membranes, as well as materials for chassis and auxiliary systems. Then, we discuss the incorporation of newly discovered protection layers and catalyst systems into the existing PEC assemblies.

11.2.1 Selection and Design Consideration of Light Absorber Materials

The primary attribute for light absorber materials, when illuminated with 1 sun AM 1.5 G solar spectrum sunlight, is that the material generates sufficient voltage and current to perform the water-splitting reaction at an appriciable rate, without the addition of external power. Although, the thermodynamic water splitting potential is 1.229 V, system losses such as electrocatalytic overpotentials and transport losses in electrolyte, for example, could significantly increase the total potential required to split water,[20,21] as shown in Figure 11.1. The operating point of the device, which determines the device efficiency, is not only dependent on the current–voltage characteristic of the light absorber, but also dependent on the overall polarization curve of the electrochemical water-splitting component. While many research materials including metal oxides and metal sulfides have shown some promise in solar water-splitting, the efficiency and stablity of these materials are far from sufficient for incorporation into a practical device. Hence, the JCAP prototyping efforts focus on the adaptation and use of commerically available and technologically relevant semiconductor materials and stacks that are often employed in photovoltatic applications.

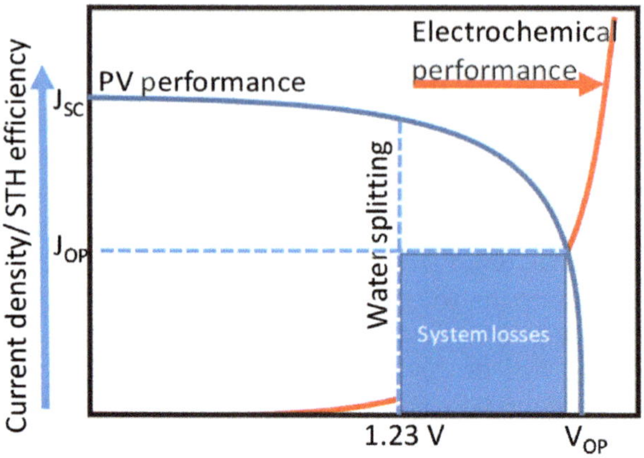

Figure 11.1 Graphical representation of the solar cell performance overlaid with the electrochemical performance. The intersection of the two curves is the potential operating point of the coupled system.

11.2.1.1 Triple-junction Amorphous Silicon

Triple-junction amorphous Si cells are one class of a few commercially available two terminal PV-grade substrates that can produce enough voltage to drive the overall water-splitting reaction. Other PV-grade substrates are often produced in a single junction cell configuration, including crystalline silicon, GaAs, CdTe, or CIGS, which do not have enough voltage to drive the overall water-splitting reaction without wiring two or more cells in series. Hence, initial JCAP efforts focused on the use of triple-junction a-Si cell for the demonstration of an integrated solar-driven water-splitting device. The goal was to understand the limitations of the material, as well as to design new cell stacks with manufacturers to improve the stability and efficiency of the overall device.

JCAP worked with Uni-Solar (Midland, Michigan) and Xunlight (Toledo, Ohio) to obtain the un-processed, bare cells without additional encapsulations for photovoltaic applications. These substrates were manufactured using roll-to-roll processes and exhibited open circuit voltages that exceed 2.1 V and short-circuit current densities that exceed 6 mA cm^{-2}. Triple-junction a-Si substrates were primarily used in two settings, one in near-neutral pH solutions and one in extreme pH solutions. In near-neutral pH solutions, the JCAP team demonstrated a membrane-free system that utilized the a-Si substrate, Pt and cobalt phosphate (Co–Pi) as electrocatalysts for the hydrogen-evolution reaction (HER) and oxygen-evolution reaction (OER), respectively, and a bulk aqueous solution buffered at pH = 9.2 by 1.0 M of boric acid/borate as an electrolyte.[22] The integrated membrane-free system exhibited a solar to fuel conversion efficiency of 3.2% in the borate solution, which exceeded the 2.5% STH conversion efficiency that has been reported previously for a comparable device,[23] due to the improved semiconductor processing and catalyst deposition protocols (Figure 11.2). However, the membrane-free system exhibited significant crossover of products, and in

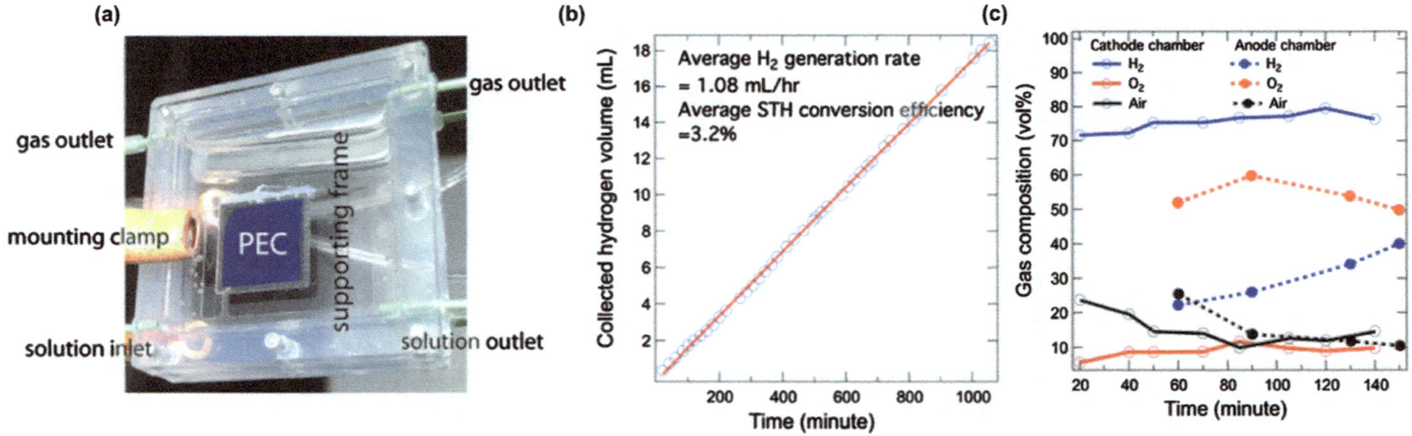

Figure 11.2 (a) Photograph of an integrated, membrane-free solar-driven water-splitting device. (b) Rate of production of hydrogen gas in the cathode compartment. The average rate of H_2 production was calculated to be 1.08 mL h^{-1} from a linear fit of the experimental data. (c) Time evolution of the gas composition (H_2, O_2 and air) from the cathode compartment (solid lines) and the anode compartment (dashed lines).
Reproduced from ref. 22 with permission from the Royal Society of Chemistry.

some measurements, up to 40% of H_2 was found in the O_2 chamber, indicating that the system was not intrinsically safe and cannot be scaled up and deployed in practice.

To build an intrinsically safe PEC device, a robust gas separation method needs to be in place. One straightforward method to achieve gas separation is to use membrane separators that are widely adapted in the water electrolysis community. Membrane separators, such as Nafion, can block diffusion and convective crossovers of gas products as well as provide necessary ionic conduction between cathode and anode chambers. JCAP's prototyping efforts centered on the development of PEC devices that operate in either 1.0 M H_2SO_4 (aq) with a cation exchange membrane or 1.0 M KOH (aq) with an anion exchange membrane. Triple junction a-Si substrates were tested under those extreme pH conditions (pH = 0 or pH = 14) and were found to be unstable. In both 1.0 M acid and base, the cells showed degradation of performance within 15 minutes. In particular, the instability of the back reflector layer (ZnO) and the top contact layer (ITO) in the triple junction a-Si corroded rapidly when exposed to the electrolyte. Various protection methods, including atomic layer deposition of TiO_2, were used to stabilize the cell without much success.

11.2.1.2 Monolithic Tandem and Triple-junction Crystalline Silicon

Due to the instability of triple junction a-Si, an alternative route for using custom-grown epitaxial tandem and triple-junction crystalline silicon cells was explored. The rationale for using crystalline silicon cells was that the Si-based photocathode would be thermodynamically stable under the hydrogen evolution conditions and thus could provide a stable light absorber for the PEC assembly. It is known that single junction crystalline Si cells cannot provide enough voltage to sustain the overall water-splitting reaction. So, the guiding principle was to trade photocurrent density for the voltage such that with proper thickness of the Si layer, an open-circuit voltage in excess of 1.5 V would be realized. Note that the multi-junction band-gap material stack has the same maximum energy conversion efficiency as a single junction material. Details of the tandem and triple junction p–i–n structures are shown below Figure 11.3.

The cell designs were based on 1 sun AM 1.5 G irradiation and the optical properties of intrinsic silicon.[24,25] The tandem junction cell was expected to produce an open circuit voltage of 1.2 V and a light limiting current density of 18 mA cm^{-2}, while the triple junction was expected to generate an open circuit voltage of 1.8 V and a light limiting current density of 10 mA cm^{-2}. The cells were configured such that the cathode was on the light-facing side for the triple junction and the dark-facing side for the tandem junction cell. The tandem-junction device was intended to be a platform for integration of photoanodes, such as tungsten oxide or bismuth vanadate, while the triple junction cell could split water without an additional light absorber.

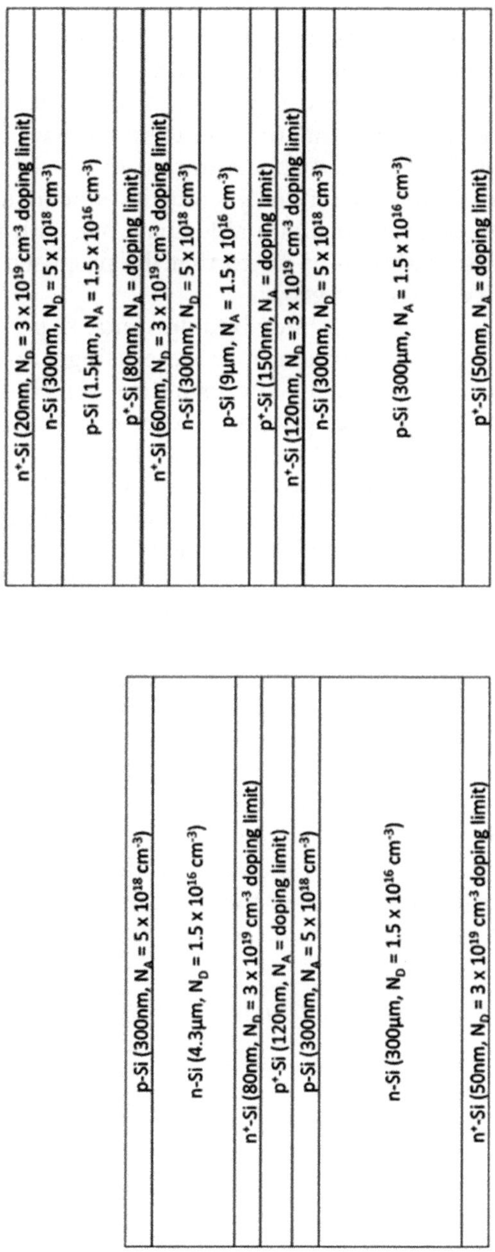

Figure 11.3 Proposed tandem and triple junction crystalline cells.

However, in practice, growing functional, high-fidelity structures proved to be an insurmountable challenge with the time and resources we had available. Initially the solar cell performance was promising for the tandem Device. However, when it was attempted to reproduce the tandem device structure, the solar cell performance of the initial wafer could not be repeated. After months of troubleshooting with LSRL, using secondary ion mass spectometery (SIMS), solar cell performance and scanning resistance probe (SRP) the conclusion was that the tunnel junction was at least an order of magnitude below the required concentration to get a functional device. The growth of functional high fidelity epitaxial silicon tandem junction structures proved to be an insurmountable challenge with the time and resources available, also it should be noted that the proposed silicon epitaxial triple junction device was never grown.

Next, we decided to design a side-by-side tandem p–n structure. This was a lower risk device concept since we were using proven components, and the manufacturing equipment and expertise were in-house. The base photoelectrode platform was a tandem silicon p–n structure, configured back-to-back through mechanical integration. The silicon tandem structure produced an open circuit voltage of approximately 1 V and a short-circuit current density of approximately 20 mA cm^{-2} when illuminated by 2 suns. Although this tandem junction device required dual illumination, it provided the flexibility to integrate either or both a photoanode and photocathode to produce the additional voltage required to split water. Using a tungsten oxide photoanode, the utility and durability of the tandem photoelectrode was proven. The tandem junction device, in conjunction with a tungsten oxide photoanode, produced a solar-to-hydrogen conversion efficiency of 0.24% as predicted by modeling and simulation. The low conversion efficiency was mainly due to the wide band-gap of the photoanode material. This is the first full device prototype demonstration in JCAP for a solar water-splitting system that produced products which were well below the explosive limits, had an operational lifetime greater than 100 hours, and employed a device configuration that is scalable. Details of the design, build, and test can be found in Walczak *et al.* 2015.[26]

11.2.1.3 Compound Semiconductor Multi-junction Photovoltaics

To demonstrate a device with greater than 10% solar-to-hydrogen (STH) conversion efficiency, silicon-based materials face significant challenges. Light absorbers based on III–V semiconductors have the potential to achieve STH conversion >20%. During the first phase of JCAP, three types of III–V substrates (Figure 11.4) were studied and employed to make full solar-driven water-splitting devices.

Figure 11.4a shows a tandem structure consisting of an InGaP top cell ($E_g = 1.84$ eV) and GaAs bottom cell ($E_g = 1.42$ eV) that was designed and modeled using Helmholtz–Zentrum Berlin's AFORS-HET software. The dual junction device was grown commercially (Sumika Electronic Materials, Inc.)

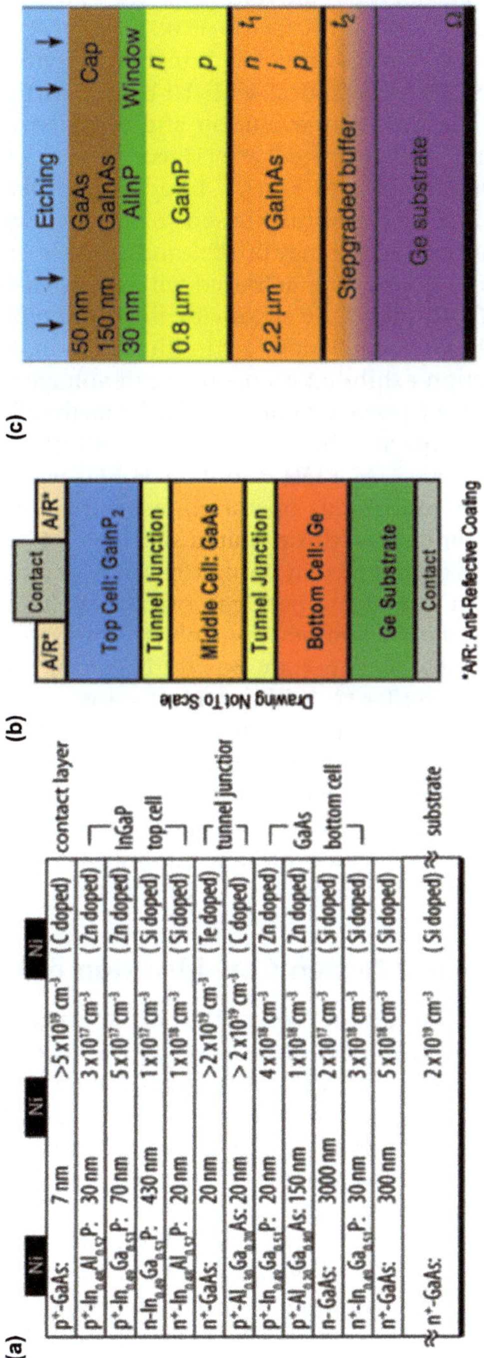

Figure 11.4 Schematic illustration of three III–V based cell structures used in JCAP.

according to specifications. The solid-state performance of this tandem structure exhibited an open circuit voltage of 2.4 V, a short-circuit current density of 7.6 mA cm^{-2} and a fill factor of 0.76 under 1 sun illumination. Using this tandem cell, a monolithically integrated device with a TiO_2 protective coating layer, in conjunction with Ni-based, earth-abundant active electrocatalysts for the hydrogen-evolution and oxygen-evolution reactions, was constructed and demonstrated a STH conversion efficiency of 8.6% under 1 sun illumination in 1 M KOH.[19] In addition, this tandem cell was also employed in a bipolar membrane based solar-driven water-splitting cell, in which a STH conversion efficiency of 10% and a device area of 1 cm^2 and a device stability of 100 hours were achieved with all earth-abundant electrocatalysts.[17] Figure 11.4b shows the schematic illustration of a triple junction cell from Spectrolab (UTJ cell). Under 1 sun illumination, this commercially available triple junction exhibited an open circuit voltage of 2.55 V, a short-circuit current density of 14.85 mA cm^{-2}, and a fill factor of 0.70.[27] Using this triple junction cell protected by a composite coating, a STH conversion efficiency >10% in both 1 M KOH and 1 M H_2SO_4 was demonstrated by the JCAP team under continuous operation, as well as under diurnal light cycling for at least four days, with simulated sunlight.[27] Figure 11.4c shows a tandem junction that consisted of a GaInP top cell with a bandgap of 1.78 eV and a GaInAs bottom cell with a bandgap of 1.26 eV. The tandem cell was grown epitaxially by metal–organic vapor phase epitaxy on a Ge substrate. The lattice mismatch to the substrate was mitigated by the use of an intermediate step-grading buffer.[28] With anti-reflection coating, the tandem cell exhibited a short-circuit current density of up to 15 mA cm^{-2} and an open-circuit voltage of 2.1 V under 1 sun illumination. Using this tandem stack, a solar-to-hydrogen efficiency of 14% was demonstrated *via in situ* surface functionalization in 1 M $HClO_4$ with RuO_2 as the oxygen evolution catalyst and Rh as the hydrogen evolution catalyst. Reproduced from ref. 29 with permission from the Royal Society of Chemistry.

11.2.2 Selection and Design Consideration of Electrolytes

Selection of electrolytes plays a critical role in defining the overall device performance (Figure 11.5). Electrolytes provide the reaction media for the HER and OER and hence affect the attainable kinetic overpotentials of the catalyst systems. For instance, relatively low overpotentials and decent stability for OER can be achieved by a range of earth-abundant, mixed metal oxides in alkaline conditions, whereas in acidic conditions, there is still a lack of earth-abundant materials for efficient and stable operation for OER.[30,31] Electrolytes also provide the ionic conduction between the cathode and the anode electrodes. The transport loss in the solution electrolytes, including ohmic resistive loss and electrodialysis, could limit the attainable conversion efficiency of the device. For instance, in an integrated solar-driven water-splitting cell, the critical dimension of the electrode needs to be less than a few centimeters in order to minimize the resistive loss, even in a

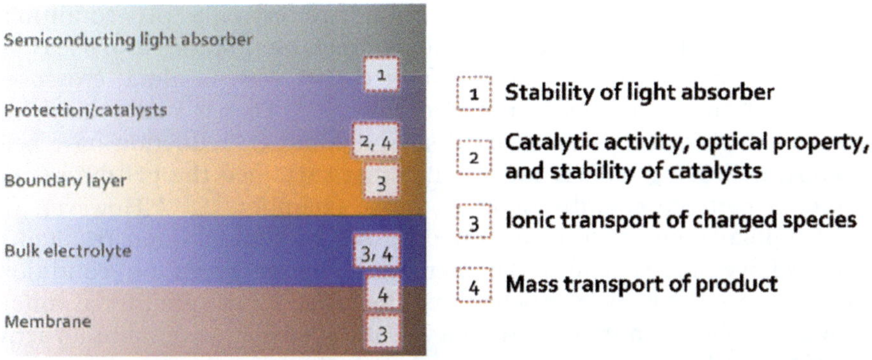

Figure 11.5 Effects of electrolyte at interfaces and on components of integrated solar water splitting devices: (1) stability of light absorbers, with failure typically at defects present in the protective coating and introduced during fabrication, deployment, or operation, (2) catalytic activity, optical properties, and stability of catalysts, (3) ionic transport of charged species, and 4) mass transport of products in separate streams and prevention of cross-over. Physical dimensions are not to scale.

highly conductive electrolyte, *e.g.*, 1.0 M H_2SO_4.[32] The electrolyte also affects the operational stability of the underlying semiconducting light absorbers.

11.2.2.1 Electrolyte Effect on Transport Losses in a Device

One of the most important roles for electrolytes in an integrated solar-driven water-splitting cell is to provide the necessary ionic transport between the cathode chamber and anode chamber to complete the overall circuit with the electron flow at the electrode/electrolyte interface. Namely, each electron that is injected into the (photo)electrodes to perform the reduction reaction at the electrode/electrolyte interface at the cathode chamber needs to be compensated by the transport of either H^+ or OH^- from or toward the anode chamber. Additional counter ion transport or steady state gradients produced during the operation would result in additional polarization loss in the system and additional energy penalties in auxiliary re-circulation or mixing to sustain the overall reaction.

To achieve near-unity transference number for either H^+ or OH^- between the cathode and anode chamber, 1 M H_2SO_4 or 1 M KOH was often used in JCAP devices in conjunction with a cation exchange membrane, such as Nafion, or an anion exchange membrane, such as Selemion.[19,27,33] Transport losses in 1 M H_2SO_4 or 1 M KOH were mainly resistive losses and the effects of resistive losses on the device designs and architectures were explored by multi-physics modeling and simulation early on in JCAP.[20,34,35] While strong base or strong acid electrolyte are highly conductive, modeling and simulation results shows that to minimize polarization loss in a macroscopic planar device, the critical length scale of the photoelectrodes needs to be less than a few centimeters at an average operating current density of

~10 mA cm^{-2}.[20,35] In addition to those two extreme pH conditions, near-neutral pH devices that incorporated bipolar membranes also use H^+ or OH^- as the majority of the ionic current carrier with minimal crossovers of other cations or anions in the electrolyte.[36–40] Un-assisted solar-driven water-splitting as well as un-assisted solar-driven CO_2 reduction have been demonstrated using bipolar membranes that sustained the pH differential between a cathode chamber and an anode chamber.[17,40,41] However, the use of bipolar membrane in near-neutral pH conditions often resulted in larger voltage losses across the membrane under operation conditions when compared to Nafion or Selemion. The cause for the additional voltage loss with bipolar membrane is still not well understood and remains an active area of research.

In addition to systems that aim to use H^+ or OH^- as the sole ionic carrier, systems that use counter ions as the major ionic current carrier have been studied by modeling and simulation with or without controlled flows at a range of device configurations.[42,43] Over the past few years, the JCAP team has evaluated the Nernstian potential loss associated with pH gradients at the electrode surfaces, the resistive loss between the cathode and anode, the product-gas crossovers, and the required pumping energy in different flow schemes as a function of the operational current density, the flow rates for the electrolyte, and the physical dimensions of the devices. Figure 11.6 shows pH distributions in a passive flow PEC system and a controlled flow PEC system under operating conditions. The trade-offs in the voltage loss, safety considerations, and energy inputs from the balance-of-systems required to produce a practical device have been evaluated and compared to membrane-free devices as well as to devices that operate at extreme pH values.[43,44] The JCAP team has provided useful insights for the general flow

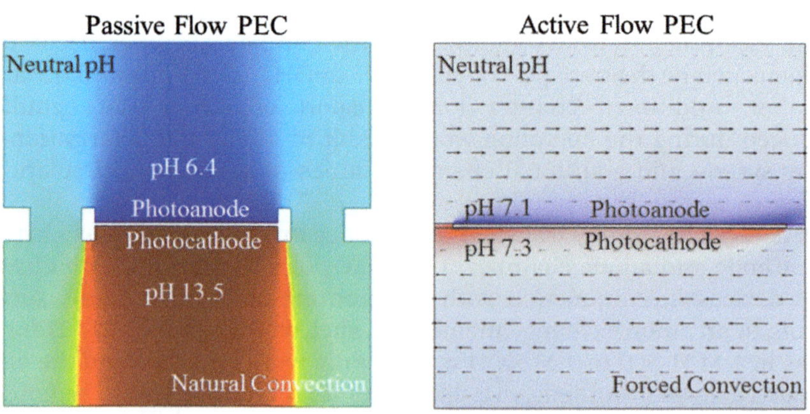

Figure 11.6 pH distributions in a passive flow PEC system and an active flow PEC system at an average current density of 18 mA cm^{-2}. The countercurrent velocity field is shown by horizontal arrows.
Reproduced from ref. 43 with permission from the Royal Society of Chemistry.

conditions and device configurations under which intrinsically safe, efficient solar-driven water-splitting cells can be operated.

11.2.2.2 Electrolyte Effect on the Stability of Semiconducting Light Absorbers

In an integrated solar fuel generator, semiconducting light absorbers, together with other components, such as electrocatalysts and membranes, are immersed in the reaction solution. To prevent significant optical loss through the catalysts or the protection layer (if present), it is necessary for overlayers in the light path to remain transparent over the spectral range absorbed by the light semiconductors and, ideally, to achieve refractive index-matching that improves light trapping. Increases of optical density in catalysts that undergo changes of oxidation state can occur and necessitate that these layers remain thin. Due to the fact that overlayers on light absorbers must be quite thin, complete isolation of the substrate from the solution is practically impossible, even with the current state-of-the-art clean room technology and especially on large scale devices. Therefore, chemical stability under idle conditions and electrochemical stability under operational conditions of the semiconducting light absorbers are highly desired.

Various protection schemes based on different materials and technologies have been demonstrated.[45,46] Success has been only achieved on self-passivating materials.[46–52] Self-passivation describes the formation of chemically inert materials from the reaction of substrate in direct contact with the electrolyte under operation conditions. This phenomenon and the evolution of this passivation layer not only depend on the physical nature of the substrate, including orientation, doping concentration, geometry and homogeneity, but also on the chemical nature of the electrolyte.[53–57] For example, for an anodically formed passivating SiO_2 on Si photoanodes, dissolution rate of the passivation layer is a function of the OH-concentration and temperature. Better passivation efficiency and stability of the self-passivation layer should be expected from Si based solar fuel generators in electrolytes with lower pH based on the Pourbaix diagram. The dissolution rate of SiO_2 drops again when OH^- concentration becomes higher than 40%. Meanwhile, when operating outdoors in the field, the passivation efficiency drops with higher temperature due to the increased dissolution rate of SiO_2. At a relatively low OH-concentration (1 M), dissolution of SiO_2 is also a function of dissolved Si concentration in solution. With the increasing concentration of Si in the electrolyte, dissolution rate of SiO_2 also drops and saturates. Moreover, the dissolution rate of SiO_2 and thus the passivation efficiency also depends on the cations and anions in the solution. It was found that smaller alkali ions (Na^+ or K^+) cause higher dissolution rate of SiO_2 compared to larger ones (Rb^+ or Cs^+). The etch rate of SiO_2 in acid containing no F-ions is much smaller than in alkaline solution.

Whether or not the semiconducting light absorber can be self-passivating in certain electrolytes can be roughly determined by the leakage current of the unprotected substrate under the operation condition within the interested potential window.

Some representative demonstrated self-passivating materials, reactions conditions, passivating layers, and performances of photoelectrodes are as follows:

- n-type crystalline Si, water oxidation, 1.0 M KOH, SiO_x
- n-type crystalline Si, water oxidation, 1.0 M K-Bi, SiO_x
- n-type amorphous Si, water oxidation, 1.0 M KOH, SiO_x
- n-type CdTe, water oxidation, 1.0 M KOH, $CdTeO_x$
- n-type InP, water oxidation, 1.0 M KOH, InO_x
- n-type InP, water oxidation, 1.0 M K-Bi, InO_x
- p-GaAs, proton reduction, 3.0 M H_2SO_4, As metal

11.2.2.3 *Electrolyte Effect on Catalytic Activity, Stability and Optical Transmittance*

11.2.2.3.1 Effect of Unintentional Cation and Anion in Electrolyte on the Catalytic Activity. Metal ions in the electrolyte, arising from manufacturing, delivery, or storage, as well as those generated during reaction, are almost impossible to avoid. Electroplated metal ions on HER catalysts during reductive reactions could affect the catalytic activity, selectivity, and Faradaic efficiency. As the most active HER catalyst, Pt is the most sensitive material to impurities in the electrolyte depending on the intrinsic impurity level and the history of the electrolyte. Impurities were often found from leached metal ions from reference electrodes, counter electrodes or epoxy. Cleanness of the glassware, stir bar, frit, *etc.* could also introduce contaminations to the system, thereby affecting the Pt activity. The use of Pt as the HER catalysts in electrolyzers or integrated solar fuel systems has been questioned due to the high cost and low earth abundancy. However, it has been pointed out that the use of Pt can be kept low without sacrificing its activity.[58,59] These studies have justified the use of Pt at the minimum amount in large scale solar fuel conversion systems. However, at low loadings, the contamination effect becomes more severe, where any plated metals on the Pt would cause the loss of active sites and thus lower its geometric activity. Low loadings would also concentrate the current from light absorber to the catalyst, which accelerates the blockage of the active sites by plated impurities. Overpotential loss thus causes the shift of the operation point and in turn affects the conversion efficiency. Large active surface area at high loadings would essentially delay the impurity effect to the catalytic activity until the coverage of the metal impurities quickly lowers the active surface area to the loading-sensitive regime.

Active research searching for non-Pt alternative HER catalysts have been widely conducted and various new materials with new structures including metal sulfides,[60–63] selenides,[64,65] phosphides,[66–68] borides,[69–71] nitrides,[72–75] carbides,[76–81] and silicides,[82–84] have been synthesized and characterized. The best HER catalysts developed so far, showing activity approaching that of

Pt, are metal phosphides, especially iron phosphide.[68] Again, in this case, the activity could be easily jeopardized by impurities, especially when the loading of catalyst is kept low. Therefore, overloading of active catalysts to lower the sensitivity to any changes on its active sites is still desired in practical systems.

For less active HER catalysts, solution impurities can affect the activity either positively or negatively, depending on the nature of the contamination from the electrolyte. For example, Pt when used as a counter electrode could be oxidized and dissolved in acid, diffused through the frit, then deposited on the HER working electrode, resulting in an improvement of geometric activity and misrepresentation of its intrinsic activity. Note that the potential window and current density level used in the characterization would also affect the deposition of metal impurities, bringing another factor to the time evolution of the activity. Experimental conditions, electrolytes, cleaning process and other details should be controlled and documented when new electrocatalysts are characterized and reported to conclude whether the catalytic activity is indeed caused by the species under consideration and not due to the presence of adventitious contaminants at low levels.

Similar to the effect of contaminants to the HER catalysts, type and concentration of contaminants affect OER catalysts. Ni-oxyhydroxide is a promising earth-abundant catalytic material for water oxidation in alkaline electrolyte. It was found that the incorporation of trace Fe impurities (<1 ppm) from commonly used KOH (reagent grade) electrolyte showed an activation process and improved activity over the pristine films. One long recognized theory behind this was the phase change between α/γ and β/β during the aging process in KOH. Recently, it was observed that a new layered double hydroxide phase, due to adventitious Fe, could be the cause of enhanced activity,[51,85–87] as discussed in Chapter 3 of this book by Bell. Adventitious Ni impurities was also reported to have promoted catalytic activity in non-purified electrolytes. Symes and colleagues find that a Ni contamination with a concentration as low as 17 nM in the solution deposited on bare FTO glass substrates acted as an active and stable catalyst for water oxidation reactions.[88] The aforementioned studies serve as proof of the important role of contamination on defining catalytic activity and highlight the importance of carefully controlled experimental conditions for evaluating and maintaining stable device operation.

11.2.2.3.2 Electrolyte Effect on Activity and Stability. Metal oxide, sulfide, phosphide, selenide, *etc.* have been used as dual functional catalysts for HER and OER reactions. However, the active species, especially on non-oxide OER reactions, are possibly in the form of metal oxy-hydroxides after leaching sulfur, selenium and phosphor into the corrosive solution under oxidation conditions. This would essentially result in a porous substrate and a high geometric activity depending on the morphology and structure of the starting materials. Similarly, HER catalysts based on metal oxides can be reduced in reductive environments. In addition, when used in acid, these oxides can dissolve and redeposit, which act as active sites for HER reactions. Chemical treatment in these examples completely altered the chemical composition at the surface. Therefore, evidence of catalytic activity from the new materials with a detailed study of its surface composition especially under operation becomes important.

OER catalysts based on earth abundant materials are more stable in strong alkaline electrolyte and therefore the stability of the catalytic activity decreases as the pH of the electrolyte decreases. When operated in bicarbonate-buffered KOH solution, a regular aging process which typically resulted in an improved OER activity in non-pure KOH, did not show an activation but a continuous degradation of activity. Instead, Ni based catalysts showed more stable activity in borate buffered KOH solution with a comparable pH to that of the bicarbonate-buffered one. Moreover, NiO_x catalysts showed much higher activity in borate buffered KOH compared to that in bicarbonate buffered one suggesting a positive influence of borate ion on the binding energies of reaction intermediates of the OER.

11.2.2.3.3 Electrolyte Effect on Light Absorption. Refractive index also needs to be considered in the design of antireflection coatings due to the slightly higher refractive index of water than air. The refractive index is also a function of the salinity. In typically used solutions (acid, alkaline or salt electrolytes), the increasing concentration leads to a small increase of the refractive index. These electrolytes with a concentration of 1 M have a refractive index of 1.34, higher than that of pure water at room temperature (1.33) and certainly higher than air (1.0). This difference must be reflected in modeling, calculation and the design of optimized light trapping structures. For example, for a single layer antireflection coating on planar polished Si, the optimum refractive index at the wavelength of 550 nm, where solar spectrum has the highest intensity, is 2.3 instead of 2.0 when designed for operation in air. Thickness in this case should be around 59 nm rather than 68 nm in air.

The electrolyte thickness also contributes to the parasitic optical losses, *i.e.*, the non-zero imaginary part of the refractive index, especially for the longer wavelength absorption in the bottom cell. The absorption coefficient of water at different temperature, ion concentration and thickness determines the solar irradiance at the surface (Figure 11.7). The absorption by pure water in the visible region of the spectrum is low in the blue and increases into the red and infrared with visible harmonics associates with state transitions. There are large variations in the measurement of pure water absorption coefficients, with the largest variations in the blue region of the spectrum, likely due to contamination by organic matter. Absorption peaks of water are consistent with the major dips in the solar spectrum (AM1.5), which is primarily filtered by water vapor, O_2, O_3, and CO_2 in the atmosphere.[45] Furthermore, the absorption coefficient at different wavelengths shows different temperature dependent behavior.[89] Calculations showed that with the increasing thickness of the water layer, transmittance losses increase especially at the longer wavelength, which affects the bottom cell

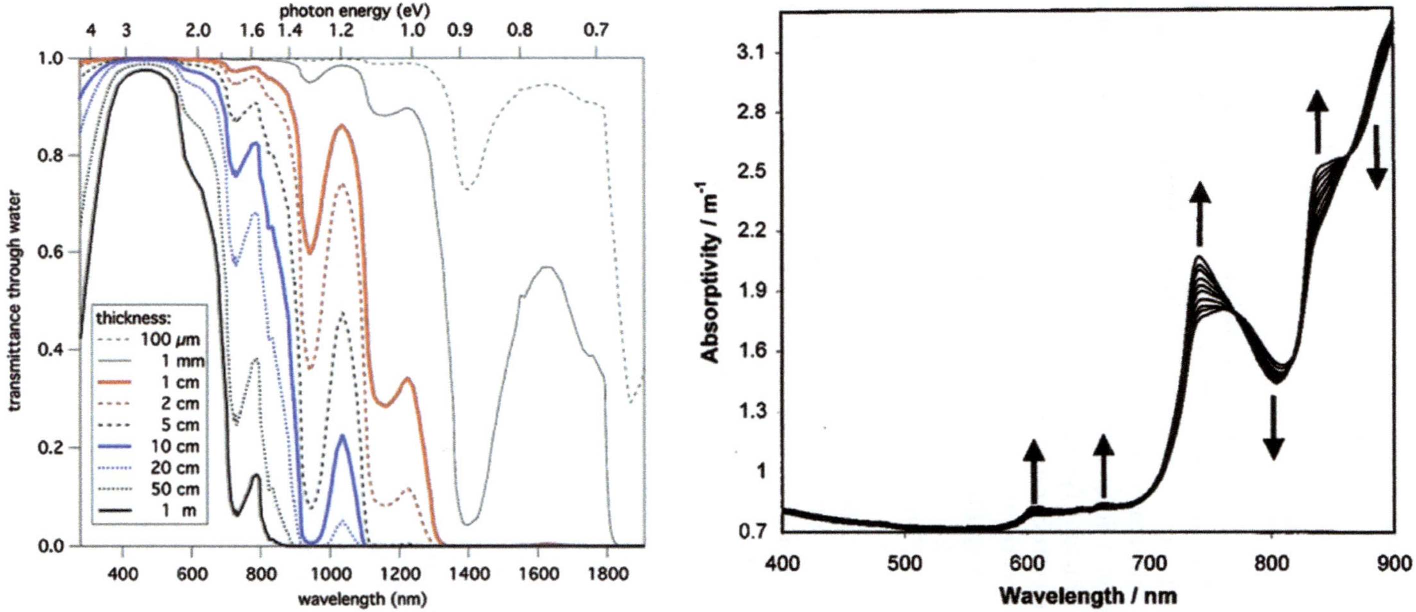

Figure 11.7 Transmittance of pure water at different water layer thicknesses (left) and the temperature dependence of the absorptivity (right). Arrows show the direction of change of the absorptivity when the temperature increases from 15 to 60 °C. Left figure reproduced from ref. 24 with permission from the Royal Society of Chemistry. Right figure reprinted with permission from V. S. Langford, A. J. McKinley and T. I. Quickenden, Temperature Dependence of the Visible-Near-Infrared Absorption Spectrum of Liquid Water, *The Journal of Physical Chemistry A*, 2001, 105, 8916–8921, Copyright 2001 American Chemical Society.

absorption in a tandem design. This parasitic absorption shifts the bandgap combination to larger band gaps and strongly affects the solar-to-hydrogen conversion efficiency.[90] Therefore, in principle, as long as the optical pathway in the water layer is kept thin (~1 cm), design of the tandem junction, as well as the efficiency, are not affected.[45]

11.2.2.3.4 Electrolyte Effect on Electrochromism of Electrocatalysts. Most OER catalysts show electrochromism, a property describing a change of optical absorption spectrum under positively applied potentials. The mechanism of the electrochromism simply relies on the intercalation/deintercalation of small ions in electrode causing a change of the oxidation state of the metal. Typically used OER catalysts, like Ni, Fe, Co, Mn, and Ir are all electrochromic with significant differences in their signature absorbance spectra under water oxidation conditions.[91] This wavelength dependent absorption spectrum is not only a material-dependent property, but also strongly affected by the morphology, loading, and total number of active sites. Therefore, in an integrated solar fuel device with the OER side as the light incident side, the electrochromism in OER catalysts will be another consideration to the design of the optimum bandgap combination for proper current matching. Meanwhile, the degradation of electrochromism as the device ages has been observed before, which is a memory effect affecting both the transmittance at the colored and bleached states and typically causing the irreversible darkening of the film. Surface species showed distinctly different valence conditions of Ni.[92] The degradation of the optical reversibility is caused by ion trapping in the structure. The optical stability of the OER catalysts, therefore, becomes another important criterion to reach a long-term operational lifetime. Permanent darkening of the OER catalysts will alter the transmitted spectrum filtered by the catalyst. Since the system is initially designed for one particular spectrum of the catalyst, *i.e.*, before aging, increased optical loss will reduce the solar-to-fuel conversion efficiency. High temperature annealing with lower density of active sites was shown to effectively enhance the optical stability.[92,93] Other techniques to improve the dynamic range, transparency of the colored state, and optical stability of catalysts have been developed, for example, by intercalation of Li ions.[94,95]

In addition, buffer ions bonding to the surface affect the catalytic activity through interaction with the active sites. The absolute number of surface active sites can be altered and thus the optical absorption spectrum is affected.

The difference in the electrochromism and optical stability of OER catalysts even in different electrolytes affect the design of the semiconductor tandem junction.

Development of semiconductor tandem junction photovoltaics is difficult. Though there is flexibility in tuning the band gap for proper absorption, this must be done in a way that maintains the electronic properties (carrier lifetime) and promotes lattice matching during integration, which could affect bulk and interfacial recombination. Therefore, direct integration of the known tandem light absorbers and electrocatalysts may involve dramatically reduced efficiency compared to the discrete PV + E counterpart, which does not require a consideration of the optical property of the electrocatalyst. This dilemma can be improved by reducing the coupling between the catalyst and semiconductor light absorbers.[58,96] Generally speaking, if the optical transmittance of the catalysts is smaller than 10–15%, integrated devices can be more efficient than the PV + E. However, if the optical loss is greater than 15%, the integrated device does not represent any merit in terms of efficiency since the typical metallization on solid state solar cells is around 10–15%. In this case, it is apparently reasonable to pattern the catalysts into grids or islands with an optical obscuration of no more than 10–15%. The advantages of this design are three-fold. First, it minimizes the effort of redesigning the light absorber based on the absorption spectrum of the catalysts. Second, it minimizes the sensitivity of the performance stability to the optical stability of the catalysts. And third, catalyst loading can be maximized for maximized efficiency which could also maximize its chemical stability. Note that this patterned catalyst design puts a strict requirement on the conductivity of the transparent TCO between the catalysts and light absorber or on the carrier diffusion length to ensure minimized lateral resistance loss and maximized carrier collection efficiency.[58]

11.2.3 Incorporation of Membrane Separators

Membrane separators are essential to the efficient, sustainable and safe operation of solar fuel devices. The main function of membrane separators includes providing ionic pathways between the cathode and anode chamber and minimizing product gas crossovers. Generally, to operate in acidic electrolyte, a cation exchange membrane (CEM) such as Nafion, sulfonated polystyrene, sulfonated polyether ether ketone, or polyimide-based membrane is often used. And to operate in alkaline electrolyte, an anion exchange membrane (AEM), in which commonly used charged side-groups include quaternary amines, imidazolium, pyridinium, and quaternary phosphonium ions, is often employed. CEM or AEM can also be employed in near-neutral pH electrolytes, in which co-ions or counter ions in the electrolyte carry the majority of the ionic current instead of proton or hydroxide. In that situation, external mixing or recirculation is required to balance the ions in the electrolyte and prevent electrodialysis. In addition to CEM or AEM, bipolar membranes can be used for sustainable solar-driven water-splitting reactions in various electrolyte pH conditions. Unlike CEM or AEM, when bipolar membranes are employed at two different pH conditions, even operating at near-neutral pH conditions, proton and hydroxide still carry the majority of the ionic currents at steady state operation.

We refer readers to Chapter 10 by Miller and Houle on membranes for solar fuels devices. This book chapter focuses on methods that were attempted and developed in JCAP for membrane incorporation in integrated solar fuel devices. These methods include, but are not limited to: mechanical compression, adhesion, and other exotic methods of fabrication and integration. Each of these approaches has their pros and cons, which we will explore further.

11.2.3.1 Mechanical Compression

Mechanical compression, like the seal made by an O-ring or gasket, can be used to create a seal between the membrane and the electrolyte system it bifurcates. Many membranes are made of thin film-like materials and therefore require a rigid support structure to be held in place against the sealing surface. This is commonly seen in clamshell style architectures in which the membrane is sandwiched between two identical and complementary electrolyte compartments with two opposing O-rings along the circumference of the sealing faces. Due to the flexible film-like structure of most membranes, O-rings with square cross-sectional profiles are typically used in this opposing geometry to ensure an adequate seal. Standard O-ring profiles can be used for almost all other purposes.

Mechanical compression is the most prevalent sealing mechanism for general purposes. The advantages of using mechanical compression to integrate membranes into assemblies include robust and repeatable sealing, chemical compatibility, and facile interchangeability. The disadvantages of using mechanical compression include excessive spatial infrastructure (body material needed to house the O-ring and its groove), restrictive design criteria (must use enough fasteners to provide an even pressure distribution), and specialized fabrication techniques.

It is important to have confidence that an assembly's seal is robust and will not fail during use. The mechanical nature of sealing a thin film using an O-ring or gasket almost guarantees a proper seal; if the clamping mechanism succeeds in providing uniform pressure the only modes of failure must be due to environmental degradation like chemical corrosion. O-rings are typically positioned in assemblies by grooves which restrict their displacement with respect to the rest of the device, so making reliable seals every time a device is assembled is nearly guaranteed.

Another advantage of using O-ring or gasket material to seal membranes is that they are available in a myriad of materials and chemical compatibilities. In many circumstances the membrane is exposed to corrosive environments that might not be compatible with standard rubber O-ring material. In these cases the O-ring or gasket material can be replaced with another material more resistant to aggressive environments, like PTFE or Kel-F.

Although O-ring and gasket material can be tailored to resist aggressive chemical environments, its inherent sealing ability can still degrade over time prompting replacement of the failed component. When this happens, it is important to be able to quickly and easily change out the old part for a new part. Mechanical compression architectures inherently take advantage of this need by virtue of their operation; replacing a single component is as easy as substituting a new one, and complementary components do not need to be replaced when the device is assembled. In other words, it is not necessary to fabricate a second full assembly if only one component fails.

While mechanical compression offers many advantages over competing sealing methods, it also brings inherent shortfalls with it. Typically, O-rings and gasket materials require a physical retention system to keep them aligned within an assembly. This is usually in the form of a channel, or groove, machined into the structure that holds the O-ring or gasket in place. The geometry of these grooves is standardized by the size and material of the O-ring or gasket being used, and modifications to these dimensions should be kept to a minimum as they are designed to provide an optimal seal. This imposes restrictions on device geometry that must be taken into consideration. In a monolithic water splitting device, for example, the path length from the cathode surface to the anode surface affects the overall device efficiency. Ideally, the path length should be minimized, thereby reducing system losses and potentially improving the device efficiency. Therefore, designs must be employed to ensure this seal is made properly while still conforming to higher level device design requirements.

Another consideration when using mechanical compression for membrane sealing is the need for uniform pressure at the sealing interface. For example, if fasteners are placed at regular intervals around the perimeter of a sealed enclosure then there will be more pressure at the location of the fastener than at a point between fasteners. If the O-ring is not compressed properly, leaks are likely to occur. This situation is unlikely to occur for most single O-ring sealing applications but arises more commonly when multiple interfaces must be compressed by the same mechanism. Because pressure must be imposed by the fastener on the O-ring or gasket material through a supporting structure, the fastener should be located as close to the sealing interface as possible to minimize structural deformations.

One must always keep in mind the fabrication method used to create the structure housing and sealing surface, as the surface finish ultimately determines the quality of the seal that can be made. In general, a seal is formed on a smooth flat surface free from deformities or irregularities. Standard applications usually require 32 to 64 micro-inches of RMS surface variation to obtain a satisfactory seal using rubber O-rings. This means that tools used to fabricate the surfaces of components to be sealed must meet the defined standards and that some methods of fabrication are not suited for mechanical compression sealing at all.

11.2.3.2 Adhesion

A second method of integrating a membrane into an assembly of components is by simply adhering it to the desired surface. Adhesion can be achieved using a variety of compounds, one of the more common being chemical resistant epoxy, such as, Hysol 9460 or 302-3m. Advantages to adhering the membrane to a support structure without the use of compression architectures include minimal spatial requirements, relaxed surface finish requirements, and the ability to seal multiple interfaces simultaneously. Some drawbacks include possible system contamination, lack of longevity, and failures associated with membrane hydrodynamics.

Adhering a membrane to a support structure using sealant or epoxy can offer improved device efficiency over alternative integration methods. It has been demonstrated experimentally and validated with modeling and simulation that maximizing STH conversion efficiency means minimizing losses.[97,98] In practice, this means minimizing the distance that charged molecular species would have to travel to get from the anode or cathode, through the membrane, and arrive at the complementary electrode. Adhesion requires only a thin strip of material to bond (note that more surface area does, however, provide a better seal, and should be maximized if possible), and can even bond directly to fuel-producing components like the PEC assembly itself in limited cases. The lack of support infrastructure needed to create a seal in this fashion, allowing for a minimal ionic path length to be established, is a major advantage of using adhesion to create a membrane seal.

We have seen that the surface finish of a sealing interface highly affects the quality of the seal that can be achieved using mechanical compression schemes. This fact is highly consequential with respect to the fabrication method used to produce the device's components, precluding many cost efficient methods from being utilized. Unlike O-ring or gasket seals, adhesion typically functions best on roughened surfaces which offer a greater surface area than similar yet smooth surfaces. High surface area finishes are associated with less refined fabrication methods which are generally more cost efficient than their alternatives, allowing functional components to be more readily available and actively developed.

Complementing their affinity for rough surfaces, adhesives are also capable of bonding multiple discontinuous surfaces simultaneously. Water splitting architectures which seek to minimize the ionic path length from anode to cathode are generally configured to have their membrane component integrated on a support structure which also positions the PEC assembly. Minimizing the ionic path length obliges the membrane to minimize its proximity to the PEC assembly, resulting in an interface of three different materials (membrane, support structure, and PEC assembly) which must all be sealed simultaneously. This is not easy to accomplish using compression mechanisms, but can be readily done with proper adhesives.

While adhesives offer unique membrane integration solutions, they also present inherent hindrances to device functionality. The electrolyte purity within a solar fuel generator is critical for device efficiency and operation, and adhesive sealants like epoxy have the potential to introduce foreign compounds to the electrolyte which may hamper fuel production activity. Therefore, their use and composition must be carefully considered before widespread adoption and implementation on larger scales.

Related to contamination concerns is the chemical compatibility of the adhesive with the electrolyte environment. Improperly matched adhesive and electrolyte systems will corrode the adhesive and otherwise degrade the integrity of the seal. This can result in adhesive compounds leeching into the electrolyte and confounding precise measurements, as well as allowing mixing of the product fuel streams resulting in potentially dangerous conditions.

Another disadvantage to using adhesives is interfacial hydrodynamics, especially when used to bond membrane material that inherently changes physical dimensions when hydrated. Membrane alteration due to hydration, or swelling, occurs when a dry membrane comes into contact with an aqueous environment that modifies the dry membrane's dimensions, usually increasing its physical volume. Changes to the physical dimensions of the membrane can affect sealing interfaces where the adhesive is rigid and non-pliable, and is most often seen with membranes that are stored dry before being introduced to the wet electrolyte environment, like Nafion. This phenomena manifests when a membrane adheres to a support structure when dry, and is then subsequently immersed in an aqueous environment after the adhesive has been set and cured. It is possible to circumvent bonding failure due to membrane dimensional changes by hydrating the membrane before adhering it to a support structure. However, adhesive that can be cured in aqueous environments must be used and the membrane continually hydrated after application. These criterions are often difficult to implement in practice, yet can afford substantial advantages over other membrane integration techniques.

11.2.3.3 Others

More exotic approaches to membrane integration have also been explored. These techniques include ultrasonic welding, encapsulation, and waxing which provide more versatility, scale-up viability, and sterility (except in the case of waxing) compared to more common methods. Their drawbacks include system contamination (in the case of waxing), lack of *post facto* modification, and high monetary barriers to entry (in the case of ultrasonic welding and encapsulation).

Ultrasonic welding is a process by which energy is locally transferred to a workpiece *via* high-frequency vibration resulting in a fusing of materials. It is used to bond dissimilar materials, most commonly plastics for packaging and medical equipment. The highly localized nature of ultrasonic welding allows for versatile designs to be utilized for membrane integration. Parameters affecting the bond strength of ultrasonic welds are dependent on the membrane material and the support structure it is to be bonded to, as well as the frequency, pressure, and knurl pattern used during the process. Successful demonstrations of this membrane integration method have been shown for Nafion–Polycarbonate joints in small scale experimental tests, and the true benefits of using this method, namely sterility and scale-up viability, have yet to be realized. Contamination, either of the fuel feedstock or in the fuel output stream, can severely limit the efficiency of a solar fuel generator. Ultrasonic welding eliminates this concern; the membrane material is bonded directly to its support structure without any foreign solvents or additives that could potentially find their way into the assembled system. The scale-up viability of ultrasonic welding cannot be overstated. It is a fast process, creating a robust bond within seconds, and is not only reliable but also readily configured for high throughput automation.

Encapsulation is another technique that can be used to integrate a membrane into an assembly. Encapsulation is the process by which the perimeter of a membrane is encased in a material while leaving an opening for ionic conduction through the membrane itself, analogous to a glass window encapsulated by a window frame allowing light to pour through the glass itself. Membrane material commonly comes in large sheets that are cut to size by the end user. More often than not the membrane material itself does not have the intrinsic properties necessary to act as a self-sealing gasket material leading to the use of traditional gaskets and O-rings. These extra materials take up valuable space and can introduce undesirable compounds to the system. These factors are mitigated by using an encapsulant material that is compatible with the system, and then bonding this frame to the assembly using traditional methods. Like ultrasonic welding, encapsulation readily lends itself to industrial-scale processes like automation giving it an advantage as solar fuel technology becomes more prevalent and its use more widespread.

Highly viscous liquids, like wax, can also be used to create a leak-free bond between membrane and support structure. Similar to encapsulation, waxing involves surrounding the perimeter of the membrane with a material more receptive to bonding, while leaving the membrane open to ionic transport. Its advantage over encapsulation is its versatility: it can be deployed in small enough amounts to create a thin layer joint between multiple discontinuous surfaces with minimal infrastructure needed to support the assembled components. Waxing is similar in practice to adhesion integration methods, but while adhesives set into a solid state after curing waxes remain liquid, albeit highly viscous, filling any overlooked surface defects that would render other sealing techniques ineffective.

For each advantage afforded, there are also potential risks that must be assessed before implementation. Ultrasonic welding, for example, is very useful for making rapid and reliable bonds between dissimilar materials. However, the disadvantages to this method are two-fold: high financial capital barriers to entry and lack of available parameterization post initial investment. Ultrasonic welding requires specialized machinery and instrumentation, as well as custom horns that must be tailored specifically to the desired materials that are to be bonded. This required infrastructure and customization significantly increases the initial investment that must be made before a single device is assembled, and can render non-industrial-scale operations untenable. A key consideration for designing a proper horn is the device geometry with respect to the membrane. Once a horn has been fabricated it absolutely defines the sealing interface. This interfacial geometry cannot be modified without substantial subsequent monetary investment into a new horn, rendering geometric parameters expensive to investigate.

Likewise, encapsulation is subject to disadvantages similar to ultrasonic welding. The most common way of utilizing encapsulation in practice is to fabricate a mold into which the pre-cut and sized membrane is placed. Encapsulant is then poured into the mold, surrounding the membrane's perimeter, and left to cure and solidify. The membrane size is thus inherently restricted to that imposed by the mold. Unless affordable and facile mold fabrication methods are used, membrane geometry is a difficult parameter to explore using encapsulation methods. In addition to geometry restrictions the encapsulant is usually a different material than the support structure it will bond to, introducing foreign compounds to the device assembly which may affect performance.

We have seen contamination concerns arise from bonding methods due to leaching, and those concerns are magnified for viscous liquid sealing solutions like waxing. The liquid nature of many wax formulations allows for sealing material to migrate away from the location where it is initially applied, sometimes completely detaching from the membrane support structure it is meant to seal. This poses more risk than just contamination as larger pieces of wax sealant could find their way to the fuel exhaust port, completely blocking the collection of fuel and causing device pressures to rise above safe levels. Table 11.1 summarizes the relative advantages and advantages of various membrane incorporation methods that were explored in JCAP.

11.2.4 Chassis and Auxiliary Components

In addition to active materials and components, chassis and auxiliary components are also important to ensure proper and sustainable function of solar-driven water-splitting devices. When selecting materials, it is important to define the desired attributes needed to meet a specific design requirement. These attributes are then evaluated against the physical properties of the material, such as its chemical, mechanical, electrical, and optical characteristics. The search involves data collected from databases, vendors, manuscripts, colleagues, and calculations. A subset of materials is identified and then evaluated under relevant operating conditions. Concurrently, the manufacturability, service life, and scalability of these materials are also analyzed.[99]

Additive manufacturing has emerged as a versatile technology that enables rapid design development. The quality of the final printed product, however, is highly cost dependent and therefore imposes a significant capital barrier to product development for independent designers. While some of the more exotic and expensive printing technologies like stereo lithography (SLA), digital light processing (DLP), and other photopolymerization techniques can provide precise models with smooth surface finishes, more cost efficient and accessible methods like fused deposition modeling (FDM) and fused filament fabrication (FFF) are limited by their filament or print head dimensions.

Below is a detailed example of how we applied this type of analysis during the initial phase of JCAP. Throughout this project an effort was undertaken

Table 11.1 Relative advantages and disadvantages of various membrane incorporation methods.

	Mechanical compression	Adhesion	Ultrasonic welding	Encapsulation	Waxing
Cell fabrication difficulty (bulk)	Medium	Easy	Hard	Easy	Easy
Cell fabrication difficulty (surface finish)	Medium	Easy	Easy	Medium	Easy
Spatial requirement	Large	Small	Small	Medium	Small
Difficulty of replacement/ repair	Easy	Medium	Hard	Medium	Medium
Frequency of replacement/ repair	Not often	Often	Not often	Not often	Often
Chemical compatibility (sealing mechanism degradation over time)	Excellent	Poor	Good	Excellent	Poor
Chemical compatibility (electrolyte fouling)	Excellent	Fair	Excellent	Excellent	Poor
Can seal discontinuous surface?	Yes	No	Yes	Yes	Yes
Effect of membrane swelling	None	Leakage	None	None	None
Post facto modification of membrane geometry	No	Yes	No	No	Yes
Cost (initial)	Medium	Low	High	Medium	Low
Cost (long term)	Low	Low	Low	Medium	Low

to identify potential device-building materials that could survive the expected environment of a long-term water splitter. Specifically, any candidate device would likely need to survive caustic acidic or basic solutions and temperatures that would repeatedly swing from 0 to 50 °C or higher. Many would need to remain transparent and rigid throughout a long working lifetime. Some materials, for instance, O-ring and other "sealing" materials, might need to maintain compliance and structural integrity while accommodating differential thermal expansion of reactor cell components.

While some of this information was readily available (see Cole–Parmer, for instance), JCAP-specific conditions weren't often described or tested in the literature. For this reason, an effort was undertaken to test some of the most universally-desirable candidate materials under our most demanding conditions, primarily long-term (over 1 day) exposure to 1 molar H_2SO_4 or 1 M KOH. Table 11.2 describes, in very general and practical terms, a variety of materials tested under a variety of conditions, depending on intended use. Some of the listings are simply a re-telling of available on-line information. The goal was to have a simple, working reference and repository for empirical information sharing.

As the work of JCAP progressed, Epotek 302-3M and 3M DP460NS became the "go-to" epoxies of choice, due to their resilience over a wide range of conditions. This prompted further survivability studies that encompassed a wider range of curing schedules and materials preparation. These tests are described in Table 11.3. For this study, "roughened" refers to a surface that has been lightly bead blasted with glass particles and subsequently solvent cleaned before further processing. Cure schedules were chosen based on the properties of polycarbonate and the specific epoxy in order to maximize cross-linking of the polymer

11.2.5 Integration of Protective Layers with Catalysts and Light Absorbers

The primary components of a solar-driven water-splitting cell are the light absorbers, electrocatalysts, membrane separators and electrolyte. One of the primary metrics for assessing the performance of these systems is the solar-to-hydrogen conversion efficiency,[15] which is a ratio of the output to input energy and assesses how well the components perform together.[100] JCAP produced solar-H_2 generators utilizing an ALD-TiO_2 protection layer on semiconductor light absorber junctions which incorporated a thin layer of sputter-deposited Ni metal as an OER catalyst precursor.[101–104] The interface formed between the Ni and the TiO_2 is integral to the charge-transport performance of the protection layer.

Significant efforts were devoted to the discovery of a new electrocatalytic materials for the oxygen evolution reaction (OER) which has resulted in the discovery of a new family of biphasic nano-structured Ce-rich catalysts by the high throughput experimentation project, as described by Chapter 9 in this book by Gergoire and co-workers.[105–108] Integration of these new electrocatalysts with existing light absorbers and protective coatings engenders new challenges from, (a) constraints imposed by the processing limitations of the light absorber material and device architecture, (b) the requirement to enable efficient optical transmission to the light absorber, and (c) the need to produce compositions, structures, and defect states at the interfaces between the light absorber, electrocatalysts and electrolyte to yield assemblies of materials which work well together to produce high-performance photoanodes.

The performance of the material interfaces depends on the specific interactions of the interfaced materials and depends on a number of factors, such as, the processing conditions, morphology, and structure. Therefore, integrating new electrocatalysts into photoanode or photocathode

Table 11.2 Caustic survivability of various materials.[a]

Material Tested	Electrolyte (type and concentration)	Notes	Survivability
3M Scotchweld epoxy # DP460NS	1 M KOH	White filled epoxy, thixotropic, highly (but not totally) opaque. 1 h work time, room temperature cure.	No visible degradation after 14 days
3M Scotchweld epoxy # DP460NS	1 M H_2SO_4		No visible degradation after 14 days
3M Scotchweld epoxy # 460 w/ 300 nm Al_2O_3 Powder mixed in (25% or so by vol.)	1 M KOH	This makes a paste that is highly "formable," does not pull back *via* surface tension, *etc.* Excellent for sealing the edges of wafers, *etc.*	No visible degradation after 14 days
3M Scotchweld epoxy # DP460NS	1 M H_2SO_4	Cured 65 °C, 3 h.	Failed
3M Scotchweld epoxy # DP460NS	1 M KOH	Cured 65 °C, 3 h.	Survived > 5 days
3M Scotchweld epoxy #DP105	1 M KOH	Water clear, higher viscosity than EpoTek 302-3M, super short work time though – 4 min. Boils under degassing vacuum. Remains somewhat flexible after curing.	No visible degradation after 14 days
3M Scotchweld epoxy #DP105	1 M H_2SO_4		No visible degradation after 6 days
3M Scotchweld epoxy #DP420NS	1 M KOH	Dark gray/totally opaque.	Fails after 2 weeks
Epo-Tek 115-129 epoxy	1 M KOH	Very slight cloudiness, higher viscosity version of 302-3M, long work times, 60 °C cure.	No visible degradation after 14 days
Epo-Tek 301-2 epoxy	1 M KOH	Water clear.	5 days (fails at adhesion surface)
Epo-Tek 301-2 epoxy	10% H_2SO_4		Fails after 7 days
Epo-Tek 301-2FL epoxy	1 M H_2SO_4	Cured 80 °C overnight (12 h).	Failed
Epo-Tek 301-2FL epoxy	1 M KOH	Cured 80 °C overnight (12 h).	Survived > 5 days
Epo-Tek 302-3M epoxy	1 M KOH	Water clear, good transmission of visible.	No visible degradation after 7 days

Epo-Tek 302-3M epoxy	10% H_2SO_4	Water clear, good transmission of visible.	No visible degradation after 7 days
Epo-Tek 302-3M black epoxy	1 M H_2SO_4	Cured 65 °C, 3 h.	Failed
Epo-Tek 302-3M black	1 M KOH	Cured 65 °C, 3 h.	Survived > 5 days
Epo-Tek 302-3M w/1% Epo-Tek AP100 Surfactant	1 M KOH	Water clear, good transmission of visible.	No visible degradation after 7 days
EpoTek 302-3M w/1% Epo-Tek AP100 Surfactant	10% H_2SO_4	Water clear, good transmission of visible.	Thin films fail after 5 days
Epo-Tek 302-3M w/1% Surfactant and fumed silica for thickening	1 M KOH	Semi opaque.	Slight visible degradation after 7 days
Epo-Tek 302-3M w/1% Surfactant and fumed silica for thickening	10% H_2SO_4		Thin films fail after 5 days
Epo-Tek 377	1 M KOH	Clear/brownish.	Fails after 7 days
Epo-Tek 377	10% H_2SO_4		No visible degradation after 7 days
Epo-Tek H70E	10% H_2SO_4		Adhesion fails in 1 day
Epo-Tek H70E	1 M KOH	Alumina filled epoxy.	Adhesion fails in 2 days
Epo-Tek OJ2145 epoxy	1 M KOH	Slightly thicker version of 302-3M. Tested on acrylic and polycarbonate. Perfect survivability.	>12 days
Epo-Tek OJ2145 epoxy	1 M H_2SO_4	Slightly thicker version of 302-3M. Tested on acrylic and polycarbonate. Perfect survivability.	>12 days
FEP	1 M KOH, 1 M NaOH	http://www.vici-jour.com	Stable
Hardman #04001, extra fast setting epoxy	1 M H_2SO_4		>11 days, No visible degradation
Kalrez	1 M KOH, 1 M NaOH	http://mykin.com	Stable
Loctite 9460 epoxy	10% H_2SO_4		Fails after 7 days
Loctite Hysol E-120HP epoxy	TBAPF6 (100 mM) in acetonitrile (CH_3CN)	Semi-opaque, moderately viscous, highly chemical-resistant.	Stable for >2 weeks

Table 11.2 (*Continued*)

Material Tested	Electrolyte (type and concentration)	Notes	Survivability
Loctite Hysol E-30CL epoxy	TBAPF6 (100 mM) in acetonitrile (CH_3CN)	Transparent, not very viscous, highly chemical-resistant.	Stable for >2 weeks
Loon Outdoors UV Clear Fly Finish (thin)	1 M KOH	Water clear, flexible, Cured with UV flashlight 5 min.	Loss of adhesion after 7 days
Loon Outdoors UV Clear Fly Finish (thin)	1 M H_2SO_4	Water clear, flexible, Cured with UV flashlight 5 min.	Survived 7 days
Neoprene	KOH, NaOH	us.burkert.com	Stable
Parylene (probably type "C"), 2 μm thick	1 M KOH	Survives well.	>5 days, no degradation observed
PEEK	1 M KOH, 1 M NaOH	www.vici-jour.com	Stable
Platinum	1 M KOH		Over 5 days (indefinite)
Plexiglass acrylic	1 M KOH, 1 M NaOH	www.eplastics.com	>30 days
Plexiglass acrylic	up to "20%" H_2SO_4	www.eplastics.com	>30 days
Polyethylene	1 M KOH, 1 M NaOH	www.vici-jour.com	Stable
Polypropylene	1 M KOH, 1 M NaOH	www.vici-jour.com	Stable
$BiVO_4$	1 M KPi buffer, pH 12.3	Unprotected $BiVO_4$ is unstable against photocorrosion at all pH, but time to failure varies.	<5 min
$BiVO_4$	1 M KPi buffer, pH 6.7	Unprotected $BiVO_4$ is unstable against photocorrosion at all pH, but time to failure varies	~30 min, immediate steady decline
EPDM	1 M KOH, 1 M NaOH	http://mykin.com	Stable
Si	1 M H_2SO_4	H-terminated Si is operationally stable in acid, but slow oxidation can occur when held in dark.	>1 h

Si	1 M KOH	Unprotected Si is readily etched in molar base.	Immediate failure
Silicone	1 M NaOH	http://mykin.com	Stable
Silicone rubber	H_2SO_4		Fail
Silicone rubber	KOH		OK for short term use. Long-term failure
Silicone Vacuum grease	1 M KOH	Dissolves and floats on the surface of the KOH.	~2 h
Silicone, Moldable, Dow Corning MS1003, 2-part	1 M KOH		Fail after 2 weeks
Silicone, Moldable, Dow Corning MS1003, 2-part	1 M H_2SO_4	Takes several days to cure. Very wet. Seemingly hydrophobic.	Stable for >2 weeks
Stainless Steel (316)	KOH		Excellent survivability
Stainless Steel (316)	10–75% sulfuric acid		Poor
Stycast 2850FT epoxy	1 M KOH		5 days (fails at adhesion surface)
Stycast 2850FT epoxy	10% H_2SO_4		Fails in 3 days
Teflon (PTFE)	1 M KOH, 1 M NaOH	www.vici-jour.com	Stable
Ti	1 M KOH, 1 M NaOH	www.parrinst.com	Stable
Torr Seal (Varian vacuum product)	1 M KOH		Adhesion fails in 1 day
Torr Seal (Varian vacuum product)	10% H_2SO_4		Adhesion fails in 1 day
Viton	10–75% Sulfuric Acid		Indefinite (excellent resistance)
Viton	KOH		Slight corrosion effects long-term

[a]These data and notes were acquired by several JCAP researchers.

Table 11.3 Curing schedule, surface preparation, and "extreme condition" tests for three specific epoxies.

Epoxy	Curing schedule	Material adhered to	Caustic	Appearance	Adherent surface Condition
Epo-Tek 302-3M	RT, 24 hours	2 minutes O_2 plasma-cleaned polycarbonate	1 M KOH	No visible degradation.	Roughened
	65 °C, 3 hours	2 minutes O_2 plasma-cleaned polycarbonate	1 M KOH	No visible degradation.	Roughened
	60 °C 20 minutes, ramped to 130 °C over 20 minutes, held at 130 °C for 3 hours	2 minutes O_2 plasma-cleaned polycarbonate	1 M KOH	No visible degradation.	Roughened
	RT, 24 hours	2 minutes O_2 plasma-cleaned polycarbonate	1 M KOH	No visible degradation.	Smooth
	65 °C, 3 hours	2 minutes O_2 plasma-cleaned polycarbonate	1 M KOH	No visible degradation.	Smooth
	60 °C 20 minutes, ramped to 130 °C over 20 minutes, held at 130 °C for 3 hours	2 minutes O_2 plasma-cleaned polycarbonate	1 M KOH	No visible degradation.	Smooth
	RT, 24 hours	2 minutes O_2 plasma-cleaned acrylic	1 M KOH	No visible degradation.	Roughened
	65 °C, 3 hours	2 minutes O_2 plasma-cleaned acrylic	1 M KOH	No visible degradation.	Roughened
	60 °C 20 minutes, ramped to 130 °C over 20 minutes, held at 130 °C for 3 hours	2 minutes O_2 plasma-cleaned acrylic	1 M KOH	No visible degradation.	Roughened
	RT, 24 hours	2 minutes O_2 plasma-cleaned acrylic	1 M KOH	No visible degradation.	Smooth
	65 °C, 3 hours	2 minutes O_2 plasma-cleaned acrylic	1 M KOH	No visible degradation.	Smooth
	60 °C 20 minutes, ramped to 130 °C over 20 minutes, held at 130 °C for 3 hours	2 minutes O_2 plasma-cleaned acrylic	1 M KOH	No visible degradation.	Smooth
	RT, 24 hours	2 minutes O_2 plasma-cleaned polycarbonate	1 M H_2SO_4	No visible degradation.	Roughened
	65 °C, 3 hours	2 minutes O_2 plasma-cleaned polycarbonate	1 M H_2SO_4	No visible degradation.	Roughened

	60 °C 20 minutes, ramped to 130 °C over 20 minutes, held at 130 °C for 3 hours	2 minutes O_2 plasma-cleaned polycarbonate	1 M H_2SO_4	No visible degradation.	Roughened
	RT, 24 hours	2 minutes O_2 plasma-cleaned polycarbonate	1 M H_2SO_4	No visible degradation.	Smooth
	65 °C, 3 hours	2 minutes O_2 plasma-cleaned polycarbonate	1 M H_2SO_4	No visible degradation.	Smooth
	60 °C 20 minutes, ramped to 130 °C over 20 minutes, held at 130 °C for 3 hours	2 minutes O_2 plasma-cleaned polycarbonate	1 M H_2SO_4	No visible degradation.	Smooth
	RT, 24 hours	2 minutes O_2 plasma-cleaned acrylic	1 M H_2SO_4	No visible degradation.	Roughened
	65 °C, 3 hours	2 minutes O_2 plasma-cleaned acrylic	1 M H_2SO_4	No visible degradation.	Roughened
	60 °C 20 minutes, ramped to 130 °C over 20 minutes, held at 130 °C for 3 hours	2 minutes O_2 plasma-cleaned acrylic	1 M H_2SO_4	No visible degradation.	Roughened
	RT, 24 hours	2 minutes O_2 plasma-cleaned acrylic	1 M H_2SO_4	No visible degradation.	Smooth
	65 °C, 3 hours	2 minutes O_2 plasma-cleaned acrylic	1 M H_2SO_4	No visible degradation.	Smooth
	60 °C 20 minutes, ramped to 130 °C over 20 minutes, held at 130 °C for 3 hours	2 minutes O_2 plasma-cleaned acrylic	1 M H_2SO_4	No visible degradation.	Smooth
Epo-Tek 302-3M Black	RT, 24 hours	2 minutes O_2 plasma-cleaned polycarbonate	1 M KOH	No visible degradation.	Roughened
	65 °C, 3 hours	2 minutes O_2 plasma-cleaned polycarbonate	1 M KOH	No visible degradation.	Roughened
	60 °C 20 minutes, ramped to 130 °C over 20 minutes, held at 130 °C for 3 hours	2 minutes O_2 plasma-cleaned polycarbonate	1 M KOH	No visible degradation.	Roughened
	RT, 24 hours	2 minutes O_2 plasma-cleaned polycarbonate	1 M KOH	No visible degradation.	Smooth
	65 °C, 3 hours	2 minutes O_2 plasma-cleaned polycarbonate	1 M KOH	No visible degradation.	Smooth

Table 11.3 (*Continued*)

Epoxy	Curing schedule	Material adhered to	Caustic	Appearance	Adherent surface Condition
	60 °C 20 minutes, ramped to 130 °C over 20 minutes, held at 130 °C for 3 hours	2 minutes O_2 plasma-cleaned polycarbonate	1 M KOH	No visible degradation.	Smooth
	RT, 24 hours	2 minutes O_2 plasma-cleaned acrylic	1 M KOH	No visible degradation.	Roughened
	65 °C, 3 hours	2 minutes O_2 plasma-cleaned acrylic	1 M KOH	No visible degradation.	Roughened
	60 °C 20 minutes, ramped to 130 °C over 20 minutes, held at 130 °C for 3 hours	2 minutes O_2 plasma-cleaned acrylic	1 M KOH	No visible degradation.	Roughened
	RT, 24 hours	2 minutes O_2 plasma-cleaned acrylic	1 M KOH	No visible degradation.	Smooth
	65 °C, 3 hours	2 minutes O_2 plasma-cleaned acrylic	1 M KOH	No visible degradation.	Smooth
	60 °C 20 minutes, ramped to 130 °C over 20 minutes, held at 130 °C for 3 hours	2 minutes O_2 plasma-cleaned acrylic	1 M KOH	No visible degradation.	Smooth
	RT, 24 hours	2 minutes O_2 plasma-cleaned polycarbonate	1 M H_2SO_4	No visible degradation.	Roughened
	65 °C, 3 hours	2 minutes O_2 plasma-cleaned polycarbonate	1 M H_2SO_4	No visible degradation.	Roughened
	60 °C 20 minutes, ramped to 130 °C over 20 minutes, held at 130 °C for 3 hours	2 minutes O_2 plasma-cleaned polycarbonate	1 M H_2SO_4	No visible degradation.	Roughened
	RT, 24 hours	2 minutes O_2 plasma-cleaned polycarbonate	1 M H_2SO_4	No visible degradation.	Smooth
	65 °C, 3 hours	2 minutes O_2 plasma-cleaned polycarbonate	1 M H_2SO_4	No visible degradation.	Smooth
	60 °C 20 minutes, ramped to 130 °C over 20 minutes, held at 130 °C for 3 hours	2 minutes O_2 plasma-cleaned polycarbonate	1 M H_2SO_4	No visible degradation.	Smooth
	RT, 24 hours	2 minutes O_2 plasma-cleaned acrylic	1 M H_2SO_4	No visible degradation.	Roughened

	65 °C, 3 hours	2 minutes O_2 plasma-cleaned acrylic	1 M H_2SO_4	No visible degradation.	Roughened
	60 °C 20 minutes, ramped to 130 °C over 20 minutes, held at 130 °C for 3 hours	2 minutes O_2 plasma-cleaned acrylic	1 M H_2SO_4	No visible degradation.	Roughened
	RT, 24 hours	2 minutes O_2 plasma-cleaned acrylic	1 M H_2SO_4	No visible degradation.	Smooth
	65 °C, 3 hours	2 minutes O_2 plasma-cleaned acrylic	1 M H_2SO_4	No visible degradation.	Smooth
	60 °C 20 minutes, ramped to 130 °C over 20 minutes, held at 130 °C for 3 hours	2 minutes O_2 plasma-cleaned acrylic	1 M H_2SO_4	No visible degradation.	Smooth
"Degassed" 3M DP460NS	RT, 24 hours	2 minutes O_2 plasma-cleaned polycarbonate	1 M KOH	No visible degradation.	Roughened
1 min, 25″ Hg vacuum	60 °C 20 minutes, ramped to 70 °C over 20 minutes, held at 70 °C, 2 hours	2 minutes O_2 plasma-cleaned polycarbonate	1 M KOH	No visible degradation.	Roughened
	RT, 24 hours	2 minutes O_2 plasma-cleaned polycarbonate	1 M KOH	No visible degradation.	Smooth
	60 °C 20 minutes, ramped to 70 °C over 20 minutes, held at 70 °C, 2 hours	2 minutes O_2 plasma-cleaned polycarbonate	1 M KOH	No visible degradation.	Smooth
	RT, 24 hours	2 minutes O_2 plasma-cleaned acrylic	1 M KOH	No visible degradation.	Roughened
	60 °C 20 minutes, ramped to 70 °C over 20 minutes, held at 70 °C, 2 hours	2 minutes O_2 plasma-cleaned acrylic	1 M KOH	No visible degradation.	Roughened
	RT, 24 hours	2 minutes O_2 plasma-cleaned acrylic	1 M KOH	No visible degradation.	Smooth
	60 °C 20 minutes, ramped to 70 °C over 20 minutes, held at 70 °C, 2 hours	2 minutes O_2 plasma-cleaned acrylic	1 M KOH	No visible degradation.	Smooth
	RT, 24 hours	2 minutes O_2 plasma-cleaned polycarbonate	1 M H_2SO_4	No visible degradation.	Roughened
	60 °C 20 minutes, ramped to 70 °C over 20 minutes, held at 70 °C, 2 hours	2 minutes O_2 plasma-cleaned polycarbonate	1 M H_2SO_4	No visible degradation.	Roughened
	RT, 24 hours	2 minutes O_2 plasma-cleaned polycarbonate	1 M H_2SO_4	No visible degradation.	Smooth
	60 °C 20 minutes, ramped to 70 °C over 20 minutes, held at 70 °C, 2 hours	2 minutes O_2 plasma-cleaned polycarbonate	1 M H_2SO_4	No visible degradation.	Smooth

Table 11.3 (*Continued*)

Epoxy	Curing schedule	Material adhered to	Caustic	Appearance	Adherent surface Condition
	RT, 24 hours	2 minutes O_2 plasma-cleaned acrylic	1 M H_2SO_4	No visible degradation.	Roughened
	60 °C 20 minutes, ramped to 70 °C over 20 minutes, held at 70 °C, 2 hours	2 minutes O_2 plasma-cleaned acrylic	1 M H_2SO_4	No visible degradation.	Roughened
				No visible degradation.	
	RT, 24 hours	2 minutes O_2 plasma-cleaned acrylic	1 M H_2SO_4	No visible degradation.	Smooth
	60 °C 20 minutes, ramped to 70 °C over 20 minutes, held at 70 °C, 2 hours	2 minutes O_2 plasma-cleaned acrylic	1 M H_2SO_4	No visible degradation.	Smooth
Elevated temperature tests					
Electrolytes heated to 65 °C for 24 hours					
Epo-Tek 302-3M	RT, 24 hours	2 minutes O_2 plasma-cleaned polycarbonate	1 M KOH	Survived, but plastic.	Roughened
	65 °C, 3 hours	2 minutes O_2 plasma-cleaned polycarbonate	1 M KOH	Became somewhat cloudy.	Roughened
	60 °C 20 minutes, ramped to 130 °C over 20 minutes, held at 130 °C for 3 hours	2 minutes O_2 plasma-cleaned polycarbonate	1 M KOH	Cloudy.	Roughened
	RT, 24 hours	2 minutes O_2 plasma-cleaned polycarbonate	1 M KOH	Survived, but plastic.	Smooth
	65 °C, 3 hours	2 minutes O_2 plasma-cleaned polycarbonate	1 M KOH	Became somewhat cloudy.	Smooth
	60 °C 20 minutes, ramped to 130 °C over 20 minutes, held at 130 °C for 3 hours	2 minutes O_2 plasma-cleaned polycarbonate	1 M KOH	Cloudy.	Smooth
	RT, 24 hours	2 minutes O_2 plasma-cleaned acrylic	1 M KOH	Perfect survival.	Roughened
	65 °C, 3 hours	2 minutes O_2 plasma-cleaned acrylic	1 M KOH	Perfect survival.	Roughened

60 °C 20 minutes, ramped to 130 °C over 20 minutes, held at 130 °C for 3 hours	2 minutes O_2 plasma-cleaned acrylic	1 M KOH	Perfect survival.	Roughened
RT, 24 hours	2 minutes O_2 plasma-cleaned acrylic	1 M KOH	Perfect survival.	Smooth
65 °C, 3 hours	2 minutes O_2 plasma-cleaned acrylic	1 M KOH	Perfect survival.	Smooth
60 °C 20 minutes, ramped to 130 °C over 20 minutes, held at 130 °C for 3 hours	2 minutes O_2 plasma-cleaned acrylic	1 M KOH	Perfect survival.	Smooth
RT, 24 hours	2 minutes O_2 plasma-cleaned polycarbonate	1M H_2SO_4	Some epoxy decomposition.	Roughened
65 °C, 3 hours	2 minutes O_2 plasma-cleaned polycarbonate	1 M H_2SO_4	Some epoxy decomposition.	Roughened
60 °C 20 minutes, ramped to 130 °C over 20 minutes, held at 130 °C for 3 hours	2 minutes O_2 plasma-cleaned polycarbonate	1 M H_2SO_4	Perfect survival.	Roughened
RT, 24 hours	2 minutes O_2 plasma-cleaned polycarbonate	1 M H_2SO_4	Some epoxy decomposition.	Smooth
65 °C, 3 hours	2 minutes O_2 plasma-cleaned polycarbonate	1 M H_2SO_4	Some epoxy decomposition.	Smooth
60 °C 20 minutes, ramped to 130 °C over 20 minutes, held at 130 °C for 3 hours	2 minutes O_2 plasma-cleaned polycarbonate	1 M H_2SO_4	Perfect survival.	Smooth
RT, 24 hours	2 minutes O_2 plasma-cleaned acrylic	1 M H_2SO_4	Some epoxy decomposition.	Roughened
65 °C, 3 hours	2 minutes O_2 plasma-cleaned acrylic	1 M H_2SO_4	Some epoxy decomposition.	Roughened
60 °C 20 minutes, ramped to 130 °C over 20 minutes, held at 130 °C for 3 hours	2 minutes O_2 plasma-cleaned acrylic	1 M H_2SO_4	Perfect survival.	Roughened
RT, 24 hours	2 minutes O_2 plasma-cleaned acrylic	1 M H_2SO_4	Some epoxy decomposition.	Smooth
65 °C, 3 hours	2 minutes O_2 plasma-cleaned acrylic	1 M H_2SO_4	Some epoxy decomposition.	Smooth
60 °C 20 minutes, ramped to 130 °C over 20 minutes, held at 130 °C for 3 hours	2 minutes O_2 plasma-cleaned acrylic	1 M H_2SO_4	Perfect survival.	Smooth

assemblies remains a significant challenge. High-throughput experimental methods have been developed to rapidly screen catalyst composition-loading libraries on light absorbers to discover synergistic interactions between catalysts and light absorbers under specific processing conditions. An example of such beneficial interactions is observed with the integration of the Ce-rich OER catalysts on $BiVO_4$ to produce photoanode assembles.[109–111] However, direct integration of Ce-rich OER catalysts to the ALD-TiO_2 [12f,36] using high-throughput inkjet printing methods, did not produce an acceptable material combination.

The Ce-rich OER catalyst discovery utilized printed inks of metal precursors on FTO (fluorine doped tin oxide) coated glass followed by calcination for several hours at 350 °C to produce the oxide catalyst. These processing conditions could be incompatible with the ALD protection layer and, more fundamentally, are potentially incompatible with doped III–V semiconductor junctions. However, Si substrates are compatible with these annealing conditions, enabling investigation of the compatibility of the catalyst and the TiO_2 layer. Utilizing dark electrodes consisting of p^+Si with a protective layer of ALD-TiO_2 calcined at 350 °C, we demonstrated that the protective TiO_2 layer is not degraded by these calcination conditions. Electrodes upon which Ni metal catalyst was deposited (after calcination) produced the same low overpotential of 300 mV at 10 mA cm^{-2} for the OER reaction at pH 14, as previously reported. However, when the Ce-rich metal oxide catalyst was prepared on the TiO_2-protected electrode by printing and then calcining precursor inks, the assemblies were inactive for the OER. This suggests that the interface between the TiO_2 and the catalyst does not have an electronic structure suitable for charge transport. Therefore, direct transfer of the precursor printing of the FTO-optimal composition ($Ni_{0.3}Fe_{0.07}Co_{0.2}Ce_{0.43}O_x$) and calcination processes used for the catalyst discovery to the TiO_2-protection layer is not feasible.

High-performing photoanode assemblies consisting of Ce-rich electrocatalysts deposited on $BiVO_4$ were discovered using high throughput experimental methods, in which the assembly performance is a product of synergistic interfacial interactions at catalyst compositions and loadings shifted from the FTO-optimal composition.[109–111] Therefore, similar catalyst composition-loading libraries were deposited onto ALD-TiO_2 coated FTO/glass and Ti metal substrates to determine if high performing OER assemblies would be produced at alternate catalyst compositions and loadings. As shown in Figure 11.8, the most active catalyst compositions in the FTO/ALD-TiO_2/catalyst assemblies were very different from those previously observed on FTO/glass. Moreover, the catalytic activity of even the best assemblies was too poor for practical implementation. In order to ascertain if the poor performance was related to the ink deposition and processing method, a (Ni–Co–Ce)O_x composition library was deposited by co-sputtering, and evaluated for OER activity without subsequent thermal processing. At all compositions, the co-sputtered catalyst ALD-TiO_2 assembly performed poorly, indicating that the direct deposition of metal oxide catalysts onto the

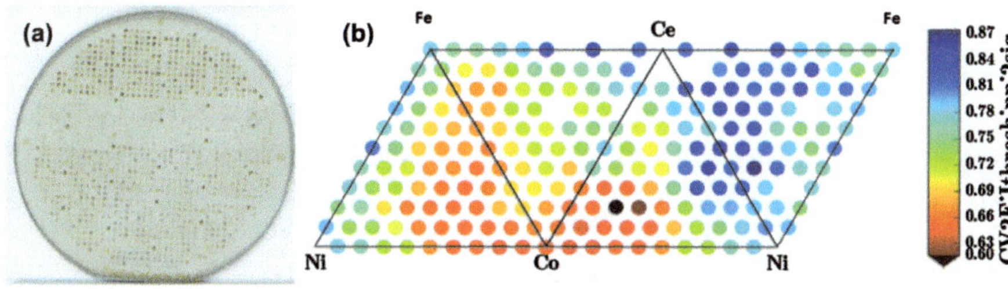

Figure 11.8 (a) Photograph of (Ni–Fe–Co–Ce)O_x catalyst composition-loading library deposited on ALD-TiO_2 coated FTO/glass substrate. (b) OER activity plot as a function of composition for one of the three catalyst loadings, indicating that the optimal composition was shifted from $Ni_{0.25}Co_{0.25}Ce_{0.5}O_x$ on FTO/glass to $Ni_{0.2}Co_{0.6}Ce_{0.2}O_x$, but with low overall activity.

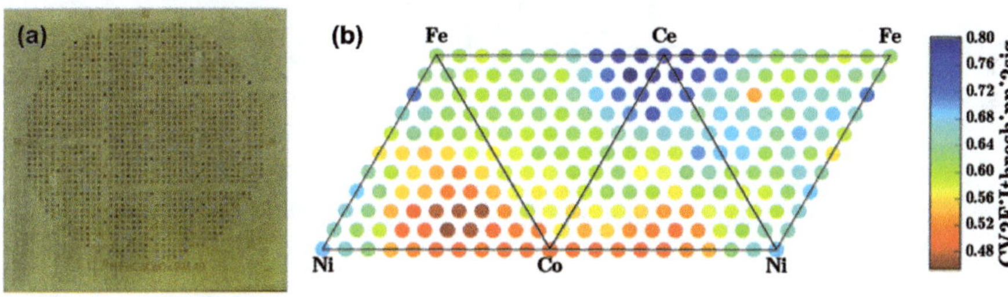

Figure 11.9 (a) Photograph of (Ni–Fe–Co–Ce)O_x catalyst composition-loading library deposited on Ti metal substrate. (b) OER activity plot as a function of composition for one of the three catalyst loadings, indicating that the optimal composition was shifted from $Ni_{0.25}Co_{0.25}Ce_{0.5}O_x$ on FTO/glass to $Ni_{0.5}Fe_{0.1}Co_{0.4}O_x$, but with low overall activity.

ALD-TiO_2 protective layer produced junctions with poor charge-transport properties. In order to determine if some Ce-rich catalyst composition would interface well with Ti metal *via* the ink deposition and thermal calcination processing conditions, a similar composition-loading library was deposited and process on Ti metal plate. As shown in Figure 11.9, the most active catalyst compositions were significantly shifted from those optimal on FTO or ALD-TiO_2. Moreover, the activity, although significantly higher than on ALD-TiO_2, was again too poor for the application.

11.3 Full Device Characterization and Evaluation

11.3.1 Different Sources of Lights and Calibration Methods

Laboratory electrochemical characterization of photoelectrochemical devices is typically conducted in an electrochemical cell containing electrolyte with the photoactive electrode facing the light. The intensity and spectrum, especially at the sample position, depends on the light source (Figure 11.10).[24] It is good practice to measure the irradiance using a

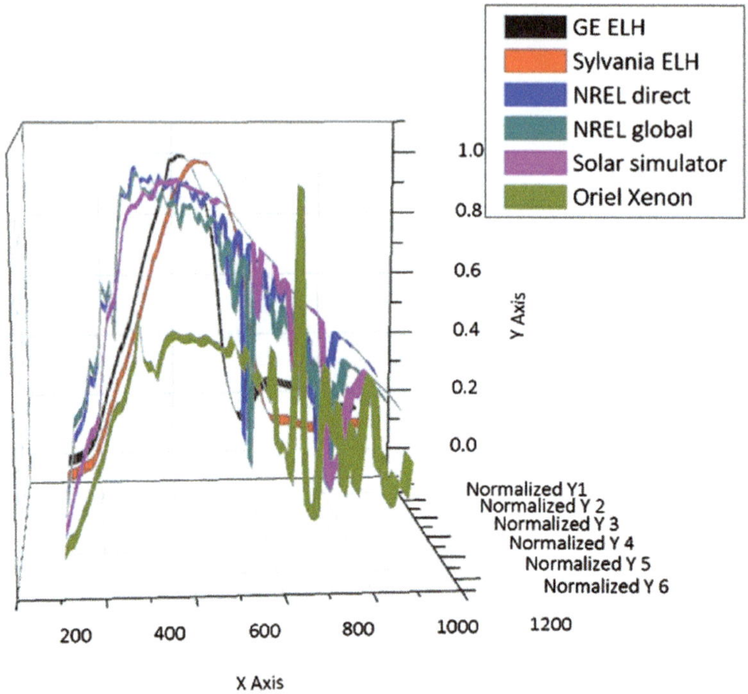

Figure 11.10 Normalized irradiance spectra of typically used broad-band light sources in photoelectrochemical research.

calibration diode for accurate assessment of the STH conversion efficiency. In short, the calibration diode is placed in-plane with the external surface of the window of the PEC cell. It can also be informative to measure the irradiance in the electrolyte, using a miniaturized photodiode with proper electrical isolation and chemical protection. The electrochemical cell has to be well designed providing minimal mechanical distortion to the optical field to maintain the optical intensity at the sample.

Quantum yield, Φ_{ext}, also known as external quantum efficiency, spectral response, or incident photon to electron conversion efficiency (IPCE), should also be measured with a calibration diode at the location as the sample position in the cell with electrolyte (or at the window). Otherwise, the reported Φ_{ext} will be inaccurate and potentially confounded by the cell design, water absorption, and configurations. Measurement of light intensity at the sample position should use a photodiode with a calibrated responsivity. The active area of the calibration diode should be comparable to the sample area to minimize the impact of hot spots.

Xenon lamps are not collimated, and the irradiance may vary spectrally, temporally, and spatially. The back reflector of a xenon lamp can be adjusted, which affects the beam profile, as well as spatial and spectral uniformity. Adjusting the power on the lamp to control its intensity creates distortion in the spectrum due to the change of the temperature and emission spectrum.

Intensity and spectrum changes during measurements can be monitored with a spectrometer. The intensity can be also monitored by a photodiode. The typical lifetime of xenon lamps is 1000 h, after which there is significant red shifting in the spectrum. After 2000 h of operation, the lamp output can be more than 30% reduced. As lamp operation hours accumulate, electrode material spatters onto the inner surface of the bulb and they obstruct radiation. This coating also changes the spectral characteristics causing distortions to its calibrated spectrum. Moreover, the gap between the anode and cathode in the arc lamp increases as the lamp ages, leading to an unstable arc during operation.

Finally, a diode with a proper bandgap which matches that of the photoelectrode under investigation should be used for the calibration of intensity. When a broad-band light source is used, the mismatch of bandgaps between the calibration photodiode and photoelectrode under investigation can lead to an unrealistic flux of above bandgap photons and incorrect values of photocurrent density.[24]

11.3.2 Product Collection

Integrated solar-driven water-splitting cells are evaluated by their ability to produce hydrogen gas. While electrical current densities were often used to quantify the device efficiency in half-cell setups, devices should be validated with direct measurement of the production rate hydrogen and oxygen. As materials improve and device designs advance the efficiency and stability of the cell in the past few years, accurate and automated product collection becomes an important task to evaluate the device performance and monitor the degradation. In this section, a method for automatically collecting and measuring the gas output of a solar fuel generator during operation is described and using this method, the performance of a monolithically integrated water splitting device was evaluated.

11.3.2.1 Experimental Setup

The instrument used to execute the automated gas collection and data analysis method is shown in Figure 11.11. It consists of two eudiometers, one for each chamber (oxygen and hydrogen) of the solar fuel generator, two digital manometers to monitor the pressure of each eudiometer, and a computer to log the data output of the manometers. Water splitting solar fuel generators take advantage of membranes to isolate gaseous products. The ideal membrane confines each gas stream to its respective chamber allowing for independent monitoring of the oxygen and hydrogen streams. The following will describe the infrastructure used to collect and monitor gas output from a single chamber, a duplicate set is used to monitor the complementary chamber independent of the first.

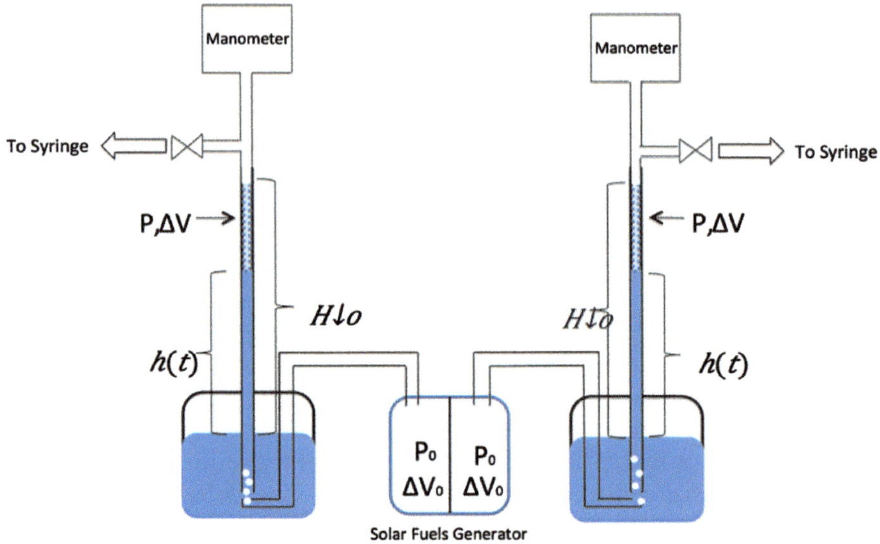

Figure 11.11 Schematic illustration of the hydrogen gas and oxygen gas collection setup.

A eudiometer is an inverted buret with one end submerged in a liquid media. The output of a gas stream is positioned in the liquid media, below the buret inlet, such that any gas leaving the stream is collected in the mouth of the buret and stored in its headspace. In this way one can collect the total gas volume output of the gas stream. The top of the inverted buret is sealed and connected to a manometer that measures pressure in the headspace in the buret, with an auxiliary valve connected to a syringe that can be toggled in order to fill the buret and prime it for measurement. The manometer measures the pressure differential between the buret headspace and the atmosphere, ensuring any changes to atmospheric pressure do not affect gas volume measurements. The manometer output is logged by a computer and further processed through an algorithm to convert temporal pressure measurements to average gas production rates. Gaseous production rates can then be used to deduce fuel cell efficiency, and eudiometer headspace aliquots can be sampled to quantify product purity.

11.3.2.2 Measurements

With the target solar fuel generator positioned in its experimental configuration, step 1, is to fit its gaseous product output ports with flexible tubing terminating beneath their respective eudiometers. If measuring product purity is necessary, then step two is to flush the eudiometer headspace with an inert gas compatible with the analytic method used for quantification. This step should be performed with the eudiometer inlet submerged to ensure no air creeps back into the headspace. With the eudiometer flushed with inert gas and sealed to a manometer, step four is to prime the eudiometer with liquid media. This is most easily accomplished by using a three-way valve to connect a syringe to the eudiometer and manometer system, and pulling vacuum to raise the height of the liquid media to a level appropriate for the manometer measurement range. Once the eudiometer is primed the three-way valve is closed to isolate the syringe from the headspace which is now filled with inert gas and its pressure monitored by the manometer. The final step is to operate the solar fuel generator and to begin collecting pressure-time data from the manometer. It is crucial to ensure that the flexible tubing leading from the device to the eudiometer does not become blocked with electrolyte or other foreign objects, as this will affect the pressure measurement leading to inaccurate efficiency calculations, as well as modifying the pressure of the device's operating condition. Other factors to consider are the tubing length, which should be minimized, and the tubing inner diameter, which should be maximized. The outlet of the tube leading to the eudiometer should also have a diffusion device allowing the outlet gas stream to enter the eudiometer in smaller bubbles that are less likely to block the eudiometer. Indeed, a tradeoff must be negotiated between maximum tube conductance and resistance due to any diffuser making the product bubbles small.

11.3.2.3 Analysis

Ideal for these measurements the device is operating in steady state. Therefore, gas is produced and enters the eudiometer at a constant rate, causing the water level to also drop at a constant rate. When measuring pressure in units of inches water column (in WC), the constant drop in water level is readily seen as a constant velocity in units of inches per second. Measuring this velocity is as trivial as fitting a straight line to the data produced by the manometer, and simply measuring the linear distance per volume of the eudiometer, which is then converted to gas production rate in units of volume per time.

The manometer measures the pressure in the headspace of the eudiometer, P as seen in Figure 11.10. From the pressure, the change in headspace volume is $\Delta V = c\Delta P$, where c (volume/length) is the measured constant unique to each eudiometer. The change in headspace volume is then related to the gas production rate of the solar fuel generator.

11.3.3 Outdoor Testing

As the materials, components and whole device design of solar-driven water-splitting cells improves, the testing environments are expanding from the lab to the outdoors in order to access the practical application of this technology. The diurnal cycling, man-made or natural, is essential to evaluate the device performance and understand the degradation pathways in the real world. Although, the primary metrics for assessing the performance of solar water splitting systems indoors and outdoors are similar, there are a number of additional variables introduced with outdoor testing. These variables, such as weather, wildlife, dust, smoke, and time of day, to name a few, are uncontrollable and need to be considered when designing these systems. The knowledge gained from outdoor testing can be used for assessing the scalability of the technology and allow for the development of robust cells that can survive the rigors of the outdoors.

Over the past few years, JCAP has demonstrated a range of device architectures for solar-driven water-splitting with STH conversion efficiency exceeded 10%, including photocathode devices and photoanode devices, as well as devices that operate in alkaline, acidic and near neutral pH conditions.[15,26,112–116] One example is an integrated photovoltaic-based electrosynthetic cell using a bipolar membrane for unassisted solar-driven water-splitting that maintains steady-state pH gradients between the anolyte and catholyte regions of the cell. A tandem-junction GaAs/InGaP photoanode coated by an amorphous TiO_2 protective layer in conjunction with Ni-based earth-abundant electrocatalyst in 0.5 M potassium borate buffer solution at Ph = 9.3 was connected to a CoP nanoparticle-coated Ti mesh cathode electrode in 1.0 M sulfuric acid solution at pH = 0. An unassisted solar-to-hydrogen (STH) conversion efficiency of 10%, with continuous operation for over 100 hours with <10% relative change in performance, were observed for >1 cm^2 area photoelectrodes, with separate product gas streams throughout the system.[114] Because of the size, stability and efficiency of the device that could produce a decent amount of hydrogen gases in an outdoor measurement, an outdoor experiment was carried out from 9 am to 4 pm (November 29–December, Caltech campus, Pasadena, CA, USA) (Figure 11.12). The illumination angle of the photoreactor prototype relative to the incident direct beam sunlight was controlled by a homemade dual-axis solar tracker (parts were obtained from www.Browndoggadgets.com). During outdoor testing, the current passing between the photoanode and cathode was monitored with no external bias applied throughout the day. The theoretical amount of $H_2(g)$ gas that would be produced was calculated from the measured current and passed charges, assuming 100% Faradaic efficiency. The temperature of the cell solution in the external reservoir was also monitored and recorded throughout the experiment, and the intensity of the solar illumination was also measured and recorded using an on-board Si diode (Thorlabs) calibrated by a NREL-calibrated Si photovoltaic reference cell.

Samples were mounted onto a single-sided Cu tape using a Ag-based conductive epoxy. Samples were then assembled into a custom-made compression flow cell 3-D Figure 11.11a that were equipped with fluorosilicone O-rings (0.96 cm^2 Teflon). The same peristaltic pump setup was also used in the outdoor measurement to ensure no bubble accumulation interfered with the optical illumination. An optical image of the outdoor measurement setup on a dual-axis solar tracker are shown in Figure 11.12b. Measurement results, including the photoanodic current density response, intensity recorded by the on-board Si photodiode and the calculated H_2 gas production based on the charge passed, are shown in Figure 11.12c.

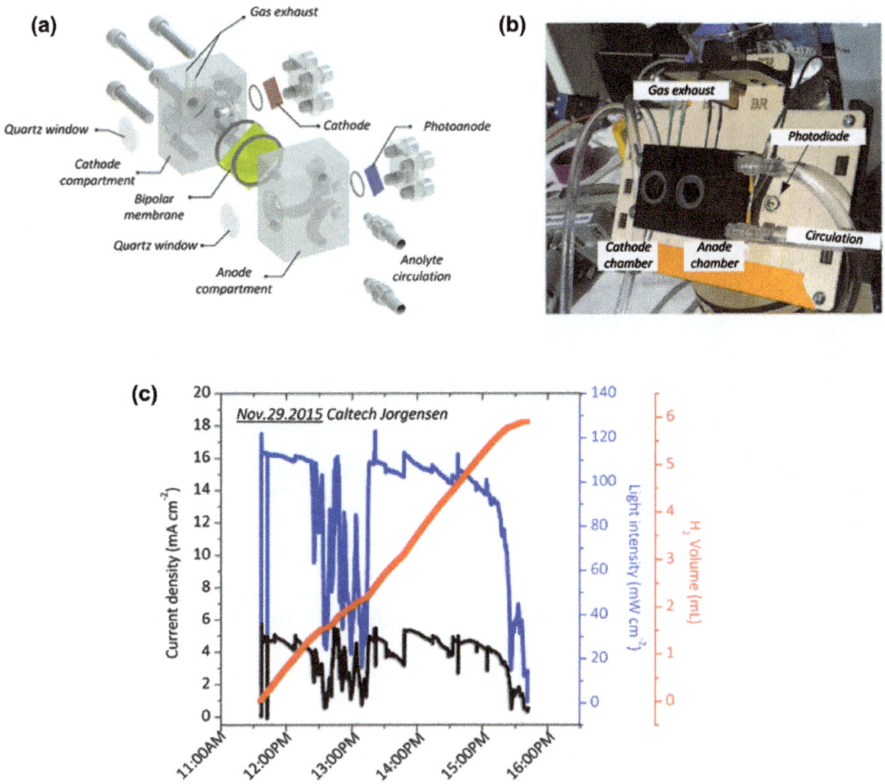

Figure 11.12 (a) A schematic illustration of the flow cell and (b) an optical image of the cell on a dual-axis solar tracker outdoor measurement setup. (c) Outdoor test results (bottom, H_2 volume calculated based on integrated charge—red curve, optical intensity—measured by an on-board photodiode from Thorlabs calibrated to the NREL photodiode—blue curve, and photocurrent response measured on the tandem substrate in the flow cell on the dual-axis solar tracker—black curve).

In addition, a mobile outdoor test station was also set up on the roof of Chu Hall at Lawrence Berkeley National Laboratory. The testing apparatus consisted of a two-axis solar tracker, a potentiostat for collecting photocurrent as a function of time, a calibrated reference Si photovoltaic cell for monitoring light intensity, thermocouples for collecting the temperature of the reference cell, and a video camera, as shown in the photograph in Figure 11.13a. The apparatus was placed next to a commercial weather station, which collected ambient temperature, wind speed and direction, and barometric pressure. One difference for outdoor *versus* indoor testing is that the incident angle of irradiation changes though out the day. To maximize light capture, the light either needs to be directed at the cell or the orientation of the cell needs to be adjusted throughout the day; ours utilized the latter. The outdoor performance of the composite-protected solar water

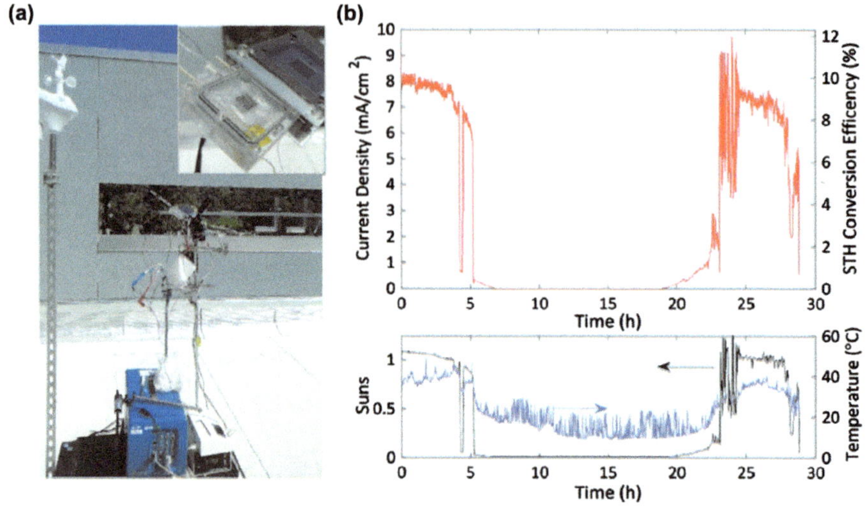

Figure 11.13 Outdoor PEC testbed: (a) the image of 2-axis solar tracker and analytical equipment, with a close-up view (inset) of the device under test adjacent to the reference diode. (b) Outdoor performance of a composite-protected solar water splitting device over 30 h with no applied bias in 1 M H_2SO_4.

splitting cell was measured by monitoring the photo-current density under illumination (Figure 11.13b). A large variation of the incident light intensity and operating temperature were observed under outdoor testing conditions. Under full illumination near mid-day, the photocurrent density from the solar water splitting cell was greater than 8 mA cm^{-2}, both on the first and second day. Overall, limited outdoor experiments have been performed in JCAP as well as in the PEC community. Standardizing the test protocols for outdoor testing and evaluating materials and assemblies at varying temperature and illumination remain an under-explored area.

11.4 System Engineering Approaches

11.4.1 Introduction to System Engineering

Integrated solar-driven water-splitting devices are a complex set of sub-systems and components that must operate concurrently and in harmony without degrading the functionality of each other. These sub-systems and components comprise a wide range of materials with often countervailing operational ranges and tolerances. For example, a low pH electrolyte may favor high membrane ionic conductivity but may also accelerate corrosion of light absorber materials. Furthermore, to design and construct such a complicated, interdependent system requires a diverse technical team with expertise in catalysis, semiconductor physics, corrosion, optics, materials science, electrochemistry, and other disciplines. To that end, using a formal system engineering approach for project management can facilitate the efficient design, fabrication, testing, and delivery of the solar fuel generator from a highly multi-disciplinary team of engineers, scientists, and managers.

System engineering is a relatively new engineering field that has been recognized as invaluable for managing the life-cycle of complex technical projects. It involves a rigorous approach described below to manage technical, schedule and budget risk, as well as formalizing the project life-cycle of experimentation, assembly, and tests. Kossiakoff and Sweet ascribe the foundation of modern-day systems engineering as a discipline that arose in response to (1) advancing technology that was accompanied by increasing risks and complexity (*e.g.*, NASA's Space Shuttle), (2) global competition requiring expert risk taking and associated risk management, and (3) scientific and engineering specialization requiring the bridging of disciplines and interfaces.[117] The somewhat generalized application of system engineering as it relates to solar fuel generator development is described below.

11.4.2 Development of Hierarchical Requirements

Central to the systems engineering methodology is the development of a set of hierarchical requirements that the final product will be tested against, for acceptance by the stakeholder. While setting and maintaining formal requirements is rather time consuming, it has a number of advantages such as:

- minimizing ambiguity in what will be expected of the hardware and the team members
- setting expectations for all stakeholders and end users
- avoiding over- and under-engineering of the product
- informing a progressively-detailed hardware design
- setting timelines and decision gates to manage technical issues
- helping engineers and scientists stay on-task, driven by the necessity of meeting specific requirements

The development of multi-level requirements begins as high-level negotiations between the end user (the stakeholder) and the project executive management. This is a critical step where all parties agree to well defined and realistic metrics for the hardware to be delivered. The output of these negotiations are termed Level 1 requirements, and they describe the general attributes and overall system-level functionality. An example of the Level 1 requirement in the development of integrated solar-driven water-splitting devices is given in Table 11.4.

The next tier of requirements, Level 2, are negotiated between the project executive management and project senior management. These requirements describe sub-system functionality to meet the parent Level 1 requirements, where sub-systems for a solar fuel generator may include product gas

Table 11.4 Level 1 requirements for a solar fuel generator.

Requirement ID	Requirement
1.1	No wires may be used
1.2	Two distinctly different prototype designs and one engineering model shall be delivered
1.3	Delivery date = July 27, 2012
1.3	Hold PDR, CDR, MRR, DR
1.4	No microbial components
1.5	No powered components
1.6	Carry 5% schedule margin
1.7	Designs shall accommodate analytical probe points in engineering model cells
1.8	Delivered prototypes shall be 10×10 cm ± 1 cm
1.9	Delivered prototypes shall appear (qualitatively) professionally prepared and attractive without suggesting impractical cost or complexity
1.10	Delivered prototypes shall allow for facile viewing of internal components and products during operation
1.11	Delivered prototypes shall generate gaseous products (during operation under illumination) that can be verified by visual means (*e.g.* bubbles) or by analysis (*e.g.* gas chromatography)
1.12	Delivered prototypes shall be robust to normal handling conditions without any loss of functionality, leakage of reactant, or breakage of internal components
1.13	Targeted operating point: 10 mA cm^{-2} current density
1.14	Shall operate under adiabatic conditions with no active heating or cooling at nominally 25 °C, 20% relative humidity
1.15	Shall be capable of wet calendar life greater than 1000 hours
1.16	Shall be capable of operational life greater than 100 hours

handling assemblies, feedstock distribution assemblies, light absorber/catalysts assemblies, *etc.* An example of Level 2 requirements is given in Table 11.5.

Finally, the Level 3 requirements are negotiated between the project senior management and cognizant engineers and scientists. These requirements describe component-level functionality and are tied to satisfying higher level requirements. An example of Level 3 requirement is given in Table 11.6.

It is important to note that as the project moves forward from a general concept to a more advanced design, the requirements will likely be updated and refined. Provisions for revising the requirements should be made early in the project, with an agreed upon, well documented approval process for changes. Any changes to Level 1 requirements must include approval from the stakeholder.

11.4.3 Testing Plan

A hierarchy of requirements is not useful without a means to confirm compliance to these requirements with unambiguous pass/no-pass outcomes. A formal testing plan should be established concurrently with

Table 11.5 Level 2 requirements for a solar fuel generator.

Requirement ID	Requirement	Design feature addressing requirement	Parent requirement
2.1A	Tilt angle shall be sized to avoid bubble buildup or gas headspace to form in channels.	Tilt angle on device x and y directions = 5°, 5°	
2.2A	Engineering model design shall allow for insertable and removable instrumentation		1.7
2.3A	Case shall be transparent		1.10
2.4A	Window shall be optically transparent to UV light	Window is UV transparent Dow Chemical EG6301 epoxy or Topas 8007X10 resin	
2.5A	Window shall be able to be molded, ground, milled, or otherwise textured for light focusing	Window is UV transparent Dow Chemical EG6301 epoxy or Topas 8007X10 resin	
2.6A	Engineering model shall allow for facile membrane mounting and removal	Clamp top and bottom frames together with o-ring seals onto membrane with nut/washer/bolt fasteners	
2.7A	No fasteners shall be exposed to acidic medium	Fasteners pass through top and bottom frames without passing through wetted compartment	
2.8A	Engineering model shall accommodate operation either by illumination mode or by electrically biased mode		
2.9A	During operation, engineering model shall be capable of measuring the following data: differential pressure, temperature, pH, voltage, O_2 concentration, H_2 concentration		
2.10A	Design shall accommodate manual water filling and atmospheric vent		
2.11A	Design shall allow for gaseous product escape without leakage of reactant liquid water	Incorporate water-tight gas vent (*e.g.* hydrophobic Goretex membrane) on case and window	1.12
2.12A	All cell components shall be compatible with sulfuric acidic medium		

Table 11.6 Level 3 requirements for a solar fuel generator.

Requirement ID	Requirement	Design feature addressing requirement	Parent requirement
3.1B	Membrane shall tolerate 20″ H_2O column pressure differential	Membrane is assembled with solution cast $SrTiO_3$ in Nafion onto polypropylene mesh	2.19B
3.2B	Membrane shall remain in place through 3-axis 360° rotation at (360° s^{-1} rate)	Incorporate membrane attachment fixture to decouple membrane attachment from cell fixtures	1.12
3.3B	Membrane matrix shall be proton conductive	Membrane matrix is Nafion	2.22B
3.4B	Membrane components including light absorber, membrane matrix, support structure, and catalyst(s) shall be stable in sulfuric acid medium	Membrane matrix is Nafion, light absorber is SrTiO3, support structure is polypropylene mesh, HER catalyst is Pt	2.22B
3.5B	Membrane catalyst(s) shall not occlude proton conductive matrix	Sputter deposit HER catalyst onto light absorber prior to membrane infiltration with proton conducting matrix	2.20B
3.6B	Gas vents shall allow product gases to escape without leakage of liquid feed at 20″ water differential	Use Goretex gaskets in countersunk gas vent fittings	2.19B
3.7B	Membrane shall be mounted such that there is no visible discontinuity in the flatness of membrane	Incorporate membrane attachment fixture to decouple membrane attachment from cell fixtures	1.12
3.8B	Membrane shall be free of edge burrs, uneven surfaces, voids, contamination, or other defects		

the development of requirements. The interdependency of testing and requirements may be visualized as a V-model shown in Figure 11.14.

In practice, an effective test plan is highly detailed to avoid any ambiguity in the test results. The plan includes written test protocols with specific instruments named, test durations fixed, allowable re-test conditions defined, and so on. In a project-oriented industrial setting, depending on the customer's specifications, quality assurance personnel may be asked to witness and certify the outcome of the tests. The test plan should require

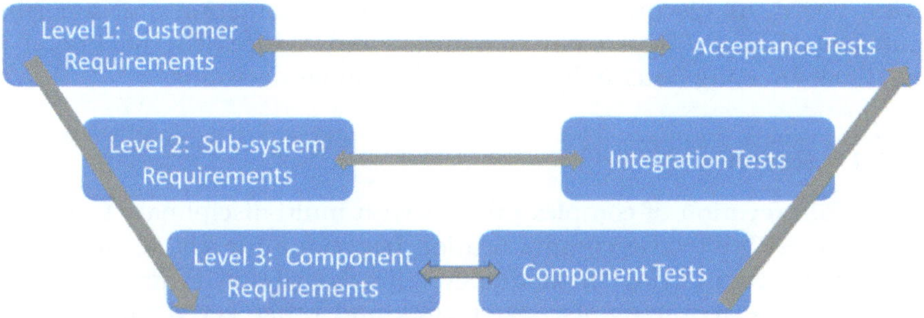

Figure 11.14 V-model of the hierarchy and interdependency of requirements and testing.

management concurrence before initiating the work to avoid post-test disputes about the efficacy or validity of the test outcome.

It is to be expected that not all the tests will have the desired outcome. However, if there is no formal process for documenting and dispositioning test issues, this can lead to significant risk and uncertainty about the entire test protocol and the test outcome. Consider the example of a team carrying out the final testing of a solar fuel generator prior to an important demonstration for a sponsor. When testing the system under illumination, the technician observes a sudden spike in the concentration of oxygen on the hydrogen side of the generator. The process engineer suspects that the technician may have improperly inserted the oxygen probe into the testing apparatus causing atmospheric ingress, rather than a much more troublesome problem of a pinhole in the gas separation membrane. To confirm this hypothesis, the engineer asks the technician to re-test again, taking more care with the insertion of the oxygen probe into the test fixture. With no oxygen detected immediately upon the start-up of the generator, the process engineer terminates the test and concludes her hypothesis of the failure mechanism was correct and the membrane was pinhole-free. She does not report the first test failure to management, and then certifies the solar fuel generator is ready for the crucial sponsor test. Was this a sound engineering decision, discounting the possibility of other potentially intermittent but much more serious design or assembly problems? Would other test engineers and management agree with this assessment? What is the risk of such a unilateral, undocumented decision? One can easily imagine that confirmation bias or other judgment errors by the process engineer could lead to profound failures impacting the entire project.

An effective way to manage such risks associated with engineering judgment on testing issues is to put in place a formal problem/failure reporting (PFR) system. In such a system, any outcome other than unequivocal success is extensively documented with details such as written descriptions of the problem or failure, post-test materials analyses, equipment calibration dates, photographs, and so on. These PFRs are then vetted before a technical review board, as discussed later. This PFR documentation and reporting approach demands rigor from all parties, and reduces the likelihood of individuals making potentially high-risk unilateral decisions.

11.4.4 Reviews

Successful execution of complex projects with multi-disciplinary teams can only occur with effective communication between all levels of the project. The systems engineering methodology relies heavily on reviews to assure that the plans, requirements, test results and other key details are well documented and agreed upon at all project levels. Reviews also serve as milestones for the project, and these help team members stay focused working towards meeting these tangible goals on schedule. The number and types of reviews carried out will depend on the complexity of the project, though in practice, reviews will include:

- Conceptual Design Review
- Preliminary Design Review
- Critical Design Review
- Hardware Certification Review

Other reviews such as Quality Assurance Audit, Manufacturing Readiness Review and Test Readiness Review can be treated as separate reviews or included within other planned reviews.

The first review, the Conceptual Design Review, is typically held with the stakeholder and project executive management. Generally, this review is less structured and informal, and is intended to include only high-level descriptions of the device and its functionality. The parties must reach agreement on Level 1 requirements, which serves as the sole element of the success criteria for the review.

In contrast, the Preliminary Design Review is much more structured. At this stage of the project, the hardware design is not complete, and functional elements are shown only in block diagram form. Level 1 and 2 requirements should be in place before holding the review, and general approaches to manufacturing and assembly should be established. A typical flow of the review is described below.

Before the review, the project management team will assemble a review board with a designated chairperson, comprising subject matter experts and experienced technical managers including members from outside the project organization. The chair will ask the board for decisions at the end of presentations, voting for a pass, preliminary pass, or fail. Board members and meeting attendees can ask for specific actions to be taken prior to the next meeting (*e.g.*, carry out a re-test, clarify a design element) through a request for action (RFA). The RFAs are collected at the end of the review, voted upon, and assigned to specific team members with a close-out date.

After collecting RFAs, the board will render the pass/fail decision based on success criteria such as whether:

- Level 1 and 2 requirements are sufficiently detailed to constrain preliminary design.
- Preliminary designs are sound and sufficiently detailed to proceed to define Level 3 requirements and comprehensive designs.
- Proposed approaches to fabrication/assembly are sound.

In the review, the team members will present Level 1 and Level 2 requirements. Next, the team will present preliminary designs and tentative fabrication approaches. The team members and board will discuss the adequacy of fabrication and assembly plans in the context of some initial "pathfinder" experimental work.

If the outcome of the Preliminary Design Review is a fail, the team will redesign and refine the Level 1 and 2 requirements per RFAs and then hold second (delta)-Preliminary Design Review shortly afterward. If the review board's decision is preliminary pass, the team will work to close out RFAs and in parallel, begin assigning Level 3 requirements. If the board's decision is a pass, the team will proceed with ramped-up experimental studies in support of the comprehensive design and set Level 3 requirements.

The next key milestone for the project is the Critical Design Review. The review board is typically asked to assign a pass/fail decision based on the following criteria:

- Open actions, issues, and RFAs that pertain to the preliminary design have been addressed.
- Design changes since the Preliminary Design Review have been reviewed and are acceptable.
- Level 1–3 requirements are verifiable and sufficiently defined to constrain the comprehensive design.
- Design related risks to going forward into manufacturing/assembly are acceptable.
- The manufacturing/assembly process and engineering plans are robust.
- The risks of proceeding into manufacturing/assembly are acceptable.
- The test plan is well understood and acceptable.
- All test equipment defined, calibrated, and demonstrated.

In this review, a top-level discussion of the schedule, workforce, and budget is included. The Level 1 and 2 requirements from the preliminary design review are presented, and the new Level 3 requirements are enumerated and defended. A key portion of the review focuses on the design packages for the hardware, with supporting experimental and computational modeling data. The design at this point should be very well defined, with hardware dimensions, specified materials, interfaces, *etc.* shown in CAD images or preferably in rapid-prototyped models. The Manufacturing/

Assembly Plan and Test Plan should be presented, with a clear understanding of the critical path steps. The team often presents a risk assessment, including potential issues and workarounds.

The final step prior to delivering the product to the customer is holding the Hardware Certification Review. This milestone provides the endpoint to all development, fabrication, and testing. The team member's task in this meeting to convince the review board that the hardware has been demonstrated to meet requirements by test, analysis, or inspection. In this review, there are usually only two primary success criteria: demonstrating that all PFRs have been appropriately addressed and the test plan has been followed with satisfactory results.

The importance of the review board in the role of project success cannot be understated. A passive, disengaged, or technically uncertain board can allow the project team members to escape scrutiny resulting in significant risk of project success. In contrast, the most effective review committee is highly engaged, comfortable in discussing multi-disciplinary topics at an advanced technical level, and respectfully adversarial. They should actively challenge the presenters and question under-supported claims. When the project team understands that the review board will take no quarter, the team members invariably rise to the challenge and conduct their project work in a much more professional and defensible manner.

It is worth mentioning in closing that if not properly managed, the unique attributes of a team of highly educated and accomplished technical managers, engineers and scientists with their particular attributes can surprisingly work against project success, indefinitely impeding progress towards the hardware delivery. These contributors with few exceptions are naturally very curious and innovative. Most are continually seeking better solutions to problems, even to those problems that were previously solved. The maxim "perfect is the enemy of good enough" describes the risk of under-managed teams continually refining designs without regard to schedule and milestones. These team members, with the admirable goal of improving the final product, often cannot see that these design improvements can come with concomitant yet unforeseen deleterious impacts to other sub-systems outside their cognizance. System engineering, with firm requirements, schedules, milestones, and deliverables, provides a framework for keeping teams on track for successful on-time delivery of the final product.

11.5 Conclusion and Outlook

The performance of integrated solar-driven water-splitting devices has advanced significantly in the past few years. Fundamental materials discovery in light absorber, electrocatalysts, membrane separators and interfacial phenomena as well as multi-physics modeling, simulation and engineering of the overall device laid the ground-work for the rapid increase of the reported device efficiencies. In particular, 10% STH conversion efficiency has changed from the ceiling of the reported device efficiencies into the floor of the reported device efficiencies in various device configurations. As the integrated solar-driven water-splitting device contains many functional materials and components as well as many auxiliary components, prototyping development plays an important role in evaluating newly discovered materials and components in the real world settings and provides a technical readiness level for practical development and deployment. Moreover, the assembly and integration of various components, many of which operate at different optimal conditions, requires significant science and engineering efforts to optimize the overall device efficiency, stability and scalability. Efficient device design in prototyping development is often guided by modeling and simulation, as well as life cycle and techno-economic analyses of the overall system. While several planar device designs have been modeled, fabricated and evaluated using PV grade materials, to further reduce the cost of the system, radically different new designs or new materials are still required in order to compete with hydrogen produced by natural gas reforming. Techno-economic analysis of such system revealed that in order to compete with traditional production methods, such as steam methane reforming, which produces hydrogen at the plant gate of ~\$1.5 per kg, the total installation of the distributed system needs to be $<$\$100 m^{-2} at a conversion efficiency of ~20%.[118] To achieve these aggressive goals, further advancement on discovery of new materials as well as on design and engineering of new device/systems architectures are required. The needed new materials include cheap metal oxides with a bandgap value between 1.7–1.8 eV and good electronic properties for efficient and stable operation of OER or HER. The new designs include the Type 1 and 2 designs (particle reactors)[14] or embedded Si microwire arrays that could be rolled out like artificial turf.[119] Significant prototyping efforts are still needed in translating those ideas into tangible prototypes so that the advantages and challenges in those new designs can be fully evaluated.

References

1. N. S. Lewis, Research opportunities to advance solar energy utilization, *Science*, 2016, **351**, 353.
2. H. B. Gray, Powering the planet with solar fuel, in *246th ACS National Meeting & Exposition*, 2013, Indianapolis, IN.
3. J. R. McKone, N. S. Lewis and H. B. Gray, Will Solar-Driven Water-Splitting Devices See the Light of Day?, *Chem. Mater.*, 2014, **26**, 407–414.
4. N. S. Lewis and D. G. Nocera, Powering the planet: Chemical challenges in solar energy utilization, *Proc. Natl. Acad. Sci. U. S. A.*, 2006, **103**, 15729–15735.
5. T. J. Jacobsson, V. Fjallstrom, M. Sahlberg, M. Edoff and T. Edvinsson, A monolithic device for solar water splitting based on series interconnected thin film absorbers reaching over 10% solar-to-hydrogen efficiency, *Energy Environ. Sci.*, 2013, **6**, 3676–3683.

6. C. R. Cox, J. Z. Lee, D. G. Nocera and T. Buonassisi, Ten-percent solar-to-fuel conversion with nonprecious materials, *Proc. Natl. Acad. Sci. U. S. A.*, 2014, **111**, 14057–14061.
7. J. S. Luo, J. H. Im, M. T. Mayer, M. Schreier, M. K. Nazeeruddin, N. G. Park, S. D. Tilley, H. J. Fan and M. Gratzel, Water photolysis at 12.3% efficiency via perovskite photovoltaics and Earth-abundant catalysts, *Science*, 2014, **345**, 1593–1596.
8. S. Licht, B. Wang, S. Mukerji, T. Soga, M. Umeno and H. Tributsch, Efficient Solar Water Splitting, Exemplified by RuO2-Catalyzed AlGaAs/Si Photoelectrolysis, *J. Phys. Chem. B*, 2000, **104**, 8920–8924.
9. K. Fujii, S. Nakamura, M. Sugiyama, K. Watanabe, B. Bagheri and Y. Nakano, Characteristics of hydrogen generation from water splitting by polymer electrolyte electrochemical cell directly connected with concentrated photovoltaic cell, *Int. J. Hydrogen Energy*, 2013, **38**, 14424–14432.
10. G. Peharz, F. Dimroth and U. Wittstadt, Solar hydrogen production by water splitting with a conversion efficiency of 18%, *Int. J. Hydrogen Energy*, 2007, **32**, 3248–3252.
11. K. Sapru; N. T. Stetson; S. R. Ovshinsky; J. Yang; G. Fritz; M. Fairlie and A. T. B. Stuart, Development of a small scale hydrogen production-storage system for hydrogen applications, in IECEC-97 - *Proceedings of the Thirty-Second Intersociety Energy Conversion Engineering Conference, 1997,* vols. 1–4, pp. 1947–1952.
12. L. G. Arriaga, W. Martinez, U. Cano and H. Blud, Direct coupling of a solar-hydrogen system in Mexico, *Int. J. Hydrogen Energy*, 2007, **32**, 2247–2252.
13. T. Lipman, *An Overview of Hydrogen Production and Storage Systems with Renewable Hydrogen Case Studies*, DOE, 2011.
14. B. D. James; G. N. Baum; J. Perez and K. N. Baum, *Technoeconomic Analysis of Photoelectrochemical (PEC) Hydrogen Production*, DOE, 2009.
15. J. W. Ager III, M. Shaner, K. Walczak, I. D. Sharp and S. Ardo, Experimental Demonstrations of Spontaneous, Solar-Driven Photoelectrochemical Water Splitting, *Energy Environ. Sci.*, 2015, **8**, 2811–2824.
16. S. Haussener, S. Hu, C. X. Xiang, A. Z. Weber and N. S. Lewis, Simulations of the irradiation and temperature dependence of the efficiency of tandem photoelectrochemical water-splitting systems, *Energy Environ. Sci.*, 2013, **6**, 3605–3618.
17. K. Sun, R. Liu, Y. K. Chen, E. Verlage, N. S. Lewis and C. X. Xiang, A Stabilized, Intrinsically Safe, 10% Efficient, Solar-Driven Water-Splitting Cell Incorporating Earth-Abundant Electrocatalysts with Steady-State pH Gradients and Product Separation Enabled by a Bipolar Membrane, *Adv. Energy Mater.*, 2016, **6**, 1600379.
18. J. L. Young, M. A. Steiner, H. Doscher, R. M. France, J. A. Turner and T. G. Deutsch, Direct solar-to-hydrogen conversion via inverted metamorphic multi-junction semiconductor architectures, *Nat. Energy*, 2017, **2**, 17028.

19. E. Verlage, S. Hu, R. Liu, R. J. R. Jones, K. Sun, C. X. Xiang, N. S. Lewis and H. A. Atwater, A monolithically integrated, intrinsically safe, 10% efficient, solar-driven water-splitting system based on active, stable earth-abundant electrocatalysts in conjunction with tandem III-V light absorbers protected by amorphous TiO2 films, *Energy Environ. Sci.*, 2015, **8**, 3166–3172.
20. S. Haussener, C. X. Xiang, J. M. Spurgeon, S. Ardo, N. S. Lewis and A. Z. Weber, Modeling, simulation, and design criteria for photoelectrochemical water-splitting systems, *Energy Environ. Sci.*, 2012, **5**, 9922–9935.
21. M. T. Winkler, C. R. Cox, D. G. Nocera and T. Buonassisi, Modeling integrated photovoltaic-electrochemical devices using steady-state equivalent circuits, *Proc. Natl. Acad. Sci.*, 2013, **110**, E1076–E1082.
22. J. Jin, K. Walczak, M. R. Singh, C. Karp, N. S. Lewis and C. Xiang, An experimental and modeling/simulation-based evaluation of the efficiency and operational performance characteristics of an integrated, membrane-free, neutral pH solar-driven water-splitting system, *Energy Environ. Sci.*, 2014, **7**, 3371–3380.
23. S. Y. Reece, J. A. Hamel, K. Sung, T. D. Jarvi, A. J. Esswein, J. J. H. Pijpers and D. G. Nocera, Wireless Solar Water Splitting Using Silicon-Based Semiconductors and Earth-Abundant Catalysts, *Science*, 2011, **334**, 645–648.
24. H. Doscher, J. F. Geisz, T. G. Deutsch and J. A. Turner, Sunlight absorption in water - efficiency and design implications for photoelectrochemical devices, *Energy Environ. Sci.*, 2014, **7**, 2951–2956.
25. M. A. Green and M. J. Keevers, Optical properties of intrinsic silicon at 300 K, *Prog. Photovolt.: Res. Appl.*, 1995, **3**, 189–192.
26. K. Walczak, Y. Chen, C. Karp, J. W. Beeman, M. Shaner, J. Spurgeon, I. D. Sharp, X. Amashukeli, W. West, J. Jin, N. S. Lewis and C. Xiang, Modeling, Simulation, and Fabrication of a Fully Integrated, Acid-stable, Scalable Solar-Driven Water-Splitting System, *ChemSusChem*, 2015, **8**, 544–551.
27. K. A. Walczak, G. Segev, D. M. Larson, J. W. Beeman, F. A. Houle and I. D. Sharp, Hybrid Composite Coatings for Durable and Efficient Solar Hydrogen Generation under Diverse Operating Conditions, *Adv. Energy Mater.*, 2017, **7**, 1602791.
28. F. Dimroth, R. Beckert, M. Meusel, U. Schubert and A. W. Bett, Metamorphic GayIn1-yP/Ga1-xInxAs tandem solar cells for space and for terrestrial concentrator applications at $C > 1000$ suns, *Prog. Photovolt.*, 2001, **9**, 165–178.
29. M. M. May, H. J. Lewerenz, D. Lackner, F. Dimroth and T. Hannappel, Efficient direct solar-to-hydrogen conversion by in situ interface transformation of a tandem structure, *Nat. Commun.*, 2015, **6**, 8286.
30. C. C. L. McCrory, S. Jung, I. M. Ferrer, S. M. Chatman, J. C. Peters and T. F. Jaramillo, Benchmarking Hydrogen Evolving Reaction and Oxygen

Evolving Reaction Electrocatalysts for Solar Water Splitting Devices, *J. Am. Chem. Soc.*, 2015, **137**, 4347–4357.

31. C. C. L. McCrory, S. Jung, J. C. Peters and T. F. Jaramillo, Benchmarking Heterogeneous Electrocatalysts for the Oxygen Evolution Reaction, *J. Am. Chem. Soc.*, 2013, **135**, 16977–16987.
32. S. Haussener, C. Xiang, J. Spurgeon, S. Ardo, N. Lewis and A. Z. Weber, Modeling, Simulation, and Design Criteria for Photoelectrochemical Water-Splitting Systems, *Energy Environ. Sci.*, 2012, **5**, 9922–9935.
33. K. Walczak, Y. K. Chen, C. Karp, J. W. Beeman, M. Shaner, J. Spurgeon, I. D. Sharp, X. Amashukeli, W. West, J. Jin, N. S. Lewis and C. X. Xiang, Modeling, Simulation, and Fabrication of a Fully Integrated, Acid-stable, Scalable Solar-Driven Water-Splitting System, *ChemSusChem*, 2015, **8**, 544–551.
34. Y. K. Chen, C. X. Xiang, S. Hu and N. S. Lewis, Modeling the Performance of an Integrated Photoelectrolysis System with 10×Solar Concentrators, *J. Electrochem. Soc.*, 2014, **161**, F1101–F1110.
35. S. Haussener, S. Hu, C. Xiang, A. Z. Weber and N. Lewis, Simulations of the irradiation and temperature dependence of the efficiency of tandem photoelectrochemical water-splitting systems, *Energy Environ. Sci.*, 2013, **6**, 3605–3618.
36. N. M. Vargas-Barbosa, G. M. Geise, M. A. Hickner and T. E. Mallouk, Assessing the Utility of Bipolar Membranes for use in Photoelectrochemical Water-Splitting Cells, *ChemSusChem*, 2014, **7**, 3017–3020.
37. M. B. McDonald, S. Ardo, N. S. Lewis and M. S. Freund, Use of Bipolar Membranes for Maintaining Steady-State pH Gradients in Membrane-Supported, Solar-Driven Water Splitting, *ChemSusChem*, 2014, **7**, 3021–3027.
38. M. Ünlü, J. Zhou and P. A. Kohl, Hybrid Anion and Proton Exchange Membrane Fuel Cells, *J. Phys. Chem. C*, 2009, **113**, 11416–11423.
39. D. A. Vermaas, M. Sassenburg and W. A. Smith, Photo-assisted water splitting with bipolar membrane induced pH gradients for practical solar fuel devices, *J. Mater. Chem. A*, 2015, **3**, 19556–19562.
40. X. H. Zhou, R. Liu, K. Sun, Y. K. Chen, E. Verlage, S. A. Francis, N. S. Lewis and C. X. Xiang, Solar-Driven Reduction of 1 atm of CO2 to Formate at 10% Energy-Conversion Efficiency by Use of a TiO2-Protected III-V Tandem Photoanode in Conjunction with a Bipolar Membrane and a Pd/C Cathode, *ACS Energy Lett.*, 2016, **1**, 764–770.
41. M. Schreier, F. Heroguel, L. Steier, S. Ahmad, J. S. Luterbacher, M. T. Mayer, J. S. Luo and M. Gratzel, Solar conversion of CO2 to CO using Earth-abundant electrocatalysts prepared by atomic layer modification of CuO, *Nat. Energy*, 2017, **2**, 17087.
42. M. R. Singh, K. Papadantonakis, C. X. Xiang and N. S. Lewis, An electrochemical engineering assessment of the operational conditions and constraints for solar-driven water-splitting systems at near-neutral pH, *Energy Environ. Sci.*, 2015, **8**, 2760–2767.

43. M. R. Singh, C. Xiang and N. S. Lewis, Evaluation of flow schemes for near-neutral pH electrolytes in solar-fuel generators, *Sustainable Energy Fuels*, 2017, **1**, 458–466.
44. M. R. Singh, K. M. Papadantonakis, C. Xiang and N. Lewis, An Electrochemical Engineering Assessment of the Operational Conditions and Constraints for Solar-Driven Water-Splitting Systems at Near-Neutral pH, *Energy Environ. Sci.*, 2015, **8**, 2760–2767.
45. K. Sun, S. Shen, Y. Liang, P. E. Burrows, S. S. Mao and D. Wang, Enabling Silicon for Solar-Fuel Production, *Chem. Rev.*, 2014, **114**, 8662–8719.
46. M. F. Lichterman, K. Sun, S. Hu, X. Zhou, M. T. McDowell, M. R. Shaner, M. H. Richter, E. J. Crumlin, A. I. Carim, F. H. Saadi, B. S. Brunschwig and N. S. Lewis, Protection of inorganic semiconductors for sustained, efficient photoelectrochemical water oxidation, *Catal. Today*, 2016, **262**, 11–23.
47. M. R. Shaner, S. Hu, K. Sun and N. S. Lewis, Stabilization of Si microwire arrays for solar-driven H_2O oxidation to $O_2(g)$ in 1.0 M KOH(aq) using conformal coatings of amorphous TiO_2, *Energy Environ. Sci.*, 2015, **8**, 203–207.
48. K. Sun, Y. Kuang, E. Verlage, B. S. Brunschwig, C. W. Tu and N. S. Lewis, Sputtered NiOx Films for Stabilization of p + n-InP Photoanodes for Solar-Driven Water Oxidation, *Adv. Energy Mater.*, 2015, **5**, 1402276.
49. K. Sun, R. Liu, Y. Chen, E. Verlage, N. S. Lewis and C. Xiang, A Stabilized, Intrinsically Safe, 10% Efficient, Solar-Driven Water-Splitting Cell Incorporating Earth-Abundant Electrocatalysts with Steady-State pH Gradients and Product Separation Enabled by a Bipolar Membrane, *Adv. Energy Mater.*, 2016, **6**, 1600379.
50. K. Sun, N. Park, Z. Sun, J. Zhou, J. Wang, X. Pang, S. Shen, S. Y. Noh, Y. Jing, S. Jin, P. Yu and D. Wang, Nickel oxide functionalized silicon for efficient photo-oxidation of water, *Energy Environ. Sci.*, 2012, **5**, 7872–7877.
51. K. Sun, F. H. Saadi, M. F. Lichterman, W. G. Hale, H.-P. Wang, X. Zhou, N. T. Plymale, S. T. Omelchenko, J.-H. He, K. M. Papadantonakis, B. S. Brunschwig and N. S. Lewis, Stable solar-driven oxidation of water by semiconducting photoanodes protected by transparent catalytic nickel oxide films, *Proc. Natl. Acad. Sci.*, 2015, **112**, 3612–3617.
52. E. Verlage, S. Hu, R. Liu, R. J. R. Jones, K. Sun, C. Xiang, N. S. Lewis and H. A. Atwater, A monolithically integrated, intrinsically safe, 10% efficient, solar-driven water-splitting system based on active, stable earth-abundant electrocatalysts in conjunction with tandem III-V light absorbers protected by amorphous TiO2 films, *Energy Environ. Sci.*, 2015, **8**, 3166–3172.
53. M. J. Madou, K. W. Frese and S. R. Morrison, Photoelectrochemical corrosion as influenced by an oxide layer, *J. Phys. Chem.*, 1980, **84**, 3423–3428.
54. E. D. Palik, J. W. Faust, H. F. Gray and R. F. Greene, Study of the Etch-Stop Mechanism in Silicon, *J. Electrochem. Soc.*, 1982, **129**, 2051–2059.

55. J. W. Faust and E. D. Palik, Study of the Orientation Dependent Etching and Initial Anodization of Si in Aqueous KOH, *J. Electrochem. Soc.*, 1983, **130**, 1413–1420.
56. E. D. Palik and V. M. Bermudez, Ellipsometric investigation of the silicon/anodic-oxide interface, *J. Phys. Colloques*, 1983, **44**, C10-179–C110-182.
57. H. G. G. Philipsen and J. J. Kelly, Anisotropy in the Anodic Oxidation of Silicon in KOH Solution, *J. Phys. Chem. B*, 2005, **109**, 17245–17253.
58. Y. Chen, K. Sun, H. Audesirk, C. Xiang and N. Lewis, A Quantitative Analysis of the Efficiency of Solar-Driven Water-Splitting Device Designs Based on Tandem Photoabsorbers Patterned with Islands of Metallic Electrocatalysts, *Energy Environ. Sci.*, 2015, **8**, 1736–1747.
59. E. Kemppainen, A. Bodin, B. Sebok, T. Pedersen, B. Seger, B. Mei, D. Bae, P. C. K. Vesborg, J. Halme, O. Hansen, P. D. Lund and I. Chorkendorff, Scalability and feasibility of photoelectrochemical H2 evolution: the ultimate limit of Pt nanoparticle as an HER catalyst, *Energy Environ. Sci.*, 2015, **8**, 2991–2999.
60. J. M. Falkowski, N. M. Concannon, B. Yan, Y. Surendranath and Heazlewoodite, Ni3S2: A Potent Catalyst for Oxygen Reduction to Water under Benign Conditions, *J. Am. Chem. Soc.*, 2015, **137**, 7978–7981.
61. C. D. Giovanni, Á. Reyes-Carmona, A. Coursier, S. Nowak, J. M. Grenèche, H. Lecoq, L. Mouton, J. Rozière, D. Jones, J. Peron, M. Giraud and C. Tard, Low-Cost Nanostructured Iron Sulfide Electrocatalysts for PEM Water Electrolysis, *ACS Catal.*, 2016, 2626–2631.
62. A. Harvey, C. Backes, Z. Gholamvand, D. Hanlon, D. McAteer, H. C. Nerl, E. McGuire, A. Seral-Ascaso, Q. M. Ramasse, N. McEvoy, S. Winters, N. C. Berner, D. McCloskey, J. F. Donegan, G. S. Duesberg, V. Nicolosi and J. N. Coleman, Preparation of Gallium Sulfide Nanosheets by Liquid Exfoliation and Their Application As Hydrogen Evolution Catalysts, *Chem. Mater.*, 2015, **27**, 3483–3493.
63. N. Kornienko, J. Resasco, N. Becknell, C.-M. Jiang, Y.-S. Liu, K. Nie, X. Sun, J. Guo, S. R. Leone and P. Yang, Operando Spectroscopic Analysis of an Amorphous Cobalt Sulfide Hydrogen Evolution Electrocatalyst, *J. Am. Chem. Soc.*, 2015, **137**, 7448–7455.
64. D. Kong, J. Cha, H. Wang, H. R. Lee and Y. Cui, First-Row Transition Metal Dichalcogenide Catalysts for Hydrogen Evolution Reaction, *Energy Environ. Sci.*, 2013, **6**, 3553–3558.
65. A. T. Swesi, J. Masud and M. Nath, Nickel selenide as a high-efficiency catalyst for oxygen evolution reaction, *Energy Environ. Sci.*, 2016, **9**, 1771–1782.
66. J. F. Callejas, C. G. Read, C. W. Roske, N. S. Lewis and R. E. Schaak, Synthesis, Characterization, and Properties of Metal Phosphide Catalysts for the Hydrogen-Evolution Reaction, *Chem. Mater.*, 2016, **28**, 6017–6044.
67. F. H. Saadi, A. I. Carim, E. Verlage, J. C. Hemminger, N. S. Lewis and M. P. Soriaga, CoP as an Acid-Stable Active Electrocatalyst for the

Hydrogen-Evolution Reaction: Electrochemical Synthesis, Interfacial Characterization and Performance Evaluation, *J. Phys. Chem. C*, 2014, **118**, 29294–29300.

68. J. F. Callejas, J. M. McEnaney, C. G. Read, J. C. Crompton, A. J. Biacchi, E. J. Popczun, T. R. Gordon, N. S. Lewis and R. E. Schaak, Electrocatalytic and Photocatalytic Hydrogen Production from Acidic and Neutral-pH Aqueous Solutions Using Iron Phosphide Nanoparticles, *ACS Nano*, 2014, **8**, 11101–11107.
69. J. Masa, P. Weide, D. Peeters, I. Sinev, W. Xia, Z. Sun, C. Somsen, M. Muhler and W. Schuhmann, Amorphous Cobalt Boride (Co2B) as a Highly Efficient Nonprecious Catalyst for Electrochemical Water Splitting: Oxygen and HydrogenEvolution., *Adv. Energy Mater.*, 2016, **6**, 1502313.
70. S. Carenco, D. Portehault, C. Boissière, N. Mézailles and C. Sanchez, Nanoscaled Metal Borides and Phosphides: Recent Developments and Perspectives, *Chem. Rev.*, 2013, **113**, 7981–8065.
71. T. Osaka, H. Ishibashi, T. Endo and T. Yoshida, Oxygen evolution reaction on transition metal borides, *Electrochim. Acta*, 1981, **26**, 339–343.
72. H. Yan, C. Tian, L. Wang, A. Wu, M. Meng, L. Zhao and H. Fu, Phosphorus-Modified Tungsten Nitride/Reduced Graphene Oxide as a High-Performance, Non-Noble-Metal Electrocatalyst for the Hydrogen Evolution Reaction, *Angew. Chem., Int. Ed.*, 2015, **127**, 6423–6427.
73. K. Xu, P. Chen, X. Li, Y. Tong, H. Ding, X. Wu, W. Chu, Z. Peng, C. Wu and Y. Xie, Metallic Nickel Nitride Nanosheets Realizing Enhanced Electrochemical Water Oxidation, *J. Am. Chem. Soc.*, 2015, **137**, 4119–4125.
74. H.-W. Liang, S. Bruller, R. Dong, J. Zhang, X. Feng and K. Mullen, Molecular metal-Nx centres in porous carbon for electrocatalytic hydrogen evolution, *Nat. Commun.*, 2015, **6**, 7992.
75. P. Chen, K. Xu, Z. Fang, Y. Tong, J. Wu, X. Lu, X. Peng, H. Ding, C. Wu and Y. Xie, Metallic Co4N Porous Nanowire Arrays Activated by Surface Oxidation as Electrocatalysts for the Oxygen Evolution Reaction, *Angew. Chem., Int. Ed.*, 2015, **54**, 14710–14714.
76. D. V. Sokolsky, V. S. Palanker and E. N. Baybatyrov, Electrochemical hydrogen reactions on tungsten carbide, *Electrochim. Acta*, 1975, **20**, 71–77.
77. W.-F. Chen, J. T. Muckerman and E. Fujita, Recent developments in transition metal carbides and nitrides as hydrogen evolution electrocatalysts, *Chem. Commun.*, 2013, **49**, 8896–8909.
78. M. D. Porosoff, X. Yang, J. A. Boscoboinik and J. G. Chen, Molybdenum Carbide as Alternative Catalysts to Precious Metals for Highly Selective Reduction of CO2 to CO, *Angew. Chem., Int. Ed.*, 2014, **53**, 6705–6709.
79. F.-X. Ma, H. B. Wu, B. Y. Xia, C.-Y. Xu and X. W. Lou, Hierarchical β-Mo2C Nanotubes Organized by Ultrathin Nanosheets as a Highly Efficient Electrocatalyst for Hydrogen Production, *Angew. Chem., Int. Ed.*, 2015, **54**, 15395–15399.

80. H. B. Wu, B. Y. Xia, L. Yu, X.-Y. Yu and X. W. Lou, Porous molybdenum carbide nano-octahedrons synthesized via confined carburization in metal-organic frameworks for efficient hydrogen production, *Nat. Commun.*, 2015, **6**, 6512.
81. Z.-Y. Wu, X.-X. Xu, B.-C. Hu, H.-W. Liang, Y. Lin, L.-F. Chen and S.-H. Yu, Iron Carbide Nanoparticles Encapsulated in Mesoporous Fe-N-Doped Carbon Nanofibers for Efficient Electrocatalysis, *Angew. Chem., Int. Ed.*, 2015, **54**, 8179–8183.
82. J. M. McEnaney and R. E. Schaak, Solution Synthesis of Metal Silicide Nanoparticles, *Inorg. Chem.*, 2015, **54**, 707–709.
83. A. B. Shein and E. N. Zubova, Electrochemical Behavior of Manganese Silicides in Sulfuric Acid Solution, *Prot. Met.*, 2005, **41**, 234–242.
84. A. K. Vijh, G. Bélanger and R. Jacques, Electrochemical reactions on iron silicide electrodes in alkaline solutions, *Mater. Chem. Phys.*, 1989, **21**, 529–538.
85. L. Trotochaud, S. L. Young, J. K. Ranney and S. W. Boettcher, Nickel–Iron Oxyhydroxide Oxygen-Evolution Electrocatalysts: The Role of Intentional and Incidental Iron Incorporation, *J. Am. Chem. Soc.*, 2014, **136**, 6744–6753.
86. S. Klaus, Y. Cai, M. W. Louie, L. Trotochaud and A. T. Bell, Effects of Fe Electrolyte Impurities on Ni(OH)2/NiOOH Structure and Oxygen Evolution Activity, *J. Phys. Chem. C*, 2015, **119**, 7243–7254.
87. S. Klaus, M. W. Louie, L. Trotochaud and A. T. Bell, Role of Catalyst Preparation on the Electrocatalytic Activity of Ni1–xFexOOH for the Oxygen Evolution Reaction, *J. Phys. Chem. C*, 2015, **119**, 18303–18316.
88. I. Roger and M. D. Symes, Efficient Electrocatalytic Water Oxidation at Neutral and High pH by Adventitious Nickel at Nanomolar Concentrations, *J. Am. Chem. Soc.*, 2015, **137**, 13980–13988.
89. V. S. Langford, A. J. McKinley and T. I. Quickenden, Temperature Dependence of the Visible-Near-Infrared Absorption Spectrum of Liquid Water, *J. Phys. Chem. A*, 2001, **105**, 8916–8921.
90. H. Doscher, J. F. Geisz, T. G. Deutsch and J. A. Turner, Sunlight absorption in water - efficiency and design implications for photoelectrochemical devices, *Energy Environ. Sci.*, 2014, **7**, 2951–2956.
91. C. G. Granquist, *Handbook of Inorganic Electrochromic Materials*, Elsevier Science B.V., 2002.
92. Y. Ushio, A. Ishikawa and T. Niwa, Degradation of the electrochromic nickel oxide film upon redox cycling, *Thin Solid Films*, 1996, **280**, 233–237.
93. L. KyoungMoo, A. Yoshio and K. Midori, Kyung Ho, K., Effects of Substrate Temperature on Electrochromic Properties of Cobalt Oxide and Oxyhydroxide Thin Films Prepared by Reactive Sputtering Using O 2 and H 2 O Gases, *Jpn. J. Appl. Phys.*, 2012, **51**, 045501.
94. R. C. Tenent, D. T. Gillaspie, A. Miedaner, P. A. Parilla, C. J. Curtis and A. C. Dillon, Fast-Switching Electrochromic Li + -Doped NiO Films by

Ultrasonic Spray Deposition, *J. Electrochem. Soc.*, 2010, **157**, H318–H322.

95. R.-T. Wen, G. A. Niklasson and C. G. Granqvist, Electrochromic nickel oxide films and their compatibility with potassium hydroxide and lithium perchlorate in propylene carbonate: Optical, electrochemical and stress-related properties, *Thin Solid Films*, 2014, **565**, 128–135.
96. M. R. Shaner, J. R. McKone, H. B. Gray and N. S. Lewis, Functional integration of Ni-Mo electrocatalysts with Si microwire array photocathodes to simultaneously achieve high fill factors and light-limited photocurrent densities for solar-driven hydrogen evolution, *Energy Environ. Sci.*, 2015, **8**, 2977–2984.
97. J. Newman, Scaling with Ohm's Law; Wired vs. Wireless Photoelectrochemical Cells, *J. Electrochem. Soc.*, 2013, **160**, F309–F311.
98. S. Haussener, S. Hu, C. Xiang, A. Z. Weber and N. S. Lewis, Simulations of the irradiation and temperature dependence of the efficiency of tandem photoelectrochemical water-splitting systems, *Energy Environ. Sci.*, 2013, **6**, 3605–3618.
99. K. A. Walczak, M. J. Hutchins and D. Dornfeld, Energy system design to maximize net energy production considering uncertainty in scale-up: A case study in artificial photosynthesis, *Proc. CIRP*, 2014, **15**, 306–312.
100. W. A. Smith, I. D. Sharp, N. C. Strandwitz and J. Bisquert, Interfacial band-edge energetics for solar fuels production, *Energy Environ. Sci.*, 2015, **8**, 2851–2862.
101. S. Hu, M. H. Richter, M. F. Lichterman, J. Beardslee, T. Mayer, B. S. Brunschwig and N. S. Lewis, Electrical, Photoelectrochemical, and Photoelectron Spectroscopic Investigation of the Interfacial Transport and Energetics of Amorphous TiO2/Si Heterojunctions, *J. Phys. Chem. C*, 2016, **120**, 3117–3129.
102. M. F. Lichterman, S. Hu, M. H. Richter, E. J. Crumlin, S. Axnanda, M. Favaro, W. Drisdell, Z. Hussain, T. Mayer, B. S. Brunschwig, N. S. Lewis, Z. Liu and H. J. Lewerenz, Direct observation of the energetics at a semiconductor/liquid junction by operando X-ray photoelectron spectroscopy, *Energy Environ. Sci.*, 2015, **8**, 2409–2416.
103. M. F. Lichterman, M. H. Richter, S. Hu, E. J. Crumlin, S. Axnanda, M. Favaro, W. Drisdell, Z. Hussain, B. S. Brunschwig, N. S. Lewis, Z. Liu and H. J. Lewerenz, An Electrochemical, Microtopographical and Ambient Pressure X-Ray Photoelectron Spectroscopic Investigation of Si/TiO2/Ni/Electrolyte Interfaces, *J.Electrochem. Soc.*, 2016, **163**, H139–H146.
104. M. T. McDowell, M. F. Lichterman, A. I. Carim, R. Liu, S. Hu, B. S. Brunschwig and N. S. Lewis, The Influence of Structure and Processing on the Behavior of TiO2 Protective Layers for Stabilization of n-Si/TiO2/Ni Photoanodes for Water Oxidation, *ACS Appl. Mater. Interfaces*, 2015, 7, 15189–15199.
105. J. A. Haber, E. Anzenburg, J. Yano, C. Kisielowski and J. M. Gregoire, Multiphase Nanostructure of a Quinary Metal Oxide Electrocatalyst

Reveals a New Direction for OER Electrocatalyst Design, *Adv. Energy Mater.*, 2015, **5**, 11.
106. J. A. Haber, Y. Cai, S. H. Jung, C. X. Xiang, S. Mitrovic, J. Jin, A. T. Bell and J. M. Gregoire, Discovering Ce-rich oxygen evolution catalysts, from high throughput screening to water electrolysis, *Energy Environ. Sci.*, 2014, **7**, 682–688.
107. J. A. Haber, D. Guevarra, S. H. Jung, J. Jin and J. M. Gregoire, Discovery of New Oxygen Evolution Reaction Electrocatalysts by Combinatorial Investigation of the Ni- La- Co- Ce Oxide Composition Space, *ChemElectroChem*, 2014, **1**, 1613–1617.
108. J. A. Haber, C. X. Xiang, D. Guevarra, S. H. Jung, J. Jin and J. M. Gregoire, High-Throughput Mapping of the Electrochemical Properties of (Ni-Fe-Co-Ce)O-x Oxygen-Evolution Catalysts, *ChemElectroChem*, 2014, **1**, 524–528.
109. D. Guevarra, A. Shinde, S. K. Suram, I. D. Sharp, F. M. Toma, J. A. Haber and J. M. Gregoire, Development of solar fuels photoanodes through combinatorial integration of Ni-La-Co-Ce oxide catalysts on BiVO4, *Energy Environ. Sci.*, 2016, **9**, 565–580.
110. A. Shinde, D. Guevarra, G. Liu, I. D. Sharp, F. M. Toma, J. M. Gregoire and J. A. Haber, Discovery of Fe–Ce Oxide/BiVO4 Photoanodes through Combinatorial Exploration of Ni–Fe–Co–Ce Oxide Coatings, *ACS Appl. Mater. Interfaces*, 2016, **8**, 23696–23705.
111. A. Shinde, G. Li, L. Zhou, D. Guevarra, S. K. Suram, F. M. Toma, Q. Yan, J. A. Haber, J. B. Neaton and J. M. Gregoire, The role of the CeO2/BiVO4 interface in optimized Fe-Ce oxide coatings for solar fuels photoanodes, *J. Mater. Chem. A*, 2016, **4**, 14356–14363.
112. J. Jin, K. Walczak, M. R. Singh, C. Karp, N. S. Lewis and C. X. Xiang, An experimental and modeling/simulation-based evaluation of the efficiency and operational performance characteristics of an integrated, membrane-free, neutral pH solar-driven water-splitting system, *Energy Environ. Sci.*, 2014, **7**, 3371–3380.
113. M. A. Modestino, K. A. Walczak, A. Berger, C. M. Evans, S. Haussener, C. Koval, J. S. Newman, J. W. Ager and R. A. Segalman, Robust production of purified H-2 in a stable, self-regulating, and continuously operating solar fuel generator, *Energy Environ. Sci.*, 2014, **7**, 297–301.
114. K. Sun, R. Liu, Y. Chen, E. Verlage, N. S. Lewis and C. Xiang, A Stabilized, Intrinsically Safe, 10% Efficient, Solar-Driven Water-Splitting Cell Incorporating Earth-Abundant Electrocatalysts with Steady-State pH Gradients and Product Separation Enabled by a Bipolar Membrane, *Adv. Energy Mater.*, 2016, **6**, 1600379.
115. E. Verlage, S. Hu, R. Liu, R. J. R. Jones, K. Sun, C. Xiang, N. S. Lewis and H. A. Atwater, A monolithically integrated, intrinsically safe, 10% efficient, solar-driven water-splitting system based on active, stable earth-abundant electrocatalysts in conjunction with tandem III-V light absorbers protected by amorphous TiO2 films, *Energy Environ. Sci.*, 2015, **8**, 3166–3172.

116. C. Xiang, Y. Chen and N. S. Lewis, Modeling an integrated photoelectrolysis system sustained by water vapor, *Energy Environ. Sci.*, 2013, **6**, 3713–3721.
117. A. Kossiakoff, W. N. Sweet, S. Seymour and S. M. Biemer, *Systems Engineering Principles and Practice*, John Wiley & Sons, 2011, vol. 83.
118. M. R. Shaner, H. A. Atwater, N. S. Lewis and E. W. McFarland, A comparative technoeconomic analysis of renewable hydrogen production using solar energy, *Energy Environ. Sci.*, 2016, **9**, 2354–2371.
119. M. R. Shaner, K. T. Fountaine, S. Ardo, R. H. Coridan, H. A. Atwater and N. S. Lewis, Photoelectrochemistry of core-shell tandem junction n-p(+)-Si/n-WO3 microwire array photoelectrodes, *Energy Environ. Sci.*, 2014, **7**, 779–790.

CHAPTER 12

High-efficiency Water Splitting Systems

MATTHIAS M. MAY,*[a] HENNING DÖSCHER*[b] AND JOHN A. TURNER*[c]

[a] University of Cambridge, Department of Chemistry, Cambridge, UK; [b] Fraunhofer ISI, Competence Center Emerging Technologies, Karlsruhe, Germany; [c] National Renewable Energy Laboratory, Golden, Colorado, USA
*Email: Matthias.May@physik.hu-berlin.de; Henning.Doescher@isi.fraunhofer.de; John.Turner@nrel.gov

12.1 The Need for High Efficiency in Solar Fuel Generation

The direct photolysis of water for hydrogen production has been deemed a Holy Grail of Chemistry.[1] First shown in 1972 with TiO_2, the photoelectrochemical splitting of water has been a dream through the decades. While the process seems to be the simplest and most straightforward approach for solar-driven water splitting, ultimately though there has not been a large-scale demonstration of a photoelectrochemical water splitting system.

For a commercially viable system, the key parameters are solar-to-hydrogen (STH) efficiency, system cost and lifetime, of which the efficiency carries the greatest weight.[2] This is not unexpected, since land area must be covered and that relates directly to the capital cost of the system. Unfortunately, the main focus of research over the decades has been on oxides due to their expected low cost and stability. In general oxide semiconductors

Energy and Environment Series No. 22
Integrated Solar Fuel Generators
Edited by Ian D. Sharp, Harry A. Atwater and Hans-Joachim Lewerenz

Published by the Royal Society of Chemistry, www.rsc.org

have very poor solid-state characteristics,[3] thus their STH efficiencies remain very low.

While the number of systems, for which spontaneous solar water splitting was demonstrated, steadily increases, the range of reported efficiencies varies greatly with very few having the efficiency necessary to be considered for a viable system. As we will focus on high efficiency systems in the following, we must discuss the definition of high efficiency and motivate the permanent struggle of the solar energy communities towards this end. Solar conversion efficiencies determine a system's production rate and, at a given capital cost, its economic feasibility. If the relation between costs for the solar absorber to the cumulated cost for module, installation and land shifts away from the absorber, efficiency becomes a key factor in determining the costs per produced energy unit. In the mature market of photovoltaic solar energy conversion, such a trend can already be observed. In solar water splitting, the overall system is more complex due to gas handling facilities and consequently, these systems will be more sensitive to the impact of conversion efficiency and thus the final price of the produced hydrogen.[2] But how to define high efficiency? One can define this *via* the ratio of realised to achievable efficiency for (a) a given material system or (b) the overall physical limit given by thermodynamics and the shape of the solar spectrum. The latter definition is more general and in line with commercial viability and will be employed here. Note that the threshold for "high" will increase with time as technological progress and economic demands shift the realised efficiencies towards the physical limit.

Physical efficiency limitations vary with the type of system that is used for solar water splitting. In general, one can distinguish two main approaches: PV-electrolysis systems and more tightly integrated photoelectrochemical systems. In the former approach, a dark electrolyser is driven by the output of a photovoltaic solar cell. Such a system may exhibit only electric coupling[4] or also mechanical coupling.[5] For maximum efficiency, the current-voltage behaviour of the solar cell and electrolyser must be matched whether directly coupled or interfaced with DC/DC converters, but apart from this restriction, the two systems can be developed independently. This advantage though must be tempered with the realization that the low capacity factor of PV (~20%) results in the high-cost electrolyzers not operating full-time, and thus for a given product output, greatly increasing the capital costs of the system and the subsequent price of the produced hydrogen.

While for approaches beyond PV-electrolysis a variety of terminologies exist, they all share the challenge that the electrochemical load, the photovoltage of the absorber, and the photocurrent cannot be optimised independently, limiting obtainable efficiencies. Before we focus on the latter approach in the following, we first discuss the various taxonomies present in the literature.

A definition from an electrochemical perspective is the term *direct water splitting*, where the semiconductor is part of the solid–electrolyte junction, and in most cases also decorated with a catalyst. This gives rise to all the effects of semiconductor photoelectrochemistry with the Fermi level in the forbidden gap, in-gap surface states that can couple to the electrolyte, and accumulation (depletion) regions. In the case of a (metallic) catalyst present on the surface, the energetic landscape can become even more complicated due to triple points, where catalyst, semiconductor, and electrolyte meet. Indirect water splitting, on the other hand, involves in this definition a semiconductor–metal contact of either predominantly Schottky- or ohmic type, followed by a metal–electrolyte contact. Here, the semiconductor is not in contact with the electrolyte, but energetically coupled to the electrolyte in an indirect manner *via* the Fermi level of the contacting metal. Other definitions of "direct" comprise the condition that the total number of junctions equals the number of charge-separating semiconductor–electrolyte junctions as opposed to buried junctions as in a photovoltaic (PV) device.[6] Such a definition does, however, not take into account that most semiconductors change their surface composition upon contact with the electrolyte forming either intrinsically a charge-separating solid–solid junction,[7,8] or experience charge-separation at the "catalyst"–absorber interface, as in the case of Co-Pi/$BiVO_4$.[9]

12.2 Efficiency Limitations and Prospects for Photoelectrochemical Energy Conversion

The design of highly efficient water splitting devices requires careful, simultaneous consideration of all efficiency-limiting factors. They comprise effects from thermodynamics, catalysis, electronic structure, materials science, and chemical engineering. Many of these challenges cannot be addressed independently, which adds to the complexity of device design. For a given set of boundary conditions such as catalyst performance, water layer thickness or solution resistivity, the optimum efficiency tandem requires an infinite fine-control of absorber bandgaps and the possibility to combine two absorbers in a tandem device. Over a wide range of bandgaps, this is possible with the III–V semiconductors.[10]

12.2.1 Fundamental Limitations: Detailed Balance Limit and Catalysis

The first and most well-known limit for the harvesting of solar energy by water splitting originates in the shape of the solar spectrum and the ability of a semiconductor to transform radiative energy into a chemical potential difference for electrons as discussed by Shockley and Queisser.[11] If we consider the steps in the photoelectrochemical water splitting process: (i) light absorption, (ii) carrier generation, (iii) carrier separation and (iv) delivery to the interface, (v) electron transfer (catalysis), and (vi) product formation, we note that the first four are in the realm of solid-state physics. Thus, the solid-state semiconducting properties (the photovoltaics) are key for an efficient solar-driven water splitting process.

In the detailed-balance limit, which was developed for photovoltaics, the main factors are radiative recombination as well as thermalisation and transmission losses due to the shape of the solar spectrum in relation to the bandgap of the semiconductor. Originally discussed for single-junction solar cells, which limits efficiencies to just above 30%, the scheme can easily be extended to multi-junction absorbers,[12] which greatly increases obtainable efficiencies as described below. To approach the single- or multi-junction Shockley–Queisser limit, the bandgap(s) of the absorber(s) have to be adapted to the solar spectrum. As we will see later, this requirement makes III–V compound semiconductors with their flexible opto-electronic properties the hitherto best absorbers for solar energy conversion.

In monolithic multi-junction absorbers, the photovoltages of the individual subcells add up to the overall device photovoltage, while the photocurrent is limited by the subcell that absorbs the least photons, as the current is equal in the whole device. Photon management, for instance by thinning a subcell to thicknesses below its photon absorption length, is a measure to alleviate the limiting photocurrent of the subjacent absorber with a smaller bandgap. The Shockley–Queisser limit can, in principle, be overcome by multi-exciton generation, where thermalisation losses are reduced by the generation of multiple excitons per absorbed photon. While this has recently been demonstrated for a water splitting system with lead sulphide quantum dots,[13] the effect is, up to now, only observed at significant quantum efficiencies for low-dimensional systems.

The second fundamental limit is catalysis, which is in the case of solar water splitting given by the two half-reactions of hydrogen and oxygen evolution. The difference in Nernst potentials equals the Gibb's free energy per electron and is the extractable energy by the combustion of hydrogen in a fuel cell. It can be considered the electrochemical load[14]—the equivalent of the energy per electron at the operating voltage of a photovoltaic cell—and depends on the type of solar fuel (1.23 eV for water splitting). We will employ the Gibb's free energy for the efficiency definition in the following, but it should be noted that some communities also use the higher heating value, which is 1.48 eV for water splitting, for the definition of the solar-to-hydrogen efficiency.[5]

The fixation of the electrochemical load in solar water splitting to 1.23 eV severely limits obtainable efficiencies: in photovoltaics, an increase of the number of absorbers decreases the photocurrent by distributing the photocurrent over more subcells, but this effect is overcompensated by an increase in operating voltage and an extension of spectral sensitivity to the infrared. Consequently, more—adequately designed—junctions benefit overall efficiency. For solar water splitting, on the other hand, a reduced photocurrent of triple- or higher-order multi-junctions combined with a fixed electrochemical load reduces the efficiency and makes already double-junction cells the optimum for water splitting. An increase of the electrochemical load to higher values, however, for instance by choice of formic acid as a product of CO_2 reduction (1.43 eV), is equivalent to an increase of the operating voltage of a photovoltaic cell. This benefits the maximum efficiencies and renders triple cells for the production of fuels with a Gibb's free energy beyond 1.6 eV more attractive than double junctions as depicted in Figure 12.1.[15] It shows the maximum obtainable solar-to-fuel efficiencies (STF) for single, double and triple absorbers with only the fundamental limits of detailed balance and catalysis considered. The top cell(s) were thinned to maximise current matching and we see that the highest efficiencies are obtained for triple junctions in the range of 2–3 eV electrochemical load. An unknown parameter here is, however, the catalytic overpotential for a given product, which depends on the actual electrochemical reaction and its charge-transfer type. For Figure 12.1, the catalyst characteristics (*i.e.* exchange current density and Tafel slope) of IrO_2 for OER, which is the dominating factor in water splitting, have been employed. Other products, however, will most likely induce higher losses from catalysis.

Catalytic overpotentials increase the required voltage to initiate the reaction as a function of the current density and can typically be described by the Tafel equation, a logarithmic voltage-current relationship.[16] In the case of water splitting, the overall catalytic overpotential is dominated by the multi-electron transfer oxygen evolution reaction (OER) and amounts—depending on the performance of the catalysts—to at least 300 mV at 10 mA cm^{-2}. This limit is dictated by thermodynamics and is unlikely to be

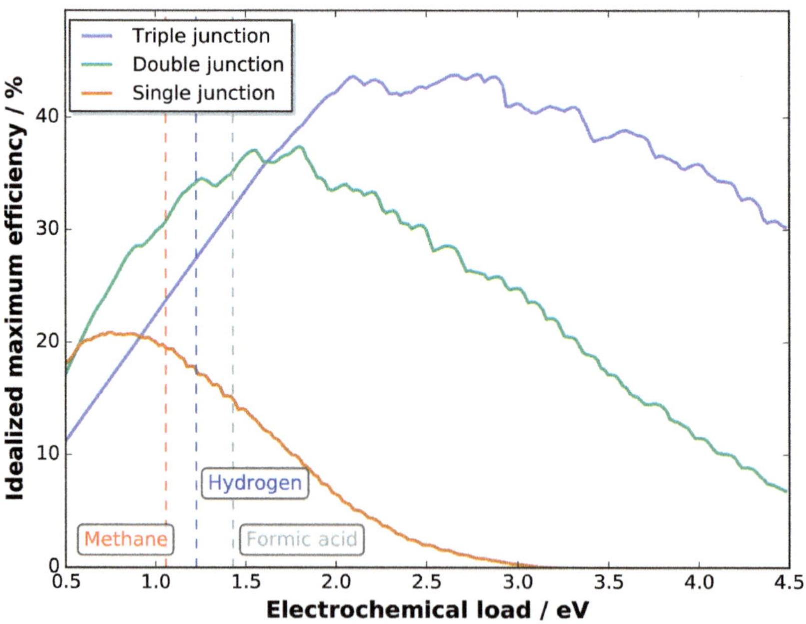

Figure 12.1 Ideal solar-to-fuel efficiencies in the detailed balance limit, with the Tafel behaviour of IrO_2 as OER catalyst. No further losses are considered; calculated with YaSoFo.[15]

overcome.[17] High-performance catalysts, especially for OER in acid, often show the drawback that they are based on noble metals.[18] The overpotentials of hitherto known catalysts for other reactions such as CO_2 reduction are significantly higher: for a recent high-efficiency system with $\eta \sim 13\%$ STF, the overpotentials of cathode and anode were almost equal and amounted to more than 800 mV.[19] If CO_2 reduction is at all viable for solar fuel production can, however, be questioned.[20]

Figure 12.2(a) shows water splitting efficiencies with an ideal tandem absorber in the detailed balance limit with 2 mm water layer and 2 Ω ohmic drop as a function of catalyst performance. For the sake of simplicity, we assume here that the overpotential is dominated by a single catalyst described by exchange current density and Tafel slope (which are, however, not fully independent). High-performance catalysts such as RuO_2 are found towards the bottom left of the graph, where theoretical efficiencies approach 30% STH. Noble-metal free catalysts typically benefit from a higher earth abundance, but they often exhibit a higher Tafel slope towards 100 mV per decade and lower exchange current densities, corresponding to a shift towards the top right of the graph. This quickly decreases maximum obtainable efficiencies to values below 20% STH. The efficiencies were obtained by optimising the bandgaps of the tandem absorber (with optimum thinning) at each combination of exchange current density and Tafel slope. Again, we observe that a change in the performance of an outer parameter requires an adjustment of the bandgaps of the absorbers. This is illustrated in Figure 12.2(b), were the resulting sum of the bandgaps for the ideal two-junction tandem absorbers is shown. With a decreasing catalyst performance, the photovoltage generated by the absorber must increase to compensate for the higher overpotentials. This is accomplished by

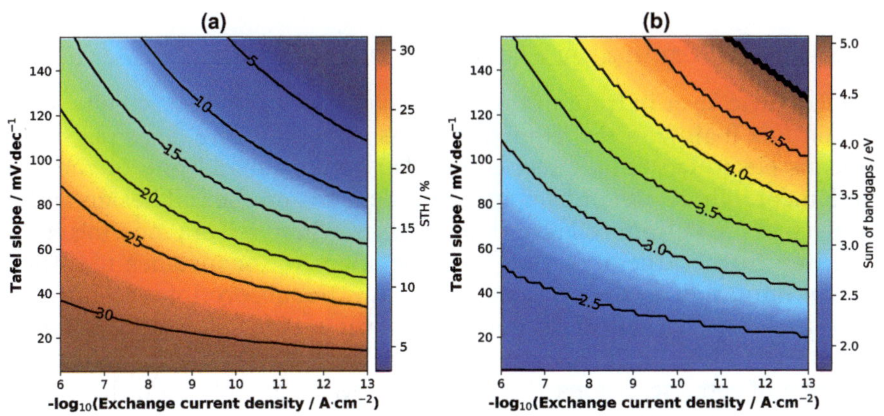

Figure 12.2 Ideal water splitting efficiencies (a) for a dual-junction tandem as a function of Tafel slope and exchange current density in the detailed balance limit as a function of catalyst performance, with 2 mm water layer and 2 Ω ohmic drop. (b) The sum of the bandgaps for the optimum tandem absorber.

increasing the bandgaps, which leads to a reduction of the photocurrent and therefore efficiency.

The above mentioned two fundamental limits are by far the most important boundary conditions for high-efficiency water splitting. The minimum photovoltage dictated by electrochemistry requires absorbers with suitable bandgaps and material properties. If the delivered photovoltage is too low to drive the reaction, the efficiency is zero. Absorbers in the intermediate bandgap range of *ca.* 1.2–1.4 eV might deliver high photocurrents,[21] but if the achieved photovoltages are in the order of only 500 mV or below, this is insufficient for water splitting. Equally, materials with high photovoltage, but high bandgap, such as TiO_2, fail because of low photocurrents due to a lack of UV photons in the solar spectrum. Multi-junction devices with adequate bandgaps are therefore imperative for efficient photoelectrochemical water splitting.

12.2.2 Further Relevant Loss Mechanisms and Mitigation Strategies

In addition to the fundamental limits, which constitute significant restrictions for solar fuel efficiencies, there are further loss mechanisms which can, in principle, be minimised. Losses such as ohmic resistivities and light absorption by the electrolyte depend on absorber materials and cell design. These are to some extent unavoidable and their magnitude depends on the characteristics of the design, such as the distance between the active areas of the half-cell reactions. Mitigation requires specific optimisation not only of the photoelectrochemical cell, but also of the absorber, which will be discussed in the following.

12.2.2.1 Absorption by Electrolyte and Catalyst

An electrolyte layer in front of the cell acts as a spectral filter, which cuts the photon flux mainly in the infrared.[22–24] As water splitting takes place in an aqueous electrolyte, this effect could also be considered one of the more fundamental limits. Its impact, however, depends on the cell design and the bandgap of the bottom cell absorber. In cases, where cell engineering cannot be driven further, the absorbers must be adapted, for instance to alleviate current limitation by the bottom cell. Finally, the absorption of light by the catalysts, which can affect photocurrents over a wide spectral range, has to be reduced, which is achieved for instance by the tailoring of size and distribution of catalyst nanoparticles.[15,25] However, depending on the catalytic properties of the bare semiconductor, co-catalysts might not be needed at all.

The effect of a water layer (Figure 12.3b), on the efficiency and thus the bandgaps, leads to a reduction of obtainable maximum efficiencies with an increasing thickness of the water layer. The optimum bandgaps of the

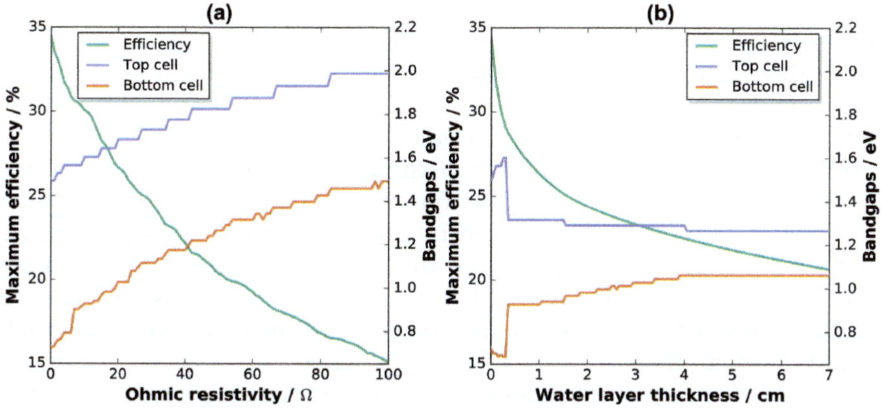

Figure 12.3 Maximum solar-to-hydrogen efficiencies and the corresponding optimum bandgaps for double junction absorbers in the detailed balance limit with additional losses: (a) ohmic drop and (b) absorption (data from ref. 26) by liquid water.

subcells, however, show a non-linear behaviour: In the limit of very thin water layers, the bottom cell bandgap must be decreased to provide more current, which has to be counter-balanced by an increase of the top cell bandgap to still provide enough photovoltage. The latter measure also reduces current limitation by the bottom cell. Around 0.3 cm thickness, however, the infrared fraction of the solar spectrum is attenuated to an extent, where a further decrease of the bottom cell bandgap does not benefit its photocurrent, which is why the optimum bandgap suddenly increases to about 0.9 eV. Current matching can still be maintained by thinning of the top cell, where the bandgap can now be reduced again with the bottom cell generating more photovoltage. Beyond 1.5 cm of water, the absorption also starts to affect the visible range, decreasing the photocurrent of the top cell. The optimum bandgap of the top cell is therefore further reduced. The loss in photovoltage is partly compensated by an increase of the bottom cell bandgap, partly by the reduced catalytic overpotential at the overall lower current density.

In the case of the metamorphic tandem discussed in Section 12.3.2.2 below, where the bottom cell absorber bandgap is larger than the optimum, a water layer thickness of 2 mm—technically certainly feasible—reduces the current by only 0.1 mA cm^{-2}. In the popular combination $BiVO_4$ (2.4 eV bandgap) with Si (1.12 eV) on the other hand, the top cell is strongly current limiting and it takes 4 cm of water for the current to drop by *ca.* 0.1 mA cm^{-2}. For higher order multi-junction devices such as triple junction cells, where the bottom absorber has a bandgap deeper in the infrared, the limiting impact of a water layer is, however, even more pronounced than in the double-junction case. This can lead to a reverse bias of the bottom absorber, leading to a "double knee"-shape in the I–V curve, where the short-circuit current is given by the top two cells, but the voltage supplied by the bottom junction is lost as soon as it turns current-limiting.[27] While for high-efficiency water-splitting, triple junctions are of limited interest, other solar fuel pathways (in an aqueous electrolyte) with higher electrochemical load could require the high photovoltage of a triple-junction and will therefore suffer more from the infrared absorption of water.

While thin water layers in front of the absorber reduce the absorption, they increase the overpotential from ion transport in the solution. The size of gas bubbles—which can be reduced by surfactants to some extent—creates another boundary condition for the thickness of the water layer. An embedding of counter-electrode and gas-conduction channels in the front window, similar to the contact fingers in photovoltaics, could potentially reduce these effects.[15]

There are two important messages from these observations for the design of efficient water splitting devices: Devices should be benchmarked under realistic conditions (see Section 12.4), as a variation of parameters such as ohmic drop or parasitic absorption by electrolyte and catalyst can significantly change optimum bandgap combinations and potentially require new absorber materials. Furthermore, the adaptation to the working conditions of a solar water splitting device necessitates a precise fine-tuning of opto-electronic properties such as bandgaps and absorber thickness, as is feasible with the III–V semiconductors.[10]

12.2.2.2 Ohmic Resistivity

Ohmic resistivities that stem from the ionic conductivity of the electrolyte, a membrane or from within the solar cell effectively increase the overvoltage required to drive the reaction.[28,29] Unlike the Tafel behaviour of the overpotential from catalysis, these resistivities show a linear increase of the overpotential with current.

While the reduction of solid-state ohmic drops, for instance from the ohmic back-contact between semiconductor and counter-electrode, is greatly reduced by the use of the appropriate contact fabrication technology,[30] the ion transport resistivity in the electrolyte can be reduced by advanced electrode geometries such as supertube arrays or flow reactors.[15,31] If ohmic drops remain high due to cell design or the use of low-conductivity electrolytes, the bandgaps of the absorbers have to be increased to provide sufficient photovoltage: Figure 12.3(a) shows maximum efficiencies for solar water splitting in the detailed balance limit (as Figure 12.1), with the introduction of an ohmic resistivity from solar cell and/or electrolyte. This loss mechanism causes a steep decrease of the obtainable efficiencies, with a drop to about half of the initial efficiency at 60Ω. To provide enough photovoltage for the effectively increased overpotential, both subcells have to increase their bandgap.

Shunt resistances in the device have to be considered as a loss mechanism as well.[32,33] While the typical origin of shunt resistance in solid-state devices, the conduction along grain boundaries, can be considered negligible in high-quality epitaxial III–V solar cells, the processing of samples, especially if the sample area is very small, can reintroduce them. In addition to solid-state shunt resistances, Seitz *et al.*[32] argue that charge-transfer from the semiconductor to the electrolyte can also be an origin of a shunt resistance, effectively decreasing obtainable efficiencies.

12.2.2.3 Non-radiative Recombination

Non-radiative recombination, which originates for instance at growth-induced defects in the semiconductor,[34,35] are a challenge for materials science and can be minimised by adequate absorber growth or the use of nanostructures. For some materials, however, high recombination rates appear to be an intrinsic property of the material, such as in the in principle very attractive dilute III–V nitrides.[36,37] For the classical III–V semiconductors, radiative recombination was greatly reduced over time by advances in epitaxial growth and the development of surface and interface passivation layers, see also Section 12.3.1. It can, however, become an issue for heteroepitaxy of compounds with a large lattice-mismatch or when growing polar III–V's on non-polar substrates. The former challenge can be tackled by metamorphic or inverted metamorphic growth concepts (Sections 12.3.2.2 and 12.3.2.3) or wafer bonding, where two wafers are brought into contact after surface sputtering in vacuum, forming covalent bonds with defects confined to a few atomic layers.[38] Recent efforts regarding the latter are directed towards defect minimisation of III–V heteroepitaxy on Si.[38–40]

Surface charge-carrier recombination in solar water splitting systems is still an issue, often reducing the photovoltage of a device working in photoelectrochemical mode compared to the voltage obtained for the same absorbers in photovoltaic mode. For InP, charge trapping states related to certain oxygen motifs on the surface were predicted by Wood *et al.*[41] Comparing water and oxygen adsorption on clean InP surfaces, it was found that these states originate in oxidation by molecular oxygen, but can be largely avoided by water, forming a different surface oxide species.[42] This was later also confirmed for GaInP by ambient-pressure photoluminescence experiments.[43] Here, a challenge for high-efficiency III–V water splitting systems remains to develop electronic surface passivation layers that are simultaneously stable in the electrolyte, reduce surface charge-carrier recombination and feature a suitable coupling to the electrocatalyst. Phosphates partially fulfil these requirements, but, at this current stage of development, lack long-term stability in the electrolyte.[44]

12.2.2.4 Energetics

Unfavourable energetic alignment of hetero-junctions within the absorber and at the interface to catalyst or electrolyte are a property of the electronic structure of a material or its (oxidized) surface and leads to photovoltage losses. For p-GaP, this photovoltage loss amounts to about 800 mV due to a conduction band offset between GaP and its oxide formed upon contact with the aqueous electrolyte.[8] In the case of the catalyst–semiconductor interface, such losses can in some cases be mitigated by using pinch-off effects at the interface of highly doped semiconductors with catalyst nanoparticles, or surface transformations that lead to a more suitable energetic alignment than the directly oxidised surface.[44,45] Also adequate doping profiles allow, to some extent, for a control of internal band alignment, as well as the energetics between semiconductor and electrolyte.[46] For efficient tunnel junctions between subcells, doping profiles have to be as abrupt as possible. The abruptness can be improved by the selection of an appropriate (*i.e.* III-rich or V-rich) surface termination before switching to the next stoichiometry during growth.[7,47] As the theoretical description of III–V semiconductors is well established, the framework of density functional theory becomes predictive regarding energetics and charge-trapping states,[41,48,49] which promotes fundamental understanding and device design.

12.2.2.5 Light Management

Photon management describes approaches to maximise the overall conversion from impinging photons to current. The main factors here are a minimisation of reflection and transmission losses and the adaptation of the absorber's thickness by thinning. Yet also the refractive indices of the surrounding medium is a parameter that finally impacts the open-circuit voltage as analysed in detail by Létay and Bett for single-junction concentrating solar cells.[50] (In the calculations presented here, we used their idealised assumption of a mirror at the back and an interface to air at the front.) The effect has recently been described as external radiative efficiency[51] and also has to be considered for solar water splitting applications,[33] especially as the surrounding medium of the top absorber is not air, but water.

To reduce reflection losses, anti-reflection coatings (ARC) such as TiO_2 combined with MgF_2 are widely used in photovoltaics. In solar water splitting, the optical interface is typically comprised of an outer cell with a transparent window, a thin layer of water, a catalyst and/or protection layer followed by the absorber. Reflective losses at the quartz-window of the outer cell to air amount to *ca.* 4% of the incident light.[23] Therefore, an ARC has to be applied to both, the outer cell window and the surface of the absorber. Ideally, anti-reflective properties can be achieved by the protection layer or the catalyst. Nanoparticles have been demonstrated to serve this purpose to some extent, reducing the reflection in the water layer, but not necessarily in air due to their different refractive indices.[15]

The above mentioned fundamental efficiency limitation imposed by the solar spectrum and absorption threshold of the semiconductors severely restricts the choice of absorber bandgaps and therefore the materials. This condition can, however, be partially weakened by photon management in the form of thinning: here, the top absorber(s) of the multi-junction cell is

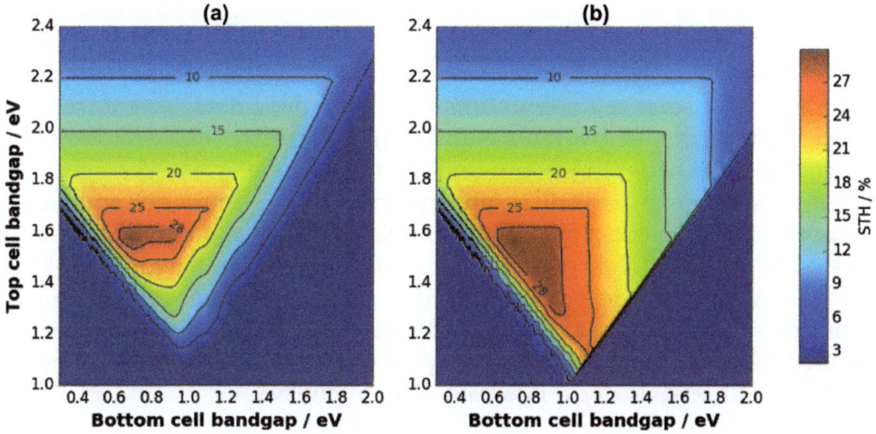

Figure 12.4 Maximum STH efficiencies of a dual junction in the detailed-balance limit with 2 mm water layer, 2 Ω ohmic drop, and IrO_2 as a catalyst. (a) No thinning, (b) with optimum thinning of the top absorber.

thinned below the absorption length to partially transmit photons above its bandgap to the subsequent absorber. Such an approach can be necessary if for instance lattice matching in an epitaxial absorber restricts feasible lattice parameters and therefore stoichiometry and bandgap. It was also used for a recent metamorphic water splitting tandem absorber.[44] Figure 12.4 plots theoretical STH efficiencies as a function of top and bottom cell bandgaps. The comparison of Figure 12.4(a), where no thinning was allowed, with Figure 12.4(b) shows that the region of high efficiencies beyond 20% STH in Figure 12.4(b) is significantly broadened. In the latter case, optimum thinning was assumed, demonstrating that it greatly extends the range of attractive bandgap combinations.

12.2.2.6 Gas Bubble Management

Especially for high-efficiency systems with their high photocurrents and corresponding gas production rates, the formation of gas bubbles becomes a challenge. They will scatter light and reduce the photon flux on the sample, while gas bubbles sticking to the surface reduce the reactive area.[52] The latter factor can be reduced by designing the surface to be hydrophilic. Another effect is mechanical and chemical stress, which was already noted in early high-efficiency systems, where corrosion appeared to be accelerated in areas, where bubble accumulation occurred.[53] This can be partially compensated by the use of surfactants, yet some surfactants such as Triton X can also act as reactants. A 1 M $HClO_4$ solution with 5 mM Triton X develops a yellowish colouring after several hours of water splitting using only Pt electrodes, which indicates that the Faradaic efficiency for water splitting in the solution is less than unity, which can also deteriorate efficiency measurements.

The direction of bubble protrusion is given by gravity, which can render the orientation of the cell, *i.e.* horizontal *versus* upright, a factor for both stability and efficiency. As already noted by Khaselev and Turner,[53] a typical, not fully upright solar cell orientation will ease bubble detachment from the surface and consequently reduce mechanical stress. This will, however, change the angle between photon and bubble path, lengthen the light path subject to scattering (as well as absorption by the electrolyte) and finally reduce efficiency. In concentrating systems, where the position of the sun is followed by means of a tracker, the tilt and therefore the magnitude of this effect will vary over the course of a day. As a consequence, gas bubble management has to be part of high-efficiency cell design, but also has to be considered for efficiency evaluation. A precise quantification of the effect on obtainable efficiencies does, however, remain elusive.

12.2.2.7 Multi-terminal Approaches

In conventional, monolithic multi-junction devices, the current is collected at two contacts at the front and back of the device, for photovoltaic operation at an ohmic solid–solid contact, for photoelectrochemical operation at a solid–liquid contact. This approach has the effect that the current is constant in the whole device and voltages add up. In multi-terminal approaches, these two restrictions are relaxed by the introduction of additional contacts in between the subcells to collect the current. This approach indeed enables high efficiencies in photovoltaics, as for instance the current limitation by the popular Ge substrate, also used as the bottom cell in a four junction device, is lifted.[54] While it could, in principle, also be used in photoelectrochemical devices, for instance by harvesting surplus voltage in the form of electrical power in a hybrid device, this would remove the advantage of the simplicity of an integrated water splitting system.[55]

12.2.2.8 Concentration

Light concentration up to 1000 suns is routinely used in high-efficiency photovoltaics, as it not only shifts the electricity generation costs away from the absorber to the overall system, which benefits pricey absorber materials, but also improves the total efficiency: the open-circuit voltage of a cell increases logarithmically with the light intensity and so does, to a lesser extent, the fill factor. At very high concentrations, charge transport becomes an issue, which is why most concentrating photovoltaic cells show a global maximum of efficiency at concentrations in the order of 200–500 suns. The requirement of direct sunlight not impeded by clouds does, however, restrict the application case for concentrating solar cells geographically.

Also, in solar water splitting, light concentration was already used in the classical GaInP/GaAs tandem (*ca.* 10 suns power equivalent).[53] Due to the benefits for expensive absorbers, such an application is considered the commercial case for III–V absorbers in economic models, with a very high sensitivity of the hydrogen price on the system efficiency.[2]

12.3 III–V Semiconductor Tandem Structures: A Testbed for High Efficiency

The reason for the widespread use of III–V semiconductors in electronic devices such as light emitting diodes, sensors, and solar cells are their highly flexible opto-electronic properties. The magnitude as well as the direct/indirect nature of their bandgap can be tuned over a wide range by combining several elements of group III and V to form binary, ternary or quaternary compounds.[10] The flexibility also holds true for the energetic positions of valence band maximum and conduction band minimum, which determines the band offset of heterojunctions, an important parameter for the design of quantum wells or tunnel junctions.

As III–V semiconductors enable absorbers providing high photovoltages at high photocurrents, photovoltaic or photoelectrochemical devices based on this material class have become a testbed for high efficiencies: single factors such as ohmic resistivity, parasitic light absorption, bubble management or the energetics of semiconductor–catalyst junctions have to be carefully addressed. Lessons learned from the design of these systems will also benefit materials that are currently not as mature, but might show a higher potential for scalability due to lower material or production costs.

12.3.1 History of the III–V Compound Semiconductors and High-efficiency Solar Cells

Earliest studies on solar cells based on the III–V material class date back to 1954,[56] the same year silicon solar cells could already demonstrate 6% conversion efficiency.[57] Only one year later, Gremmelmaier reported 4% efficiency for a single-junction GaAs photovoltaic cell, albeit on a very small area and suffering from a—in comparison to Si—lower available material quality.[58] The potential of binary III–V semiconductors covering a wide range of bandgaps to exploit the solar spectrum more efficiently than Si-based solar cells was quickly discovered,[59] yet bulk material quality as well as surface charge carrier recombination were still an obstacle. The latter was tackled by Alferov *et al.* in 1970, developing the first III–V heterojunction solar cell, where they terminated the GaAs with an $Al_xGa_{1-x}As$ layer.[60] It took another five years until the first concentrating GaAs solar cells equipped with an $Al_xGa_{1-x}As$ window layer surpassed with about 23% reported efficiency at 10 suns the efficiency of Si solar cell technology.[†,61] In the same year, III–V semiconductors appeared for the first time in the context of solar water

[†]It has to be noted, however, that photovoltaic efficiency reports up to the mid/end of the 1970s are prone to relatively large errors due to non-standardised benchmarking routines.

splitting, when Yoneyama *et al.* proposed the use of a p-GaP photocathode in combination with a TiO_2 photoanode for unassisted water splitting.[62] Due to the high bandgap of the TiO_2 absorber, however, already the theoretical limit of this wired multi-junction approach was intrinsically limited to below 2%. A few years later, the first photovoltaic III–V tandem cells were developed which provided an open-circuit voltage of 2 V and would therefore in principle have already afforded unbiased water splitting.[63]

One of the main driving factors for the further development of III–V multi-junction cells was—and still is—high-efficiency photovoltaics in space applications, where the material cost is less relevant. While this potential had been recognised quite early, the heavy weight of GaAs solar cells based on GaAs wafer substrates did, however, initially outweigh the efficiency benefits of III–V solar cells in space.[64] It took until the 1990s for thin-film III–V solar cells to be deployed as power supply in satellites.[65] Advances in metal-organic vapour phase epitaxy, the preparation technique that started to replace liquid phase epitaxy in the 1980s, are nowadays routinely enabling efficiencies of triple-junction solar cells for space beyond 30%.[66] These cells are at the moment mostly based on epitaxial thin films grown monolithically on a Ge substrate. The latter acts as the bottom cell and has the advantage of reduced weight and cost compared to GaAs wafers. Recent trends here replace the Ge with the more abundant and less expensive Si,[38] targeting the contribution of substrates to III–V PV cell manufacturing costs, presently amounting to more than 50%.[67]

Reaching maturity in space, high-efficiency photovoltaics came back to earth in the form of terrestrial concentrator photovoltaics, where the higher sensitivity of the market to materials cost is (partially) compensated by the use of a light concentrating element with a small-area absorber. Here, wafer-bonded quadruple junction cells currently enable efficiencies of 46%,[68] and multi-junction cells of higher order will soon break the 50% mark.

Still lacking a commercial application case, the further development of III–V-driven water splitting was progressing in a comparatively abrupt, non-evolutionary manner. Already in the 1980s, InP photocathodes were demonstrated to be relatively stable and efficient, but only with applied bias due to the low bandgap of InP resulting in a low photovoltage.[69] The first breakthrough for highly-efficient, immersed water splitting came in 1998, when a GaInP/GaAs cell with one buried and one solid–liquid junction, developed at NREL, enabled 12.4% unbiased water splitting under concentrated illumination, a record, which was to remain unchallenged for 17 years.[53] The first full device equipped with a gas-separating membrane, designed at JCAP, employed a commercial GaInP/GaAs photovoltaic tandem from AZUR Space equipped with a TiO_2 protection layer and enabled 10% efficiency.[70] A further step towards the 15% STH efficiency milestone was achieved by a photovoltaic double-junction GaInP/GaInAs core, grown metamorphically on a Ge substrate. Here, photoelectrochemical *in situ* functionalisation provided (electro)chemical and electronic passivation of the immersed surface that was decorated with Rh electrocatalyst nanoparticles leading to 14% STH efficiency under one sun.[44] Less than two years later, the highest STH efficiency benchmark is now 16%, continuing the trend to adopt buried charge-separating junctions, here with an inverted metamorphic GaInP/GaInAs tandem, completed by a sputtered PtRu co-catalyst layer.[46] Up to now, however, all of these systems suffer from a limited stability in the order of 10–100 hrs.

For non-immersed systems, where typically a double or triple junction photovoltaic absorber is coupled to a single electrolyser or an array thereof, 16 and 18% STH efficiency under one sun were already reported in 2001.[71,72] The current benchmarks here are III–V concentrator photovoltaic cells, that are with around 30% STH efficiency under 40–200 suns about twice as efficient as the immersed systems.[5,73]

12.3.2 Highly Efficient III–V Tandem Structures in Solar Water Splitting

12.3.2.1 Classical III–V Tandem Photoelectrochemistry

As high efficiency solar converters are necessary for efficient photoelectrochemical water splitting, it follows that III–Vs would be excellent candidates for water splitting devices. The first reported use of a III–V for water splitting was a tandem consisting of a p-GaP photocathode and an n-TiO_2 photoanode with a reported water splitting efficiency of 0.25%.[74] Later a separated tandem of p-InP as the photocathode and n-GaAs as the photoanode was used and a water splitting efficiency of 8% was reported.[75] However, these devices utilized two semiconductors illuminated in parallel, thus each of their electrodes operated separately. This is not an ideal configuration as the maximum possible efficiency cannot be reached with this design.

In 1994,[76] NREL described the first monolithic III–V tandem solar cell consisting of a lattice matched $GaInP_2$/GaAs p/n, p/n tandem cell. Based on that design, in 1998 the monolithic photoelectrochemical (PEC)/PV device was developed with a p-$GaInP_2$ photoelectrochemical top cell, backed by a p/n GaAs cell. A water splitting efficiency of 12.4% was reported.[53]

12.3.2.2 Metamorphic Device Concepts

Lattice-matched epitaxial growth, where the lattice constant of the epitaxial layer matches the substrate, avoids growth defects originating in relaxation of the top absorber. This restriction to a lattice constant does, however, limit the material choice as ternary or quarternary III–V compounds exhibit stoichiometry-dependent lattice constants with associated bandgaps. The boundary condition can, however, be relaxed by metamorphic growth, where the lattice constant is changed stepwise from the substrate to the subsequent absorber in so-called grading layers. Defects do still arise, but they are limited to dedicated layers with a thickness of the order of 1 μm that are not part of the photoactive region.[77] The cells above the grading layer as well as the window layer are then typically lattice-matched.

With such a metamorphic photovoltaic tandem core, based on GaInP (bandgap: 1.78 eV) and GaInAs (1.26 eV) grown pseudomorphically on a Ge substrate, photoelectrochemical *in situ* modification recently enabled 14% STH.[44] This value was later, in addition to the efficiency measurements at Helmholtz-Zentrum Berlin, confirmed for a different sample at the benchmarking facilities of JCAP.

Here, the highly n-doped AlInP window layer that terminated the GaInP top cell, served as the starting point for further surface treatment. After photoelectrochemical etching, it was functionalised in a single aqueous $RhCl_3$ solution to avoid unfavourable[42] exposure of the III–V surface to ambient oxygen. A pulsed, stroboscopic photoelectrochemical deposition of Rh nanoparticles simultaneously provided the HER electrocatalyst and formed a phosphate-rich surface, which appears to passivate the surface both electronically and (electro)chemically. The resulting device structure is sketched in Figure 12.5.

Originally intended for concentrating photovoltaic operation, the device was optimized to AM 1.5D illumination, with a thinning of the GaInP top cell to alleviate the current-limitation by the bottom cell. Theoretically, the bandgap combination of 1.78 and 1.26 eV could feature 16.9 mA cm^{-2} without thinning and 18.5 mA cm^{-2} with optimum thinning under AM 1.5G, equivalent to 20.8 and 22.8% STH.

Initial potentiostatic catalyst deposition adapted from previous work on InP photocathodes[78] induced strong catalyst absorption due to a

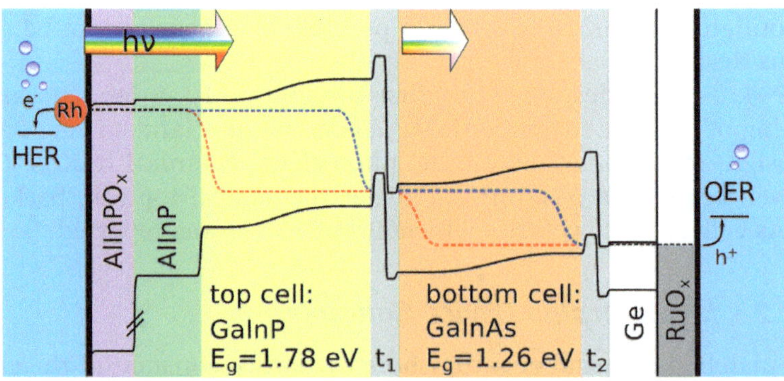

Figure 12.5 Energy schematic of the metamorphic tandem layer structure under illumination. The ohmic back contact is connected to a sputtered RuO_x counter electrode for the oxygen evolution reaction (OER). The hydrogen evolution (HER) takes place at the Rh nanoparticles. Contacts between subcells are facilitated by tunnel junctions, where switching between majority-charge-carrier types occurs. Black, dashed lines represent the Fermi level, blue (red) the quasi-Fermi levels of electrons (holes) and arrows indicate the illumination.
Adapted from ref. 44.

nanoparticle morphology featuring rather large particles. A pulsed (50 ms) deposition with the addition of stroboscopic illumination improved the resulting catalyst morphology and its associated transparency, enabling saturation current densities closer to what was expected for photovoltaic operation with anti-reflection coating.[44] Nanoparticle sizes showed a rather wide distribution, with an average diameter of *ca.* 20 nm.

Fill factor and open-circuit potential were, however, below the values expected from the photovoltaic device, indicating non-ideal energetic alignment and surface charge-carrier recombination. The photocurrent for an unbiased operation under simulated AM 1.5G (Wacom WXS-50S solar simulator, vertical configuration) with a wired RuO_x OER counter-electrode was 11.5 $mA\,cm^{-2}$, with a saturation current of 14 $mA\,cm^{-2}$ (Figure 12.7, green curve). A further decrease of the photovoltage losses associated with photoelectrochemical operation by 250 mV could unlock access to this photocurrent, which would result in 17.2% STH.

The solution for efficiency determination was 1 M perchloric acid with no surfactants and sample sizes were typically around 0.5 cm^2, as defined by the black, high-viscosity Electrolube ER 2162 epoxy. Faradaic efficiency was confirmed to be (near) unity by long-term eudiometric gas collection, finding the expected gas volumes for H_2 and O_2 evolution. The stability of the device was, however, limited with a 50% decrease of the photocurrent over a period of 40 h. One problem here appeared to be the mechanical stress induced by the hydrogen bubbles on the catalyst nanoparticles, as a horizontal configuration showed a tendency towards longer lifetimes.

Figure 12.6(a) shows external quantum efficiencies (EQE) of the tandem operated in photovoltaic mode with ohmic Au front-contacts. After etching of the cap layer, the EQE of the tandem was significantly below the values of the fully processed photovoltaic device (*cf.* ref. 5) due to the lack of an anti-reflection coating (ARC). The deposition of a very high loading of Rh catalyst nanoparticles decreases these values even more due to the introduction of light scattering (in air) and absorption by the catalyst. The scattering in air is evidenced by an optically greyish-turbid appearance.

The former effect is, however, reversed by the introduction of a water layer on top of the surface that, with its higher refractive index, turns the catalyst layer into an ARC, partially compensating for the catalyst absorption.[15,79] An *in situ* measurement of reflectance in the electrolyte before and after catalyst deposition confirms these ARC properties (Figure 12.6b), the reflectivity is strongly reduced. The combined effect of ARC and catalyst absorption is, however, not spectrally uniform and therefore changes the solar spectrum impinging on the absorber. An analysis of relative EQE before and after catalyst deposition suggests that the top cell becomes current-limiting, which should be compensated by a reduction of the top absorber thinning.

Recent progress in the tandem absorber epitaxy led to an increased transparency of the tunnel junctions.[5] This improved the efficiency of the device, which was designed for a concentrating photovoltaic operation, *i.e.* a maximisation of the electric power output at AM 1.5D and not for a

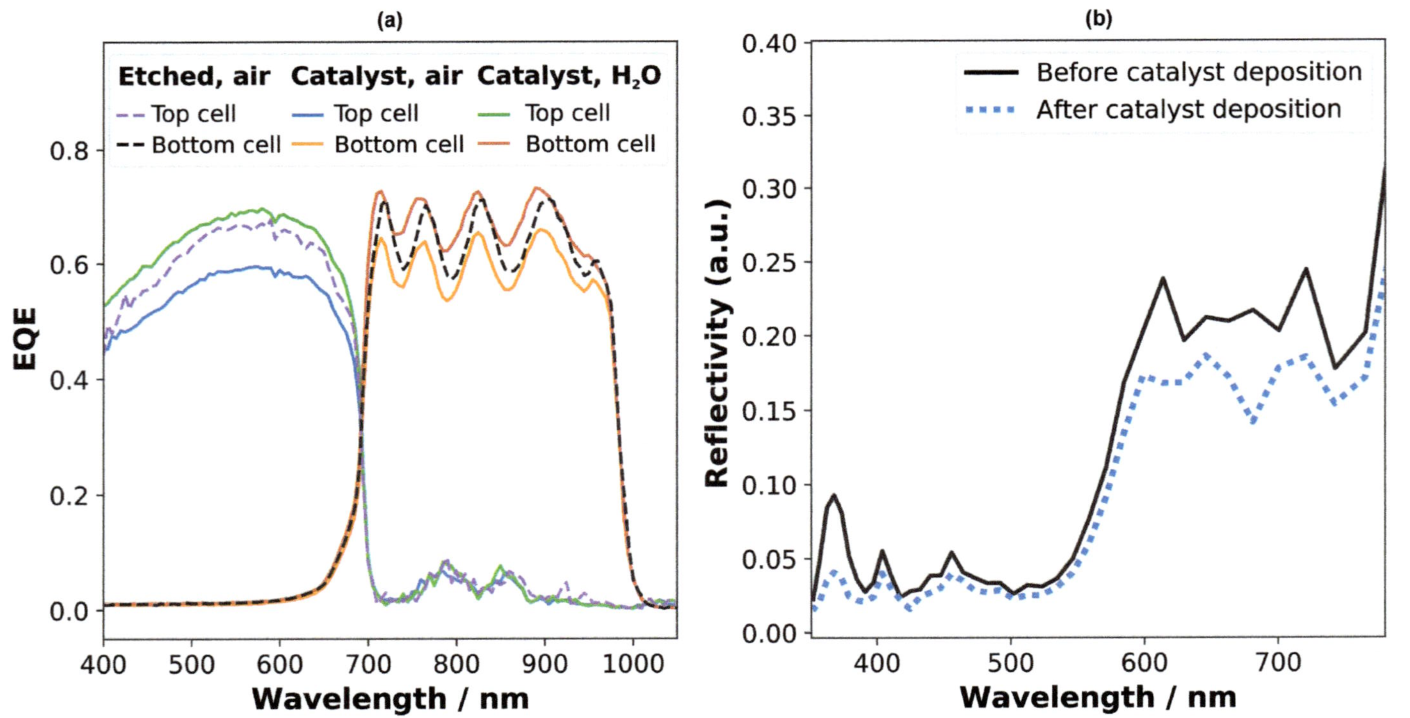

Figure 12.6 (a) External quantum efficiencies of the metamorphic device in photovoltaic operation at different stages of device processing. (b) Reflectance of the tandem in $RhCl_3$ solution before and after Rh nanoparticle deposition.[15]

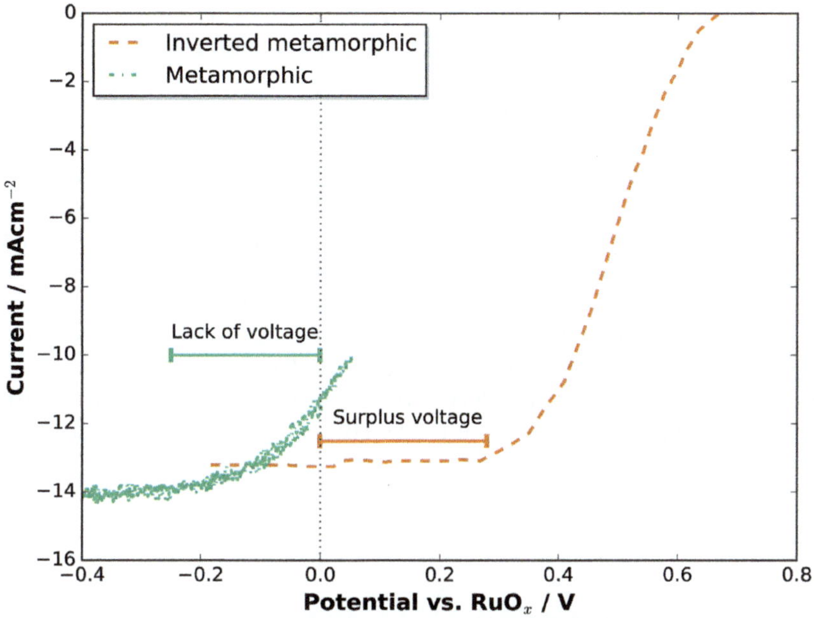

Figure 12.7 Comparison of the I–V curves of the metamorphic (green curve) and the inverted metamorphic (orange curve, Section 12.3.2.3) tandem against a RuO_x counterelectrode.[44,46] While the metamorphic tandem cannot realize its full photocurrent due to a lack of photovoltage, the inverted metamorphic device generates surplus voltage, which would make the use of more earth-abundant catalysts feasible but suffers from higher reflection losses.

maximisation of the photocurrent at a certain threshold voltage as required for solar water splitting. A first step towards an absorber customised for solar water splitting at AM 1.5G comprised an adapted thinning derived from EQE measurements.[15] Combined with a further optimised catalyst deposition on a TiO_2 protection layer, this already promises significant efficiency improvements for this device concept.[80] Further enhancements of the efficiency require the improvement of surface energetics to reduce photovoltage losses as well as the reduction of surface charge-carrier recombination, which then would allow for a further reduction of absorber bandgaps to increase the photocurrent.

12.3.2.3 Inverted Metamorphic Device Concepts

Engineering solutions, such as top-absorber thinning as described above and in Section 2.2.5, aim to mitigate practical STH efficiency limitations imposed by limited design flexibility with a given set of materials. In case of a too small difference in bandgap energy, *i.e.* between the classical GaInP (1.81 eV) and GaAs (1.42 eV) tandem absorbers, the remainder of the solar photon flux reaching the bottom junction does not nearly suffice to match the top absorber current.[81] Thinning represents a practical work-around when limited materials design capabilities are available, though it is inferior to the options of either increasing the top absorber bandgap (for higher voltage at equal current) or decreasing the bottom absorber bandgap (higher current, but decreased voltage).

Figure 12.8 illustrates these design options based on a theoretical detailed balance contour plot (a) and the bandgap *vs.* lattice constant relation for III–V semiconductors (b), which limits their implementation in practice. Here, efficiencies were calculated using a water layer of 2 mm, no thinning, and a constant overpotential of 800 mV to represent losses from catalysis and ohmic drop. Note that the latter condition changes the shape of the isoefficiency contourlines when compared to Figure 12.4(a). The classical GaInP/GaAs tandem features STH prospects of only 15% efficiency (black dot). Both higher voltage and about 19% efficiency would be possible on the

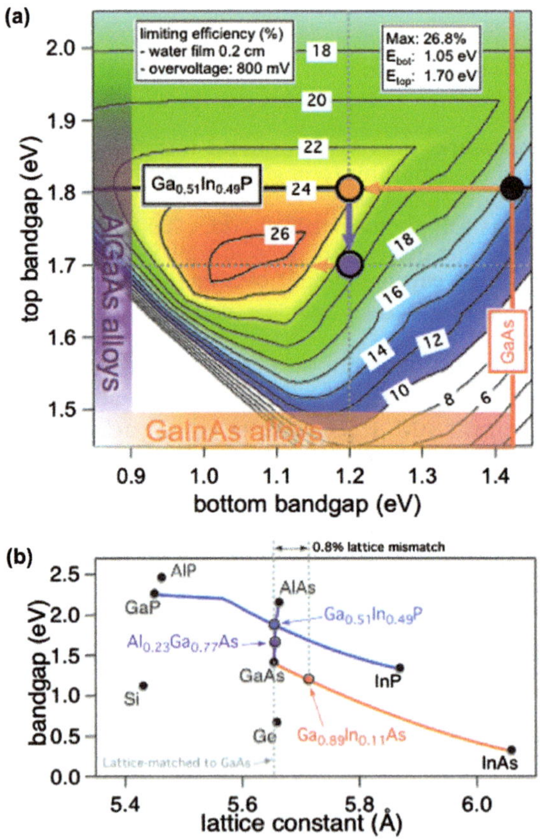

Figure 12.8 (a) Solar-to-hydrogen isoefficiency contour plot and (b) bandgap *vs.* lattice constant for various III–V alloys.
Reproduced with permission from ref. 46, Copyright 2017, Springer Nature.

same GaAs substrate (red line) when implemented with a higher top absorber bandgap. In practice, excessive dislocation nucleation upon reducing the lattice constant during epitaxial growth promotes non-radiative charge carrier recombination and prohibits implementing more Ga-rich GaInP top absorbers to approach this optimum. Note that the calculation in Figure 12.8(a) is based on the assumption of optically thick absorbers. Of course, top absorber thinning may enable a similar redistribution of sub-cell current and 19% efficiency prospects, but with the sacrifice of surplus potential, which is problematic for using abundant catalysts and for generating higher value reaction products. In contrast, a reduction of the bottom absorber band gap (orange arrow) represents the most effective design modification, towards 24% STH prospects given sufficient surplus potential compensating the implied loss of voltage.

The development of metamorphic growth techniques (Section 12.3.2.2) eventually enabled graded buffer layers to change III–V lattice constants during epitaxial growth without prohibitive sacrifice of material quality. Both growth cost and material performance goals still limit the design flexibility. In practice, a single graded buffer stack increases the lattice constant to that of a particular GaInAs composition before growth of the entire tandem structure. Overall, lowered bandgaps harvest more of the solar flux than the classical GaInP/GaAs tandem design, but bandgap combinations remain linked (to vertical combinations of red and blue lines in Figure 12.8b). The principle trade-off described above remains. For instance, the 1.78 eV/1.26 eV combination (Section 12.3.2.2) features off-optimum efficiency prospects of about 20.8% for thick absorbers. Prospects of over 22% are accessible theoretically with a 1.85 eV top junction and practically by top-absorber thinning, with and without the ability to harvest additional potential.

In contrast, full flexibility to combine arbitrary III–V bandgaps according to theoretical prediction represents a major milestone for advancing solar fuel research. Once achieved, theory and hereof derived device-level models, that also include heterogeneous charge-transfer and catalysis, could serve as valuable guidance for advanced device design. Beyond enabling new STH conversion records and structural advances, the transfer of inverted metamorphic multi junction (IMM) growth techniques from PV context[82] finally enables dedicated PEC device design independent of the plain utilization of pre-existing solar cell structures.[46] In contrast, a novel GaInAsP/GaInAs PEC tandem structure transferred back to PV set a new landmark efficiency record of 32.6% for dual-junction solar cells operated without sunlight concentration.[83]

IMM devices are grown inverted with the eventual illuminated side grown first followed by the graded layers and the bottom junction. Substrate removal during post-growth processing allows for the device to be re-oriented. This strategy has a distinct advantage over a more traditional upright growth. Because the top junction generates a higher voltage than the bottom junction—and therefore, a larger fraction of the total power—good material quality in the top junction is essential to high overall efficiency. The inverted-growth architecture allows the lattice-matched top junction to be grown with very few defects and excellent material quality. Deleterious effects, if any, from the dislocations that result from the lattice mismatch are largely confined to the lower-power-producing bottom junction.

After growth, a 1-μm-thick gold film is deposited on the back surface of the device to act as both a rear ohmic/electrical contact and a photon reflector. In the latter capacity, the gold film allows the bottom cell to be thinned by about a factor of two while maintaining the same effective absorption volume, thus enabling a cost reduction. If the junction quality is high enough, the reflector can also enhance the photon recycling to increase the voltage.[84] The device is then bonded to a flat, rigid "handle" to provide mechanical support (such as a silicon wafer[46]) with epoxy and the GaAs substrate is removed, leaving the wide-bandgap top junction as the uppermost layer. For research purposes, the substrate is usually removed by selective chemical etching, but emerging techniques such as spalling[85] or epitaxial lift-off[86] are intended to enable reuse of the GaAs substrate, which is ~100× thicker than the IMM device and one of the largest cost drivers for III–V epitaxy.[87]

Figure 12.9 shows a schematic of the ground-breaking 1.8/1.2 eV GaInP/GaInAs IMM PEC device.[46] The incident solar flux is split between the top GaInP and bottom GaInAs junctions that are series connected *via* a transparent tunnel junction. At the PtRu-modified semiconductor/electrolyte interface, conduction-band electrons reduce protons to evolve hydrogen gas while water is oxidized at a metal anode, producing oxygen and protons. The transparent, AlGaInP-based composition graded buffer layer stack (CGB) is indicated in the Figure 12.9(b) transmission electron microscopy (TEM) cross-section. Dislocations are necessary to change the lattice constant and access lattice-mismatched material, and are observed throughout the transparent CGB. However, the CGB largely confines the dislocations and prevents them from penetrating into the lower junction, thus maintaining device performance.

Figure 12.10 summarizes recent progress towards optimum STH conversion with IMM PEC devices,[46] indicating relevant device structures (Figure 12.10a), associated IPCE (Figure 12.10b), and valid on-sun[81] current density–voltage (J–V) measurements (Figure 12.10c). Six monolithic, III–V tandem water-splitting devices (characterized while coupled to a RuO_x anode) represent several structural advances achieved:

- LM-upright: classical GaInP/GaAs lattice-matched, upright PEC/PV tandem.[53,81]
- LM-inverted: inverted implementation of an analogue GaInP/GaAs device confirming the inverted growth and processing and demonstrating the advantages of the Au back reflector.
- IMM-p: IMM GaInP/GaInAs device with 1.2 eV ($Ga_{0.89}In_{0.11}As$) replacing the GaAs bottom junction.
- IMM-pn: A p–n architecture implemented by a thin (25 nm or less), electrolyte-adjacent n-GaInP layer optimizes photovoltage.

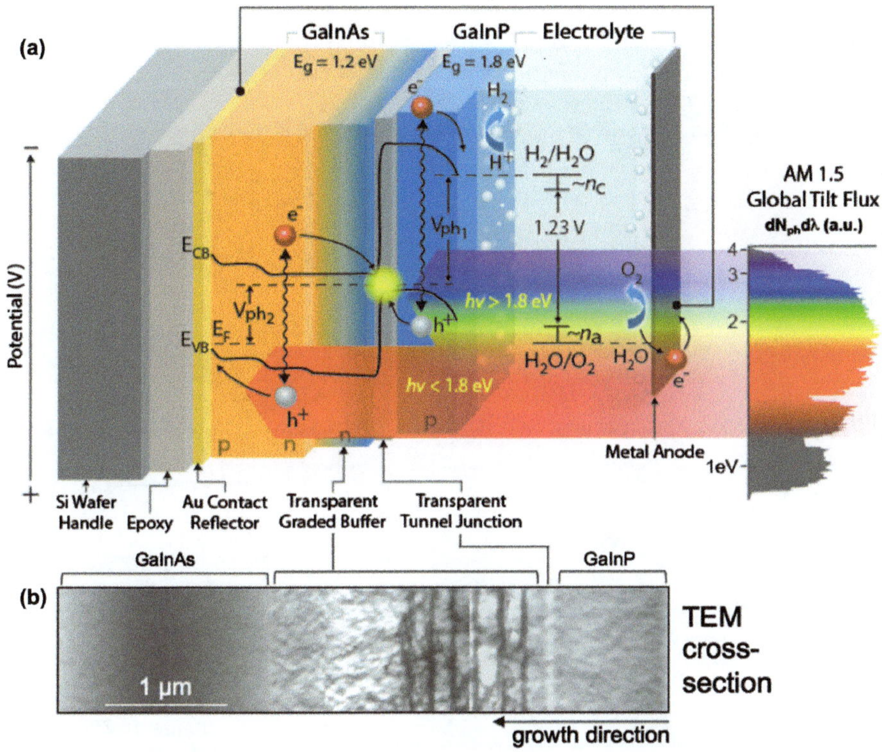

Figure 12.9 (a) Schematic of IMM configured for water splitting, (b) TEM cross-section of IMM device highlighting the transparent, compositionally graded buffer (CGB) layer and the transparent tunnel junction. Reproduced with permission from ref. 46, Copyright 2017, Springer Nature.

- IMM-pnw: Addition of 20 nm-thick n-AlInP surface passivation and 10 nm n-GaInP capping layers.
- IMM-AlGaAs: Proof-of-principle IMM AlGaAs/GaInAs device (replacing the GaInP with 1.7 eV $Al_{0.23}Ga_{0.77}As$) as a first step towards yet higher STH conversion (see Figure 12.8).

The LM-upright structure[81] basically reproduces the classical design[53] and serves as a reference. IPCE characterization (Figure 12.10b) confirms decent external quantum efficiency (grey line) on the order of 60% throughout the relevant parts of the solar spectrum for both GaInP top (≥1.8 eV) and GaAs bottom (≥1.4 eV) junctions. Reflection off the device (and PEC cell) surface already diminishes the solar flux before reaching the absorber layers by about 20% (thin dashed red line), but some room for improvement remains. Integration of the quantum efficiency over the solar irradiance spectrum (light grey area) shows a significant current mismatch between top and bottom junction (11.9 *vs.* 7.8 mA cm^{-2}), in line with unbiased on-sun water-splitting

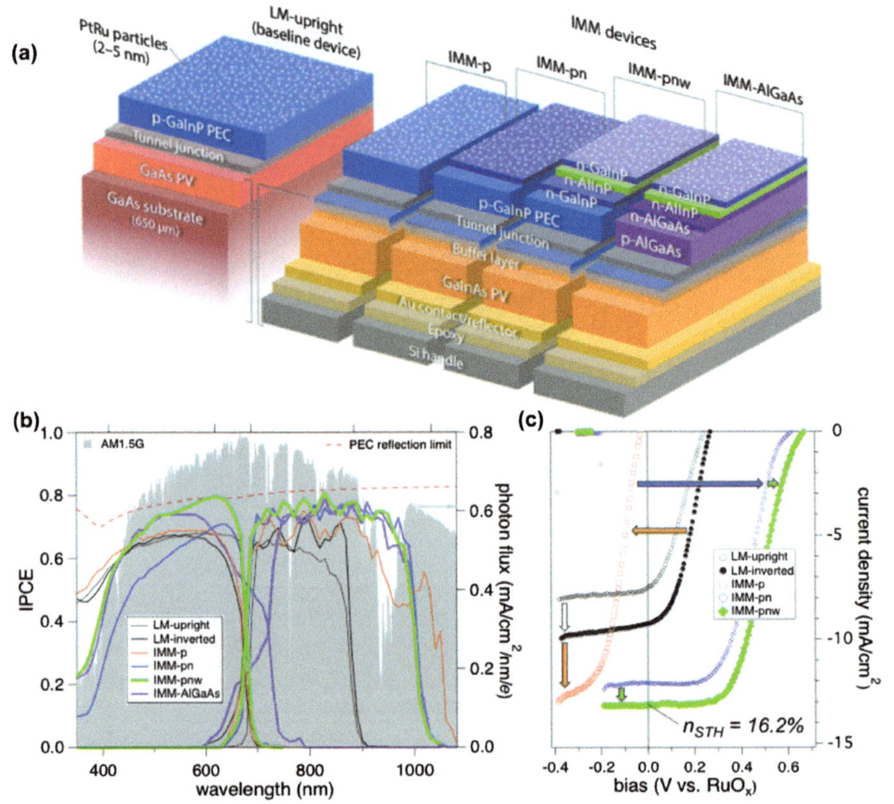

Figure 12.10 (a) Schematic of tandem structure advances, (b) IPCE, and (c) on-sun J–V characterization.
Reproduced with permission from ref. 46, Copyright 2017, Springer Nature.

operation at 7.6 mA cm^{-2} (Figure 12.10c) or about 9.3% STH efficiency. The LM-inverted design consists of identical materials, but highlights a major advantage of the inversion procedure. Substrate removal enables Au back contact deposition directly on the epitaxial GaAs absorber material, where it also serves as an efficient back reflector, effectively preventing photon losses. IPCE (Figure 12.10b, black line) shows significant improvements of the long-wavelength response and interference oscillations. In combination with some current re-distribution by slightly thinning the top absorber, an on-sun performance of 9.3 mA cm^{-2} (Figure 12.10c, black line) or 11.4% STH could be achieved.

The IMM-p device represents a first step towards utilizing the design flexibility enabled by inverted metamorphic III–V epitaxy. The novel bandgap pairing between GaInP (1.8 eV) and GaInAs (1.2 eV) successfully lifts the current limitation by the bottom absorber. Its spectral response extends significantly towards longer wavelengths (Figure 12.10b, orange line) with an integrated current density of 14.8 mA cm^{-2} for the bottom junction. The light-limited photocurrent of the tandem increases to above 12 mA cm^{-2} (Figure 12.10c, orange). However, the photocurrent-for-photovoltage tradeoff (orange arrows) inherent to the lowering of the bottom bandgap resulted in insufficient photovoltage to drive the water splitting reaction (0% STH). This highlights the necessity of understanding the critical design criteria for maximum water-splitting efficiency: photocurrent density (and thus efficiency) is maximized with lower-bandgap energy pairings with the critical constraint of maintaining sufficient photovoltage for water splitting.

Apparently, all previously discussed devices suffer from excessive overvoltage losses. Among various mechanisms the band-edge energetics of the PEC junction contributes a significant share. The conduction band of GaInP aligns about 0.6 V above the hydrogen evolution reaction potential.[88] In principle, hydrogen evolution at high rates requires only about 50 mV overpotential.[89] Due to the corresponding offset in the valence band and bulk Fermi levels, the holes reaching the electrolyte *via* the anode come with reduced (and in the IMM-p case insufficient) oxidative potential for the oxygen-evolution reaction.

The implementation of a p–n homojunction serves to utilize the excess overpotential at the GaInP electrolyte interface for an effectively enhanced and tunable photovoltage. Actually, the IMM-pn device produces a near-optimized photovoltage that is 0.55 V higher (Figure 12.10c, blue arrow) compared to IMM-p. Earlier studies indicate that such a "buried junction" device forms a simple ohmic contact at the electrolyte, but do not consider band bending of the n-type layer.[90] The n-GaInP in the IMM-pn produces upward band bending toward the electrolyte that lowers the conduction band systematically with thickness and doping density, allowing the excess overpotential to be reduced. The IMM-pn device shows unbiased water splitting at 11.5 mA cm^{-2} (Figure 12.10, blue) or 14.1% STH. The IPCE analysis (Figure 12.10b, blue line) uncovers insufficient quantum conversion efficiency of the current limiting GaInP top junction, in particular towards shorter wavelengths. Presumably, the upward band bending in the n-GaInP layer hinders the transport of holes generated near the surface, which are then subject to increased non-radiative recombination.

Integration of a thin n-type AlInP window layer on top of the IMM-pn device results in much better surface passivation, as demonstrated by much improved top junction IPCE of the IMM-pnw device (Figure 12.10b, green line). AlInP has a relatively wide, indirect bandgap and is highly transparent, with its conduction band aligned to that of GaInP to ensure facile electron transport to the electrolyte. The deep valence band provides passivation by blocking minority-carrier holes from surface recombination sites while also counteracting the upward band bending of n-GaInP that occurs when in contact with electrolyte. Because aluminium-containing semiconductors such as n-AlInP are unstable in contact with aqueous electrolyte, the device also includes another thin n-GaInP capping layer for surface protection. Despite some parasitic absorption occurring in the capping layer that mostly affects shorter wavelengths, the design enables a significant net improvement of the unbiased water-splitting performance to 13.2 mA cm^{-2} (Figure 12.10c, green line) or 16.2% STH efficiency. The on-sun characterization of the IMM-pnw device clearly demonstrates its excellent photovoltage, letting it achieve its light-limited photocurrent more than 200 mV earlier than necessary.

The design flexibility of the IMM PEC device concept enables a dedicated trade-off of such surplus photovoltage for higher photocurrent by designing a device with lowered top and bottom junction bandgaps. The IMM-AlGaAs structure represents a proof-of-principle for a first step in that direction. As outlined in Figure 12.8(a), AlGaAs compositions replacing the top junction enable a bandgap reduction targeting 1.7 eV (purple arrow). In a second step, further reduction of the GaInAs bottom absorber bandgap by incorporation of additional In (orange arrow) may restore current-matching and effectively exploit additional solar flux at longer wavelengths. Initially, top absorber thinning serves for current matching instead (Figure 12.10b, purple lines). Due to stability issues, the top junction material might need to be replaced by GaInAsP[91] with, a composition enabling the target bandgap of 1.7 eV.

12.4 Efficiency Measurement and Characterization Strategies

As described above, tandem (or multi-junction) absorber structures constitute a requisite for high efficiency solar energy conversion. In the case of unbiased solar water splitting operation, both modelling (Section 12.2) and experimental implementation (Section 12.3) clearly demonstrate the need for specifically designed dual-junction structures for achieving optimum performance. The increased structural complexity (multiple absorbers as well as their electronic interconnect) triggers a chain of challenges beginning with the dedicated design of optimum device structures and the ability for their experimental implementation—quickly leading to the question of adequate and accurate characterization. Most prominently, solar-to-hydrogen (STH) conversion efficiency serves as the primary figure of merit for both scientific progress and potential commercialization options. Therefore, our discussion of regularly overlooked measurement issues and advisable techniques centres on the ability of determining reliable, realistic, and valid STH efficiencies. Of course, other quantitative characterization methods equally benefit from enhanced measurement accuracy. Without, even qualitative analyses based on sample-to-sample comparison can produce misleading conclusions.

12.4.1 Standard Solar Irradiance *vs.* Laboratory Light Sources

We have already discussed the critical impact of flux distribution within the illumination spectrum on the efficiency prospects and design guidelines for solar fuel devices (Section 12.2), where the parasitic sunlight absorption in aqueous electrolyte and catalyst shaped the effectively available irradiance on the absorber level. In practice, laboratory characterization requires artificial illumination, but the available light sources strongly deviate from the solar spectrum, even in the case of so-called solar simulators. Figure 12.11 displays optical emission spectra of typical illumination sources used for PEC characterization (acquired with a StellarNet spectrometer) compared to solar irradiance (global, air mass 1.5, ASTM G173-3).[81] Note that by analogy with a classical pyranometer (or thermopile) calibration, intensities are set to equivalent power within the relevant wavelength regime (280–1800 nm), which comprises more than 95% of terrestrial solar power (1 kW m^{-2}). Global intensity adjustment remains the only calibration parameter for a given light source, while the mismatch of flux distribution strongly depends on the relevant spectral region, *i.e.* the bandgap and absorption characteristics of the specimen. Both multi-exciton generation and hot carrier extraction remain negligible in classical absorber materials, where each absorbed photon eventually generates a pair of thermalized charge carriers. Hence, the use of photon flux units (rather than power units) in Figure 12.11 provides for direct proportionality to current generation potential at the given wavelength. Spectral integration then yields the theoretical current limit (for unity light absorption and charge-carrier extraction) associated

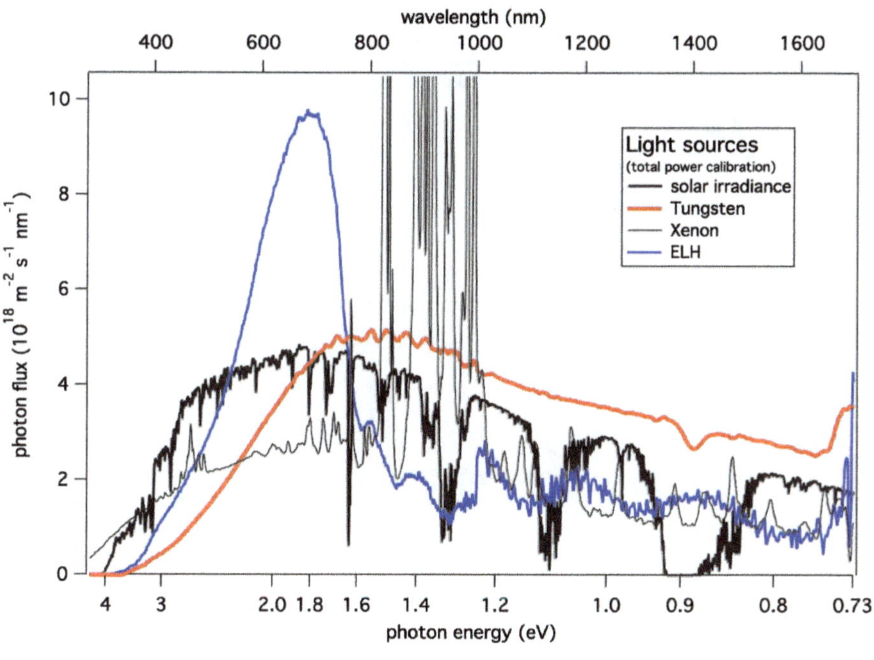

Figure 12.11 Spectral distribution of flux for AM1.5 global irradiance compared to typical PEC laboratory white-light sources (adjusted to provide solar equivalent illumination power).
Reproduced from ref. 81 with permission from the Royal Society of Chemistry. ELH: Tungsten halogen lamp with dichroic reflector.

with each absorber bandgap—an important measure for sub-cell current matching in tandem device design and for overall advances in STH conversion efficiency.

Despite being calibrated to provide solar-equivalent power, all laboratory white-light sources (*e.g.* those shown in Figure 12.11) deviate from the standard solar flux distribution and introduce significant error in absorber current limits for most bandgaps, though recent developments of multi-LED solar simulators promise advances here. For tandem absorbers, this effect is smallest for bandgap combinations that are ideally adapted to the (real) solar spectrum and show perfect current matching. With the number of junctions and the degree of non-ideality, the magnitude of this error will typically increase.

Laboratory light-source calibration approaches often rely on calibrated reference PV devices instead of a spectral power measurement. Adjusting the light-source intensity to achieve the calibration current with the reference solar cell placed at the measurement position removes systematic deviation of the current limit—when, and only when, testing absorbers with equal bandgap. Figure 12.12 exemplifies the approach using a GaInP reference solar cell. At its bandgap energy of 1.81 eV, all light sources produce a

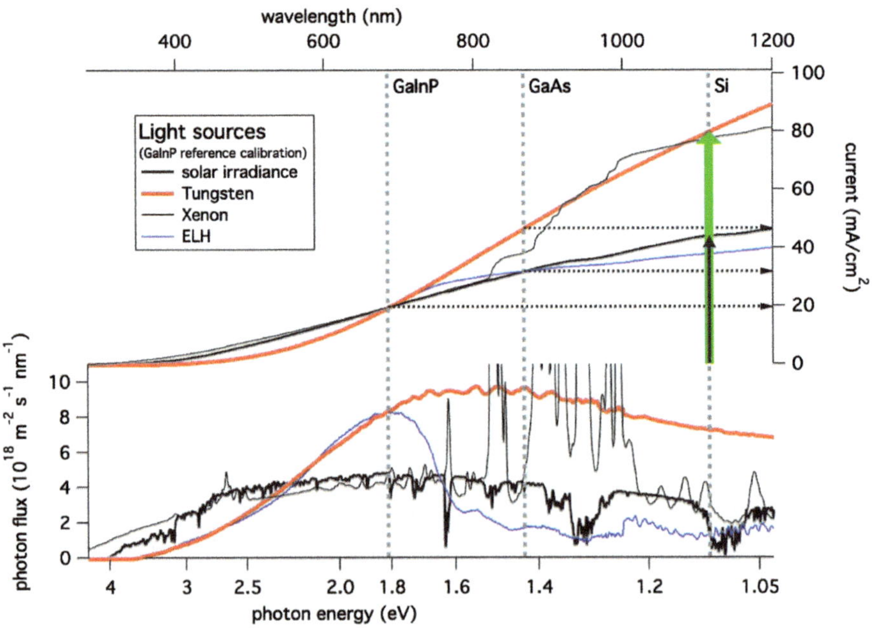

Figure 12.12 Current limitation *vs.* absorber bandgap for AM1.5 global illumination as well as for various laboratory white-light sources (top) based on reference-cell calibration exemplified for an externally calibrated GaInP reference solar cell.
Reproduced from ref. 81 with permission from the Royal Society of Chemistry.

solar-equivalent current-density limit of 19.4 mA cm^{-2}, but the spectral distribution of flux still strongly depends on the source type. The calibrated emission flux of the tungsten lamp actually only equals AM1.5 global solar irradiance around 2.2 eV; the source systematically lacks emission in the ultraviolet (UV) range, but is compensated by an adequate excess of intensity in the visible range. In contrast, the xenon source produces a significant surplus in UV emission as well as excessive characteristic emission lines in the near infrared (NIR) region, particularly in the range of 800–1000 nm (1.2–1.6 eV).

In general, the light-source calibration implies substitution of photon count between different emission wavelengths. Results are only valid in the case where photon absorption and charge-carrier extraction probabilities are independent of the excitation energy. Both the reference cell and the tested material must comply with this idealized absorbed condition (IAC), which is essentially equivalent with an experimental finding of constant IPCE above the respective bandgaps. Explicit IAC violations include advanced absorber concepts (multi-exciton generation, upconversion), molecular materials (dye sensitization, organic materials), and light management (photonic coupling, absorber thinning). In general, non-abrupt absorption edges affect virtually any semiconductor to some extent, but in particular Si and other indirect transition materials. In contrast, high-efficiency water-splitting devices generally require nearly ideal absorber materials, and their particular implementation with advanced III–V semiconductors goes along with good IAC compliance. Nevertheless, the requisite for tandem absorber designs still impedes the ability of straightforward STH efficiency characterization with typical laboratory light sources due to their solar-inequivalent flux distribution.

12.4.2 Tandem Device Characterization: A Case Study on Common Practice *vs.* Result Validation

The impacts of solar-inequivalent laboratory light sources include both absolute misjudgment (often tremendous overestimation, even for single-junction absorbers[23]) of critical performance parameters including STH efficiency and precarious misinterpretation of relative results. For exemplification of these trends in the case of a tandem absorber, we discuss measurements of the classical GaInP/GaAs tandem structure design. In detail, we report on an improved upright epitaxial III–V tandem PEC structure grown by metalorganic vapour-phase epitaxy[76] (Figure 12.13) that closely matches the three main features of the classical GaInP/GaAs design:[53] (i) an optically thick, epitaxial, 4 μm p-type GaInP top absorber ($Ga_{0.51}In_{0.49}P$ composition, 1.81 eV bandgap energy) forming the hydrogen-evolving PEC junction with the electrolyte at its surface; (ii) a buried GaAs bottom junction on a single-crystalline GaAs(100) substrate as joint growth template providing a PV bias to drive the water-splitting reaction without external voltage

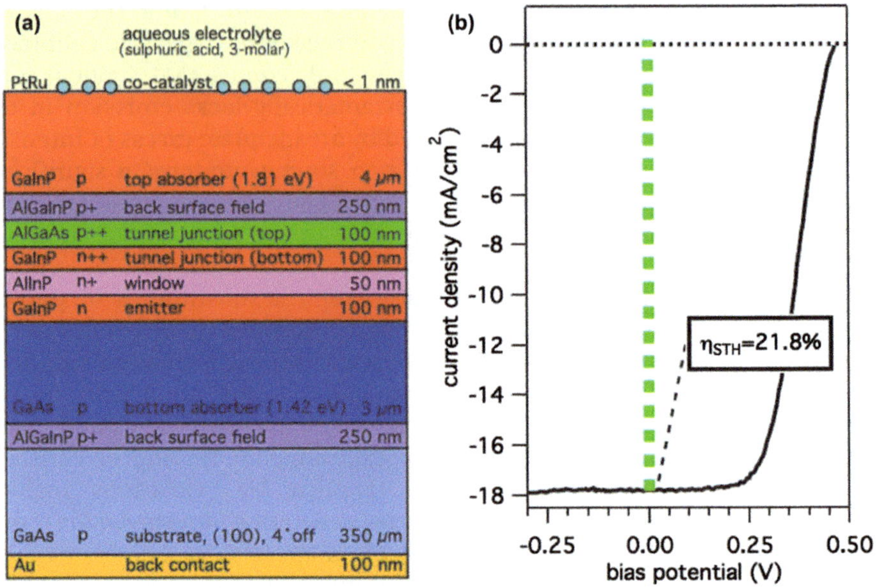

Figure 12.13 (a) Improved upright epitaxial GaInP/GaAs tandem PEC device structure and (b) current density *vs.* voltage characteristic of a 0.085 cm^2 sample under tungsten white-light illumination adjusted with GaInP reference. Here, "η_{STH}" is subject to errors, as discussed in the text. Reproduced from ref. 81 with permission from the Royal Society of Chemistry.

supply; and (iii) a tunnel junction for electrical series-connection of both sub-devices in a tandem configuration. Structural advances of the current design include aluminium incorporation in the tunnel junction, window layer, and back-surface field for better performance and lower parasitic absorption, as well as a heterojunction GaAs bottom cell and a sputtered PtRu co-catalyst. Figure 12.13(b) demonstrates traditional laboratory-based STH conversion-efficiency characterization. In a two-electrode configuration measured *vs.* an IrO_2 counter electrode, we observe a direct water-splitting operation driven by a tungsten white light source (with 3″ water filter) set at a 1 sun intensity (at position A of Figure 12.18) employing a calibrated GaInP reference solar cell. At short-circuit condition (referring to the absence of an external bias potential), we observe a current density of 17.7 mA cm^{-2} (on an 0.085 cm^2 sample), indicating an "STH" (inflated, as discussed below) conversion efficiency of 21.8%.

In comparison, dotted horizontal arrows in Figure 12.12 mark the expected GaInP (19.4 mA cm^{-2}) and GaAs (31.7 mA cm^{-2}) light levels for AM1.5 global illumination. The optically thick top absorber (Figure 12.12a) provides sufficient absorption length to filter all sunlight above its bandgap energy. In principle, the residual bottom-cell illumination light level of 12.3 mA cm^{-2} should then constrain the series-connected device performance as the current-limiting junction. Although the historic result[53] still appears compatible (82%) with that fundamental limitation, our current data indicated a clear violation (144%).

Critical evaluation of the illumination light-source used (Figure 12.12) quickly reveals a significant systematic error in our experimental configuration, causing a drastic overestimation of STH conversion efficiency: the spectral emission shape of the tungsten source causes a vast relative over-illumination of the GaAs bottom junction, associated with an effective bottom sub-cell light level of 27.3 mA cm^{-2} (instead of 12.3 mA cm^{-2} for AM1.5 global; both values subsequent to GaInP filter) when using a GaInP reference cell for intensity adjustment at 1 sun level. In effect, the much lower light level of the GaInP top absorber (19.4 mA cm^{-2}) then imposes the current limitation of the tandem device for the laboratory measurement—in contrast to the situation in actual sunlight. In principle, the light-level ratio of an ELH source as used in ref. 53 appears much better (Figure 12.12), but the spectral distribution strongly deviates from solar irradiance, and we cannot exclude significant calibration error affecting the historic measurements. Note that our tandem result (Figure 12.13) still requires 91% quantum conversion of the incident light in the GaInP top junction.

Even for simpler single-junction configurations, PEC characterization is particularly susceptible to spectral mismatch issues: the high voltage requirement for unassisted water-splitting particularly focuses on high-bandgap materials and, accordingly, the high photon-energy onset of the solar spectrum—despite the minor contribution of the UV region to the total solar current-generation potential. The fundamental deviation of white-light-source emission shapes (Figures 12.11 and 12.12) accentuates the mismatch, with only little chance for suppression through intensity calibration. For instance, xenon-emission-based illumination systems fundamentally over-supply the UV range, whereas tungsten sources fall short in that region. Popular solar simulators consist of xenon arc lamps with specific optical filters for spectral modification. Figure 12.14 displays the transmission spectrum of the widespread Oriel AM1.5G filter (yellow), featuring a steep UV cut-off around 3.6 eV, a broad attenuation feature in the near IR (between 1.0 and 1.6 eV), and more or less constant transmission (close to unity) anywhere else. Its application to the native xenon lamp spectrum (grey, various calibrations) creates a somewhat closer match (green) to the AM1.5 global solar irradiance standard (black): still, flux distributions only agree well in the visible range, but both UV cut-off and near IR attenuation induce a certain balance of photon surpluses and deficits within either region. Among a pyranometer, Si and GaInP references, intensity calibration deviates only marginally (by a few %). Also, the light levels for individual absorbers (*e.g.* 102.6% for GaInP or 98.6% for GaAs, both based on Si reference) appear fairly reasonable. In contrast, the light source configuration fixes incorrect light level ratios for tandem devices, such as 1 : 0.895 for our exemplary GaInP/GaAs structure. Intensity calibration may only decide whether to over-/undersupply the top/bottom absorber by this factor, but the overall performance of tandem

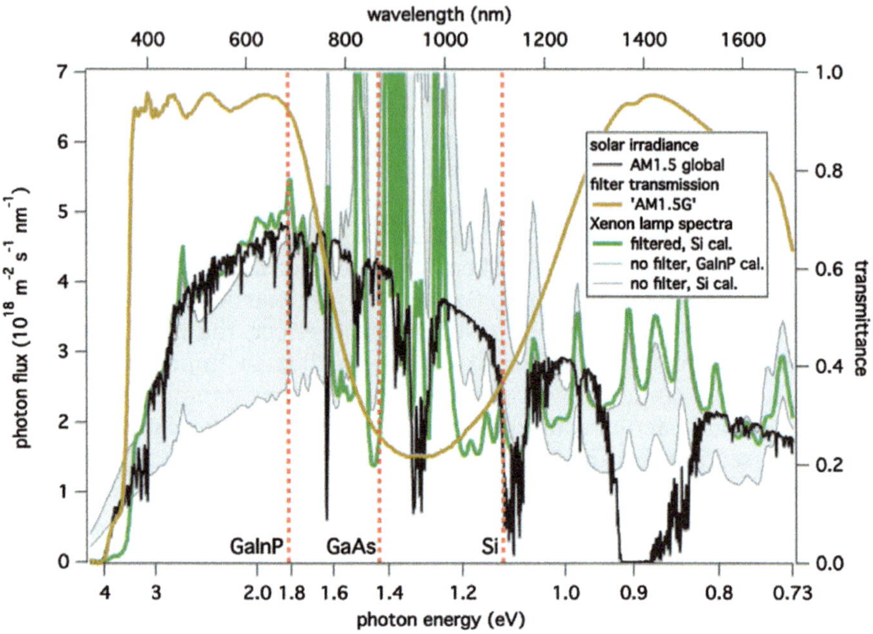

Figure 12.14 Transmission spectrum of an Oriel AM 1.5G optical filter (yellow) as well as filtered (green) and unfiltered (grey) xenon arc lamp spectra (various calibrations) in comparison to AM 1.5 global solar irradiance. Reproduced from ref. 81 with permission from the Royal Society of Chemistry.

absorbers critically depends on current matching between the sub-devices. Hence, both absolute STH efficiency figures and qualitative insights for advanced tandem device development may be flawed significantly.

12.4.3 Utilization of Natural Sunlight and Secondary Illumination Errors

For minimizing the spectral-mismatch error, outdoor PEC characterization was conducted under sunlight illumination. Of course, both the intensity and spectral composition of the terrestrial sunlight spectrum depend strongly on factors such as season, angle of incidence, air pollution, and altitude. All results shown in Figure 12.15 were measured in Golden, CO, USA, in the direct vicinity of the NREL Solar Radiation Research Laboratory, where precise data on actual solar irradiance are continuously recorded and published.[92] All the data were normalized following the established PV convention of assuming 1 kW m^{-2} intensity as a 1 sun illumination level regardless of the applied spectrum.[93] The spectral mismatch between the applied sunlight and the solar irradiance standards was mitigated by appropriate timing (to ensure nearly AM1.5 global conditions). It still remains a

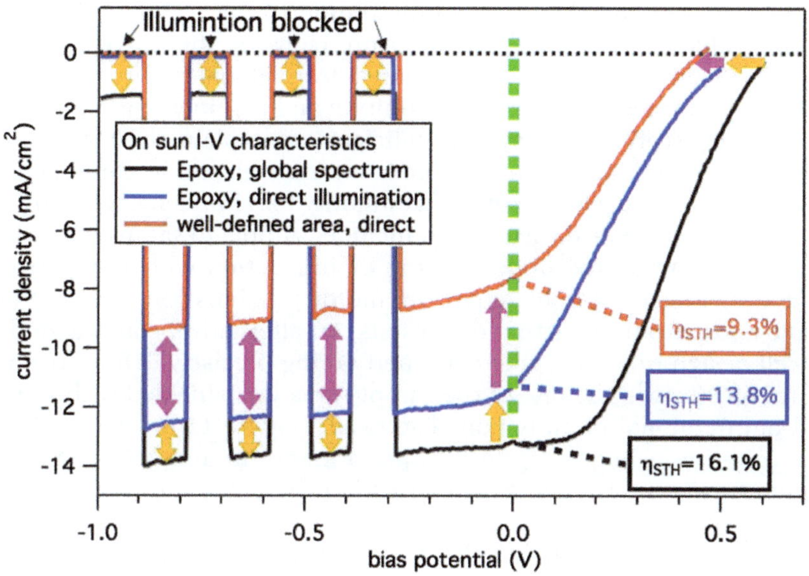

Figure 12.15 Water-splitting characteristics measured with actual sunlight (outdoors) for standard epoxy PEC tandem cathodes *vs.* IrO_2 anodes with normalized global (black) and direct (blue) illumination in comparison to proper area definition (red); current density at zero bias (green dashed line) is considered as an indication of STH conversion efficiency η_{STH}, whereas only the red curve is a real η_{STH} value. Reproduced from ref. 81 with permission from the Royal Society of Chemistry.

potential source of error (less than 5%[23]), but of negligible magnitude at the present accuracy level of PEC STH efficiency determination.

In a first attempt for the structure shown in Figure 12.13, we found unbiased water-splitting operation driven only by sunlight to occur beyond 13 mA cm^{-2} current density for a calculated 16% STH conversion (Figure 12.15, black line). The values fell below the initial laboratory characterization result (Figure 12.13b), but still greatly exceeded experimental limits derived from quantum conversion reference characterization obtained with incident photo to current efficiency measurements (IPCE, see below). Apparently, additional factors contributed to STH overestimation beyond an inadequate choice of illumination source. Significant current levels while blocking the direct light path (Figure 12.15, black line) confirmed indirect device illumination that is part of the global solar spectrum, but the effect could be artificially enhanced by the glassware instrumentation surrounding the PEC sample. Indirect light paths were excluded by using a dark compartment around the PEC cell, where the device is exclusively illuminated through a collimating tube[94] designed to restrict the incident sunlight to the direct and circumsolar (5° field of view) portion. In that configuration, the current density above 11 mA cm^{-2} (Figure 12.15, blue line) indicates almost

14% STH conversion. Note that normalization to the 1 kW m^{-2} 1 sun convention[93] enables the direct comparison between the current densities shown in Figure 12.15, despite the change from global to direct irradiance.

Inappropriate active-device-area definition and confinement remained the last area of possible experimental deficiency to explain the residual overestimation of PEC performance. Epoxy encased electrode construction is common throughout the scientific community for simplicity, flexibility, and corrosion resistance. Downsides can include: sample-to-sample variation of device area, unexpected under-etching or interaction with the electrolyte, and optical impact of light reflection and/or transmission. In an effort to exclude epoxy from impacting our results, an alternative, compression-type PEC cell design was used, where an inert O-ring precisely defines the active area (of 0.185 cm^2); the rest of the sample area was shielded with Al foil to avoid unintentional illumination. The result (Figure 12.15, red line) might be perceived as a huge drop in PEC performance to about 7.6 mA cm^{-2} in current density or 9.3% in STH conversion. Actually, the values show remarkable consistency with experimental current limitations estimated by IPCE (Figure 12.16). Pragmatically, only that final number truly represents the STH energy-conversion efficiency (unity Faradaic efficiency provided, *i.e.*, absence of sacrificial reactions, to be confirmed by consistent gas chromatic

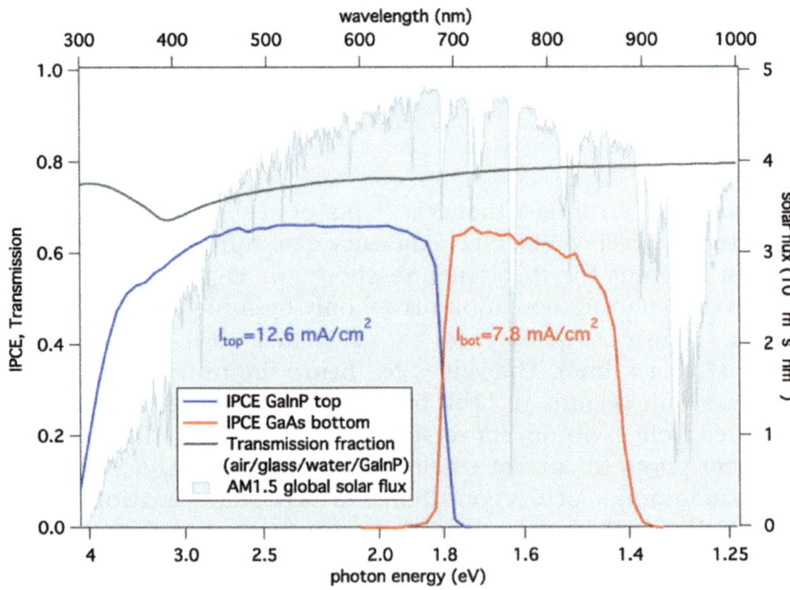

Figure 12.16 Independent IPCE characterization of GaInP top (blue) and GaAs bottom (red) absorbers of our PEC tandem device acquired with appropriate bias illumination; modelled transmission through air/glass/electrolyte/semiconductor interface system (dark grey); and AM1.5 global solar flux (light grey area).
Reproduced from ref. 81 with permission from the Royal Society of Chemistry.

product analysis), whereas the device performance was significantly overrated in all earlier measurements due to uncontrolled systematic errors.

In total, typical—but inappropriate—PEC testing techniques led to a major inflation of the estimated STH performance—in this case, to more than double (17.7 *vs.* 7.6 mA cm^{-2}) the correct value, as obtained with advanced techniques inspired by multi-junction PV characterization. Beyond primary illumination errors (regarding the light source spectral mismatch and intensity calibration), secondary deficiencies may also contribute significantly to overrated performance observations. In order to understand the insufficiencies related to epoxy-encased electrodes, we measured the spectral transmission properties of our exemplary Loctite HySol 9462 epoxy (Figure 12.17) that we used for laboratory-scale PEC electrode assembly. The inset in Figure 12.17 illustrates the assembly concept: rectangular semiconductor absorbers usually receive an external electrical back-contact with Ag paint and Cu wire before the entire device, except the central area of the front surface, is coated by a chemically inert epoxy to prevent contact with the reactive electrolyte during operation. Epoxy layer thicknesses of the order of 1 mm ensure proper illumination-area definition because the thinnest

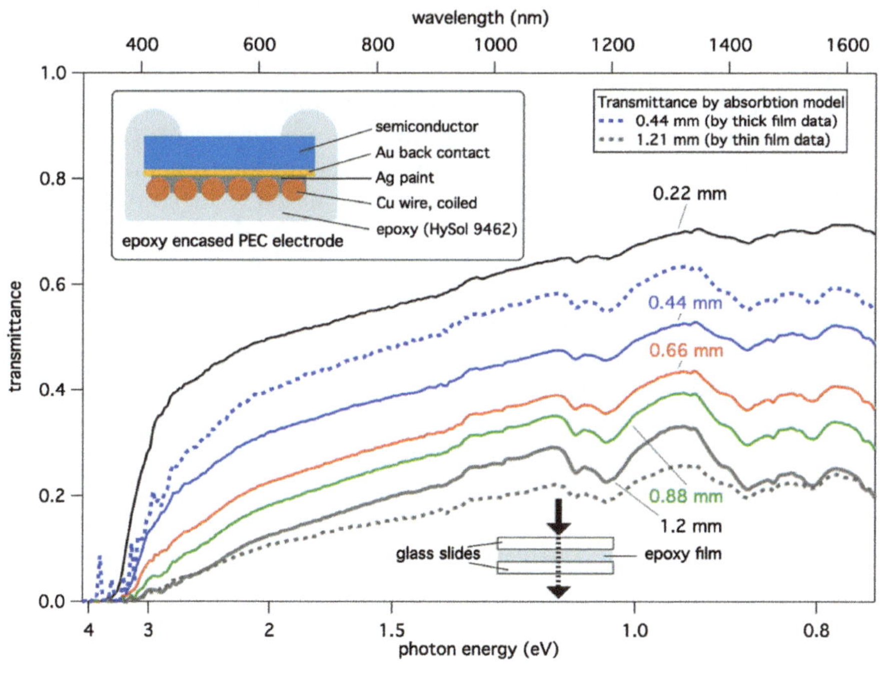

Figure 12.17 Spectral transmission through Loctite HySol 9462 epoxy films confined between glass slides, as well as schematic diagram (inset) of the epoxy-encased PEC electrode manufacturing concept.
Reproduced from ref. 81 with permission from the Royal Society of Chemistry.

layers (0.1 mm) appear optically semi-transparent. The optical transmission spectra of defined epoxy films confined between two glass slides (Figure 12.17) demonstrate significant penetration even through thicker layers. In a simple absorbance model (Figure 12.17, dashed lines), transmission of the air/glass/epoxy/glass/air layer stacks should only depend on reflection of interfaces (constant) and absorption in the epoxy (exponential decay). Calculations do not reproduce the data, but indicate sub-exponential behaviour consistent with significant translucence of the epoxy, probably based on scattering of incident light.

The consequence is an imprecise area definition for PEC devices, where the epoxy only defines the chemically active surface area, but also allows for partial illumination of the covered area (Figure 12.17, inset). The contribution to device performance remains largely unpredictable because epoxy thickness and the fraction of covered area vary. Additionally, the lateral collection of charge carriers plays a significant role. Scanning light-spot analysis could provide the means to quantify this error,[95] but analysis must take into account that the reaction area is larger than the illuminated area, which benefits catalysis. Based on higher light transmission and better charge-carrier transport, the limiting GaAs bottom junction might benefit more than proportionally in epoxy-encased electrode configurations. Both effects supposedly contribute most for small, research-scale samples. An active area of about 0.085 cm^2 was used for epoxy electrodes represented in this study. A brief statistical survey over the range of 0.02–0.06 cm^2 samples (~20 pieces) from the same growth run did not show an evident correlation between active area and observed light-limited photocurrent. Other epoxy types provide more desirable light-transmission properties, but complete opacity to the full solar-irradiance spectrum remains to be confirmed. The error for the area definition can be greatly reduced by the use of black high-viscosity epoxy (*e.g.* the Electrolube ER 2162 or 1455 used in ref. 44), which enables steep edges to minimise imprecise areas due to edge creep. Larger-area samples, where only a small fraction of the absorber surface is covered by the epoxy, further reduces the error of area associated with a partially transparent epoxy. The latter approach also reduces the dark current, which benefits photovoltage and consequently efficiency. For more discussion on various epoxy-related issues, see Chapter 11, "Prototyping development of integrated solar-driven water-splitting cells" in the present book.

In general, the light path between the chosen illumination source and the defined illumination error constitutes another source of potential measurement error and possible efficiency overrating. The contribution of unintended indirect light paths inducing an illumination (and according photocurrent) surplus was described above. Our solution (dark compartment and collimating tube) implies the switch to direct irradiance in good consistency with the most likely application scenario of highest efficiency water splitting devices. Operation under mild to intermediate sunlight concentration may help to compensate rather high materials and production costs. The outdoor characterization approach with natural sunlight circumvents not only the spectral mismatch of laboratory light sources, but also issues related to the divergent nature. In combination with the specific PEC operation scenario (within an electrochemical cell, made from glass or with glass windows, filled with aqueous electrolyte) potential error mechanisms regarding the light source intensity calibration and unintentional illumination concentration require consideration.

In principle, light level adjustment based on an appropriate PV reference device promises the highest accuracy when done in place, but in the absence of the PEC cell (position A in Figure 12.18). In contrast, intensity calibration in dry (air-filled) or wet (electrolyte-filled) *outer* PEC cells (position B in Figure 12.18) is subject to additional inaccuracies and possible systematic performance inflation. Neglect of the reflective loss at the additional air–glass glass interfaces (approximately 4% per pass) during light source intensity adjustment (based on achieving the calibration current of the PV reference cell) will systematically produce a higher (not solar-equivalent) flux level. The error can, however, be reduced by employing filters in

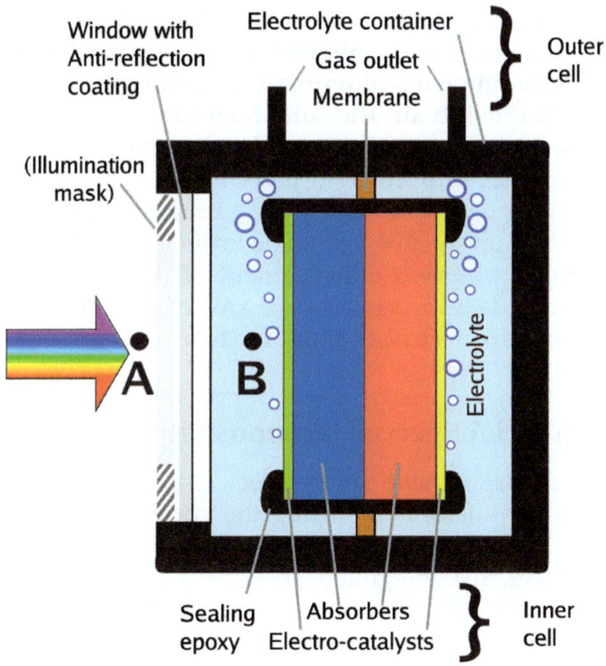

Figure 12.18 Sketch of a photoelectrochemical cell for efficiency validation with the two points for spectral calibration, A and B. While B reduces errors from global illumination, A is the more realistic point of reference as long as cell components do not act as light concentrating elements. If effects of global illumination and diverging light cannot be excluded, a mask for the definition of the illuminated area has to be used. Adapted from ref. 15 with permission from the Royal Society of Chemistry.

combination with the reference cell, as discussed in ref. 96. For tandem cells, this requires a number of calibrations equal to the number of subcells, each with a suitable combination of filters in front of the reference cell to mimic the transmission of the overlying solar cells as well as a filter for the spectral range between the absorption onset of the reference cell and the bottom solar cell. This multi-step calibration requires a tuning of the illumination source irradiance in the spectral regions of the subcells independently. While this is relatively straightforward for multi-source (*e.g.* LED-based) solar simulators, it requires the introduction of additional filters for single lamp sources, where only the overall intensity can be adjusted.

Potential unintentional performance record inflation might be highest when filling the PEC cell after calibration, as the lowered refractive index offset will effectively reduce the reflective loss then. Neglect of the parasitic light absorption in the electrolyte (see Section 12.2) may affect calibrations done in the presence of the electrolyte. An artificially overrated light level would compensate the lost photon flux after a simple adjustment to the calibration current. Common Si-based PV reference devices appear most susceptible, as light absorption in water is most relevant for longer wavelengths. Despite an appropriate intensity calibration, the divergent nature of artificial light sources in combination with a typical cell might finally introduce a slight unintended illumination concentration. Compared to the calibration ideally done in air, the subsequent installation of a typical PEC set-up introduces the glass window and the electrolyte materials with much higher refractive index into the light path. In consequence, the whole configuration effectively acts as a weak lens.[96] Even high-end solar simulators typically specify a 4° divergence half angle artificially inflating the light level and resulting in an observed enhancement of the photocurrent by about 10% in our example (assuming an active area of 0.16 cm^2, a window thickness of 0.2 cm and an electrolyte film of 0.2 cm).

12.4.4 Differential Spectral Responsivity

Another approach for reliable solar-to-fuel efficiency evaluation is to use multi-source sun simulators employing the spectrometric characterisation technique, the most common approach used in multi-junction photovoltaics.[15,97,98] Here, the integrated spectral responsivity s_j of a subcell j under a light source i with the spectrum e_i has to deliver the same current as under the absolute spectral irradiance of reference spectrum e_{ref} (AM 1.5G), subject to a correction factor A_i:

$$\sum_{i=1}^{N} A_i \int_{\lambda} s_j(\lambda)e_i(\lambda)\mathrm{d}\lambda = \int_{\lambda} s_j(\lambda)e_{\mathrm{ref}}(\lambda)\mathrm{d}\lambda; \quad j=1,2\ldots N \tag{12.1}$$

The spectral responsivity is measured by a variation of the light source, which in turn is calibrated with a suitable reference cell. The correction factor A_i, obtained from the linear equation system in eqn (12.1), gives us then the correction we have to apply to the ith of the N subcells.[97,98] Another option would be the extension of the differential spectral responsivity scheme[99] to multi-junction absorbers replacing the white bias light with monochromatic light sources addressing each subcell. This has become feasible with the availability of sun-simulators that employ a large number of independently adjustable LED's. To probe $s_j(\lambda)$ of a subcell, the bias light of the complementary subcells is set to render the probed cell current-limiting, yet still at near-operational overall current. Then, the differential spectral responsivity (DSR), $\tilde{s} = \Delta I/\Delta$, is measured with a monochromatic, modulated probe beam employing a lock-in amplifier. An integration of $\tilde{s}$ allows us to obtain $s(\lambda)$. With this factor, the response of a subcell is again corrected for the mismatch between actual and simulated AM 1.5G spectrum according to eqn (12.1).

For solar water splitting, conditions near the operating current are necessary because of the non-linearity of catalyst overpotentials and as the optical properties of the device are likely to change because of gas evolution and potential changes in electrolyte composition. The approach has the advantage, that violations of IAC (see above) do not deteriorate the result. Filtering out the signal of the modulated light source at high bias current is, however, a difficult task with the commonly used monochromated lamp sources because of their low spectral irradiance that renders signal recovery almost impossible. Slow charge-transfer kinetics in a water splitting system, which can increase the response time of the system to a change in illumination, aggravate the difficulty. In a recent attempt for photovoltaics, the lamp source has been replaced by a femtosecond laser system with an optical pulse-to-CW converter to increase the spectral irradiance by a factor of about 1000.[100] Even with such a light source, the measurement of the spectral responsivity in a photoelectrochemical environment remains a challenge.

12.4.5 Solar-to-hydrogen Conversion Reference Laboratories

We propose applying the following standards for future PEC performance reporting: (i) traceable disclosure of the illumination-source configuration (lamp, filters, optics, PEC configuration, calibration routine) and/or its measured spectral distribution; (ii) thorough device-area definition (including confinement of the illumination area and avoidance of indirect light paths); (iii) complementary IPCE confirmation of the solar-generation potential or, alternatively, DSR by use of multi-source solar simulators; and (iv) proper consideration of Faradaic efficiency. In the long term, only standardized and validated PEC testing and STH efficiency determination techniques will provide a credible, objective base for scientific progress and technological deployment of solar water-splitting devices for solar fuel generation. The formation of acknowledged reference laboratories analogous to the PV community should be considered.

Stability reporting (see also Chapter 8 "Degradation of semiconductor electrodes in photoelectrochemical devices") forms an intrinsic part of PEC performance evaluation, as reported efficiencies are always a time-average, with most devices exhibiting only a limited lifetime. One way to include stabilities in an easily accessible way could be to report the time, *t*, over which the efficiency, STH(*t*), was averaged, in the form of an index, *i*, giving the logarithm of the averaging time in seconds:[15]

$$\mathrm{STH}_i = \frac{1}{10^i}\int_0^{10^i} \mathrm{STH}(t)\mathrm{d}t$$

Here, an efficiency of STH_0 corresponds to an averaging over 1 s, while the ultimate goal for applications would be several years, *i.e.* STH_8.

In benchmarking η STH (or STF for products beyond hydrogen), the protocols should identify and mitigate critical influence factors that introduce measurement errors. We emphasise that only unbiased (0 V *vs.* the counter-electrode), *i.e.* spontaneous fuel-forming conditions, can be considered for the reporting of η STH/STF efficiencies. Sample areas should be large enough to minimise errors from imprecise definition of the electrochemically active/illuminated area as well as to realistically account for solution resistivity and the sensitivity of the device performance to pinholes. Our benchmarking protocol includes: (1) on-sun (*i.e.* using natural sunlight), direct-only (diffuse excluded) solar illumination as necessary for accurate and precisely defined illumination, that also mitigates optical concentration effects inherent to diverging-source solar simulators, (2) incident photon-to-current efficiency (IPCE) or differential spectral responsivity (DSR) measurements for spectral correction, translation of performance measurements to reference conditions such as AM1.5G, and η STH validation including active-area definition, and (3) confirmation of Faradaic efficiency.

12.5 Summary and Outlook

In summary, we have shown that high-efficiency water splitting systems require tandem absorbers specifically adapted to the necessary boundary conditions such as catalysis and cell design. With their flexible optical and electronic properties, combined with mature manufacturing technology, the III–V semiconductors allow for the implementation of highly efficient photoelectrochemical systems. Current work is dedicated to pushing experimental demonstrations further towards the theoretical limits. Depending on the system, this requires a conversion of surplus voltage to current by shifting bandgap combinations, a reduction of reflective losses or photovoltage losses related to energetics and recombination, or—in most cases—a combination of all of them.

Efforts in light concentrating cells, heteroepitaxy of thin-film III–V on Si substrates or inverted metamorphic growth with substrate reuse have the potential to significantly reduce the impact of absorber costs on the overall solar hydrogen price. Also wafer bonding—combined with substrate lift-off—is another emerging technique to remove the restriction of lattice constant-matching or the need for grading layers and therefore promises to benefit efficiencies due to an enhanced flexibility. Yet even if other absorber systems turn out to be more cost-competitive, the III–V systems currently serve as a testbed for high-efficiency water splitting in general, with lessons to be learned for catalyst requirements, cell design, and efficiency validation.

Acknowledgements

MMM acknowledges funding from the fellowship programme of the German National Academy of Sciences Leopoldina, grant LPDS 2015-09. HD acknowledges support by an EU Marie Curie Fellowship (IOF no. 300971). JAT acknowledges support from the US Department of Energy, Office of Science, Office of Basic Energy Sciences, Solar Photochemistry Program under Contract Number DE-AC36-08GO28308.

References

1. A. J. Bard and M. A. Fox, *Acc. Chem. Res.*, 1995, **28**, 141–145.
2. B. A. Pinaud *et al.*, *Energy Environ. Sci.*, 2013, **6**, 1983–2002.
3. I. Austin and N. Mott, *Adv. Phys.*, 1969, **18**, 41–102.
4. A. Nakamura, Y. Ota, K. Koike, Y. Hidaka, K. Nishioka, M. Sugiyama and K. Fujii, *Appl. Phys. Express*, 2015, **8**, 107101.
5. J. Ohlmann, J. F. M. Sanchez, D. Lackner, P. Förster, M. Steiner, A. Fallisch and F. Dimroth, *AIP Conf. Proc.*, 2016, **1766**, 080004.
6. A. C. Nielander, M. R. Shaner, K. M. Papadantonakis, S. A. Francis and N. S. Lewis, *Energy Environ. Sci.*, 2015, **8**, 16–25.
7. T. Hannappel, M. M. May and H.-J. Lewerenz, in *Photoelectrochemical Water Splitting: Materials, Processes and Architectures*, ed. H.-J. Lewerenz and L. Peter, The Royal Society of Chemistry, 2013, pp. 223–265.
8. B. Kaiser, D. Fertig, J. Ziegler, J. Klett, S. Hoch and W. Jaegermann, *ChemPhysChem*, 2012, **13**, 3053–3060.
9. D. K. Zhong, S. Choi and D. R. Gamelin, *J. Am. Chem. Soc.*, 2011, **133**, 18370–18377.
10. I. Vurgaftman, J. R. Meyer and L. R. Ram-Mohan, *J. Appl. Phys.*, 2001, **89**, 5815–5875.
11. W. Shockley and H. J. Queisser, *J. Appl. Phys.*, 1961, **32**, 510–519.
12. P. Baruch, A. D. Vos, P. Landsberg and J. Parrott, *J. Sol. Energy Mater. Sol. Cells*, 1995, **36**, 201–222.
13. Y. Yan, R. W. Crisp, J. Gu, B. D. Chernomordik, G. F. Pach, A. R. Marshall, J. A. Turner and M. C. Beard, *Nat. Energy*, 2017, **2**, 17052.
14. M. T. Winkler, C. R. Cox, D. G. Nocera and T. Buonassisi *Proc. Natl. Acad. Sci. U. S. A.* 2013, **110**, E1076–E1082.

15. M. M. May, D. Lackner, J. Ohlmann, F. Dimroth, R. van de Krol, T. Hannappel and K. Schwarzburg, *Sustainable Energy Fuels*, 2017, **1**, 492–503.
16. A. J. Bard and L. R. Faulkner, 2nd edn, Wiley, New York, 2001.
17. M. T. Koper, *J. Electroanal. Chem.*, 2011, **660**, 254–260.
18. C. C. L. McCrory, S. Jung, I. M. Ferrer, S. M. Chatman, J. C. Peters and T. F. Jaramillo, *J. Am. Chem. Soc.*, 2015, **137**, 4347–4357.
19. M. Schreier, F. Heroguel, L. Steier, S. Ahmad, J. S. Luterbacher, M. T. Mayer, J. Luo and M. Grätzel, *Nat. Energy*, 2017, **2**, 17087.
20. B. Parkinson and J. Turner, in *Photoelectrochemical Water Splitting: Materials, Processes and Architectures*, ed. H.-J. Lewerenz and L. Peter, The Royal Society of Chemistry, 2013, pp. 1–18.
21. A. Azarpira, *et al.*, *Adv. Energy Mater.*, 2015, **5**, 1402148.
22. B. Parkinson, *Acc. Chem. Res.*, 1984, **17**, 431–437.
23. A. Murphy, P. Barnes, L. Randeniya, I. Plumb, I. Grey, M. Horne and J. Glasscock, *Int. J. Hydrogen Energy*, 2006, **31**, 1999–2017.
24. H. Döscher, J. F. Geisz, T. G. Deutsch and J. A. Turner, *Energy Environ. Sci.*, 2014, **7**, 2951–2956.
25. Y. Degani, T. Sheng, A. Heller, D. Aspnes, A. Studna and J. Porter, *J. Electroanal. Chem. Interfacial Electrochem.*, 1987, **228**, 167–178.
26. L. Kou, D. Labrie and P. Chylek, *Appl. Opt.*, 1993, **32**, 3531–3540.
27. K. A. Walczak, G. Segev, D. M. Larson, J. W. Beeman, F. A. Houle and I. D. Sharp, *Adv. Energy Mater.*, 2017, **7**, 1602791.
28. J. K. Dohrmann and N.-S. Schaaf, *J. Phys. Chem.*, 1992, **96**, 4558–4563.
29. K. Walczak, *et al.*, *ChemSusChem*, 2015, **8**, 544–551.
30. M. Murakami and Y. Koide, *Crit. Rev. Solid State Mater. Sci.*, 1998, **23**, 1–60.
31. M. R. Singh, C. Xiang and N. S. Lewis, *Sustainable Energy Fuels*, 2017, **1**, 458–466.
32. L. C. Seitz, Z. Chen, A. J. Forman, B. A. Pinaud, J. D. Benck and T. F. Jaramillo, *ChemSusChem*, 2014, **7**, 1372–1385.
33. K. T. Fountaine, H. J. Lewerenz and H. A. Atwater, *Nat. Commun.*, 2016, **7**, 13706.
34. L. Forbes, *Solid-State Electron.*, 1975, **18**, 635–640.
35. L. M. Peter and K. G. Upul Wijayantha, *ChemPhysChem*, 2014, **15**, 1983–1995.
36. M. Barroso, S. R. Pendlebury, A. J. Cowan and J. R. Durrant, *Chem. Sci.*, 2013, **4**, 2724–2734.
37. S. Almosni, *et al.*, *J. Appl. Phys.*, 2013, **113**, 123509.
38. R. Cariou, *et al.*, *IEEE J. Photovoltaics*, 2017, **7**, 367–373.
39. H. Döscher, B. Kunert, A. Beyer, O. Supplie, K. Volz, W. Stolz and T. Hannappel, *J. Vac. Sci. Technol., B: Microelectron. Nanometer Struct.*, 2010, **28**, C5H1.
40. O. Romanyuk, O. Supplie, T. Susi, M. M. May and T. Hannappel, *Phys. Rev. B*, 2016, **94**, 155309.
41. B. C. Wood, E. Schwegler, W. I. Choi and T. Ogitsu, *J. Phys. Chem. C*, 2014, **118**, 1062–1070.

42. M. M. May, H.-J. Lewerenz and T. Hannappel, *J. Phys. Chem. C*, 2014, **118**, 19032–19041.
43. J. L. Young, H. Döscher, J. A. Turner and T. G. Deutsch, *J. Phys. Chem. C*, 2016, **120**, 4418–4422.
44. M. M. May, H.-J. Lewerenz, D. Lackner, F. Dimroth and T. Hannappel, *Nat. Commun.*, 2015, **6**, 8286.
45. W. Calvet, E. Murugasen, J. Klett, B. Kaiser, W. Jaegermann, F. Finger, S. Hoch, M. Blug and J. Busse, *J. Phys. Chem. Chem. Phys.*, 2014, **16**, 12043–12050.
46. J. L. Young, M. A. Steiner, H. Döscher, R. M. France, J. A. Turner and T. G. Deutsch, *Nat. Energy*, 2017, **2**, 17028.
47. B. E. Sagol, U. Seidel, N. Szabo, K. Schwarzburg and T. Hannappel, *Chimia*, 2007, **61**, 775–779.
48. A. C. Meng, J. Cheng and M. Sprik, *J. Phys. Chem. B*, 2016, **120**, 1928–1939.
49. T. A. Pham, Y. Ping and G. Galli, *Nat. Mater.*, 2017, **16**, 401–408.
50. G. Létay and A. W. Bett, *Proceedings of the 17th European Photovoltaic Solar Energy Conference*, 178–181 (2001).
51. M. A. Green, *Prog. Photovoltaics*, 2012, **20**, 472–476.
52. S. Hernández, G. Barbero, G. Saracco and A. L. Alexe-Ionescu, *J. Phys. Chem. C*, 2015, **119**, 9916–9925.
53. O. Khaselev and J. A. Turner, *Science*, 1998, **280**, 425–427.
54. X. Sheng, *et al.*, in *2014 IEEE Photonics Conference* (2014), 433–434.
55. M. Weber and M. Dignam, *Int. J. Hydrogen Energy*, 1986, **11**, 225–232.
56. H. Welker, *Physica*, 1954, **20**, 893–909.
57. D. M. Chapin, C. S. Fuller and G. L. Pearson, *J. Appl. Phys.*, 1954, **25**, 676–677.
58. R. E. Gremmelmaier, *Z. Naturforsch., A: Phys. Sci.*, 1955, **10**, 501–502.
59. J. J. Loferski, *J. Appl. Phys.*, 1956, **27**, 777–784.
60. Z. Alferov, V. Andreev, M. Kagan, I. Protasov and V. Trofim, *Sov.* Phys. *Semicond.*, 1971, **4**, 2047–2048.
61. L. W. James and R. L. Moon, *Appl. Phys. Lett.*, 1975, **26**, 467–470.
62. H. Yoneyama, H. Sakamoto and H. Tamura, *Electrochim. Acta*, 1975, **20**, 341–345.
63. S. M. Bedair, M. F. Lamorte and J. R. Hauser, *Appl. Phys. Lett.*, 1979, **34**, 38–39.
64. R. C. Knechtli, R. Y. Loo and G. S. Kamath, *IEEE Trans. Electron Devices*, 1984, **31**, 577–588.
65. C. Flores, R. Campesato, F. Paletta, G. L. Timo, E. Rossi, L. Brambilla, A. Caon, R. Contini and F. Svelto, in *Conference Record of the Twenty Third IEEE Photovoltaic Specialists Conference* (1993), 1369–1374.
66. G. F. X. Strobl, L. Ebel, D. Fuhrmann, W. Guter, R. Kern, V. Khorenko, W. Koestler and M. Meusel, in *Conference Record of the 40th IEEE Photovoltaic Specialists Conference* (2014), 3595–3600.
67. K. A. W. Horowitz, M. Woodhouse, H. Lee and G. P. Smestad, *AIP Conf. Proc.*, 2015, **1679**, 100001.

68. Press Release, Fraunhofer ISE (Dec. 2014).
69. A. Heller, E. Aharon-Shalom, W. A. Bonner and B. Miller, *J. Am. Chem. Soc.*, 1982, **104**, 6942–6948.
70. E. Verlage, S. Hu, R. Liu, R. J. R. Jones, K. Sun, C. Xiang, N. S. Lewis and H. A. Atwater, *Energy Environ. Sci.*, 2015, **8**, 3166–3172.
71. O. Khaselev, A. Bansal and J. Turner, *Int. J. Hydrogen Energy*, 2001, **26**, 127–132.
72. S. Licht, B. Wang, S. Mukerji, T. Soga, M. Umeno and H. Tributsch, *Int. J. Hydrogen Energy*, 2001, **26**, 653–659.
73. J. Jia, L. C. Seitz, J. D. Benck, Y. Huo, Y. Chen, J. W. D. Ng, T. Bilir, J. S. Harris and T. F. Jaramillo, *Nat. Commun.*, 2016, **7**, 13237.
74. A. J. Nozik, *Appl. Phys. Lett.*, 1976, **29**, 150–153.
75. R. C. Kainthla, B. Zelenay and J. O. Bockris, *J. Electrochem. Soc.*, 1987, **134**, 841–845.
76. K. A. Bertness, S. R. Kurtz, D. J. Friedman, A. E. Kibbler, C. Kramer and J. M. Olson, *Appl. Phys. Lett.*, 1994, **65**, 989–991.
77. F. Dimroth, R. Beckert, M. Meusel, U. Schubert and A. W. Bett, *Prog. Photovoltaics*, 2001, **9**, 165–178.
78. A. G. Muñoz, C. Heine, M. Lublow, H. W. Klemm, N. Szabo, T. Hannappel and H.-J. Lewerenz, *ECS J. Solid State Sci. Technol.*, 2013, **2**, Q51–Q58.
79. J. D. Porter, A. Heller and D. E. Aspnes, *Nature*, 1985, **313**, 664–666.
80. W.-H. Cheng, M. H. Richter, M. M. May, J. Ohlmann, D. Lackner, F. Dimroth, T. Hannappel, H. A. Atwater and H.-J. Lewerenz, arXiv:1706.01493, 2017.
81. H. Döscher, J. L. Young, J. Geisz, J. Turner and T. Deutsch, *Energy Environ. Sci.*, 2016, **9**, 74–80.
82. J. F. Geisz, *et al.*, *Appl. Phys.·Lett.*, 2007, **91**, 023502.
83. NREL Efficiency Chart. https://www.nrel.gov/pv (2017).
84. M. A. Steiner, J. F. Geisz, I. Garcia, D. J. Friedman, A. Duda and S. R. Kurtz, *J. Appl. Phys.*, 2013, **113**, 123109.
85. D. Shahrjerdi, *et al.*, *Appl. Phys· Lett.*, 2012, **100**, 053901.
86. N. J. Smeenk, J. Engel, P. Mulder, G. J. Bauhuis, G. M. M. W. Bissels, J. J. Schermer, E. Vlieg and J. J. Kelly, *ECS J. Solid State Sci. Technol.*, 2013, **2**, P58–P65.
87. M. Woodhouse and A. Goodrich, *Tech. Rep.* (NREL, 2013).
88. A. Bansal and J. A. Turner, *J. Phys. Chem. B*, 2000, **104**, 6591–6598.
89. J. R. Bolton, S. J. Strickler and J. S. Connolly, *Nature*, 1985, **316**, 495–500.
90. S. W. Boettcher, *et al.*, *J. Am. Chem. Soc.*, 2011, **133**, 1216–1219.
91. N. Jain, *et al.*, in *Conference Record of the 43rd IEEE Photovoltaic Specialists Conference* (2016), 0046–0051.
92. NREL-SRRL. https://www.nrel.gov/midc/srrl bms.
93. ASTM, in G 173 – 03, ASTM International, West Conshohocken, 2003, DOI: 10.1520/g0173-03r1.
94. T. Cannon, *Sol. Cells*, 1986, **18**, 233–241.

95. T. E. Furtak, D. C. Canfield and B. A. Parkinson, *J. Appl. Phys.*, 1980, **51**, 6018–6021.
96. H. J. Snaith, *Energy Environ. Sci.*, 2012, **5**, 6513–6520.
97. G. Siefer, C. Baur, M. Meusel, F. Dimroth, A. W. Bett and W. Warta, in *Conference Record of the Twenty-Ninth IEEE Photovoltaic Specialists Conference*, 2002, 836–839.
98. C. Baur and A. W. Bett, in *Conference Record of the Thirty-First IEEE Photovoltaic Specialists Conference* (2005), 583–586.
99. Metzdorf, *J. Appl. Opt.*, 1987, **26**, 1701–1708.
100. S. Winter, T. Fey, I. Kröger, D. Friedrich, K. Ladner, B. Ortel, S. Pendsa and F. Witt, *Measurement*, 2014, **51**, 457–463.

CHAPTER 13

Continuum-scale Modeling of Solar Water-splitting Devices

MEENESH R. SINGH,[a] SOPHIA HAUSSENER[b] AND ADAM Z. WEBER*[c]

[a] Department of Chemical Engineering, University of Illinois at Chicago, IL 60607, USA; [b] Laboratory of Renewable Energy Science and Engineering, EPFL, Station 9, 1015 Lausanne, Switzerland; [c] Joint Center for Artificial Photosynthesis, Lawrence Berkeley National Laboratory, Berkeley, CA 94720, USA
*Email: azweber@lbl.gov

13.1 Introduction

A solar-fuel generator is a complex device involving different species in multiple phases distributed under varying temperature and pressure over a broad range of scales of dimensions from nanometers to 100s of centimeters. Due to the coupling of a multitude of processes, any change in one property or component may affect the solar-fuel production reaction and consequently overall device efficiency and operation. It is difficult to comprehend intuitively these typically nonlinear effects and consequences on device operation. In this aspect, mathematical modeling is ideally suited to play a significant role in the research and development of devices, cells, and their components.

An integrated solar-hydrogen generator that uses light-capturing semiconductors, combined with electrocatalysts in contact with an electrolyte, to generate oxygen and hydrogen *via* the electrolysis of water,

$$2H^+ + 2e^- \rightarrow H_2 \qquad 0\ V \tag{13.1}$$

Energy and Environment Series No. 22
Integrated Solar Fuel Generators
Edited by Ian D. Sharp, Harry A. Atwater and Hans-Joachim Lewerenz

Published by the Royal Society of Chemistry, www.rsc.org

$$2H_2O \rightarrow O_2 + 4H^+ + 4e^- \qquad 1.229\ \text{V} \tag{13.2}$$

is a promising route for the direct production of fuels from sunlight. Such a scheme is the simplest of all solar-driven fuel-forming devices, as well as demonstrating the highest efficiency, stability, and scalability.[1–4] The overall solar-to-hydrogen (STH) conversion efficiency of such a system depends on the performance and materials properties of all the individual components as well as the system design. Over the past few years, significant advances have been made in materials discovery and development for individual components as well as in the design and implementation of solar-driven water-splitting devices at the system level, many of which are mentioned or described in other chapters of this book.

Figure 13.1 shows a schematic illustration of various photoelectrochemical processes in an integrated solar water-splitting cell. Key processes include light absorption and photoexcited charge transport in semiconductor materials, interfacial charge transport, and electrocatalysis for the hydrogen evolution reaction (HER) and the oxygen evolution reaction (OER), ion and solvent transport in electrolytes and membranes, energy transport, and product gas transport. Due to the heavily coupled nature of these interactions, one cannot consider or model them independently in order to predict overall cell performance and behavior. However, one can use different pseudo steady-state and/or equilibrium expressions depending on the focus of the study since the phenomena occur on vastly different time scales.

Although the phenomena are complex, they can roughly be separated into two intrinsic curves as shown in Figure 13.2. The first is the light absorber power or diode curve, which represents the overall performance of the semiconductor components including light absorption, charge carrier generation and separation, *etc.* The second is the performance or polarization

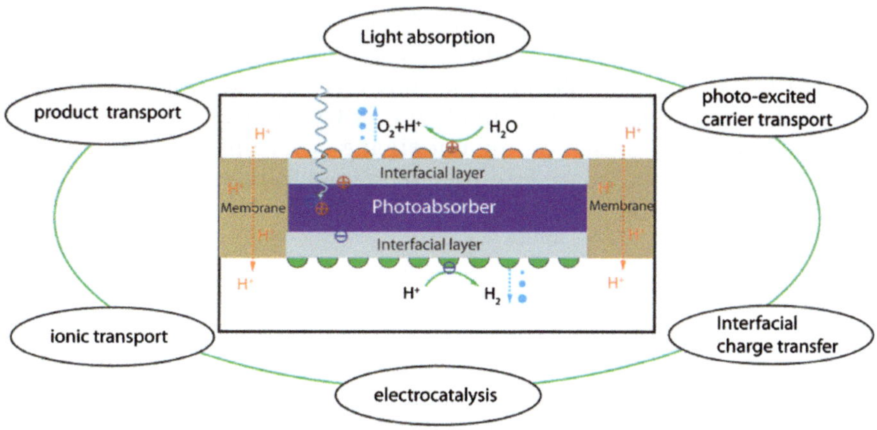

Figure 13.1 Schematic illustration of various coupled photoelectrochemical processes in an integrated solar-driven water-splitting cell.

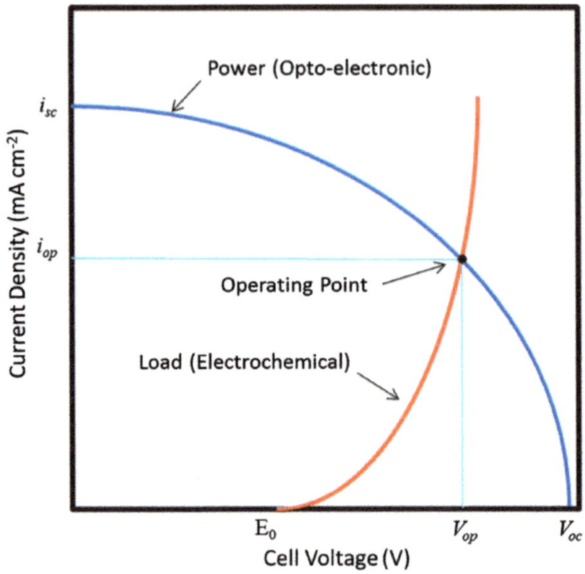

Figure 13.2 Light absorber power and electrochemical load curves, where the intersection represents the solar-fuel device operating point.

curve of the electrolyzer, which comprises the various mass- and ion-transport losses, the catalyst activity, *etc.* The intersection of these two curves represents the operation point of the integrated device. It is readily apparent that depending on the intersection point, changes to light insolation or catalysis may have larger or smaller impacts. For example, if it is on the steeper part of the power curve, changing the catalyst activity to reduce the operating potential could have a very large impact on overall cell performance; however, changing the open-circuit voltage or fill factor of the semiconductor might not have as large an impact. Furthermore, if these curves do not intersect due to large overpotentials or insufficient fluxes or voltage from the semiconductor, then operation is not possible. Thus, the goal of device and component modeling is to predict each of the curves (potentially coupled) based on the underlying governing physics.

For either curve, the work performed or gained by the system is simply the power as expressed by

$$P = IV \tag{13.3}$$

where I is the current and V is the cell voltage.

In general, the focus of the continuum modeling effort in this field as initiated by JCAP is on using the models to help understand the different physical phenomena in order to enable technology development. In particular, modeling is used to optimize physical geometries of given prototype designs,[5] evaluate the viability and efficiencies of design concepts,[6–10] define operational conditions and constraints,[11,12] guide materials development in terms of design targets and specific phenomena,[13] and conduct *in silico* placements and analysis of cells under real-world conditions. Such ambitions necessitate a multiphysics and perhaps multiscale approach that virtually integrates the various components and phenomena.

The scope of this chapter is to discuss the modeling approaches and governing equations that are utilized for continuum-level modeling of integrated devices. In such a framework, this chapter will focus more on transport than complex modeling of reaction mechanisms or semiconductor physics that are handled at the component level but have yet to be fully integrated into a device-level model; this approach is also in agreement with where the field was when JCAP started. In the discussion of equations and physics, concerns and issues that arise will be noted and test cases will be discussed. As much of the work on the actual solutions and specific cases are described in the various peer-reviewed literature and are typically much more specific, they are not discussed in detail throughout this chapter; furthermore, a recent review article discusses many of these findings.[3] Thus, this chapter is a companion to that article in many ways and provides more perspective and details of approach. Finally, while one can consider a photovoltaic in series with an electrolyzer to provide solar water splitting (*i.e.*, separating the two curves in Figure 13.2), as mentioned throughout this book, such an architecture is outside the scope of this chapter, which focuses on integrated devices (in agreement with the initial JCAP primary mission), and thus will not be discussed; it should be noted though that many of the concepts and approaches discussed in this chapter equally apply to each subset of such devices respectively.[14]

13.1.1 Definitions

Before examining the modeling approach, some definitions and baseline formation are required. As discussed recently,[3,15] there is a taxonomy of various designs. In fact, the initial modeling approach was to examine classes of cell architectures and not necessarily specific designs *per se*. This taxonomy includes microwire, side-by-side, louvered, microfluidic, recirculating, vertically integrated, *etc.* In these designs, the governing physics and underlying phenomena are the same, although additional effects may or may not occur (*e.g.*, no membranes in microfluidics, light tunneling and trapping in microwires, *etc.*).

Figure 13.3 shows an example of a vertically integrated design. As one can see, this device is composed of a light absorber that may be spatially integrated in the middle or at one electrode, electrocatalyst particles for hydrogen and oxygen evolution reactions, a membrane for ion conduction and simultaneous product separation, and flow fields and layers that could be either gas or liquid fed. The device effectively presents no direct semiconductor/electrolyte junction and can be described as buried photovoltaic

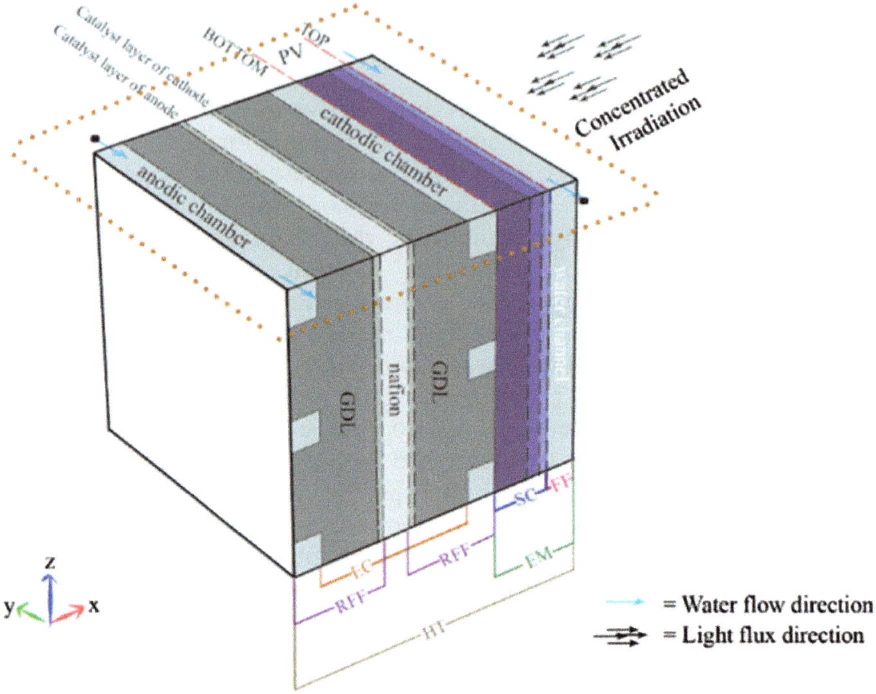

Figure 13.3 Schematic representation of the modeling domains relevant for an integrated PEC device utilizing concentrated irradiation.[16] EC: electrochemical component, (R)FF: (reactive) fluid flow, HT: heat transfer, EM: electromagnetic wave propagation, SC: semiconductor physics (with dashed lines showing the junction layers), PV: photovoltaic components, GDL: gas diffusion layer.

elements connected by integrated conductors to the anodic and cathodic catalysts.[1,15]

In terms of thermodynamics, the minimal cell potential corresponds to the Gibbs free energy,

$$U^{\theta} = \frac{\Delta G}{nF} \tag{13.4}$$

where n is the number of electrons transferred and F is Faraday's constant. Usually, most devices are not operated at the reference conditions. To account for this effect, one can determine the change in Gibbs free energy for the operating conditions, which can be calculated from handbooks[17] for the electrochemical generation of hydrogen and oxygen from water.

To compare different designs, it is useful to have a figure or parameter of merit, which can be the overall solar-to-hydrogen (STH) efficiency, η.

This efficiency can be defined as the power derived from the net hydrogen production relative to that of the incident photons,[18]

$$\eta_{\mathrm{STH}} = \frac{\text{Power Generated}}{\text{Power Available}} = \frac{J \times \dfrac{\mathrm{LHV}}{nF}}{P_{\mathrm{s}}} \tag{13.5}$$

where J is the solid-state operating current density, LHV is the lower heating value of hydrogen, and P_s is the solar insolation (W m^{-2}) at a specific location, date, and time. Such a definition is readily measured experimentally. However, one must also be cognizant of defining the efficiency in terms of not just the enthalpy but also the total free energy, which is a more rigorous thermodynamic definition that accounts for entropy and thus follows the second law of thermodynamics. From this standpoint, the efficiency can be written in terms of the fluxes and intrinsic efficiencies of the processes occurring,

$$\eta_{\mathrm{STH}} = \frac{J \cdot U^{\theta}}{P_{\mathrm{s}}} \eta_{\mathrm{F}} \eta_{\mathrm{pc}}, \tag{13.6}$$

where η_F is the Faradaic efficiency of the electrode reaction, and η_{pc} is the product collection efficiency. If parasitic reactions at the electrode are negligible, then $\eta_F = 1$. The product collection efficiency is defined as

$$\eta_{\mathrm{pc}} = \frac{I - \int_{A_{\mathrm{sep}}} nFN_{\mathrm{fuel}} \mathrm{d}A}{I}, \tag{13.7}$$

where N_{fuel} is the flux of fuel crossover through the membrane of area A_{sep}, and A is the area of electrodes. In this fashion, one does not explicitly model the recombination reactions of the crossover species at the electrodes. If such crossover occurs, then the device current used for the relevant reaction is reduced (essentially η_F decreases) but η_{pc} would increase and only represent the gas crossover that is not consumed at the other electrode. Also, for most water-splitting devices, one worries about the hydrogen crossover as a loss of fuel and does not consider the oxygen crossover explicitly, although depending on the hydrogen catalyst and operating potential it could recombine with some of the generated hydrogen to produce water (*i.e.*, $\eta_F < 1$).

13.2 Modeling Methodology and Approach

As shown in Figure 13.2, there are two main physical phenomena in a solar-fuels device. For light capture, one must absorb the incident photons that are available at each individual device location. These photons then generate holes and electrons when their energy is above the bandgap energy. Next, the generated carriers are separated and transported through the semiconductor to the reaction sites where they are utilized to electrochemically generate hydrogen and oxygen (*i.e.*, intersection point in Figure 13.2). At the reaction site, the overall driving force is the electrochemical potential difference between the ions and the electrons and its deviation from the equilibrium potential, where the ion potential is dominated by transport phenomena in the electrolytes. Thermal management can be used to affect this balance, where increased temperatures support ionic charge transport and kinetic processes but negatively impact charge recombination and transport in the light absorber. The objective of a mathematical model is to find the balance between the competing processes and, consequently, describe the state of the system, with an eye towards optimizing performance.

A robust model should predict device behavior accurately under different operating conditions (*e.g.*, temperature, irradiation, *etc.*). Such robustness requires a comprehensive description of all the major phenomena constituting operation. In practice, a comprehensive model describing operation under all the operating conditions is impossible because of the computational cost and limited understanding of the coupling of various phenomena. The general approach is to describe specific phenomena under specific limiting-case conditions and gain a better understanding of the system and then iteratively improve the coupling of the different phenomena. The computational efficiency balances the tradeoff between comprehensiveness and accuracy of model predictions. Therefore, the accuracy of model predictions is always limited by the choice of governing equations, input system parameters, and assumptions relaxing the complexity of the model.

While individual processes, such as light absorption and carrier transport in semiconductors or electrocatalysts for water-splitting reactions, have been extensively modeled and simulated, only a few reports modeled the whole cell operation that coupled more than two processes.[5–7,16,19–21] Because of this, the effects that reaction product transport has on light absorption and ionic transport are often overlooked and have not been quantitatively treated.

Therefore, advanced multiphysics, multidimensional modeling efforts are based on detailed component models but require an important additional focus on the accurate definition of the boundary conditions between the components of the device and exchange of information in the different components coupling the physical phenomena. Conservation equations (*e.g.* for energy, charge, momentum, and mass) and transport equations (*e.g.* for electromagnetic waves and species) are solved with the accurate interface conditions for the component coupling. The interface can simply require continuity in fluxes or can also account for complex physical phenomena (*e.g.* electrochemical reactions or charge transport at the semiconductor–liquid interface). This coupling introduces an additional layer of complexity because detailed component models accounting for a subset of physical phenomena might rely on the solution of another subset of equations (*e.g.* information about temperature distribution to provide detailed temperature-dependent material properties).[16,22]

An example of a program flow is shown in Figure 13.4, that contains six physical modules: (i) electromagnetic wave propagation, (ii) fluid flow, (iii) reactive fluid flow, (iv) electrochemistry and charge transfer, (v) heat transfer, and (vi) semiconductor physics. The wave propagation simulation

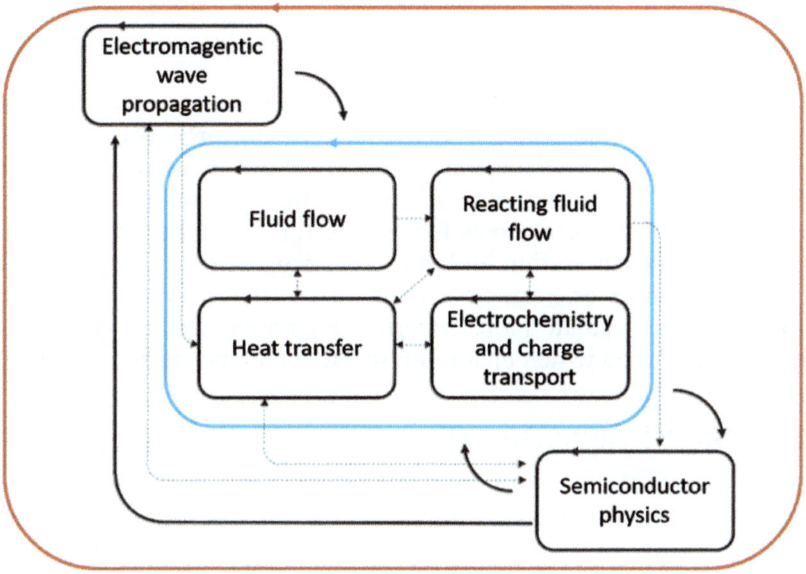

Figure 13.4 Schematic simulation flow diagram indicating three iterative loops that have to converge in order for the model to provide a unique solution. The dashed lines show interactions between the various modules.

starts with an initial guess on the temperature field (locally resolved). The heat source term from this simulation will contribute to the energy conservation module. The heat transfer module interacts with all other models to provide the temperature field, which is used to estimate properties and fluxes. The fluid flow module interacts with the heat transfer module and the reactive fluid flow module, providing velocity vector fields and receiving temperature fields. The electrochemistry module provides current density vector fields to the reactive flow module and an estimate of the heat source or sink term due to the chemical reaction and ohmic losses. In parallel, the semiconductor device simulation is performed starting with an initial temperature. The resulting surface averaged current densities and electric potential from the semiconductor and electrolyzer charge transfer equations are used to predict the operating current and potential of the device. The temperature field from the heat transfer module is compared to the temperature of the semiconductor module and an intermediate iteration loop is started repeating the semiconductor physics module until the temperature fields of this module and the heat transfer module converge. Finally, the converged temperature field of the five sub modules is compared to the temperature in the electromagnetic wave propagation module, and this global loop is repeated until those two temperature fields converge.

The six modules are not necessarily solved in the complete computational domain. Therefore, proper interface conditions have to be developed, which couple the sub-components of the domain to the complete simulation domain. Similarly, the boundary conditions for the various modules might not necessarily coincide with the boundary conditions of the complete simulation domain.

Challenges arise from the computational expenses in this scheme incorporating three nested loops, which have to converge. Furthermore, underrelaxation (*i.e.*, minimizing variable value changes iteration to iteration) in the different iteration algorithms has to be applied in order to ensure a robust scheme that converges for a wide range of operating parameters (*i.e.* irradiation concentration and mass flow rates of the reactant). Additionally, the mesh must be adapted dependent on the operating temperatures, *i.e.* for every intermediate loop. Criteria for automatized, adaptive meshing have to be developed which ensure mesh independence of the solution.[16,22]

13.2.1 Model Dimensionality

As discussed throughout this chapter, the various phenomena occur at different lengthscales and with different possible symmetries; thus models can likewise exist in different dimensionality according to the questions being explored. Typically, as one scales up the dimensionality, the physics and phenomena being solved are simplified in order for the problem to remain tractable. For example, detailed microkinetic reaction mechanisms can be explored at the 0-D scale, although if one requires full dimensionality for the different inputs, such kinetics are normally simplified to something akin to a Butler–Volmer expression.

Lower dimensional models also provide a quick means to understand the integration of different physics. For example, the work of Berger and Newman[23] is 1-D but outlines and explores the coupling of semiconductor physics and reactions. Once identified and validated, the equations and approach can be ported to more complicated models. Similarly, 1-D models may be sufficient for examining specific component properties and their relationships (*e.g.*, separator conductivity and permeability tradeoff) within the system,[13] but where the system is simplified often in terms of 0-D expressions or boundary conditions. Also, 0-D and 1-D models can be used as quick checks of possible device feasibility, where more in-depth multidimensional models can follow to fine-tune the predictions.

For the different models and approaches, different software packages have been utilized, both commercial and open source. Figure 13.5 shows some of the packages used as a function of modeling dimensionality and general phenomena explored within JCAP and the solar-fuels community. Included in the figure is a theory, which is taken separately from the continuum models described in this chapter. In this sense, the theory has its own dimensionality as it looks more at the atomic scale and provides perhaps upscaled expressions that can be used in the continuum models, especially for kinetic and possibly bandgap parameters. An ultimate model goal is to couple the different software or equations at the same scale (*i.e.*, coupling of the multidimensional packages), but this remains elusive.

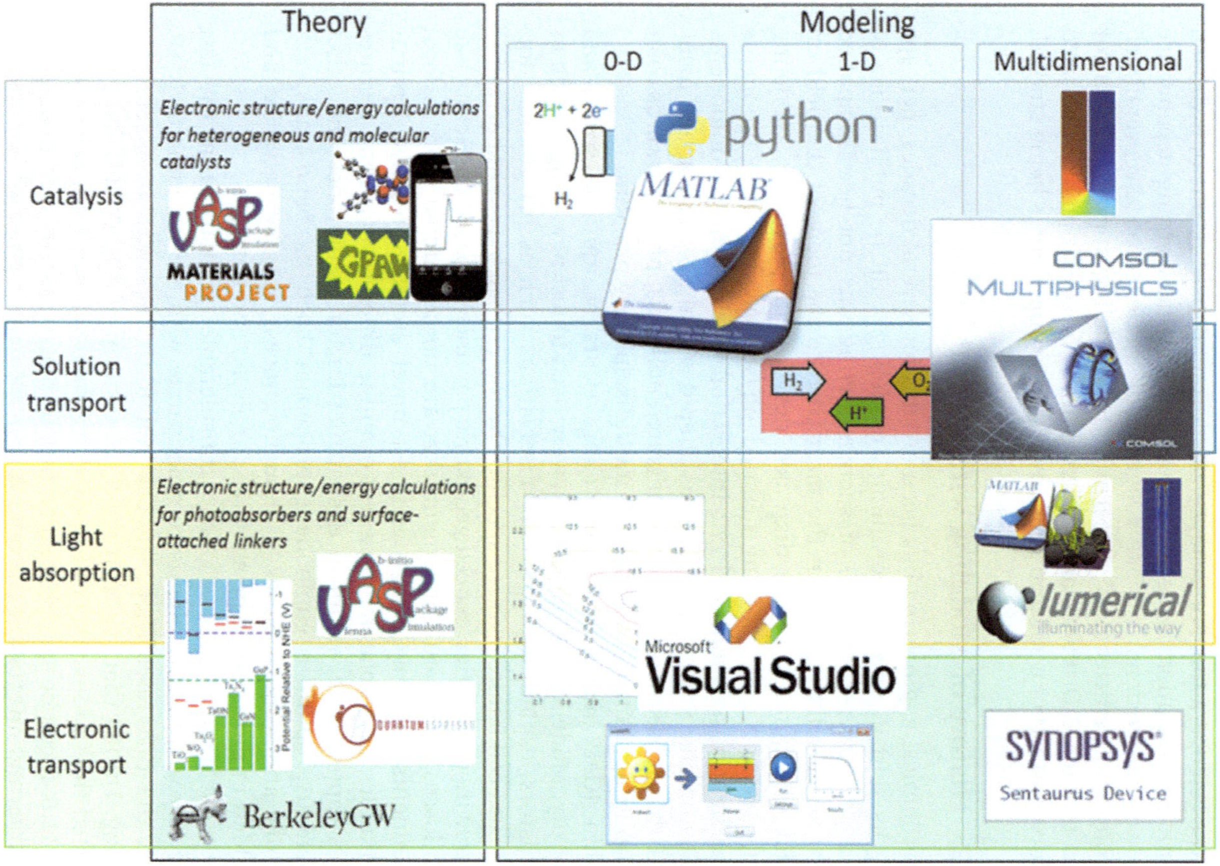

Figure 13.5 Model software and sample results as a function of dimensionality and physical phenomena being explored as utilized at the Joint Center for Artificial Photosynthesis (note that the diagram is meant to be illustrative and not exhaustive). Figure supplied by Alan Berger and all trademarks are owned by their respective companies.

In the literature, most studies have examined 0-D (analytical) or 1-D solar-hydrogen systems.[24] While the low dimensional simulation studies provided important trends and guidelines for solar-fuels research, the 2-D or 3-D effects in the cell operation play a critical role in understanding the real-life prototype cell operation. For instance, the current density or rate of hydrogen generation distribution along the photoelectrodes has a strong dependence on the cell dimensions and has a significant influence on the overall cell STH conversion efficiency.[5,10,20,21] Thus, the modeling approach taken for devices is often 2-D unless appropriate symmetry conditions can be used to reduce it to 1-D or a specific phenomenon or component is being studied.

13.2.2 History of Modeling within JCAP

Electrochemical engineering has been around for several decades, perhaps having its start in terms of continuum modeling with the publications of Newman.[25] However, the use of such modeling was not heavily applied to solar water splitting due, in the most part, to the focus of the field on material development rather than cell designs. In particular, modeling typically focused on examining detailed physics of singular components such as light absorption, and not integration of the various components and phenomena. Such an attitude was altered with the formation and focus of JCAP that subsequently took an integrated holistic approach towards solar water splitting, as benefiting a large multidisciplinary program.

Within JCAP, mathematical modeling occurs at multiple scales and complexity. Modeling and simulation were used to test new designs such as recirculators, explain experimental observations such as pH gradient formation and cell failure, explore specific phenomena such as light harvesting in microwire geometries, set cell geometric parameters based on factors like ohmic losses, elucidate phenomena that are unmeasurable such as local environments, virtually design and test novel architectures like laminar microfluidics or bipolar membranes, examine different operating regimes such as vapor phase, establish design metrics for membranes and other components, optimize component properties such as band gaps, *etc.* Modeling and simulation as a thought process is now pervasive and embedded in most prototype activities, analysis, and designs.

While such modeling has gained considerable traction within the last couple of years, its founding is commensurate with the founding of JCAP. In fact, the first researcher visit between JCAP sites was for macroscale modeling. During that multiday visit, the concept of a JCAP Simulator that integrated the various physics to enable a virtual cell was born as shown in Figure 13.6. It can be seen that an ambitious plan was developed to take 3-D geometries and integrate at the boundaries the various programs for specific phenomena (see Figure 13.5) that transfer the relevant variables among the different programs. Such a vision remains unrealized due to the difficulty of convergence of the physics and geometries and the tradeoff between dimensionality and feasibility; thus, simplifications to certain physics are

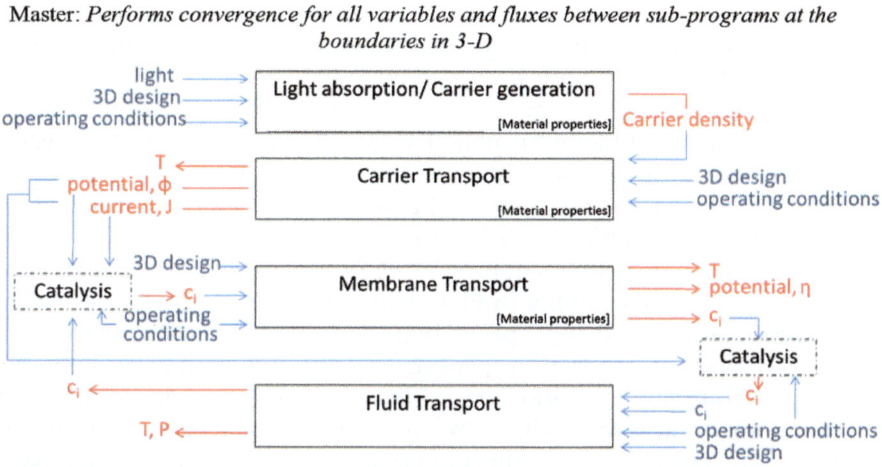

Figure 13.6 Initial program flow chart and structure for the JCAP Simulator.

required depending on the question being answered. For the most part, many of the modeling efforts focused more on transport related issues as this was lacking in the field compared to the more well-trodden semiconductor physics. Although ambitious and not fully realized, parts of the JCAP Simulator were developed and released both internally and externally to the community to aid anyone who had new designs or properties to explore virtually.

As mentioned, modeling with JCAP was present since its inception. Furthermore, the original focus was not only on isolated component issues but also those at the balance of cells including thermal management, bubble management, the flow of the gases and liquids throughout multiple cells, including product collection, *etc.* These nascent efforts were greatly scaled back as other issues dominated the research agenda. After a few years of preliminary work and some conflicting activities and personalities, simulation again became a higher priority with the founding of the crosscutting Modeling and Simulation Team (MaST) within the center. This team acknowledged the contributions of modeling and simulation across the various JCAP projects and helped to establish a single vision for these activities, which was translated to one of the four thrust areas in the JCAP renewal. Overall, one of the main successes of JCAP in its first phase was the use of modeling and concomitant prototype design to demonstrate how one needs to consider integration and cell-level issues such as product separation in order to formalize designs for water splitting that are scalable. Since the initial studies of JCAP, multiphysics modeling has been undertaken by various research groups to help examine critical issues if the technology is to transition from the lab to the field.

13.3 Governing Equations

Continuum-scale modeling of solar water-splitting devices requires describing the load and power curves (see Figure 13.2). The power curve is dominated by the semiconductor physics and light capture as discussed in Section 3.2. Relative to the design consideration of semiconductors in a photovoltaic device, the optical and/or electronic interactions among the semiconductors, protective layers, electrocatalysts layers, solution electrolytes and product bubbles also play a vital role in optimizing the overall photoelectrochemical cell performance. For the load curve, the potential is controlled by the various transport and kinetic-related overpotentials. In the sections below, we examine each in turn.

Before doing so, it is worth commenting on the typical model structure. As can be seen in the example in Figure 13.7, models are composed of the governing equations and appropriate boundary conditions. Each variable is required to have a corresponding governing equation with each derivative requiring a boundary condition. In addition to the governing differential equations, physical laws such as the sum of mole fractions and electroneutrality are considered in the equation count (*e.g.*, only $N-1$ mass balances are independent due to the sum of the mole fractions equaling unity). Throughout the sections below, various governing equations and boundary conditions are introduced for the different physics. To construct the model, one must decide which are necessary and which physics are being solved and then construct a diagram similar to Figure 13.7 to ensure a properly specified simulation, with its interfaces and boundary conditions.

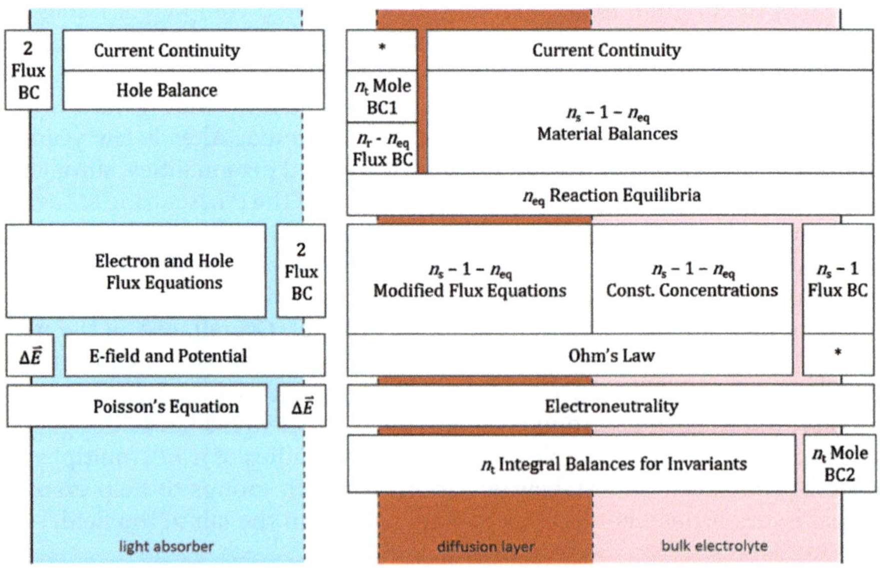

Figure 13.7 Modeling equation diagram showing the main governing equations and boundary conditions used for a complete simulation.
Figure adapted from ref. 23, Copyright 2014 The Electrochemical Society.[23]

13.3.1 Transport in Electrolytes

Mathematical modeling of species transport in electrolytes requires physical description of species flux, along with mass, charge, momentum, and energy conservation. The objective of such a model is an accurate prediction of the distribution of electrolyte potential and species concentrations in the electrolytic solution.

13.3.1.1 Mass Conservation

The conservation of mass requires,

$$\frac{\partial c_i}{\partial t} + \nabla \cdot \mathbf{N}_i = R_i, \tag{13.8}$$

where R_i is the net rate of formation of ith species participating in the homogeneous (typically acid–base) reactions, c_i and $\mathbf{N}_i$ are the concentration and superficial flux of species i, respectively. The rate of change of species concentration is due to the divergence of species flux and homogeneous reactions. The divergence of flux is the dominating term of eqn (13.8) in bulk electrolyte, where the reactions are close to equilibrium. However, the reaction layer near the electrode will also have contributions from the reaction terms.

A similar charge-conservation can be written since the current is defined as

$$\mathbf{i}_l = \sum_i z_i F\,\mathbf{N}_i \tag{13.9}$$

where $\mathbf{i}_l$ is the current density in the electrolyte, z_i is the charge number of species i, and F is Faraday's constant. Since there is no current generation in the solution (only at the boundary reaction sites), one can write the current conservation as

$$\nabla \cdot \mathbf{i}_l = 0 \tag{13.10}$$

13.3.1.2 Acid–Base Equilibria

As noted, the homogeneous reactions are dominated by acid/base reactions in the electrolytes. These are typically water- and buffer-dissociation kinetics. The water-dissociation reaction is given as

$$H_2O \rightleftarrows H^+ + OH^- \tag{13.11}$$

The forward rate constant and equilibrium constant of water-dissociation are 2.4×10^{-5} mol L^{-1} s^{-1} and 1×10^{-14} mol^2 L^{-2}, respectively. Various types of pH buffer such as acetate, phosphate, borate, and carbonate buffers are used in solar water-splitting systems for operation at neutral or near-neutral pHs. Buffers can readily ionize/de-ionize to balance pH changes in the electrolyte due to consumption/production of protons at the electrodes. The dissociation reaction of buffer is given as

$$HA^{z-} \rightleftarrows H^{+} + A^{(z+1)-} \tag{13.12}$$

This reaction will be under equilibrium if the rate of buffer dissociation is much faster than the rate of proton formation at the electrode. As the current density increases, the rate of proton formation can become comparable to or higher than the rate of buffer dissociation. In this case, there will be the formation of a layer of electrolyte attached to the electrode, where the concentrations of buffer species are not in equilibrium. This layer of electrolyte is known as the reaction layer, which is usually smaller than the Nernst diffusion layer.

13.3.1.3 Mass-species Fluxes

The electrolytes used for water splitting contain ionic species (protons, hydroxides, buffer ions, co-cations, and co-anions) and neutral species (water, dissolved hydrogen and dissolved oxygen). The flux of these species is described differently for three classes of electrolytes such as infinitely dilute (0–1 mol%), moderately dilute (1–10 mol%) and concentrated electrolytes (10–100 mol%), which are discussed as follows.

13.3.1.3.1 Infinitely Dilute Electrolyte. In the infinitely dilute electrolyte, every species moves independently from each other and they interact only through with the solvent (*i.e.*, water). One can assume unity activity coefficients and concentrations for activity. There are three mechanisms of species transport including diffusion, migration, and convection where the driving forces are concentration gradient, electric field, and velocity field, respectively. Thus, the net flux of species in the electrolyte can be written as a Nernst–Planck equation,

$$\mathbf{N}_i = \underbrace{-D_i \nabla c_i}_{\text{Diffusion}} \underbrace{- z_i u_i F c_i \nabla \phi_l}_{\text{Migration}} \underbrace{+ \mathbf{v} c_i}_{\text{Convection}}, \tag{13.13}$$

where ϕ_l is the electrolyte potential, D_i and u_i are the diffusion coefficient and mobility of ion i, respectively, and $\mathbf{v}$ is the typically mass-averaged velocity,

$$\mathbf{v} = \frac{\sum \mathbf{N}_i M_i}{\rho} \tag{13.14}$$

where M_i is the molecular weight of species i and ρ is the density of the entire solution. The ionic mobility and diffusion coefficient are related by the Nernst–Einstein equation (at infinite dilution)

$$u_i = \frac{D_i}{RT} \tag{13.15}$$

where R is the ideal-gas constant and T is the absolute temperature. Here, the electrolyte is defined by two properties such as diffusion coefficients and mobilities, which can be functions of local temperature, pressure, and concentrations. For example, the Stokes–Einstein relationship provides temperature- and concentration-dependent diffusion coefficients.

13.3.1.3.2 Moderately Dilute Electrolyte. The actual driving force for diffusion and migration of ions in the electrolyte is the gradient of electrochemical potential ($-\nabla\mu_i$), such that the flux of ionic species can be written as

$$\mathbf{N}_i = -c_i u_i \nabla\mu_i + c_i \mathbf{v} \tag{13.16}$$

The product of mobility and gradient of electrochemical potential represents the velocity of ions due to diffusion and migration. The electrochemical potential of species can be written as

$$\mu_i = RT\ \ln(\gamma_i c_i) + z_i F\phi \tag{13.17}$$

where γ_i is the activity coefficient of species i, which depends on temperature and concentration. It should be noted that the activity of a single ion cannot be truly measured and thus one is required to assume a reference ionic species.[26] Substituting eqn (13.17) into (13.16) yields the more familiar expression for ionic flux,

$$\mathbf{N}_i = -D_i \nabla c_i - D_i c_i \nabla \ln\ \gamma_i - z_i u_i F c_i \nabla\phi + c_i \mathbf{v} \tag{13.18}$$

This equation is similar to eqn (13.13), except the inclusion of the second term on the right, which reduces the net flux of species due to the non-ideal behavior of electrolyte. It can be seen that the flux expression reduces to eqn (13.13) for infinitely dilute electrolytes ($\gamma_i = 1$).

13.3.1.3.3 Concentrated Electrolyte. Dilute-solution theory fails in multicomponent concentrated electrolytes, as it does not account for interactions between different species (only between the solvent and the species). The flux expression for concentrated electrolyte can be obtained from the conservation of energy between two interacting species. From the force balance, the driving force acting on a species is balanced by the drag forces due to interactions with other species, such that

$$c_i \nabla\mu_i = \sum_j K_{i,j}(\mathbf{v}_j - \mathbf{v}_i) = \sum_j \frac{RTc_i c_j}{c_T D_{i,j}}(\mathbf{v}_j - \mathbf{v}_i) \tag{13.19}$$

where $K_{i,j}$ is the friction coefficient, $\mathbf{v}_i$ is the mass-average velocity of species i, $D_{i,j}$ is the binary diffusion coefficient, and c_T is the total concentration of all species. The expression for flux of species in concentrated electrolyte can be obtained by inverting the above equation (which for more than three species requires numerical methods)

$$\mathbf{N}_i = c_i \mathbf{v}_i = \left(\sum_j \frac{c_j}{D_{i,j}}\right)^{-1} \left(-\frac{c_T c_i \nabla\mu_i}{RT} + \sum_j \frac{c_i c_j \mathbf{v}_j}{D_{i,j}}\right) \tag{13.20}$$

For use in many modeling and simulation, one can utilize effective properties defined as

$$D_{e,i}=\frac{c_T}{\sum_j \frac{c_j}{D_{i,j}}}=\frac{\sum_j c_j}{\sum_j \frac{c_j}{D_{i,j}}};\qquad \mathbf{v}_{e,i}=\frac{\sum_j \frac{c_j\mathbf{v}_j}{D_{i,j}}}{\sum_j \frac{c_j}{D_{i,j}}} \tag{13.21}$$

The effective diffusion coefficient is the weighted-harmonic mean of the binary diffusion coefficients, whereas the effective velocity is the weighted-arithmetic mean of the species velocity. We can now re-write the flux-expression using these averaged properties and electrochemical potential definition (eqn (13.16)) to a more familiar form,

$$\mathbf{N}_i=-D_{e,i}\nabla c_i - D_{e,i}c_i\nabla \ln \gamma_i - z_i u_{e,i}Fc_i\nabla\phi + c_i\mathbf{v}_{e,i} \tag{13.22}$$

The above flux-expression for concentrated electrolyte looks very similar to the flux-expression for moderately-dilute electrolytes (eqn (13.17)). However, they differ in the value and nature of effective diffusion coefficients and velocities. Eqn (13.22) will reduce to eqn (13.18) for $c_i \ll c_0$, where c_0 is the concentration of solvent. Eqn (13.22) is a more accurate representation of species flux that can be used for any type of electrolytes. The simplified eqn (13.18) and (13.13) will yield overestimated values of flux as compared to the flux predicted using eqn (13.22).

Although the focus has been on the gradient of electrochemical potential for mass flux, the temperature, pressure, and body sources (*e.g.*, gravity or magnetic field) can also contribute to the transport of species. The pressure gradient is not an independent driving force. However, the temperature gradient can cause a mass flux, which is known as the Soret effect. The Soret effect can be more significant when the solar-fuel generator is exposed to varying solar insolation or has significant solar irradiation concentration. A more general flux expression accounting for these effects can be written as

$$c_i\left(\nabla\mu_i+\overline{S_i}\nabla T-\frac{M_i}{\rho_i}\nabla p\right)=\sum_j\frac{RTc_ic_j}{c_TD_{i,j}}(\mathbf{v}_j-\mathbf{v}_i)+\sum_j\frac{RTc_ic_j}{c_TD_{i,j}}(\alpha_j-\alpha_i)\nabla\ln T \tag{13.23}$$

where $\overline{S_i}$ is the partial molar entropy, ρ_i is the density, and α_i is the thermal diffusion coefficient. A flux-expression similar to eqn (13.22) can be obtained by defining effective properties.

13.3.1.4 Membrane Transport

The membrane serves key roles in separating products, electrolyte compartments, as well as perhaps some structural integrity. For porous separators, the above flux equations in terms of transport within electrolytes holds with just changing properties for porosity and tortuosity, *e.g.*,

$$\psi^{\text{eff}} = \frac{f_v}{\tau}\psi = f_v^{1.5}\psi \tag{13.24}$$

where, ψ_k represents any property, f_v is the volume fraction of the fluid phase, τ is the tortuosity of the fluid phase, and the last equality stems from the often-used Bruggeman expression. For ionic or charged membranes, however, additional complications are due to the fixed ionic sites, which alter the governing equations slightly since one ion cannot move and has a fixed concentration, as well as the boundary condition of the membrane with the electrolyte.

For the latter, the material balance typically ignores any interfacial mass-transfer resistance, meaning one can equate the electrochemical potential for each species across the interface as a boundary condition. For polymer membranes, this leads to a discontinuity in the potential (called the Donnan potential) due to the background charge caused by the fixed charged groups in the polymer, which should be included in the model.[23] Care should be taken when defining the potential to be used along with the reference states for electrochemical potentials, if the quasi-electrostatic potential is selected, the condition becomes

$$[RT\ \ln\ (\gamma_i c_i) + z_i F\phi]_{\text{electrolyte}} = [RT\ \ln\ (\gamma_i c_i) + z_i F\phi]_{\text{membrane}} \tag{13.25}$$

In most cases, the Debye length is short enough that membranes can be modeled using the electroneutrality assumption, but the Poisson and Nernst–Planck equations can be used if necessary.

Finally, one should note that more complicated expressions and models can be used such as those for bipolar membranes, but this is beyond the scope of this chapter. Concentrated-solution theory can typically be applied as long as one is careful about the definition of ionic and electrochemical potentials and transport properties.

13.3.1.5 Charge Transport and Conservation

As noted in eqn (13.9), one can write an overall charge conservation equation for the total ionic current. The divergence of this is typically zero since the electrochemical reactions occur at the boundaries. In addition, the variation in the electric field is directly proportional to the charge distribution in the medium (*i.e.*, Poisson's equation),

$$\nabla^2\phi = \frac{F}{\varepsilon}\sum_i z_i c_i \tag{13.26}$$

where, ε is the dielectric constant of the medium. Since the value of F/ε is very large, any charge separation in the electrolyte requires a very large variation in the electric field. Therefore, one can often assume electroneutrality,

$$\sum_i z_i c_i = 0 \tag{13.27}$$

This relationship adds to the equation count (similar to the sum of the mole fractions equals 1) and serves as the governing equation for the ionic potential. This also means that convection of ionic pairs cannot be used to carry charge (see eqn (13.9))

Finally, for ionic charge transport, one can also obtain the more familiar Ohm's law,

$$\mathbf{i} = -\kappa \nabla \phi \tag{13.28}$$

where the conductivity can be correlated to the various mobilities

$$\kappa = F^2 \sum_i z_i^2 c_i u_i \tag{13.29}$$

This expression is often used in models that ignore speciation and are focused on other aspects than detailed ionic transport.

13.3.1.6 Electron Transport

If additional conductive layers such as protection layers or FTO coatings exist, then one may need to use Ohm's law to adjust the electric field at the catalyst interface,

$$\mathbf{J} = -\sigma \nabla \phi \tag{13.30}$$

where, σ is the electronic conductivity.

13.3.1.7 Product Gases

The gases produced at the electrodes dissolve and then bubble-out from the electrolyte. This can be handled numerically by utilizing mass transfer coefficients and more complicated bubble models, although the latter is not typically used for solar-fuel modeling. For transport through the membrane, one can use a permeability or mass-transfer coefficient,

$$R_{\text{degas}, i} = k_L a \, (c_{i,l} - c_{i,g}) \tag{13.31}$$

where $k_L a$ is the volumetric mass-transfer coefficient, $c_{i,l}$ is the concentration of dissolved gas in the counter chamber and $c_{i,g}$ is the concentration of product gas in the bubbles in the counter chamber. The rate of gas crossover through the membrane together with the rate of degassing affects the total crossover of product gases and hence efficiency in a solar-fuel generator (see eqn (13.6) and (13.7)). The rate of degassing can be added to the right side of mass-balance eqn (13.8).

13.3.1.8 Momentum Conservation

Depending on the operating conditions, a velocity field can be developed in a solar-fuel generator than can affect species transport (*e.g.*, see eqn (13.13)). Evolution of bubbles at the electrode can create natural convection in the electrolyte, which increases with decreasing distance between electrodes. The maximum velocity of the electrolyte is always smaller than the terminal velocity of bubbles.

The gas bubbles and temperature gradients produced during electrolysis can generate natural convection in the electrolyte, which enhances the transport of species. The temperature gradient creates density gradient in the electrolyte that induces a velocity field. The effect of rising bubbles on the electrolyte can be discrete and computationally challenging. Alternatively, the velocity field can be obtained by defining a surface velocity at the electrode–electrolyte interface. The bubbles can quickly attain a terminal velocity, whose value depends on the size of the bubble after it detaches from the surface. The terminal velocity of bubbles can be averaged over the volume of a slice of electrolyte near the electrode. The velocity of the electrolyte at the electrode surface is given by the volume-averaged terminal velocity, which yields the velocity field in the electrolyte. It was found that the intensity of mixing due to natural convection strongly depends on the presence of membrane.[11] The membrane-less electrochemical cells could develop natural convection to support 10 mA cm^{-2} of current density in the near-neutral electrolyte, whereas the membrane-containing cells could not induce significant convection, thereby resulting in a maximum current density of <1 mA cm^{-2}.

Another scenario for momentum and convective flow is when the electrolyte is externally pumped through the solar-fuel generator, *i.e.* when forced convection is present. For the above cases, the Navier–Stokes equation can be used to obtain velocity and pressure field due to forced/natural convection[27]

$$\frac{\partial(\rho\mathbf{v})}{\partial t}+\mathbf{v}\cdot\nabla(\rho\mathbf{v})=-\nabla p+\mu\nabla^2\mathbf{v}+S_m \tag{13.32}$$

where, μ and $\mathbf{v}$ are the viscosity and mass-averaged velocity, respectively. The transient term in the momentum conservation equation represents the accumulation of momentum with time and the second term describes convection of the momentum flux (which is often small). The first two terms on the right side represent the divergence of the stress tensor and the last term represents other sources of momentum, typically other body forces like gravity or surface tensions, which can become important for free convection or during bubble formation.

13.3.2 Transport in Photoabsorbers

The semiconductor component generates the photovoltage and photocurrent required to drive the photoelectrochemical reactions by absorbing photons with energy greater than its bandgap to generate energized electrons and holes. Lower energy photons might still be absorbed but they will not contribute to charge generation but only to increase thermal energy. The collected fraction of absorbed photons with energies larger than the bandgap energy of the semiconductor that is actually converted to electron–hole pairs is described by the optical quantum yield. In general, there is either a junction between the semiconductor and liquid electrolyte to generate the photovoltage, or a buried solid/solid junction (*i.e.*, a photovoltaic). Surface charges are critical in a liquid-junction cell to form the depletion region that affects charge transport. This is an important distinction from a buried-junction cell with an electrocatalyst on the surface, where high rates of charge transfer between absorber-catalyst and catalyst-electrolyte have the potential to decouple the rate of charge transfer from the properties of the semiconductor–electrolyte junction. Non-continuum, quantum modeling is required in order to predict the quantum yield.

For solar water-splitting, most light absorbers are composed of multiple materials with different bandgaps to absorb enough photons to generate the required photovoltage. The selection of the optimal bandgap combinations of semiconductor materials in a solar-hydrogen device has different guiding principles than for a solid-state photovoltaic device. Under the Shockley–Queisser detailed balance limit for photovoltaic efficiency,[7] the photovoltage and photocurrent vary in the opposing trend with the varying bandgap combinations, which results in an optimal semiconductor bandgap combination that maximizes the solar-to-electricity conversion efficiency. However, in a solar-hydrogen device, the electrochemical potential difference between the HER and the OER (*e.g.*, 1.23 V under standard conditions) sets the lower bound for the required photovoltage, and therefore multijunction cells are often required since they can generate higher photovoltages. The optimal bandgap combinations for the tandem cell structure in a solar-driven water-splitting cell have been evaluated by others.[7,9,20,24,28,29] The theoretical framework linking semiconductor/liquid junctions, molecular donor–acceptor systems, and heterogeneous semiconductor/metal systems has long been established.[30]

13.3.2.1 Light Capture

Over the light absorber domain, light capture is determined based on the intrinsic bandgaps of the materials and the optical density and geometry of the light absorber and systems (antireflection coatings, lenses, *etc.*). Some designs allow for preferential pathways and waveguiding of light such that internal reflections occur and increase capture of incident photons. Theoretical limits can be used as a first approximation to calculate the absorbed photons. The detailed balance limit[7] describes the amount of incident solar irradiation that is absorbed and produces an electron–hole pair, *i.e.* charge or current, in the semiconductor, minus the current lost due to blackbody radiative recombination

$$J = J_{\mathrm{ph}} - J_{\mathrm{rr}} = q\int_0^{\lambda_g} n_{\mathrm{ph,sol}}(\mathbf{x}, \lambda)\mathrm{d}\lambda - q\int_0^{\lambda_g} n_{\mathrm{ph,b}}(\lambda, T)\mathrm{d}\lambda \tag{13.33}$$

where, J_{ph} describes the photocurrent density (electron and hole generation fluxes), J_{rr} is the current density lost due to radiative recombination, λ_g is the optical wavelength corresponding to the band-gap energy of the semiconducting light absorber, $n_{\mathrm{ph,sol}}$ is the spectral photon flux arriving at the surface of the earth at a specific location $\mathbf{x}$, and $n_{\mathrm{ph,b}}$ is the spectral photon flux due to blackbody radiation at a temperature T.

In the case of a dual-absorber tandem cell, the irradiation is partially absorbed by the top cell and the remaining above-bandgap radiation, $\lambda > \lambda_{\mathrm{g,top}}$, is absorbed by the bottom cell. In solar water splitting, a dual or triple-absorber tandem cell is typically used because such a structure can provide a sufficient potential difference to drive the electrochemical reaction; the system can produce an open-circuit potential difference that is greater than the equilibrium potential difference for production of H_2 and O_2 from water under standard conditions (1.23 V).

The detailed balance limit does not account for the fact that the absorption of a photon requires a finite length of the semiconductor. In a simplified approach, the Beer–Lambert law can be used to more accurately predict the actual number of photons absorbed. Depending on the geometry, one can also account for spatial variations or absorption through Beer–Lambert expression

$$\psi_{y,\lambda} = \psi_{\lambda,\mathrm{AM1.5}} \exp(-\omega_{\mathrm{eff},\lambda} y) \tag{13.34}$$

where $\psi_{\lambda,\mathrm{AM1.5}}$ is the incident AM1.5G photon flux and $\omega_{\mathrm{eff},\lambda}$ is the effective spectral absorption coefficient of the materials the light is penetrating (*e.g.*, catalyst or electrolyte). One can then integrate the above expression through the light absorber to get the generation rate of holes or electrons (similar to (13.33)), *e.g.*,

$$G = \int\int \omega_{\mathrm{eff},\lambda}\, \psi_{\lambda,\mathrm{AM1.5}} \exp(-\omega_{\mathrm{eff},\lambda} y)\, \mathrm{d}\lambda\, \mathrm{d}y \tag{13.35}$$

To be rigorous, one should solve the governing Maxwell equations in the geometry.[16,31] This computationally expensive approach allows one to accurately consider the morphology of the solar absorber and to incorporate wave effects (for example resonances), potentially leading to higher efficiencies.

13.3.2.2 Hole and Electron Transport

As noted, the absorption of light above the bandgap for each light absorber results in the generation of holes and electrons. Also, for most cases, this is a buried junction and thus treated as such. The governing material balances for these species (accounting for only band-to-band recombination) can be written as

$$\frac{\partial p}{\partial t} = -\nabla \cdot \mathbf{N}_h + G + k_{\text{rec}}\left(n_i^2 - np\right) \tag{13.36}$$

where $\mathbf{N}_h$ is the flux of holes, n is the electron concentration, p is the hole concentration, k_{rec} is the radiative recombination rate, and n_i is the equilibrium constant written as an intrinsic charge carrier concentration. The accumulation term on the left is often ignored in water-splitting simulations as it occurs on a much faster time scale than the other processes. Within the semiconductor, the current density can be written as

$$\mathbf{J} = F(\mathbf{N}_h - \mathbf{N}_e) \tag{13.37}$$

where $\mathbf{N}_e$ is the flux of electrons. One can then define the potential and electric field using Poisson's equation

$$\nabla \cdot \Phi_s = -\mathbf{E} \tag{13.38}$$

$$\nabla \cdot \mathbf{E} = \frac{F}{\epsilon_0 \epsilon_R}(p - n + N_b) \tag{13.39}$$

where N_b is the concentration of dopant atoms that provide a background charge to the local charge density, ε_0 is the permittivity of free space, and ε_R is the dielectric constant of the light absorber.

In terms of transport equations for the electrons and holes, one can use Nernst–Planck equations (see (13.13))

$$\mathbf{N}_h = -D_h \nabla p + u_h p \mathbf{E} \tag{13.40}$$

$$\mathbf{N}_e = -D_e \nabla n + u_e n \mathbf{E} \tag{13.41}$$

where one can relate the diffusivities and mobilities (drift term) by Nernst–Einstein (13.15). The above form the set of governing equations to calculate the current in the light absorber that is used at the boundary with the catalyst to promote the reactions (see Section 3.3). The kinetic equations serve to provide boundary conditions for the above equations (*i.e.*, set fluxes due to continuity) as well as the equality between electron electrochemical potentials.

13.3.2.3 Diode Equation

Although the equations and boundary conditions from the previous two sections can be used to simulate light capture and semiconductor physics, often one uses a simplified ideal-diode equation for these processes when modeling water-splitting devices. The main reason is that most of the continuum models have focused on understanding the interplay between the transport phenomena and load rather than detailed light absorbers and photovoltaics, which have already been studied extensively. Thus, the performance curve of a typically light absorber (see Figure 13.2) can be approximated by a fitted diode equation of the form

$$J = J_{\text{ph}} - J_0\left[\exp\left(\frac{qV + qiR_{\text{ser}}}{kT}\right) - 1\right] - \frac{V}{R_{\text{sh}}}, \tag{13.42}$$

where J_0 describes the dark saturation current density and R_{ser} and R_{sh} are area-normalized series and shunt resistances ($\Omega\,\text{m}^2$), respectively, and account for resistive losses through the emitter and base, as well as accounting for contact resistances and current shunting. For the detailed balance limit, the quantity R_{ser} is zero and R_{sh} goes to infinity. The resulting J–V performance of the light absorbers is characterized by the short-circuit current density, J_{sc}, the open-circuit voltage, V_{oc}, and the fill factor, FF,

$$\text{FF} = \frac{J \cdot V|_{\max\{J \cdot V\}}}{J_{\text{sc}} V_{\text{oc}}} \tag{13.43}$$

It should be noted that the above definition is valid for buried junction devices in which the voltage generated is between the anode and cathode; for semiconductor/liquid junctions, FF an V_{oc} can only be determined from a reference measurement using an analogous dark system to define the specific point on the electrochemical potential scale. Eqn (13.42) captures the phenomena discussed in this section empirically, and is a transcendental equation that is solved with the transport equations to determine the overall operation point and current of the device (see Figure 13.2).

13.3.3 Kinetics

For most solar water-splitting cells, interface treatment of the charge-transfer reactions rather than a porous electrode is utilized. Thus, the kinetic expression becomes essentially a boundary condition for the ionic current, which is current-matched to the generation of electrons and holes as discussed below. This transfer current depends on the electrochemical kinetics, which are expressed by use of Butler–Volmer expressions for the oxygen evolution reaction (OER) and the hydrogen evolution reaction (HER),

$$i_{\text{R,OER/HER}} = i_{0,\text{OER/HER}} \times \left[\left(\frac{c_{\text{red}}}{c_{\text{red},0}}\right)^{\gamma_{\text{red}}} \exp\left(\frac{\alpha_{\text{a,OER/HER}} F \eta_{\text{op}}}{RT}\right) - \left(\frac{c_{\text{ox}}}{c_{\text{ox}_0}}\right)^{\gamma_{\text{ox}}} \exp\left(\frac{-\alpha_{\text{c,OER/HER}} F \eta_{\text{op}}}{RT}\right)\right] \tag{13.44}$$

where $i_{0,\text{OER}}$ and $i_{0,\text{HER}}$ are the OER and HER exchange current densities, respectively, and $\alpha_{\text{a},i}$ and $\alpha_{\text{c},i}$ are the OER and HER anodic and cathodic transfer coefficients, respectively. While the Butler–Volmer expression is only rigorously accurate for single electron, outer-sphere transfer reactions, it often is used for multi-electron reactions, where the parameters are fit empirically or derived from mechanisms, as it fits the experimental data. The overpotential is defined as

$$\eta_{\mathrm{op}} = \phi_{\mathrm{s}} - \phi_{\mathrm{l}} - U^{\theta} \tag{13.45}$$

The concentration terms in the kinetic equations are unity for a well-stirred, or continuously flushed, reactor.

The electronic or solid potential is derived from the semiconductor physics modeling as described above, and the ionic potential is often what remains to be determined. This is because the transfer current is solved for from matching this expression to that of the semiconductor physics (*i.e.*, the intersecting point in Figure 13.2). Therefore, the kinetics couples the relevant variables at the interface and thus connects the different types of models. Finally, if one wishes, one can also account for effects such as recombination using a similar expression to (13.36) or Butler–Volmer expressions (13.44) if the holes recombine with electrons in the catalyst for example. Accounting for trap states, and related phenomena or using microkinetic expressions can be used for the kinetics, but this is beyond the scope of this chapter. Similarly, the impact of band energetics and surface states on the kinetic properties of the semiconductor electrodes are also outside the scope, however, for modeling purposes, one can incorporate microkinetics and/or the above interface states as reaction expressions that impact the electrolyte or surface potentials.[23]

Finally, the electrostatic charges on the surface of the electrode can produce a double layer in the electrolyte. The thickness of the double layer is of the order of 1 nm,[32] which is significantly smaller than typical electrode separation distances. Therefore, electroneutrality (13.27) can be assumed everywhere in the electrolyte including the electrode interface.

13.3.4 Heat Transfer

For heat transfer in the solid components, one can solve the simplified heat conservation equation,

$$\frac{dT_i}{dt}\rho_i Cp_i + \rho_i Cp_i \mathbf{v} \cdot \nabla T_i = \nabla \cdot (k_{\mathrm{th}} \nabla T_i) + Q, \tag{13.46}$$

where $\frac{dT_i}{dt}$ is the rate of change of the temperature of component i with respect to time, ρ_i is that component's density, Cp_i is its heat capacity, k_{th} is its thermal conductivity and Q is the volumetric rate of heat generation. Typically, heat conduction (the first term on the right) dominates for heat transfer although bubble formation, wind flow, and other influences can play a role in heat convection (second term on the left).

While this equation is relatively trivial, the critical part is related to the various source terms resulting from the energy generation or use in the various domains. The various source terms are described in detail by Tembhurne *et al.*[16,22,33] and are summarized below. In the semiconducting domain, three sources contribute to the energy equation: the thermalization energy of high-energy photons that are absorbed and generated electron–hole pairs, and the resistive and magnetic losses resulting from low-energy photon absorption:

$$Q_{EM_\lambda}^{PV} = Q_{TH}|^{\lambda < hc/E_g} + Q_R|^{\lambda > hc/E_g} + Q_M|^{\lambda > hc/E_g}, \tag{13.47}$$

$$Q_\mathrm{R} = \tfrac{1}{2}(\mathfrak{R}(i_x \cdot E_x) + \mathfrak{R}(i_y \cdot E_y) + \mathfrak{R}(i_z \cdot E_z)), \tag{13.48}$$

$$Q_M = \Re\left(\tfrac{1}{2} i\omega((B_x \cdot H_x^\star) + (B_y \cdot H_y^\star) + (B_z \cdot H_z^\star))\right). \tag{13.49}$$

The resistive and magnetic losses (calculated based on the electric and magnetic fields, **E** and **H**, respectively) are also calculated in all other domains of the device, including reactant channel, electrolyte, and the electrode.

In the solid and liquid conductors (electrodes and electrolytes), ohmic resistances and the electrochemical reaction will contribute to the heat source term:

$$Q_\mathrm{EC} = -i_\mathrm{s} \cdot \nabla\phi_\mathrm{s} - i_\mathrm{l} \cdot \nabla\phi_\mathrm{l} + \left(\phi_\mathrm{s} - \phi_\mathrm{l} - E_\mathrm{eq} + T\frac{\partial E_\mathrm{eq}}{\partial T}\right) i_\mathrm{loc}. \tag{13.50}$$

The reaction can be endothermic or exothermic depending on whether the operative voltage is below or above the thermoneutral potential (1.481 V), respectively; consequently, the Q-term is actually a sink or a source term, respectively.

Finally, heat generation can occur due to absorption of photon energy in the reflective layers, electrolyte, catalyst, protection layers, chassis, and lens, *etc.* In particular, a significant portion of the infrared spectrum is absorbed by the electrolyte. The optical heat generation is given by

$$Q_\mathrm{Op} = 1/V_L \sum_{\lambda_0,\mathrm{Min}}^{\lambda_0,\mathrm{Max}} \alpha(\lambda_0) I(\lambda_0) A_L,$$

where $\alpha(\lambda_0)$ is the absorption coefficient of the incident flux within that element at the light wavelength in a vacuum, λ_0, $I(\lambda_0)$ is the incident flux intensity, A_L is the area exposed to insolation of that element, and V_L is the volume of element. The bounds of the summation are placed by the radiation flux spectrum (280 to 4000 nm for the incident solar spectrum).

In terms of boundary conditions for the energy balance, these can be perhaps weather related (forced convection from the wind) or radiation to the atmosphere, or even conduction through scaffolding or heat exchangers.

13.4 Sample Cases

Polarization loss is an important metric to assess the performance of a solar-fuels generator. Polarization loss is the energy loss due to the transport of species (by migration and diffusion) and concentration gradients, which can be represented as a sum of (i) ohmic loss, (ii) diffusion loss, and (iii) Nernstian loss. The ohmic loss is due to the resistance of the electrolyte, and the diffusion loss originates from the ionic gradient in the boundary layer near each electrode due to the applied current density. The ohmic and diffusion losses can be combined into the solution loss such that

$$\Delta\phi_{\text{solution}} = \underbrace{\int \frac{i_l}{\kappa} d\mathbf{x}}_{\Delta\phi_{\text{ohmic}}} + \underbrace{\sum_i \int \frac{F z_i D_i \nabla c_i}{\kappa} d\mathbf{x}}_{\Delta\phi_{\text{diffusion}}} \tag{13.51}$$

where, $\mathbf{x}$ is the position coordinate, and $d\mathbf{x}$ is the differential volume given as $dx_1 dx_2 dx_3$. The ionic gradients shift the concentrations of reacting species next to the electrode surfaces (*e.g.*, protons, and hydroxide anion) away from those present in the bulk. This causes an increase in the equilibrium potential of the oxygen evolution reaction (OER) and the hydrogen evolution reaction (HER), which are referred to collectively as the Nernstian loss. The sum of losses due to differences in pH at the two electrodes is given by

$$\Delta\phi_{\text{Nernstian}} = \underbrace{\frac{2.303RT}{F}(\text{pH}_{\text{cathode}} - \text{pH}_{\text{bulk}})}_{\Delta\phi_{\text{cathode pH}}} + \underbrace{\frac{2.303RT}{F}(\text{pH}_{\text{bulk}} - \text{pH}_{\text{anode}})}_{\Delta\phi_{\text{anode pH}}} \tag{13.52}$$

The Nernstian loss due to pH change at the cathode is referred to as the *cathode pH loss*, and that due to the pH change at the anode is referred to as the *anode pH loss*. The losses above are due to the transport of species in the electrolyte, which, in turn, depend on the applied current density, electrolyte composition, electrolyte hydrodynamics, membrane composition, and catalyst selectivity. The kinetic overpotentials for the OER and HER also contribute to the total losses in the electrochemical cell.

13.4.1 Simplifications to the Flux-expressions

The mathematical modeling of aqueous electrolytes is treated only to a limited extent in the simulation of solar-fuel generators. Aqueous electrolytes have been modeled using a dilute solution theory, assuming ideal solution, isothermal and isobaric conditions. However, the realistic electrolytes are non-ideal, multi-component solutions operating under varying temperature and pressure conditions. These over-simplified models of electrolytes yield under-estimated values of polarization losses and current density. Although rigorous models of electrolytes such as concentrated solution theory or generalized solution theory as described above can provide an accurate prediction of concentration, potential, and current distribution, they are computationally expensive. Therefore, it is important to reduce the model by employing physically realistic assumptions.

13.4.1.1 Shockley–Queisser

As noted, the ideal-diode (eqn (13.42)) is often assumed. This makes the problem much more tractable in terms of solving mainly for the electrochemical load curve and then just matching the point to the power curve. Such an approach is good when one is not as worried about the detailed semiconductor phenomena and is looking at mainly different operating regimes in terms of electrolyte or cell designs.

13.4.1.2 Butler–Volmer

Similar to just using the ideal-diode equation for the semiconductor, one can just use the Butler–Volmer expression (eqn (13.44)), which simplifies the kinetic description, but with perhaps ohmic terms incorporated for the electrochemical load (*i.e.*, adding in the potential drop explicitly for the ionic potential in the overpotential (*i.e.*, eqn (13.47) in eqn (13.45)). In this fashion, one can quickly scan and optimize band gaps and arrangement and explore the gross impact of different properties without having to model the transport phenomena in detail.

13.4.1.3 Strongly Acidic or Alkaline Electrolytes

Transport of species in strong acid ($pH < 1$) or strong base ($pH > 13$) is primarily due to migration, as the concentration gradients are negligible. In this case, the flux expression (eqn (13.13)) only requires the migration term and the current density is given by the Ohm's law (eqn (13.28)). The concentration polarization and the associated losses such as diffusion loss and concentrated-related (*i.e.*, Nernstian) losses are also negligible in strongly acidic and alkaline electrolytes. The Ohmic loss is the only transport-related source of energy loss in this system; kinetic losses still occur as well.

13.4.1.4 Supporting Electrolyte

A supporting electrolyte contains species that do not react at the electrodes but can help in charge conduction, and thus decrease the electric field and ion migration behaviors. When the concentration of supporting electrolyte is sufficiently large, the migration term can be neglected in the flux expression. The diffusion and reactant concentration losses are the dominant sources of polarization loss.

13.4.1.5 Buffered Electrolyte

To operate away from strong acid and alkaline environments, where there are durability concerns for the components, one can use pH buffers, such as weak acids, which can produce or consume protons as the pH near the electrodes increases or decreases, respectively. Therefore, the pH buffers modulate the pH of the electrodes and promote conductivity in the electrolyte. The kinetics of acid–base reactions in the buffered electrolytes are generally faster than the rates of electrochemical reactions, so that they can be assumed to be in quasi-equilibrium. The assumption of acid–base equilibrium reduces the degree of freedom for electrolyte species and hence the numbers of mass conservation equations. As a consequence of this assumption, the thickness of the reaction layer will be zero and the diffusion loss associated with the reaction layer will be not be captured by the model. A more complicated concentrated solution approach is recommended as there are multiple species in the electrolyte and effects such as electrodialysis (*e.g.*, due to transport through an ionomer separator) can limit performance over time (*i.e.*, even with buffered solutions, concentration gradients, and thus polarization losses, can build up with time).

13.4.1.6 1-D Approximation to 2-D and 3-D Solar-fuels Generator

The solar-fuels generators are planar 3-D devices, where species transport is uniform along the length of the membrane, so that the species are distributed in two dimensions. Since the streamlines of electrolyte potential and current density do not cross each other, the species distribution is a function of the effective distance between electrodes. Therefore, the species and potential distributions in planar cells can be reconstructed in a 1-D cell with an effective separation distance. The velocity field in a 2-D flow can also be approximated in a 1-D geometry by including reaction terms in eqn (13.13), such that rate of generation of species is equal to the ratio of its concentration to the residence time of the flow. Such approximation to 1-D geometry significantly reduces the computational load, while still solving for all of the essential physics.

13.4.2 Case Study 1: Design Guidelines and Cell Scaling from 2D PEC Device Models

Multi-dimensional device models are crucial for feasibility studies of PEC device designs and architectures, and for the understanding of how the scaling of PEC devices has to be envisaged. Few multi-dimensional PEC cell models exist in literature, with almost none prior to the work described herein. Recently, Carver *et al.*[34] developed a 2-D model of a PEC cell in which the purpose was to operate in non-saturating conditions in order to suppress the bubble evolution and corresponding losses. Tembhurne *et al.*[16,22] developed a 2-D model of a PEC cell in order to understand how heat transfer can affect the performance and to identify and omit heterogeneities in the temperature, velocity and current density distributions. Hankin *et al.*[35]

developed 2-D and 3-D models of a PEC cell incorporating a detailed interface model of the hematite photoelectrode, and observed a non-uniform current distribution along the electrode surface. The latter was identified as critical when scaling the PEC cell.

Here, we discuss the findings of the 2-D device model by Haussener *et al.*,[5,20] which aimed to answer questions related to the design and architecture of a PEC cell and its scale up. The model was applied to two characteristics designs, defined based on a survey of the various laboratory-scale device demonstrations and design ideas in the literature: design A representing a side-by-side configuration of the anodic and cathodic channels, and design B representing a top-bottom configuration of the anodic and cathode channels as shown in Figure 13.8. Focus was put on the quantification of the ohmic losses in the system, consequently the buried photovoltaics was approximated by a diode-like behavior with sufficiently large open circuit potential and the electrocatalysts used were rare but efficient and selective candidates. This ensured that in the majority of conditions the photocurrent was constant for a large potential range, and the overpotential required to drive the catalysts was

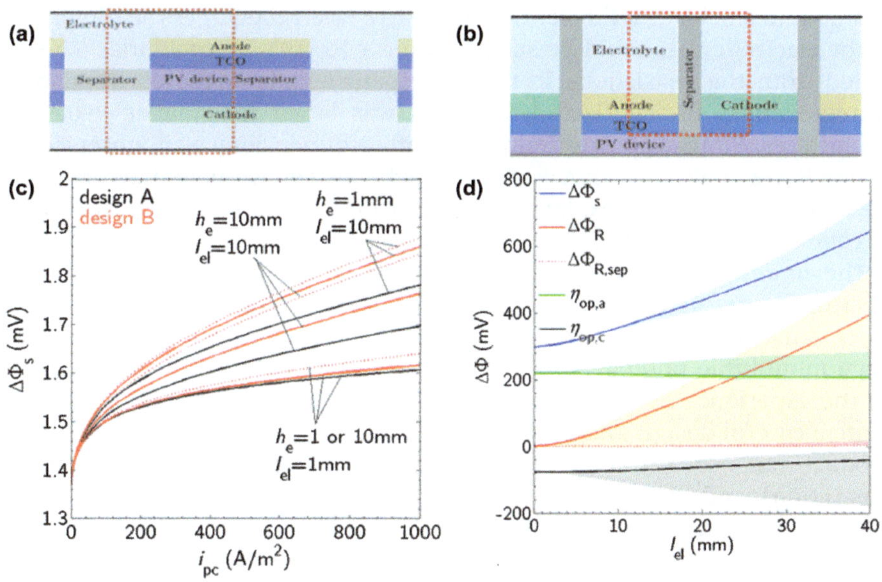

Figure 13.8 Schematic diagrams for (a) design A and (b) design B where the box denotes the modeling domain. (c) Overall cell polarization curves—average cell potential *vs.* photocurrent density—for design A (black) and design B (red) at $l_{el}/l_d = 0.5$ (solid red) and $l_{el}/l_d = 0.9$ (dotted red) for various electrode lengths and electrolyte heights, for $t_{sep} = 10$ μm. (d) Average (lines) and variation along electrode (bands around lines) for the total device overpotential ($\Delta\Phi_s$), ohmic losses ($\Delta\Phi_R$), potential losses over separator only ($\Delta\Phi_{R,sep}$), and anode/cathode overpotentials ($\eta_{op,a/c}$) for design A at $h_e = 1$ mm.
Adapted from ref. 5 with permission from the Royal Society of Chemistry.

minimized to 300 to 400 mV. Concentration variations in the electrolyte were neglected. Charge transport and conservation in the solid and liquid electrodes and electrolytes was solved, with a charge source and sink term linking the two phases, related to the electrochemical reactions. The oxygen and hydrogen evolution reactions were modeled using Butler–Volmer type expression, strictly valid only for metal-electrolyte interfaces. Transport of the dissolved, non-charged product species (hydrogen and oxygen) was solved using diffusion-advection transport equations, where it was assumed that bubbles did not affect the behavior. The convection term was related to pressure differentials between the anodic and cathodic chamber, and was shown only to be an issue with crossover if porous separators were used.

Arguing that an additional ohmic overpotential of only 100 to 200 mV could be sustained by the photoabsorbers, it became apparent that the ionic path length had to be limited to few tens of millimeters (see Figure 13.8). Detailed simulations of design A showed that—in the case where no lateral conduction in the photoabsorber or its protection layer was present—the electrode dimensions had to be limited to 10 to 38 mm for a 100 mV overpotential, dependent on the thickness (or volume) of the electrolyte (varying between 1 and 50 mm thick). The thicker the electrolyte layer, the larger the dimension of the electrode that could be sustained for a 100 mV overpotential. This resulted from the restrictions in the ionic path length or the corresponding current density in the electrolyte, which was larger for smaller electrolyte thicknesses. Similar observation and limiting dimensions were observed for design B. Generally, design A worked at slightly lower overpotential for similar dimensions, which was explained by the orthogonalization of the irradiation direction and the main ionic-transport direction.

The understanding that the path length of the ion transport and the electrolyte conductivity can be limiting makes it apparent that a practical (large-scale) PEC device or PEC plant cannot be envisaged as an extremely large monolithic photoabsorber device, as often proposed, or as suggested by the experience of scaling photovoltaic devices. Instead, PEC devices require a careful design and optimization of the coupled mass (ion transport, product transport and separation), charge (charge generation, transport and separation), and heat transfer (radiation absorption) phenomena as shown in Figure 13.8d. Potentially, a checkerboard like design with alternating paths for ion transport and photo absorption/electrochemistry could work for a scaled device.

Enhancing the lateral conduction can help to allow for larger electrode lengths at reduced ohmic losses. This lateral conduction can result from (i) a well conducting photoabsorber, (ii) a well conducting catalyst layer, and/or (iii) an additional transparent conducting layer. Lateral conduction allows for the charge to be redistributed, resulting in shorter path lengths of the ions in the relatively low-conducting electrolyte at the expense of longer path lengths of the electron or holes in the relatively high-conducting solid components. The larger the electrode, the more significant this effect. For example for a transparent conductive oxide (TCO) layer with a conductivity of 10 Ω/□ and design

A, the ohmic overpotential of a 15 mm electrode could be reduced by 40% (from 196 mV to 117 mV). However, the advancement of the TCO layer could be further increased if the kinetics of both reactions could be enhanced. For the reference conditions, the conductivity of the TCO layer is non-limiting, while the reaction rate is limiting. The simulations showed that if the exchange current densities could be increased by a factor 10^9 (for example by rougher surfaces or the use of a better catalyst), then the conductivity of the TCO would be limiting. Consequently, a careful and simultaneous analysis of the lateral solid conductivity, the reaction kinetics, and also the electrolyte conductivity has to be done in order to optimize this combination for reduced ohmic overpotentials. Obviously, the addition of a conducting interlayer will affect the band alignment and has to be carefully engineered to allow proper charge transport across the different interfaces.

The conductivity of the electrolyte was varied in order to understand if less aggressive environments than a 1 M sulfuric acid electrolyte could be used without affecting the performance. Decreasing the electrolyte conductivity by one order of magnitude (roughly equivalent to one pH point), significantly affected the overpotential, especially for larger electrode dimensions. For example for design B, operating the system with a dual absorber (1.0 eV and 1.6 eV) which would allow for a maximum efficiency of 28%, the reduction of the conductivity by one order of magnitude decreased the efficiency by 1/3 for a 2 cm electrode and by 2/3 for a 4 cm electrode. Reduction of the conductivity by two orders of magnitude reduced the efficiency to almost 0%. Consequently, it is difficult to imagine that a PEC device could work well in near-neutral environments unless supporting electrolyte is added in order to increase the conductivity. However, in such a case, concentration buildup of the charge transport ions has to be carefully monitored in order to suppress additional overpotentials due to concentration gradients as discussed in the previous section of this chapter.

Understanding how the choice of the bandgaps of a multiple photoabsorber can affect the performance of the device has been investigated with simplified equivalent circuit models.[29] Here, it was investigated how the overpotential can affect the optimal choice of bandgaps in a dual absorber architecture in design B. The larger the overpotential—as a result of slow catalysts or large ohmic losses—the larger the two bandgaps needed for highest efficiency. This maximal possible efficiency dropped as a function of this overpotential. Consequently, a proper design of a PEC device requires a consideration of not only the dimensions but also the behavior of the photoabsorbing and the electrochemical components in order to have them work optimally together.

13.4.3 Case Study 2: Multiphysics Modeling in Concentrated PEC

A particular challenge is added to the photoelectrochemical energy conversion when utilizing concentrated irradiation. Concentrated irradiation becomes an interesting approach for PEC energy conversion as it can help to reduce the mass and volume of costly components since the dilute solar irradiation (instantaneous irradiation at about 1000 $W m^{-2}$) is firstly concentrated using relatively low cost and low tech components such as mirrors or lenses before being incident on the costly photoelectrochemical components. Transient techno-economic-sustainability modeling of conceptual PEC device designs provided evidence that utilizing concentrated irradiation can profit in two ways:[4] (i) increased economic competitiveness, and (ii) ability to still use higher performing but costly photoelectrochemical components. However, utilization of concentrated irradiation introduces operational challenges related to the increased thermal energy input resulting from thermalization energy of high-energy photons and from the absorbed energy of a broad spectrum of absorbed photons in all components of the device potentially increasing the temperature of the device and increasing the charge density in the device. The increased temperature leads to increased charge recombination in semiconductors and reduced stability of materials. However, generally, transport and chemistry are enhanced and operate at larger rates at increased temperatures, potentially benefiting the performance of the device. Essential for higher temperature operation is the knowledge of the temperature heterogeneity in a PEC device and, subsequently, the use of smart thermal management to shuffle the heat between the different components.[16,22,36]

A sample test case is done by modeling the device as shown in Figure 13.3. It should be noted that the photovoltaics and electrolyzer components are in direct contact and consequently allow for heat transfer between the components. This is in contrast to physically separated PV-electrolyzer systems, utilizing wires for the charge transfer and for which no smart thermal management strategies can really be applied. The model solves simultaneous the equations described in Section 13.3. A detailed description of the results and interesting predictions on how smart thermal management can enhance the performance can be found elsewhere.[22] The model allows for a multi-dimensional resolution of the variable fields, and provides information on the feasibility of photoelectrochemical devices operating under concentrated irradiation. Here we focus on some challenges and inconsistencies in the results, which result from incomplete knowledge of physics. Using the model, one can obtain a Sankey diagram of the energy flows in the device as shown for an example device and material combinations in Figure 13.9. Such an approach and the model provides information on the repartitioning of the energy streams, allowing identification of irreversibilities and losses, and consequently directing optimization.

When evaluating the energy conservation, the overall energy is conserved in the computational domain, but locally for two sub-components (the semiconductor and the electrochemical components), there is energy lost. This is not surprising: we do not explicitly solve for energy conservation in these two subcomponents but only for the charge conservation. This results in the appearance of these two heat loss terms, shown in Figure 13.9 as a darker color of the respective arrows. Consequently, the heat sources

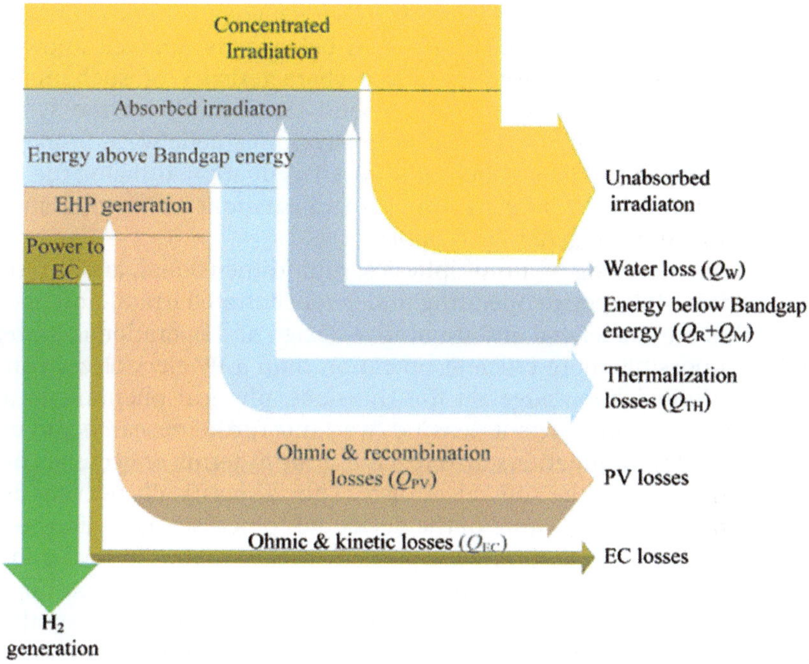

Figure 13.9 Sankey diagram depicting the distribution of energy at various stages of its utilization for a concentration irradiation of 18 suns, *i.e.* 18 000 kW m^{-2}. The width of the arrows shows approximate re-partitioning for a device utilizing a dual-junction cell made of III–V based materials. The darkened areas of the arrows of the PV and EC losses are not considered in the total heat source of the energy equation.
Reproduced from ref. 16, http://dx.doi.org/10.1149/2.0311610jes, © 2016 The Authors. Published under the terms of the CC BY 4.0 license, https://creativecommons.org/licenses/by/4.0/.[16]

provided from the semiconductor and electrochemistry module to the heat transfer module are underestimated. The lack of physical information and interpretation of the lost energy in the two components needs to be addressed in the future by developing dedicated energy conservation models.

The simplest way to transfer heat from the photoabsorbers to the catalyst is through convection where the reactant stream (liquid water in the case of water electrolysis) cools the photoabsorbers while being heated. The reactant at elevated temperature is then used in the (photo)electrochemical reaction. Potentially this reactant/cooling stream might interfere with the incoming, concentrated radiation (as is the case in our representative design) and absorb radiation in the range relevant for the photoabsorber. This would be undesired and the reactant channel needs to be designed accordingly. Knowledge of the radiative properties of the various relevant components is necessary for accurate modeling. These properties include the temperature-dependent band gap of the photoabsorber, and the spectrally-resolved absorption coefficient of the photoabsorber as well as the reactant. Often, the solar absorber is not made of thin films of pure materials but rather heterogeneous, nanostructured films for which the volume-averaged absorption behavior is not well known and understood. Reported apparent radiative characteristics of such materials cannot generally be applied as they depend significantly on the synthesis and morphology of the film. Instead, the electromagnetic wave propagation has to be predicted *via* the solution of Maxwell's equation, utilizing the locally resolved refractive index information. Such calculations are computationally very expensive and not generally feasible.

Overall, the modeling of a multiphysics, multidimensional, and multiscale device such as a PEC device operating under concentrated irradiation provides new insights into beneficial and innovative design and operational strategies, including a possibly more efficient operation than a PV-electrolyzer combination. The limited knowledge on the transient, physical phenomena at increased irradiation and current densities warrants further research and model development. The simulations also reach a limit in terms of accuracy of predicted performance in the solid electrolyte. This is specifically relevant as it is related to the multiscale nature of the device and processes. Questions related to the validity of electroneutrality at a particular scale, to the physical transport mechanisms at elevated temperatures and current densities (proton hopping, electro-osmosis, *etc.*) need to be addressed by incorporating small-scale models, potentially considering molecular-scale simulations.

13.5 Summary

Modeling and simulation-guided development of integrated solar-driven water-splitting devices has made significant advances in recent years. Multidimensional multiphysics models that account for various photoelectrochemical processes have provided design guidelines for semiconductors, interfacial layers, electrocatalysts, electrolytes and membrane separators. Modeling and simulation have also revealed a range of viable prototype architectures that can accommodate efficient, stable, scalable and safe solar-driven water-splitting reactions. This chapter discussed modeling approaches, history, and governing phenomena and equations.

In terms of outlook and challenges, the impact of active fluid flow as well as multiphase flow due to bubble evolutions on the cell performance has yet to be assessed by modeling and experimental studies. There is still a need for coupling multiphase flow and fluid dynamics models with the mass-transport equations to predict the accurate distribution of species, current and potential. There is also a need for coupling energy balance equation to more rigorously compare the cross-term coupled phenomena. In addition, a complete coupling of the various physical models is the holy grail in the community and not yet reality since there is a need to couple across lengthscales and time scales the more complicated semiconductor physics and light capture with the transport phenomena. Finally, utilizing more complicated kinetics expressions derived from careful experimentation or *ab-initio* modeling is also a possible area of future work.

Acknowledgements

The authors thank those researchers within JCAP, both past and present, who have helped guide and design modeling of solar water-splitting devices. In particular, we thank Alan Berger, John Newman, Carl Koval, and C. X. Xiang. We thank Saurabh Tembhurne, EPFL, for leading the development of the coupled multi-physics model for the concentrated PEC test case. MRS acknowledges financial support from the University of Illinois at Chicago. Part of this material is based upon work performed in the framework of the projects Solar Hydrogen Integrated Nano Electrolysis (SHINE) and Strategic Computation and Optimization of Unified Templates for Solar fuels (SCOUTS), funded by the Nano-Tera.ch Initiative (grant #145936) and the Swiss National Science Foundation (grant #155876), respectively. This material is partly based upon work performed by the Joint Center for Artificial Photosynthesis, a DOE Energy Innovation Hub, supported through the Office of Science of the US Department of Energy under Award Number DE-SC0004993.

References

1. M. G. Walter, E. L. Warren, J. R. McKone, S. W. Boettcher, Q. X. Mi, E. A. Santori and N. S. Lewis, *Chem. Rev.*, 2010, **110**, 6446.
2. J. W. Ager, M. R. Shaner, K. A. Walczak, I. D. Sharp and S. Ardo, *Energy Environ. Sci.*, 2015, **8**, 2811.
3. C. X. Xiang, A. Z. Weber, S. Ardo, A. Berger, Y. K. Chen, R. Coridan, K. T. Fountaine, S. Haussener, S. Hu, R. Liu, N. S. Lewis, M. A. Modestino, M. M. Shaner, M. R. Singh, J. C. Stevens, K. Sun and K. Walczak, *Angew. Chem., Int. Ed.*, 2016, **55**, 12974.
4. M. Dumortier, S. Y. Tembhurne and S. Haussener, *Energy Environ. Sci.*, 2015, **8**, 3614.
5. S. Haussener, C. X. Xiang, J. M. Spurgeon, S. Ardo, N. S. Lewis and A. Z. Weber, *Energy Environ. Sci.*, 2012, **5**, 9922.
6. Y. K. Chen, S. Hu, C. X. Xiang and N. S. Lewis, *Energy Environ. Sci.*, 2015, **8**, 876.
7. S. Hu, C. X. Xiang, S. Haussener, A. D. Berger and N. S. Lewis, *Energy Environ. Sci.*, 2013, **6**, 2984.
8. H. Doscher, J. F. Geisz, T. G. Deutsch and J. A. Turner, *Energy Environ. Sci.*, 2014, 7, 2951; A. Berger, R. A. Segalman and J. Newman, *Energy Environ. Sci.*, 2014, 7, 1468; C. X. Xiang, Y. K. Chen and N. S. Lewis, *Energy Environ. Sci.*, 2013, 6, 3713.
9. L. C. Seitz, Z. B. Chen, A. J. Forman, B. A. Pinaud, J. D. Benck and T. F. Jaramillo, *ChemSusChem*, 2014, **7**, 1372.
10. Y. K. Chen, C. X. Xiang, S. Hu and N. S. Lewis, *J. Electrochem. Soc.*, 2014, **161**, F1101–F1110.
11. J. Jin, K. Walczak, M. R. Singh, C. Karp, N. S. Lewis and C. X. Xiang, *Energy Environ. Sci.*, 2014, **7**, 3371.

12. M. R. Singh, K. M. Papadantonakis, C. Xiang and N. S. Lewis, *Energy Environ. Sci.*, 2015, **8**, 2760–2767.
13. Y. K. Gaudy and S. Haussener, *J. Mater. Chem. A*, 2016, **4**, 3100–3114.
14. M. A. Modestino and S. Haussener, *Annu. Rev. Chem. Biomol. Eng.*, 2015, **6**, 13.
15. A. C. Nielander, M. R. Shaner, K. M. Papadantonakis, S. A. Francis and N. S. Lewis, *Energy Environ. Sci.*, 2015, **8**, 16.
16. S. Tembhurne and S. Haussener, *J. Electrochem. Soc.*, 2016, **163**, H988.
17. R. H. Perry and D. W. Green, *Perrys's Chemcial Engineers' Handbook*, McGraw-Hill, New York, 1997; *CRC Handbook of Chemistry and Physics*, CRC Press, Boca Raton, FL, 1979.
18. R. H. Coridan, A. C. Nielander, S. A. Francis, M. T. McDowell, V. Dix, S. M. Chatman and N. S. Lewis, *Energy Environ. Sci.*, 2015, **8**, 2886.
19. Y. K. Chen, K. Sun, H. Audesirk, C. X. Xiang and N. S. Lewis, *Energy Environ. Sci.*, 2015, **8**, 1736; M. R. Singh, K. Papadantonakis, C. X. Xiang and N. S. Lewis, *Energy Environ. Sci.*, 2015, **8**, 2760.
20. S. Haussener, S. Hu, C. Xiang, A. Z. Weber and N. Lewis, *Energy Environ. Sci.*, 2013, **6**, 3605.
21. C. Xiang, Y. Chen and N. S. Lewis, *Energy Environ. Sci.*, 2013, **6**, 3713.
22. S. Tembhurne and S. Haussener, *J. Electrochem. Soc.*, 2016, **163**, H999.
23. A. Berger and J. Newman, *J. Electrochem. Soc.*, 2014, **161**, E3328.
24. J. R. Bolton, S. J. Strickler and J. S. Connolly, *Nature*, 1985, **316**, 495.
25. J. Newman and K. E. Thomas-Alyea, *Electrochemical Systems*, John Wiley & Sons, New York, 2004.
26. Y. Ma, M. Doyle, T. F. Fuller, M. M. Doeff, L. C. De Jonghe and J. Newman, *J. Electrochem. Soc.*, 1995, **142**, 1859.
27. R. B. Bird, W. E. Stewart and E. N. Lightfoot, *Transport Phenomena*, John Wiley & Sons, Inc., New York, 2002.
28. H. Döscher, J. F. Geisz, T. G. Deutsch and J. A. Turner, *Energy Environ. Sci.*, 2014, 7(9), 2951–2956.
29. K. T. Fountaine, H. J. Lewerenz and H. A. Atwater, *Nat. Commun.*, 2016, 7, 13706.
30. N. S. Lewis, *Annu. Rev. Phys. Chem.*, 1991, **42**, 543; H. Gerischer, *Advances in Electrochemistry and Electrochemical Engineering*, Interscience Publishers, 1961, vol. 1.
31. K. T. Fountaine and H. A. Atwater, *Opt. Express*, 2014, **22**, A1453.
32. J. N. Israelachvili, *Intermolecular and Surface Forces*, Academic Press, Waltham, MA 2011.
33. S. Y. Tembhurne, M. Dumortier and S. Haussener, *"Heat transfer modeling in integrated photoelectrochemical hydrogen generators using concentrated irradiation"*, presented at *15th International Heat Transfer Conference*, Kyoto, Japan, 2014.
34. C. Carver, Z. Ulissi, C. Ong, S. Dennison, G. Kelsall and K. Hellgardt, *Int. J. Hydrogen Energy*, 2012, **37**, 2911.
35. A. Hankin, F. Bedoya-Lora, C. Ong, J. Alexander, F. Petter and G. Kelsall, *Energy Environ. Sci.*, 2017, **10**, 346.
36. J. C. Stevens and A. Z. Weber, *J. Electrochem. Soc.*, 2016, **163**, H475.

Subject Index